Lecture Notes in Computer Science 988

Edited by G. Goos, J. Hartmanis and J. van Leeuwen

Advisory Board: W. Brauer   D. Gries   J. Stoer

T0189434

Springer
Berlin
Heidelberg
New York
Barcelona
Budapest
Hong Kong
London
Milan
Paris
Santa Clara
Singapore
Tokyo

Andrew U. Frank   Werner Kuhn (Eds.)

Spatial
Information Theory

A Theoretical Basis for GIS

International Conference COSIT '95
Semmering, Austria, September 21-23, 1995
Proceedings

 Springer

Series Editors

Gerhard Goos, Karlsruhe University, Germany

Juris Hartmanis, Cornell University, NY, USA

Jan van Leeuwen, Utrecht University, The Netherlands

Volume Editors

Andrew U. Frank
Werner Kuhn
Dept. of Geoinformation, Technical University of Vienna
Gusshausstr. 27-29, A-1040 Vienna, Austria

Cataloging-in-Publication data applied for

Die Deutsche Bibliothek - CIP-Einheitsaufnahme

Spatial information theory : a theoretical basis for GIS ;
international conference ; proceedings / COSIT '95,
Semmering, Austria, September 21 - 23, 1995. Andrew U. Frank
; Werner Kuhn (ed.). - Berlin ; Heidelberg ; New York ;
Barcelona ; Budapest ; Hong Kong ; London ; Milan ; Paris ;
Tokyo : Springer, 1995
  (Lecture notes in computer science ; Vol. 988)
  ISBN 3-540-60392-1
NE: Frank, Andrew U. [Hrsg.]; COSIT <1995, Semmering>; GT

CR Subject Classification (1991): I.2-3, H.2, H.5, E.1-2, I.5-6, J.2

ISBN 3-540-60392-1 Springer-Verlag Berlin Heidelberg New York

© Springer-Verlag Berlin Heidelberg 1995
Printed in Germany

Typesetting: Camera-ready by author
SPIN 10485731   06/3142 – 5 4 3 2 1 0   Printed on acid-free paper

Foreword

COSIT, the Conference On Spatial Information Theory, has become a regular event for scientists who are interested in the understanding and representation of spatial information. It is a truly interdisciplinary meeting, where common problems are approached from the viewpoints of many disciplines. It is not only multi-disciplinary, as results from many different disciplines are necessary to achieve a complete picture; but interdisciplinary in the sense that an approach from one discipline must be combined with approaches from other disciplines to allow for new questions to be posed.

Spatial Information Theory brings together three fields of research of enormous importance for the Geographic Information System (GIS) technology:

- spatial reasoning,
- representation of space, and
- human understanding of space.

These three strands of research, two more technical and the third more cognitive, are concerned with Geographic or Environmental Space, i.e., the space which is populated with immovable objects larger than human beings and in which humans move around to learn about it.

Spatial reasoning is oriented towards the logical deduction of spatial information from spatial facts; it is related to formal logic and deduction. *Representation of space* has its roots in the tradition of Artificial Intelligence, but it is also practically important for the design of spatial data structures. Spatial Information Theory brings these efforts together with the discussion of *human spatial cognition*. How do people think about space? What logical (or illogical) conclusions do they draw? What spatial aspects can be represented in natural languages? What are the components of spatial information? These questions connect efforts that focus on the understanding of space from the cooperating fields of philosophy (spatial ontology), linguistics, and psychology.

COSIT'95 documents the advances in the field of spatial information theory during the last five years since the NATO Advanced Study Institute on Cognitive and Linguistic Aspects of Geographic Space [Mark and Frank 1991]. After a conference in Pisa (Italy) in 1992, with a similar title [Frank and others 1992], the first COSIT conference was organised in 1993 on the Island of Elba (Italy) [Frank and Campari 1993]. Many references to papers from these meetings have appeared in the literature, demonstrating the continuous and growing interest in the subject of spatial information theory.

This year, 78 papers were submitted to COSIT by scientists from all over the world, slightly more than half of them from Europe and most of the rest from North-America. The submissions showed great variety, ranging from contributions improving on previous results, to reports on new observations and hypotheses, to reviews of the 'state of the art' and discussions of new approaches. The authors came from a wide variety of disciplines, namely Computer Science (20), Geography (20), Spatial Reasoning (10), Cognitive Science (8), Linguistics (3), Surveying (3), Artificial Intelligence (3), Planning (3), Psychology (3), Engineering (2), Cartography (1), Philosophy (1), and Management (1).

Each paper was reviewed by three experts, at least two from the same or a closely related discipline as the authors. The reviewers were generally very diligent and provided high quality assessment of the papers, often with extensive comments to the authors for possible improvement of the presentation. The program committee and the additional experts who willingly took on the burden of careful review made a strong contribution to the quality of the conference and of this book. The choice of the 36 papers which could be presented in the single track conference program was extremely difficult, but we are confident to have succeeded in selecting a stimulating and balanced program.

Sabine Timpf has organized a 'Doctoral Consortium' immediately following the COSIT'95 conference. It provides an opportunity for young scientists to present their research topics to their peers and to discuss them with experienced researchers. For this consortium, we are preparing a set of papers on work in progress as a separate volume (available from the Department of Geoinformation at the Technical University Vienna).

The conference organization was in the hands of the staff of the Department of Geoinformation. Sabine Timpf contributed much to the entire organization and Roswitha Markwart, Irene Orchard, Edith Unterweger, and Rebecca Winn worked long hours to deal with the administrative side. Their contribution is gratefully acknowledged. The conference was only made possible by the combined efforts of all the authors who submitted papers, and of the reviewers who helped to select the best contributions and to improve the presentation. We gratefully acknowledge their contribution.

Vienna, July 1995

Andrew U. Frank

Werner Kuhn

References

Frank, A.U., I. Campari, and U. Formentini, ed. *Theories and Methods of Spatio-Temporal Reasoning in Geographic Space*. Lecture Notes in Computer Science 639. Springer-Verlag, 1992.

Frank, A.U. and I. Campari, ed. *Spatial Information Theory - a Theoretical Basis for GIS (Proceedings of the European Conference on Spatial Information Theory COSIT'93)*. Lecture Notes in Computer Science. Springer-Verlag, 1993.

Mark, D.M. and A.U. Frank, ed. *Cognitive and Linguistic Aspects of Geographic Space*. NATO ASI Series D: Behavioural and Social Sciences, vol. 63. Kluwer Academic Publishers, 1991.

Program Committee Chairs

Andrew U. Frank (Austria)
Werner Kuhn (Austria)

International Program Committee Co-Chair

David M. Mark (USA)

Scientific Committee

Marc Armstrong (USA),
Norbert Bartelme (Austria),
Patrick Bergougnoux (France),
Ralf Bill (Germany),
Peter Burrough (The Netherlands),
Barbara Buttenfield (USA),
Irene Campari (Italy),
Kai-Uwe Carstensen (Germany),
Roberto Casati (France),
Costancio Castro (Spain),
Jean Paul Cheylan (France),
Nick Chrisman (USA),
Tony Cohn (United Kingdom),
Helen Couclelis (USA),
David Cowen (USA),
Max Craglia (UK),
Leila De Floriani (Italy),
Pierre Dumolard (France),
Max Egenhofer (USA),
Carola Eschenbach (Germany),
Giacomo Ferrari (Italy),
Manfred Fischer (Austria),
Peter Fleissner (Austria),
Wolfgang Förstner (Germany),
Georg Franck (Austria),
Wm. Randolph Franklin (USA),
Christian Freksa (Germany),
Anthony Galton (United Kingdom),
Chris Gold (Canada),
Reg Golledge (USA),
Mike Goodchild (USA),
Georg Gottlob (Austria),
Oliver Günther (Germany),
Ralph Hartmut Güting (Germany)
Thanasis Hadzilacos (Greece),
Daniel Hernandez (Germany),

John Herring (USA),
Klaus Hinrichs (Germany),
Stephen Hirtle (USA),
Erland Jungert (Sweden),
Marinos Kavouras (Greece),
Fritz Kelnhofer (Austria),
Karl Kraus (Austria),
Benjamin Kuipers (USA),
Robert Laurini (France),
Gerard Ligozat (France),
Duane Marble (USA),
Matt McGranaghan (USA),
Robert McMaster (USA),
Martien Molenaar (The Netherlands),
Mark Monmonier (USA),
Dan Montello (USA),
Jean-Claude Mueller (Germany),
Harlan Onsrud (USA),
Peter vanOosterom (The Netherlands),
Dimitris Papadias (USA),
Enrico Puppo (Italy),
Jan van Roessel (USA),
Tapani Sarjakoski (Finland),
Hansgeorg Schlichtmann (Canada),
Michel Scholl (France),
Roberto Scopigno (Italy),
Timos Sellis (Greece),
Richard Snodgrass (USA),
Joseph Strobl (Austria),
Erik Stubkjaer (Denmark),
A Min Tjoa (Austria),
Andrew Turk (Australia),
Barbara Tversky (USA),
Michael Wegener (Germany),
Robert Weibel (Switzerland),
Michael Worboys (UK)

Additional Referees

Frank Bethge (Germany)
Bud Bruegger (Italy)
Felix Bucher (Switzerland)
Adrijana Car (Austria)
Eliseo Clementini (Italy)
Andrew Curtis (USA)
Andreas Dieberger (USA)
Peter Haunold (Austria)
Annette Herskovits (USA)
Clifford Kottman (USA)
Mirjanka Lechthaler (Austria)
Jayant Sharma (USA)
Heinz Stanek (Austria)
Eva-Maria Stephan (Switzerland)
Sabine Timpf (Austria)
Andrej Vekovski (Switzerland)
Annette van Wolff (Germany)

Sponsorship

Ingenieurkonsulent Dipl.-Ing. Josef Angst, Austria
Bank Austria Aktiengesellschaft, Austria
European Commission, COMMETT Programme
Vermessungsbüro Dipl.-Ing. Peter Schmid, Gruppe Geoinformation, Austria
Technische Universität Wien, Austria

Organizing Committee

Chair: Sabine Timpf (Austria)
           Roswitha Markwart (Austria)
           Irene Orchard (Austria)
           Edith Unterweger (Austria)
           Rebecca Winn (Austria, UK)

Table of Contents

Wayfinding

Multiple Representations

Qualitative Spatial Reasoning II

Spatial Analysis

Temporal Reasoning

Cultural, Social, and Linguistic Aspects of Space

Spatial Relations

Posters

Authors' Index

Naive Geography*

Max J. Egenhofer
National Center for Geographic Information and Analysis
and
Department of Spatial Information Science and Engineering
Department of Computer Science
University of Maine
Orono, ME 04469-5711, U.S.A.
max@mecan1.maine.edu

David M. Mark
National Center for Geographic Information and Analysis
and
Department of Geography
State University of New York at Buffalo
Buffalo, NY 14261-0023, U.S.A.
dmark@geog.buffalo.edu

Abstract. This paper defines the notion and concepts of *Naive Geography*, the field of study that is concerned with formal models of the common-sense geographic world. Naive Geography is the body of knowledge that people have about the surrounding geographic world. Naive Geography is envisioned to comprise a set of theories that provide the basis for designing future Geographic Information Systems that follow human intuition and are, therefore, easily accessible to a large range of users.

1   Introduction

Naive Geography is the field of study that is concerned with formal models of the common-sense geographic world. It comprises a set of theories upon which next-generation Geographic Information Systems (GISs) can be built. In any case, Naive Geography is a necessary underpinning for the design of GISs that can be used without major training by new user communities such as average citizens, to solve day-to-day tasks. Such a scenario is currently a dream. Most GISs require extensive training, not only to familiarize the users with terminology of system designers, but also to educate them in formalizations used to represent geographic data and to derive geographic information. Naive Geography is also the basis for the design of intelligent GISs that will act and respond as a person would, therefore, empowering people to utilize GISs as reliable sources, without stunning surprises when using a system. This paper defines the notion[1] and concepts of Naive Geography.

---

*   This work was partially supported by the National Science Foundation (NSF) for National Center for Geographic Information and Analysis (NCGIA) under grant number SBE-8810917. Max Egenhofer's work is further supported by NSF grant IRI-9309230, and grants from Intergraph Corporation, Space Imaging Inc., Environmental Systems Research Institute, and the Scientific and Environmental Affairs Division of the North Atlantic Treaty Organization.

[1]   A poem by Waddington (1993) used the same term in a different context.

Although various aspects of Naive Geography have been studied for at least 40 years in a piecemeal fashion, Naive Geography has never been addressed comprehensively as a theory of its own. Occasionally, different terms have been used to describe certain aspects of it—Spatial Theory (Frank 1987), Geographical Information Science (Goodchild 1992), Spatial Information Theory (Frank and Campari 1993), Environmental Psychology, or plain Artificial Intelligence. Aspects of Naive Geography have been also considered within academic geography, and can be found in books by Bunge (1962) or Abler *et al.* (1971). By labeling Naive Geography, and distinguishing it from related areas in spatial information theory, geographic information science, and Naive Physics, we intend to catalyze and focus work on some very central issues for these fields, and for artificial intelligence and GIS in general.

Central to Naive Geography is the area of spatial and temporal reasoning. Many concepts of spatial and temporal reasoning have become important research areas in a wide range of application domains such as Physics, Medicine, Biology, and Geography. Particularly the field of Naive Physics (Hayes 1979; 1985a) addresses concerns that appear at a first glance to be very similar to Naive Geography. We will, however, be more specific on the domain, and the types of representation and reasoning by focusing on common-sense reasoning about geographic space and time; subsequently called geographic reasoning. We argue that such a focus is necessary to treat appropriately the ontological and epistemological differences among the different application domains of spatio-temporal reasoning—their data and their reasoning methods, the way people use these data and interact with them.

Much of Naive Geography should employ *qualitative reasoning methods*. Note that this notion of qualitative reasoning is distinct from the notion of qualitativeness as it is occasionally used in geography to allude to descriptive rather than analytical methods. In qualitative reasoning a situation is characterized by variables that can only take a small, predetermined number of values (De Kleer and Brown 1984) and the inference rules use these values *in lieu* of numerical quantities approximating them. Qualitative reasoning enables one to deal with partial information, which is particularly important for spatial applications when only incomplete data sets are available. It is important to find representations that support partial information. Qualitative and quantitative approaches have significantly different characteristics. While quantitative models use absolute values, qualitative models deal with magnitudes, which can sometimes be seen as abstractions from the quantitative details; therefore, qualitative reasoning models can separate numerical analyses from the determination of magnitudes of events which may be assessed differently, depending on the context in which the particular situation is viewed. This is not to be confused with fuzzy reasoning, which is frequently applied to dealing with imprecise information (Zadeh 1974). Qualitative spatial reasoning is exact, as is its outcome; yet, the resulting qualitative spatial information may be underdetermined, i.e., there is a set of possible values, one of which is the correct result (Morrissey 1990). Qualitative information and qualitative reasoning are not seen as substitutions for quantitative approaches, they are rather complementary methods, which should be applied whenever appropriate. For many decision processes qualitative information is sufficient; however, occasionally quantitative measures, dealing with precise numerical values, may be necessary and that would require the integration of quantitative information into qualitative reasoning. Qualitative approaches allow the users to abstract from the myriad of details by establishing *landmarks* (Gelsey and

McDermott 1990) when "something interesting happens"; therefore, they allow them to concentrate on a few but significant events or changes (De Kleer and Brown 1984).

The remainder of this paper continues with a brief review of Naive Physics (Section 2), and then defines Naive Geography in more detail (Section 3). Section 4 discusses an approach that promises progress toward the development of a Naive Geography. In Section 5, we lay out a sampling of ingredients of a Naive Geography. Section 6 presents our conclusions and points out some directions for further research.

2   Naive Physics

"Naive Physics is the body of knowledge that people have about the surrounding physical world. The main enterprises of Naive Physics are explaining, describing, and predicting changes to the physical world." (Hardt 1992, p. 1147). The term *Naive Physics* was coined by Patrick Hayes, and introduced in his *Naive Physics Manifesto* (Hayes 1978), a passionate and visionary statement that provided a catalyst for much research into qualitative methods for spatial and temporal problem solving. It was motivated by the recognition that Artificial Intelligence was—in the late 1970s—full of toy problems: "Small, artificial axiomatizations or puzzles designed to exercise the talents of various problem-solving programs or representational languages or systems" (Hayes 1978, p. 242). To overcome this limitation, Hayes proposed that researchers should concentrate on modeling common-sense knowledge.

Related terms and concepts include Intuitive Physics, Qualitative Physics, and Common-Sense Physics—some of these terms are more or less synonymous with Naive Physics, whereas others treat similar problems using different approaches. *Intuitive Physics* (McClosky 1983) addresses people's thinking about such tasks as dropping an object on a target while walking. Many people demonstrated poor performance in predicting when to release an object, which indicated that their intuitive models of physics may deviate from our current text-book examples of Newtonian Physics. Similarly, Naive Geography may follow Intuitive Physics as it may contradict many of our currently employed models for geographic space and time. *Qualitative Physics* (De Kleer and Brown 1984; De Kleer 1992) describes models of small-scale space in which objects undergo mechanical operations. A well-investigated example is the attempt to replicate the behavior of an analog clock (Forbus *et al.* 1991). While Qualitative Physics employs some methods that may be relevant to Naive Geography, it differs because Qualitative Physics usually focuses on the mechanics of a system and excludes human interaction.

Naive Physics by no means excludes geographic spaces. Indeed, Hayes's (1978) seminal paper on the topic contains examples of lakes and other geographic features; however, the great majority of the work in naive, common-sense, qualitative, and intuitive physics deals with spaces and objects manipulable by people, perceived from a single view point. There is strong evidence, from a variety of sources, that people conceptualize geographic spaces differently from manipulable, table-top spaces (Downs and Stea 1977; Kuipers 1978; Zubin 1989; Mark 1992a; Montello 1993; Pederson 1993; Mark and Freundschuh 1995). Thus, we think the new term, *Naive Geography*, is appropriate as part of an attempt to focus the research efforts of theoretical geographers and other spatial information theorists, on formal models of common-sense knowledge of geographic spaces.

3 Naive Geography: the Notion

In this paper, we are using the notion and concepts of *Naive Geography* to refer to what might otherwise have been called the *Naive Physics of Geographic Space*. Modifying Hardt's (1992) definition of Naive Physics:

> *Naive Geography is the body of knowledge that people have about the surrounding geographic world.*

Naive Geography captures and reflects the way people think and reason about geographic space and time, both consciously and subconsciously. *Naive* stands for instinctive or spontaneous.

Naive geographic reasoning is probably the most common and basic form of human intelligence. Spatio-temporal reasoning is so common in people's daily life that one rarely notices it as a particular concept of spatial analysis. People employ such methods of spatial reasoning almost constantly to infer information about their environment, how it evolves over time, and about the consequences of changing our locations in space. Naive geographic reasoning can be, and has to be, formalized so that it can be implemented on computers. As such Naive Geography will encompass sophisticated theories.

Naive geographic reasoning may actually contain "errors" and will occasionally be inconsistent. It may be contrary to objective observations in the real, physical world. These are properties that have been dismissed by the information systems and database communities. The principle of databases has been storage of non-redundant data to avoid potential inconsistencies. Information systems are supposed to provide one answer, one and only one. Naive Geography theories give up some of these restricted views of an information system.

3.1 The Essence of Naive Geography: Geographic Space

Geographic space is large-scale space, i.e., space that is beyond the human body and that may be represented by many different geometries at many different scales. Occasionally, geographic space has been defined as space that cannot be observed from a single viewpoint (Kuipers 1978; Kuipers and Levitt 1988). The intention of this definition was to describe the fact that geographic space comprises more than what a person sees. Of course, this definition falls short the moment one considers hills, towers, skyscrapers, hot-air balloons, airplanes, and satellites from which one can gain a view of much larger portions of space than by standing in a parking lot. A better definition of geographic space might be the space that contains objects that we humans do not think of being manipulable objects.

Geographic space is larger than a molecule, larger than a computer chip, larger than a table-top. Its objects are different from an atom, a microscopic bacterium, the pen in your hand, the engine that drives your car. Geographic space may be a hotel with its many rooms, hallways, floors, etc. Geographic space may be Vienna, with its streets, buildings, parks, and people. Geographic space may be Europe with mountains, lakes and rivers, transportation systems, political subdivisions, cultural variations, and so on. Within such spaces, we constantly move around. We explore geographic space by navigating in it, and we conceptualize it from multiple views, which are put together (mentally) like a jigsaw puzzle. This makes geographic space distinct from small-scale space, or table-top space, in which objects are thought of as being manipulable and whenever an observer lacks some information about these objects, he or she can get this information by moving the object into such a position that one can see, touch, or measure the relevant parts.

3.2 Naive Geography for GIS Design

In addition to the scientific motivation of trying to get a better understanding of how people handle their environments, there is the need to incorporate naive geographic knowledge and reasoning into GISs. The concepts and methods people use to infer information about geographic space and time become increasingly important for the interaction between users and computerized GISs. While many spatial inferences may appear trivial to us, they are extremely difficult to formalize so that they could be implemented on a computer system. Current methods to derive spatial and temporal information about geographic space are limited; therefore, we see a big gap between what a human user wants to do with a GIS, and the spatial concepts offered by the GIS. Today's GISs do not sufficiently support common-sense reasoning; however, in order to make them useful for a wider range of people, and in order to allow for prediction or forecasting, it will be necessary to incorporate people's concepts about space and time and to mimic human thinking; therefore, we will focus on common-sense geographic reasoning, reasoning as it is performed by people, reasoning whose outcome makes intuitive sense to people, reasoning that needs little explanation.

In the past, geographic reasoning has been limited to calculations in a Cartesian coordinate space; however, Euclidean geometry is not a good candidate for representing geographic information, since it relies on the existence of complete coordinate n-tuples. Likewise, pictorial representations are inadequate since they overdetermine certain situations, e.g., when drawing a picture representing a cardinal direction, a sketch also includes information about the sizes of the objects and some relative distances. Formalized spatial data models have been extensively discussed in the context of databases and GISs; however, to date there are, for instance, no models for a comprehensive treatment of different kinds of spatial concepts and their combinations that are cognitively sound and plausible. More flexible and advanced methods are needed to capture the results from cognitive scientists' studies, such as the fact that the nature of errors in people's cognitive maps is most often metrical and only rarely topological (Lynch 1960), or how topological structure (Stevens and Coupe 1978) or gestalt are used for spatial reasoning. Researchers have identified different types of spaces with related inference methods (Piaget and Inhelder 1967; Golledge 1978; Couclelis and Gale 1986). GISs need to include such intelligent mechanisms to deal with often complex spatial concepts. If GISs can achieve geographic reasoning in a manner similar to a human expert, these systems will be much more valuable tools for a large range of users—family members who are planning their upcoming vacation trip, scientists who want to analyze their data collections, or business people who want to investigate how they performed in various geographic markets.

3.3 What Naive Geography is Not

Naive Geography is neither arm-chair science, nor does it employ Mickey-Mouse research. Likewise, Naive Geography is neither childish nor stupid geography, nor is it the geography of ignorant or simple-minded people. It is not geography by the uneducated nor for the uneducated. Despite the attempts to capture human performance, naive geographic reasoning does not aim at being descriptive, neither in its methodologies nor in its results and interpretations. And it is not just another term for fuzzy reasoning, nor is fuzzy reasoning a substitute for Naive Geography—it might have its value as one of several methods for naive geographic reasoning, though. Finally, Naive Geography is not a replacement for GIS.

3.4 Naive Geography and Related Disciplines

Naive Geography is not a completely new discipline. Quite the opposite, it is closely related to several of our current scientific and engineering disciplines, and builds upon them. Geography is the most obvious discipline—it is part of the name Naive Geography. Geography is the science concerned with relationships, processes, and patterns of our surrounding world, and as such it addresses at a coarse level the kind of issues we are concerned with. At a more detailed level, the domain-specific fields contribute to Naive Geography. They include geology, archeology, economics, and transportation as they describe particular domain knowledge that shapes the users' and analysts' mental models and therefore, often enable inference that is otherwise impossible.

These geographic disciplines are not the only relevant fields for Naive Geography. Naive Geography has to employ concepts and principles of cognitive science and linguistics to ensure a linkage with the way people perceive geographic space and time, and the ways they communicate about them. Naive Geography is associated with anthropology as it has to accommodate regional and cultural particularities in how people deal with geographic space and time. There is the field of psychology upon which Naive Geography builds. And philosophy may contribute to Naive Geography as Aristotle's, Kant's, or Leibnitz's views of space frame many of the discussions about the nature of Naive Geography.

Finally, there are the fields that provide us the tools to express and formalize naive geographic knowledge: engineering as it pertains to the modeling of geographic information, from measurements about the Earth to GIS user interface design, as well as computer science and mathematics.

This scanning of relevant fields is certainly incomplete, and there may be many others whose findings and influences may be even more dramatic than those listed here. There are many who contribute—as there are many who will benefit.

4 Towards the Development of Naive Geography

Naive Geography has to bridge between different scientific perspectives; therefore, in order to investigate naive geographic concepts, researchers have to combine different research methodologies. It will be the *interplay* between the different approaches that will provide the exciting and useful results.

The framework for developing Naive Geography consists of two different research methodologies: (1) the *development of formalisms* of naive geographic models for particular tasks or sub-problems so that programmers can implement simulations on computers; and (2) the *testing and analyzing of formal models* to assess how closely the formalizations match human performance. For Naive Geography, the two research methods are only useful if they are closely integrated and embedded in a *feedback loop* to ensure that (1) mathematically sound models are tested (bridging between formalism and testing) and (2) results from tests are brought back to refine the formal models (bridging between testing and implementable formalisms). The outcome of such a complete loop leads to refined models, which in turn should be subjected to new, focused evaluations. In an ideal scenario, this leads to formal models that ultimately match closely with human perception and thinking. From the refinement process we may gain new insight into common-sense reasoning and we may actually derive certain reasoning patterns. The latter—the generic rules—would manifest *naive geographic knowledge*.

Research in the area of spatial relations provides an example in which the combination and interplay of different methods generates useful results. The treatment of spatial relations within Naive Geography must consider two complementary sources: (1) the cognitive and linguistic approach, investigating the terminology people use for spatial concepts (Talmy 1983; Herskovits 1986; Retz-Schmidt 1988) and human spatial behavior, judgments, and learning in general; and (2) the formal approach concentrating on mathematically based models, which can be implemented on a computer (Egenhofer and Franzosa 1991; Papadias and Sellis 1994; Hernández 1994). The formalisms serve as hypotheses that may be evaluated with human-subject testing (Mark *et al.* 1995).

5   Some Elements of Naive Geography

The mere identification of a comprehensive set of elements of Naive Geography comprises a major research task, and its completion would provide a big step towards the successful manifestation of Naive Geography. As a starting point, we present an *ad hoc* collection of elements that would contribute to a Naive Geography. The list is by no means exhaustive, and some of the following may turn out to be false, or at least uncommon and/or limited to specific cultures, primarily those of the authors. We present these elements to give the reader a flavor of what we intend should be included in Naive Geography.

5.1   Naive Geographic Space is Two-Dimensional

Manipulable objects on a table-top are essentially three-dimensional. Even a sheet of paper has a thickness. Furthermore, in everyday-object (manipulable) space, the three dimensions are all about equal. Objects are easily rotated about any axis, or obliquely. When an object is moved, we expect its properties, spatial and non-spatial, to remain unchanged.

Geographic space under Naive Geography is, in contrast, essentially two-dimensional. There is considerable evidence that the horizontal and vertical dimensions are decoupled in geographic space. For example, people often grossly over-estimate the steepness of slopes, and the depths of canyons compared to their widths. So, instead of parsing a three-dimensional space into three independent one-dimensional axes, geographic space seems to be interpreted as a horizontal, two-dimensional space, with the third dimension reduced more to an attribute (of position) rather than an equal dimension. This is very much like the 2 1/2-D representations used in computational vision (Marr 1982). That GISs have succeeded in the marketplace with little or no capabilities to do three-dimensional analysis is testimony to the nature of geographic space. A two-dimensional system for CAD (computer-aided design) would not likely be successful.

5.2   The Earth is Flat

This is a different point than the one about two-dimensionality. In most of our large-scale reasoning tasks, this is a common simplification. It is not a discussion as to whether it is admissible, or not. People do it. When traveling from Boston to New York, one disregards the Earth's curvature. This is independent of the mode of transportation. Trans-Atlantic air travelers often ask why the flight path goes all the way up over Greenland, rather than going straight across—the great circle, shortest path between two points across the surface of a sphere, is not part of common-sense knowledge for most people.

5.3 Maps are More Real Than Experience

Perhaps this point should be, "Maps are more faithful to the reality of geographic space than are our direct experiences of such spaces." Many times, we hear statements like, "When I get home, I want to look at the route on a map, to see where I went." This seems to be based on a naive assumption that the truth about where one is in geographic space is better represented by a map-based, map-like, or configurational view of geographic space, than it is by our memories of our experiences with that space from within.

5.4 Geographic Entities are Ontologically Different from Enlarged Table-Top Objects

As geographic space differs from table-top space, so are the properties and the behavior of many entities in geographic space different from those on a table top. The issue is not just mere size. In his paper *Ontology of Liquids*, Hayes (1985b) gave an excellent example with a detailed discussion of how the ontology of lakes is different from that of many other objects composed of liquids. He showed how a phenomenon/entity in geographic space has an ontology that is not simply an enlarged version of the table-top manipulable world.

5.5 Geographic Space and Time are Tightly Coupled

The linkage between space and time is an aspect of Naive Geography that deserves special attention. The term *geographic space and time* is understood such that *geographic* distributes over *space and time*—formalists would tend to write geographic (space and time). As there is geographic space, we want to argue that there is geographic time, i.e., time that is inherently linked to geographic concepts (Egenhofer and Golledge 1994). We select one of several examples to underline this claim:

Many cultures have pre-metric units of area that are based on effort over time (Kula 1983). The English *acre* (Jones 1963; Zupko 1968; 1977), the German *morgen* (Kennelly 1928), and the French *arpent* (Zupko 1978) all are based on the amount of land that a person with a yoke of oxen or a horse can plow in one day or one morning. There have been similar measures for distance, such as how far a person can walk in an hour, or how far an army can march in a day. We know of no such "effort-based" units of measure for manipulable (table-top) space.

5.6 Geographic Information is Frequently Incomplete

Another setting for geographic reasoning is given by the constraint that reasoning in geographic space must typically deal with incomplete information. Nevertheless, people can draw sufficiently precise conclusions, e.g., by completing information intelligently or by applying default rules, frequently based on common sense. A number of cognitive studies have provided evidence that people may employ hierarchically organized schemes to reason in geographic space and to compensate for missing information (Hirtle and Jonides 1985; McNamara *et al.* 1989).

5.7 People use Multiple Conceptualizations of Geographic Space

When thinking about geographic space, people typically employ several different concepts, and change between them frequently. Such conceptualizations of space may reflect the differences between perceptual and cognitive space (Couclelis and Gale 1986), or may be based on different geometrical properties, such as continuous vs.

discrete (Egenhofer and Herring 1991; Frank and Mark 1991). The dependency on scale, or difference in the types of operations people would typically employ, has been raised as another motivation for distinguishing different types of spaces (Zubin 1989).

5.8 Geographic Space has Multiple Levels of Detail
This aspect of representing geographic space is orthogonal to multiple conceptualizations of geographic space. A conceptualization of geographic space may have several levels of granularity, each of which will be appropriate for problem solving at different levels of detail. In cartographic applications, this aspect has been considered to be part of *scale* (Buttenfield 1989). The naive view of geographic space implies that processing a query against a more detailed representation would not provide a more precise query result.

5.9 Boundaries are Sometimes Entities, Sometimes Not
The fact that Naive Geography models geographic space as it is perceived by people, is strongly reflected in the way boundaries are represented. There is no uniform view of what a boundary is and how it is established—even if one could agree on a model for the physical entities. Such simple configurations as national boundaries may have diverse interpretations, even if the countries involved agree over the extent of their territories. Conventionally, political subdivisions are modeled as a partition of space in which a boundary separates one nation's land from its neighbor. Each of the neighbors may actually have a different perspective, namely that the boundary belongs to their country. As such, the boundary between two neighboring countries may be considered a pair of boundaries. Smith (1994) argues, from a philosophical point of view, that there may be geographic situations in which the boundary between two adjacent areas is even asymmetric. As examples he cites situations in which one country did not recognize the existence of a national boundary with its neighbor, while the other country considered it a valid boundary. Political subdivisions are certainly not the only cases in which such multiple views of boundaries may occur. The same case could be made for land parcels and the question as to who owns the boundary between two adjacent parcels.

5.10 Topology Matters, Metric Refines
In geographic space, topology is considered to be first-class information, whereas metric properties, such as distances and shapes, are used as refinements that are frequently less exactly captured. There is ample evidence that people organize geographic space such that topological information is retained fairly precisely, capturing such relationships as inclusion, coincidence, and left/right (Lynch 1960; Stevens and Coupe 1978; Riesbeck 1980).

5.11 People have Biases Toward North-South and East-West Directions
People's mental maps of directions and distances are frequently quite gross simplifications, with particular preferences for alignments in North-South and East-West directions. Despite exposure to maps and satellite images, we often ignore *geographic reality*. For instance, at a global scale, South America often is considered to be due south of North America. Likewise, most people misjudge latitudes when trying to compare cities in North America and Europe (Tversky 1981). While such misconceptions are similar to those found by Stevens and Coupe (1978), they cannot be explained with a hierarchical conceptualization of geographic space. A potential source for some of these errors are climate comparisons, and the equation (for the

Northern hemisphere) that colder means further North, and warmer equates to further South, may indicate that factors other than geographic location may influence estimations of directions.

Biases toward strict cardinal directions appear also in judgments about coastlines—the U.S. East coast is frequently believed to be due North-South (Mark 1992b). Such misconceptions may have surprising consequences when people interact with information systems. For example, most people requesting the satellite image South of the State of Maine from an image archive, would expect to receive an image that covers parts of New Hampshire and Massachusetts (Frank 1992). They would be puzzled to get nothing but water!

People tend to have similar biases towards North-South directions and right angles in navigation, where they may be irritated by slight deviations from the norm and consequently perform poorly in wayfinding.

5.12 Distances are Asymmetric

Euclidean geometry includes the axiom that a distance from point A to point B is equal to the distance from B to A. In naive geographic space, this premise is frequently violated. Distances are not only thought of as lengths of paths on the Earth's surface, but frequently seen as a measure for how long it takes to get from one place to another (Kosslyn *et al.* 1978). The *shortest* path may have multiple interpretations, e.g., in terms of distance, time, fuel consumption, or toll. Even if the same path, in opposite directions, is chosen between two points, the *distance* as people perceive it may not be the same (Golledge *et al.* 1969): terrain may influence how fast one can travel or traffic during rush hours may slow down travel in one direction.

While distance applies as a measure between positions in geographic space, it extends to abstract concepts where it captures *conceptual closeness*. For example, among water bodies, a pond is conceptually closer to a lake than to the sea, because one can find more conceptual differences between a pond and the ocean than between a pond and a lake. The shorter the distance is, the more similar the instances are. Again, such distances among concepts are frequently asymmetric, implying that the induced similarity is asymmetric as well (Papadias 1995), i.e., if A is similar to B, then B is not necessarily similar to A.

5.13 Distance Inferences are Local, Not Global

Geographic distances are thought of as local, i.e., covering the neighborhood between the two points of interest, without involving locations remote to both objects. Common coordinate systems, however, have their origins at the equator, and distance differences are calculated as differences of lengths from the equator and from Greenwich. How far it is from Bangor, Maine to Orono, Maine is based on how distant Bangor and Orono are from the equator, and how remote Bangor and Orono are from Greenwich, U.K. (Goodchild 1994). In a similar way, any distinction about North, South, East, and West is related on the reference frame's (remote) origin. Despite the convenience of such coordinate calculations, alternative spatial reference systems are needed in support of Naive Geography. Such reference systems should pay attention to neighborhood relations, as demonstrated in measurement-based systems (Buyong *et al.* 1991), or use coordinate-based calculations as a last resort of inference, as supported by deductive geographic databases (Sharma *et al.* 1994).

5.14 Distances Don't Add Up Easily

Reasoning about distances along networks in geographic space underlies formalisms that differ considerably from standard calculus. Usually, one adds up lengths of segments along a path, irrespective of their values, to obtain the length of the entire path. This method provides unreasonable results in cases where the values to be added differ by large amounts. For instance, the distance between the airports in Bangor, Maine and Santa Barbara, California is approximately 5,000 kilometers. When computing the travel distance from the University of Maine to UC Santa Barbara, it would make little sense to add the relatively short legs between the campuses and the respective airports—10 kilometers and 1.5 kilometers—to the overall distance and claim that it took 5,011.5 kilometers to get from one campus to the other.

6 Conclusions

This paper described the notion and concepts of Naive Geography. Naive Geography establishes the link between how people think about geographic space and how to develop formal models of such reasoning that can be incorporated into software systems. Such intelligent GISs—one or two generations down the road—would be intuitive to use and would provide powerful reasoning capabilities and some limited methods to make predications of human behavior. Like Patrick Hayes in his *Naive Physics Manifesto*, we consider our framework as a start of a discussion, to be revised in the future.

Common-sense reasoning is difficult, and if there are formalizations that appear to be common-sensical, then they are excellent results. Unfortunately, our scientific communities frequently consider such formalizations as "too simplistic"—because everyone understands them, and science should have some complexity to be considered science. We disagree with this attitude at the level of common-sense reasoning. *If it is simple and solves the problem, then it is good.*

7 Acknowledgments

Andrew Frank, Pat Hayes, and Barry Smith deserve special credit for discussions and comments that helped us shape the concepts of Naive Geography. We are also grateful to many others—impossible to be named explicitly—who have contributed as well.

8 References

Abler, R., J. Adams, and P. Gould (1971) *Spatial Organization—The Geographer's View of the World.* Englewood Cliffs, NJ: Prentice-Hall.

Bunge, W. (1962) *Theoretical Geography.* Lund: C.W.K. Gleerup.

Buttenfield, B. (1989) *Multiple Representations: Initiative 3 Specialist Meeting Report.* National Center for Geographic Information and Analysis, Santa Barbara, CA, Technical Report 89-3.

Buyong, T., W. Kuhn, and A. Frank (1991) A Conceptual Model of Measurement-Based Multipurpose Cadastral Systems, *URISA Journal* 3(2):35-49.

Couclelis, H. and N. Gale (1986) Space and Spaces. *Geografiska Annaler* 68(B):1-12.

De Kleer, J. (1992) Physics, Qualitative. in: S. Shapiro (ed.), *Encyclopedia of Artificial Intelligence*. Second Edition. New York: John Wiley & Sons, Inc., 2:1149-1159.

De Kleer, J. and J. Brown (1984) A Qualitative Physics Based on Confluences. *Artificial Intelligence* 24:7-83.

Downs, R. and D. Stea (1977) *Maps in Minds: Reflections on Cognitive Mapping*. New York: Harper and Row.

Egenhofer, M. and R. Franzosa (1991) Point-Set Spatial Topological Relations. *International Journal of Geographical Information Systems* 5(2):161-174.

Egenhofer, M. and R. Golledge (1994) *Time in Geographic Space: Report on the Specialist Meeting of Research Initiative 10*. National Center for Geographic Information and Analysis, Santa Barbara, CA, Technical Report 94-9.

Egenhofer, M. and J. Herring (1991) High-Level Spatial Data Structures for GIS. in: D. Maguire, M. Goodchild, and D. Rhind (eds.), *Geographical Information Systems, Vol. 1: Principles*. London: Longman, pp. 147-163.

Forbus, K., P. Nielsen, and B. Faltings (1991) Qualitative Spatial Reasoning: The CLOCK Project. *Artificial Intelligence* 51:417-471.

Frank, A. (1987) Towards a Spatial Theory. in: *International Geographic Information Systems (IGIS) Symposium: The Research Agenda*. Arlington, VA, pp. 215-227.

Frank, A. (1992) Personal communication.

Frank, A. and I. Campari, Eds. (1993) *Spatial Information Theory, European Conference, COSIT '93. Lecture Notes in Computer Science* Vol. 716. New York: Springer-Verlag.

Frank, A. and D. Mark (1991) Language Issues for GIS. in: D. Maguire, M. Goodchild, and D. Rhind (eds.), *Geographical Information Systems, Vol. 1: Principles*. London: Longman, pp. 147-163.

Gelsey, A. and D. McDermott (1990) Spatial Reasoning About Mechanisms. in: S. Chen (Ed.), *Advances in Spatial Reasoning*. 1:1-33, Norwood, NJ: Ablex Publishing Corporation.

Golledge, R. (1978) Learning about Urban Environments. in: T. Carlstein, D. Parkes, and N. Thrift (Eds.), *Timing Space and Spacing Time*. London: Edward Arnold.

Golledge, R., R. Briggs, and D. Demko (1969) The Configuration of Distances in Intra-Urban Space. *Proceedings of the Association of American Geographers*, pp. 60-65.

Goodchild, M. (1992) Geographical Information Science. *International Journal of Geographical Information Systems* 6(1):31-45.

Goodchild, M. (1994) Personal communication.

Hardt, S. (1992). Physics, Naive. in: S. Shapiro (Ed.), *Encyclopedia of Artificial Intelligence*. Second Edition. New York: John Wiley & Sons, Inc., 2:1147-1149.

Hayes, P. (1978) The Naive Physics Manifesto. in: D. Michie (Ed.), *Expert Systems in the Microelectronic Age*. Edinburgh, Scotland: Edinburgh University Press, pp. 242-270.

Hayes, P. (1985a) The Second Naive Physics Manifesto. in: J. Hobbs and R. Moore (Eds.), *Formal Theories of the Commonsense World*. Norwood, NJ: Ablex, pp. 1-36.

Hayes, P. (1985b) Naive Physics I: Ontology of Liquids. in: J. Hobbs and R. Moore (Eds.), *Formal Theories of the Commonsense World*. Norwood, NJ: Ablex, pp. 71-108.

Hernández, D. (1994) *Qualitative Representation of Spatial Knowledge*, Lecture Notes in Computer Science, Vol. 804, New York: Springer-Verlag.

Herskovits, A. (1986) *Language and Spatial Cognition—An Interdisciplinary Study of the Prepositions in English*. Cambridge, MA: Cambridge University Press.

Hirtle, S. and J. Jonides (1985) Evidence of Hierarchies in Cognitive Maps. *Memory and Cognition* 13(3):208-217.

Jones, S. (1963) *Weights and Measures: An Informal Guide*. Washington, D.C.: Public Affairs Press

Kennelly, A. (1928) *Vestiges of Pre-Metric Weights and Measures Persisting in Metric-System Europe*, 1926-1927. New York: The Macmillan Company.

Kosslyn, S., T. Ball, and B. Reiser (1978) Visual Images Preserve Metric Spatial Information: Evidence from Studies of Image Scanning. *Journal of Experimental Psychology: Human Perception and Performance* 4:47-60

Kuipers, B. (1978) Modeling Spatial Knowledge. *Cognitive Science* 2:129-153.

Kuipers, B. and T. Levitt (1988) Navigation and Mapping in Large-Scale Space. *AI Magazine* 9(2):25-46.

Kula, W. (1983) *Les Mesures et Les Hommes. Paris: Maison des Sciences de L'Homme*. [Translated from Polish by Joanna Ritt; Polish edition 1970.]

Lynch, K. (1960) *The Image of a City*. Cambridge, MA: MIT Press.

Mark, D. (1992a) Spatial Metaphors for Human-Computer Interaction. *Fifth International Symposium on Spatial Data Handling*. Charleston, SC, 1:104-112.

Mark, D. (1992b) Counter-Intuitive Geographic "Facts:" Clues for Spatial Reasoning at Geographic Scales. in: A. Frank, I. Campari, and U. Formentini (Eds.), *Theories and Methods of Spatio-Temporal Reasoning in Geographic Space*. Lecture Notes in Computer Science No. 639, Berlin: Springer-Verlag, pp. 305-317.

Mark, D., D. Comas, M. Egenhofer, S. Freundschuh, M. Gould, and J. Nunes (1995) Evaluating and Refining Computational Models of Spatial Relations Through Cross-Linguistic Human-Subjects Testing, *COSIT `95*, Semmering, Austria, *Lecture Notes in Computer Science*, Springer-Verlag.

Mark, D. and S. Freundschuh (1995) Spatial Concepts and Cognitive Models for Geographic Information Use. in: T. Nyerges, D. Mark, R. Laurini, and M.

Egenhofer (Eds.), *Cognitive Aspects of Human-Computer Interaction for Geographic Information Systems*. Dordrecht: Kluwer Academic Publishers.

Marr, D. (1982) *Vision*, San Francisco, CA: W.H. Freeman.

McClosky, M. (1983) Intuitive Physics. *Scientific American* 248(4):122-130.

McNamara, T., J. Hardy, and S. Hirtle (1989) Subjective Hierarchies in Spatial Memory, *Journal of Environmental Psychology: Learning, Memory, and Cognition* 15(2):211-227.

Montello, D. (1993) Scale and Multiple Psychologies of Space. in: A. Frank and I. Campari (Eds.), *Spatial Information Theory: A Theoretical Basis for GIS*. Lecture Notes in Computer Sciences No. 716, Berlin: Springer-Verlag, pp. 312-321.

Morrissey, J. (1990) Imprecise Information and Uncertainty in Information Systems. *ACM Transactions of Information Systems* 8(2): 159-180.

Papadias, D. (1995) Personal communication.

Papadias, D. and T. Sellis (1994) Qualitative Representation of Spatial Knowledge in Two-Dimensional Space. *VLDB Journal* 3(4):479-516.

Pederson, E. (1993) Geographic and Manipulable Space in Two Tamil Linguistic Systems. in: A. Frank and I. Campari (Eds.), *Spatial Information Theory: A Theoretical Basis for GIS*. Lecture Notes in Computer Sciences No. 716, Berlin: Springer-Verlag.

Piaget, J. and B. Inhelder (1967) *The Child's Conception of Space*. New York: Norton.

Retz-Schmidt, G. (1988) Various Views on Spatial Prepositions. *AI Magazine* 9:95-105.

Riesbeck, C. (1980) You Can't Miss It: Judging the Clarity of Directions. *Cognitive Science* 4:285-303.

Sharma, J., D. Flewelling, and M. Egenhofer (1994) A Qualitative Spatial Reasoner. in: T. Waugh and R. Healey (Eds.) *Sixth International Symposium on Spatial Data Handling*. Edinburgh, Scotland, pp. 665-681.

Smith, B. (1994) The Formal Ontology of Space: An Essay in Mereotopology. in: L. Hahn (Ed.), *The Philosophy of Roderick Chisholm*. Chicago and LaSalle: Open Court (in press).

Stevens, A. and P. Coupe (1978) Distortions in Judged Spatial Relations. *Cognitive Psychology* 10:422-437.

Talmy, L. (1983) How Language Structures Space. in: H. Pick and L. Acredolo (Eds.), *Spatial Orientation: Theory, Research, and Application*. New York: Plenum Press, pp. 225-282.

Tversky, B. (1981) Distortions in Memory for Maps. *Cognitive Psychology* 13:407-433.

Waddington, M. (1993) Naive Geography. *Queen's Quarterly* 100(1):149.

Zadeh, L. (1974) Fuzzy Logic and Its Application to Approximate Reasoning. in: *Information Processing*. North-Holland Publishing Company.

Zubin, D. (1989) Untitled, in: D. Mark, A. Frank, M. Egenhofer, S. Freundschuh, M. McGranaghan, and R. M. White (Eds.), *Languages of Spatial Relations: Initiative Two Specialist Meeting Report*. Technical Paper 89-2, National Center for Geographic Information and Analysis, Santa Barbara, CA, pp. 13-17.

Zupko, R. (1968) *A Dictionary of English Weights and Measures*. Madison, WI: The University of Wisconsin Press.

Zupko, R. (1977) *British Weights and Measures: A History from Antiquity to the Seventeenth Century*. Madison, WI: The University of Wisconsin Press.

Zupko, R. (1978) *French Weights and Measures Before the Revolution: A Dictionary of Provincial and Local Units*. Bloomington, IN: Indiana University Press.

Qualitative Spatial Reasoning Using Algebraic Topology

Boi Faltings

Laboratoire d'Intelligence Artificielle (DI)
Swiss Federal Institute of Technology (EPFL)
IN-Ecublens, 1015 Lausanne, Switzerland
Tel. +41-21-693-2738, FAX: +41-21-693-5225
E-Mail: faltings@lia.di.epfl.ch

Abstract. Many propositions in spatial and qualitative reasoning can be modelled as regions in phase or configuration spaces. Deciding the consistency of k propositions then amounts to deciding the question of whether there exists a point which simultaneously falls into all k corresponding regions. A more difficult problem is to decide whether there is a point which falls *only* within the given k regions. I call this *feasibility* of the set of propositions.

In this paper, I present a method for deciding consistency and feasibility for convex regions using only topological inference. It uses Helly's theorem to decide consistency of any set of k propositions based on information about consistency of small subsets. Using methods of algebraic topology, I show a sufficient method to compute a minimal skeleton of *feasible* places which accurately models the connectivity between feasible environments.

The method has been implemented. I show how to formulate and solve the piano-movers problem, an important problem in spatial reasoning, using the framework.

1   Introduction

Consider the two example problems shown in Figure 1. Example a) might arise in a mystery story: during construction work, a body was found at a spot x which is known to have been visible from the north side of the house which used to stand there. The previous owner vaguely remembers once having seen the gardener dig a suspicious hole on the east of the house. The question is: might the gardener be the killer, i.e. could the hole he dug be the place where the body was found? In Example b), the problem might be that we know certain operating regions of a chemical plant, and have to shut it down safely from a current operating point x without provoking either a boiler overheat or a fire.

Both problems can be formulated qualitatively using a set of propositions involving the position of the hole (a) or the operating point of the plant (b). More formally, let the state of a system be expressed by a continous variable x and its model be a set of qualitative propositions $\{p_i(x)\}$. Each proposition is then associated with a region $\{r(p_i)\}$ such that $p_i(x)$ is true if and only if the state $x \in$

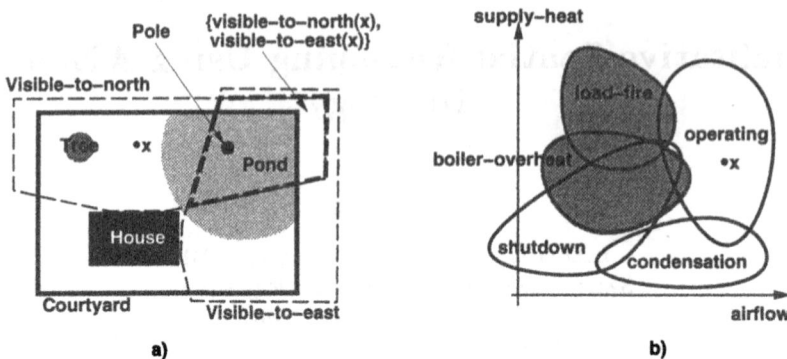

Fig. 1. *Examples of problems expressible by convex regions in two-dimensional space: a) position of a point, b) operating regions.*

$r(p_i)$. Thus, in example a), the proposition `visible-to-north(x)` corresponds to a region outlined by dashed lines, and in example b), `boiler-overheat(x)` corresponds to another region. In most previous work, regions are intervals of real numbers. In this paper, I consider general regions in d-dimensional space, but with the restriction that the individual regions are *convex*.

I assume that the precise geometry of regions is unknown, i.e. the pictures shown in Figure 1 are not available as such. Instead, through natural language descriptions or measurements, *overlap relations* among small subsets of regions are given. In d-dimensional space, I require knowledge of all simultaneous overlaps of $d+1$ regions and call these *(d+1)-relations*; for example, in 2-dimensions, all simultaneous overlaps of 3 regions are given. In the example of the pond, we might know that there is a pole in the middle of the pond which is visible both to the north and to the east of the house. This means that there must be a simultaneous intersection between the 3 regions: `visible-to-north`, `visible-to-east` and **pond**. Further relations might be given by the fact that the pond touches both the northeast corner of the house and the east and north courtyard walls, and other information about the relative locations of points. In the example of the chemical plant (b), experience might indicate that some situations in the shutdown regime also make the boiler overheat and the load catch fire, so that there is an overlap between these operating regions.

I define an *environment E* as a conjunction of a set $\{p_i\}$ of unnegated propositions. Associated with the environment is a place P such that the environment is true whenever the state falls within P. P is the intersection of all regions corresponding to propositions in E: $P(E) = \bigcap_{p_i \in E} r(p_i)$. In example a), the environment $E = \{\texttt{visible} - \texttt{to} - \texttt{north}(x), \texttt{visible} - \texttt{to} - \texttt{east}(x)\}$ is true of all positions x which fall within the intersection of the regions `visible-to-north` and `visible-to-east` (shown by the bold dashed lines in Figure 1).

An environment and its associated place is called *consistent* if there exists a point satisfying all propositions in E, which is equivalent to the condition that the associated place is a nonempty region. In example a), E is consistent, but the environment $E' = \{\texttt{visible} - \texttt{to} - \texttt{east}(x), \texttt{tree}(x)\}$ does not correspond to any place and is therefore inconsistent. An environment is called *feasible*, a stronger condition than consistency, if there is a point satisfying *only* the propositions in it. In example a), E is not feasible, since all positions x which satisfy E also satisfy the proposition $\texttt{pond}(x)$. In most applications, feasibility is more important than consistency. In our mystery story, my theory tells us that it is not possible to have dug one and the same hole both to the north and to the east of the house, i.e. the gardener is maybe not the killer after all. In process control, feasibility is important to know that the plant can be safely shut down by passing first through the `condensation` mode.

Somewhat surprisingly , it is possible to decide consistency and often also feasibility without knowledge of the precise shapes of the regions. I first give a simple algorithm for deciding consistency of any environment based on (d+1)-relations only. The main result of this paper is an algorithm for finding a minimal *skeleton* of *feasible* environments such that (i) every feasible environment is a superset of an environment in the skeleton, and (ii) any feasible trajectory is modelled by a sequence of adjacent feasible places. The skeleton is again obtained using only topological inference on the (d+1)-relations. In process control, (b) in (Figure 1), this skeleton could be used to find a path for bringing the process to a certain state while avoiding dangerous situations. In section 4, I show how a robot motion planning problem can be formulated and solved using convex regions and their minimal skeleton.

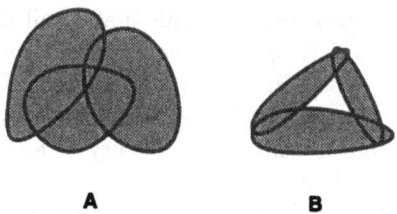

A   B

Fig. 2. *Situation A and B have the same pairwise relations, but the intersection of all three regions is consistent only in A.*

This work has been inspired in part by work on the logic of binary spatial relations ([2, 3, 4]). In these approaches, situations are modelled as conjunctions of *binary* relations among regions. However, in situations with more than 2 dimensions, such a representation is insufficient to answer queries about the consistency and feasibility of environments of propositions, as shown by the example of Figure 2. However, reasoning with binary relations can often be done in polynomial time, and may also be used for tasks such as qualitatively simulating a dynamic process [3].

Section 2 of this paper describes the theory and algorithms for deciding consistency of a set of regions. Section 3 gives an algorithm for computing homology groups which is required for the method of Section 2. Section 4 formulates the piano-movers problem as a problem of qualitative spatial reasoning, and shows the solution obtained using the methods I describe.

2 Deciding Consistency, Feasibility and Connectedness

In this section, I present three algorithms for (i) finding all maximal consistent (and feasible) environments, (ii) finding all minimal feasible environments, and (iii) finding adjacencies between environments.

Maximal Consistent Environments. An environment is consistent if the corresponding regions have a simultaneous intersection. This can be decided from the $(d+1)$-relations using Helly's theorem ([1]):

Theorem 1. *A set $\mathcal{R}$ of n convex regions r_i in d-dimensional space has a common intersection if and only if all subsets of $d + 1$ regions $\in \mathcal{R}$ have an intersection.*

The set of all consistent environments can be summarized by the set of *maximal* environments, consistent environments for which there is no superset which is also consistent. Note that any maximal environment is also *feasible*, i.e. there must exist a point where only the propositions in E hold. All subsets of a maximal environment are also consistent, but not necessarily feasible.

Minimal Feasible Environments. Recall that I call an environment E *feasible* if there is a point satisfying *only* the propositions in E and no others. Contrary to consistency, feasibility of an environment E implies neither that any superset nor that any subset of E is also feasible. Furthermore, I will show below that it is impossible to compute feasibility of an arbitrary environment using only local $(d+1)$-relations.

Topological inference is sufficient to compute all *minimal* feasible environments using only $(d+1)$-relations. In many reasoning tasks where feasibility is important, the minimal environments are in fact the most interesting ones, as they represent those solutions which require a minimal set of assumptions. Furthermore, the minimal feasible environments are equivalent to the maximal consistent environments of negated propositions: if E is minimal feasible, $\bigcup_{p \notin E} \neg p$ is a maximal consistent set of negated propositions.

The algorithm for computing all minimal feasible environments works under the following assumptions:

- the union of regions in the universe $\mathcal{U}$ is a simply connected region without holes.
- the union of regions is bounded by a *frame* of regions which are never part of any query, but for which all $(d+1)$-relations are known.

When these two assumptions hold, the presence of a feasible environment can be detected by considering the *changes* the topology of $\mathcal{U}$ undergoes as parts are removed. Figure 3 shows the principle for computing minimal environments. It

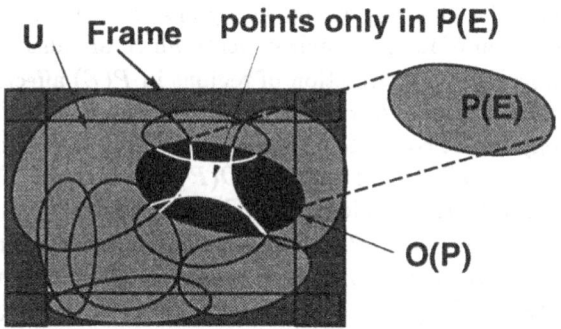

Fig. 3. *The Universe $\mathcal{U}$ is a simply connected region in configuration space, bounded by a frame. An environment E is feasible if removing its associated place $P(E)$ leaves a "hole" in $\mathcal{U}$ This can be detected considering the part of $\mathcal{U}$ which overlaps $P(E)$, shown in black and called the* overlap set *of E.*

is based on the *Alexander duality* ([7]) in algebraic topology. For the purposes of this problem, the Alexander duality states that removing a simply connected interior region P of dimension d from a simply connected space $\mathcal{U}$ of dimension d will leave a "hole" characterized by the existence of a homology group ([7]) of rank $d - 1$. Because of the frame, all places which might be considered will always be interior, and because of convexity of the regions, they will be simply connected. For such places and environments, the following theorem holds:

Theorem 2. *An environment E is feasible if and only if (1) it is consistent, (2) removal of the associated place $P(E)$ (the intersection of all regions corresponding to propositions in E) from $\mathcal{U}$ changes the topology of $\mathcal{U}$ from simply connected to a space with a hole, i.e. $\mathcal{U} - E$ has a non-empty homology group of rank $d - 1$.*

Proof. if: Suppose that E was not feasible, i.e. any point within $P(E)$ also satisfies some other proposition p_x. Then removal of $P(E)$ cannot affect the connectivity of $\mathcal{U}$, since all points in $P(E)$ would be part of some other region in $\mathcal{U}$. Thus, the conditions imply that E is feasible.

only if: Any feasible environment E must be consistent by definition, and $P(E)$ is interior because of the existence of a frame. Suppose that removal of $P(E)$ does not leave a hole. Then, all points in $P(E)$ also satisfy some other proposition, and E is not feasible.

$\square$

However, I do not know of any algorithm for determining the topology of the space $\mathcal{U} - P(E)$. The algorithm for deciding whether a region is simply connected, which I present in section 3, is only applicable to a union of a set of convex regions. Thus, the feasibility criterion can be applied computationally only when region $\mathcal{U} - P(E)$ can be represented as a union of convex regions. This is the case when environments are minimal, since the boundary of a minimal environment is formed exclusively by convex regions. Therefore, the method only applies to the detection of feasible environments which are also minimal.

Since the removal of the intersection of regions in $P(E)$ affects the topology of $\mathcal{U}$ only through the parts of $\mathcal{U}$ which actually overlap it, the way its removal affects the topology can be computed by only considering the regions of $\mathcal{U}$ which actually overlap $P(E)$, called the *overlap set* $O(P)$ (see Figure 3). When removal of P creates a hole in $\mathcal{U}$, the topology of $O(P)$ must also contain this hole.

In general, an overlap set $O(P)$ whose topology is different from that of P cannot completely cover all points in P so that P must be feasible. Since each of the individual regions is convex, the topology of a place is always simply connected. Thus, all places whose overlap sets have a nonempty homology group of rank $< d - 1$ are also feasible. In fact, I show below that these environments are *join* places which make up the connections between minimal feasible environments. In Section 3, I shall give an algorithm which allows deciding whether the overlap set of a place is simply connected or not.

Connectivity. In many cases, we are considering a state space where trajectories correspond to behaviors of a device. I define a *qualitative* trajectory between two environments x and y to be a sequence of feasible environments such that there exists a precise trajectory requiring only the propositions in these environments in the given order. Furthermore, I require the trajectory to be *minimal* in the sense that each environment specifies the minimal set of propositions required to traverse the space. I define the *feasible skeleton* of the space to be the graph whose set of paths is exactly the set of minimal feasible qualitative trajectories.

When two points x and y are within the same minimal feasible place P_{min}, but possibly in different subregions, a minimal trajectory between x and y is the composition of a trajectory passing from x to a point $z \in P_{min}$ and from z to y. If place $P0$ is contained in place P_{min}, any pair of points $x \in P0$ and $z \in P_{min}$ can be connected by a path through a sequence $(P0, P1, P2, ..., P_{min})$ of places such that for all Pi, $E_{min} \subseteq Ei \subseteq E0$, i.e. the path never traverses any regions not already contained in $P0$. Thus, there is always a trajectory from any feasible place to a minimal place such that regions are monotonically removed. For two points in the same minimal feasible environment, the only assumptions required for a trajectory between them are those which hold at the points themselves.

When the two points x and y are in different minimal environments, the trajectory between them is composed of three parts: the trajectories from x to a minimal place and to y from another minimal place, and the trajectory between minimal places. This is illustrated in Figure 4.

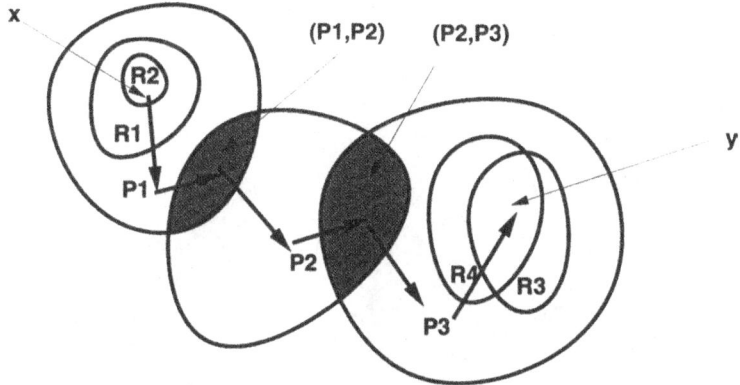

Fig. 4. *Connectivity of places. A path from x and y is composed of a path from x to the minimal place P1, a path along the minimal skeleton from P1 to P3, and a path from P3 to y. Note that all environments on the path from x to a point in P1 are subsets of E1 ∪ R1 ∪ R2 and require only the propositions which hold at x itself. The path from P1 to P3 passes through two join places (shaded), in this case P1 ∩ P2 and P2 ∩ P3.*

Two minimal places $P1$ and $P2$ are adjacent if they overlap each other, as shown in Figure 4. Moving from a point in $P1$ to a point in $P2$ is possible only by moving through a *join* place J, as illustrated in Figure 5, which contains at least all regions in $E1 \cup E2$. Note that the join place is included in both $P1$ and $P2$ and thus provides a minimal feasible path between the two places. Join places thus complete the minimal skeleton of the space.

Join places are identified by the following property:

Theorem 3. *The overlap set $O(J)$ of a join place connecting two adjacent minimal places $P1$ and $P2$ has a homology group of rank $d-2$ or smaller.*

Proof. I refer to the situation shown in Figure 5. As $P1$ and $P2$ are connected via a join place J, removing $P1$, $P2$ and J_i from $\mathcal{U}$ leaves a single "hole". Thus, the overlap set $O(P1 \cup P2 \cup J)$ contains exactly one homology group of rank $d-1$. As $P1$ and $P2$ are only connected through the join place, removing just $P1$ and $P2$ from $\mathcal{U}$ leaves two separate holes. Thus, the overlap set $O(P1 \cup P2)$ has exactly two homology groups of rank $d-1$. Since $J \in (P1 \cup P2)$, the difference between the overlap sets $O(P1 \cup P2)$ and $O(P1 \cup P2 \cup J)$ is exactly J, and removing J from $O(P1 \cup P2)$ removes one homology group of rank $d-1$. Following the theory of Mayer-Vietoris sequences (see section 3), this is possible only if the intersection contains a homology group of rank $d-2$. But this intersection is just the overlap set $O(J)$. When $d > 2$, there may be a more complex structure where join places are themselves connected by other join places. The same reasoning then applies recursivly to show that the overlap set of these places has a homology group of rank $d-3, d-4, \dots$. □

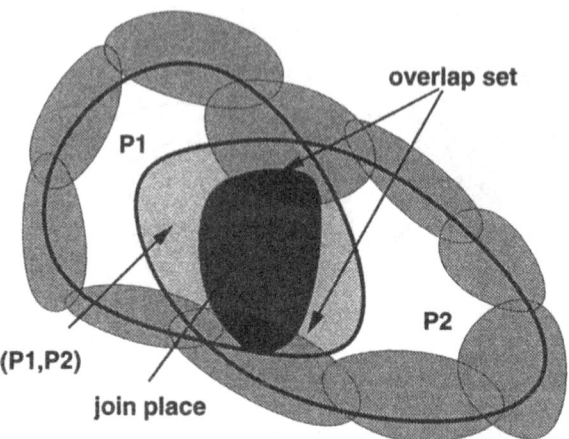

Fig. 5. *Any trajectory between two adjacent minimal places must pass through a* join *environment. The overlap set of any* join place *(shown in black) must have a homology group of rank* $d - 2$ *or smaller. In this case, d=2 and so a homology group of rank d-2 means that the overlap set is disjoint.*

Since a join place is always characterized by the presence of homology groups of rank less than $d - 1$, it can be found using a similar criterion as that used for finding minimal feasible places themselves.

Adjacencies passing through join places are now simple subset/superset relations: minimal place P is directly adjacent to join place Q if and only if $P \subset Q$ and there is no other Q' such that $P \subset Q' \subset Q$. This is the basis for computing the feasible skeleton of the universe $\mathcal{U}$, a graph whose nodes are the feasible minimal, maximal and join places, and whose arcs are the adjacencies given by the subset/superset relations.

Using Minimal and Maximal Environments to Decide Feasibility. Knowledge of the minmal and maximal environments partitions the set of possible environments into three sets:

- **feasible:** minimal and maximal feasible environments found by the two algorithms above.
- **infeasible:**
  any environment E such that $(\exists E_{max})E_{max} \subset E$ or $(\exists E_{min})E_{min} \supset E$, i.e. which is a superset of a maximal environment or a subset of a minimal feasible environment.
- **possibly feasible:**
  any environment E between a minimal and a maximal environment, i.e. $(\exists E_{min}, E_{max})E_{min} \subset E \subset E_{max}$.

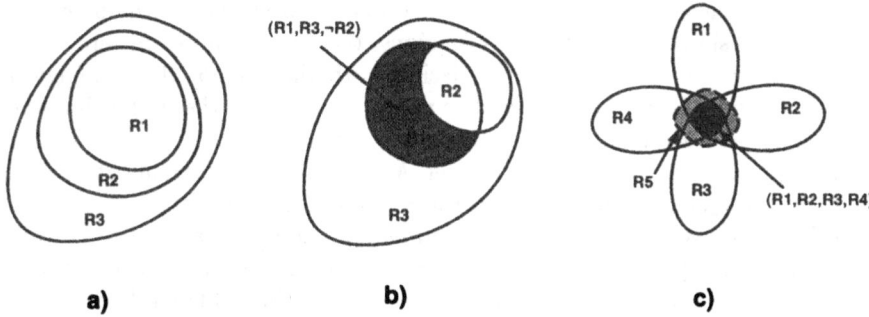

Fig. 6. *The environment* $(R1, R3, \neg R2)$ *is inconsistent in a) but consistent in b) (shaded region). The two situations can be distinguished by characterizing the overlaps more closely. However, these relations cannot be composed. In c), no relation between 3 regions can indicate that the overlap of* $(R1, R2, R3, R4)$ *(black) exists only with region* $R5$, *i.e. that* $(R1, R2, R3, R4, \neg R5)$ *is inconsistent.*

While consistency can be decided unambiguously, feasibility can only be decided for a subset of the environments. Figure 6 shows that the information present in the (d+1)-relations is insufficient to infer more: situations a) and b) have the same (d+1)-relations, but the set of feasible environments is different. This is due in part to the fact that (d+1)-relations only express consistency, but not feasibility. By considering (d+1)-relations based on *feasibility*, more precision can be obtained. These relations would distinguish, for example, the case where the overlap between $A \cap B$ is completely contained in C from the case where the overlap $A \cap B$ is also feasible without C. These inclusion relations are in fact a generalized version of some of the relations used in [2, 3] for *pairs* of regions.

However, as Figure 6 c) shows, these relations are not sufficient to rule out all inconsistent environments, as it can happen that the intersection of k regions falls entirely within a region R_{k+1} even though the intersection of no subset of the k regions had any containment relation with R_{k+1}. Thus, topological inference is guaranteed to find only the *minimal* feasible environments.

3 Computing Topologies

As was shown in the preceeding section, the skeleton of all minimal feasible environments consists of places such that their overlap set has a topology other than simply connected. Following homology theory ([7]), the topology of a space can be represented by the dimensionalities of its homology groups. More precisely, I express the topology of a d-dimensional set of regions by a vector $(k_0, k_1, ..., k_{d-1})$ whose i-th component is the dimensionality of the i-th homology group H_i of the space. This number is equal to the number of i-dimensional "holes" in the space: 0-dimensional holes correspond to disconnected pieces, 1-dimensional ones cut out simply connected disks, 2-dimensional holes simply connected spheres, and so on. For a simply connected set of regions, the vector will be all zero.

I now present an algorithm for deciding whether or not the topology of a union of regions is simply connected or not. It decomposes the set by eliminating one region at a time until only a single one remains, whose topology is known to be all zero since it is a convex region. Assume that the current set of regions is $X = \{r_1, r_2, ..., r_k\}$ and the algorithm eliminates r_j to obtain $Y = X - r_j$. Consider the overlap of r_j with the remaining regions in Y, which we call Z. Homology theory (more precisely, the theory of Mayer-Vietoris sequences [7]) shows that for every homology group of rank l eliminated from X, the overlap Z must contain a homology group of rank $l - 1$. For example, if X is a torus and Y simply connected, the removed region must have two disjoint intersections with Y. Thus, if at each elimination Z is simply connected, no homology groups are ever eliminated. Since none are present at the end, the original region was simply connected. Conversely, when X is simply connected, there is a greedy elimination order such that at every step, the intersection is simply connected.

Using the same principles, it is possible to construct an algorithm which determines the exact topology of a set of regions. However, because of possible ambiguities this algorithm might have to search different elimination orders and is unnecessarily inefficient.

To compute whether the intersection of a base place B with the union of a set of regions $R = \{r_1, r_2, ..., r_n\}$ is simply or multiply connected, I apply the following recursive algorithm $\texttt{homology}(B, R)$:

1. $X \leftarrow R$
2. while $X \neq \{\}$ do
   (a) if the intersection $B \cap \bigcap_{i=1..n} r_n$ is nonempty, i.e. a subset of a maximal consistent place, return $\texttt{simply-connected}$
   (b) find $r_j \in X$ such that the overlap $r_j \cap (X - r_j)$ is simply connected, i.e. has no homology groups of any rank. The topology of the intersection is computed by calling the algorithm recursively: $\texttt{homology}(B', R')$ with $B' = B \cap r_j$ and $R' = O(B') \cap X$.
   (c) if such an r_j is not found: return $\texttt{multiply-connected}$
   (d) set $X \leftarrow X - r_j$
3. if B is consistent, return $\texttt{simply-connected}$, otherwise return $\texttt{multiply-connected}$

If all maximal consistent places are known, the consistency tests required by the algorithm amount to simple subset relations, and the algorithm is then very efficient, in spite of the fact that its worst-case complexity is clearly exponential.

Computing the Minimal Feasible Skeleton.

1. compute the set of maximal consistent environments $\mathcal{M}$.
2. compute the set of candidates for minimal feasible and join environments as all subsets shared by at least 2 environments in $\mathcal{M}$.

3. $m \leftarrow$ those candidates E such that $O(P(E))$ is not simply connected.
4. $\{Adj\} \leftarrow$ all pairs $E_1, E_2 \in (\mathcal{M} \cup m)$ such that $E_1 \subset E_2$, but there is no E' such that $E_1 \subset E' \subset E_2$.
5. return $(\{nodes, undirectededges\}) = (\mathcal{M} \cup m, \{Adj\})$.

4 Example: the Piano Movers Problem

As an example of an application of the formalism I developed, consider the *piano movers problem* [8], a problem which has drawn much attention in robotics and spatial reasoning. In the piano movers problem, the goal is to find a path for moving a single rigid *moving object* from an initial to a final position such that it does not collide with any of the fixed and rigid obstacles.

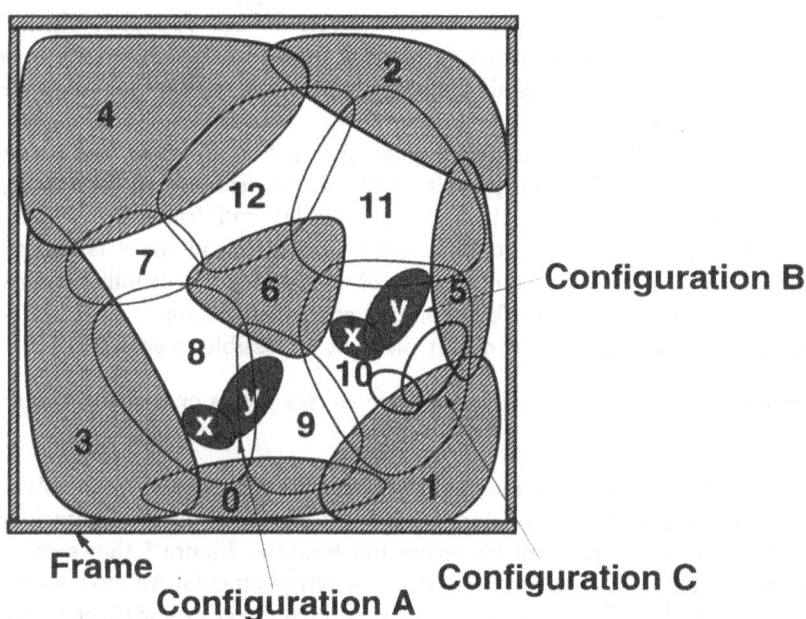

Fig. 7. *Input representation of an instance of the piano movers problem. Pieces, hatched, and cavities, shown as transparent, are numbered sequentially. The moving object consists of two pieces, x and y, and is shown in three different configurations. I represent its qualitative configuration by an environment of overlaps written as a/b, where a denotes the part of the moving object and b the region it overlaps. Thus, configuration $A = \{x/8, y/8, y/9\}$, $B = \{x/10, y/10, y/11\}$ and $C = \{x/1, y/1, y/5, x/10, y/10\}$. Configurations A and B contain only bubbles and are thus legal, while C contains three obstacles and is thus not legal.*

In a qualitative version of the problem, positions of the moving object are given by regions within which all points are considered equivalent. As an example, consider the situation shown in Figure 7. It shows a *moving object*, in grey, and a set of obstacles, hatched. The moving object consists of two convex regions, labelled x and y. The free space available for moving the object is covered with convex regions called *bubbles*. I define a qualitative configuration of the moving object by the combination of regions, obstacles or bubbles which the moving object overlaps.

A *configuration* is a particular position and orientation of the moving object and defines a point in a *configuration space* ([6]), which is spanned by these parameters. Configuration space consists of blocked configurations where the moving object would overlap a fixed one, called *blocked space*, and its complement of legal positions, called *free space*. I model both spaces by a universe of two different types of convex regions: *obstacles* and *bubbles*. Each possible overlap between a part of the moving object and part of a fixed object defines a configuration space region (c-region) of illegal configurations, called an *obstacle* o_i. Blocked space B is the union of all obstacles: $B = \bigcup_i o_i$. Each possible overlap between a part of the moving object and a cavity defines a c-region which I call a *bubble* b_j. Note that in contrast to blocked space, free space is only a *subset* of the union of all bubbles, as all configurations falling within blocked space are excluded from it: $\mathcal{F} = \bigcup_j b_j - B$

I enclose all objects by a rectangular *frame FR*, a set of regions which bounds all objects, whose purpose is to ensure consistency of the topological computation. The union of all obstacles, bubbles and frame is the universe $\mathcal{U} = \bigcup_i O_i \cup \bigcup_j B_j \cup FR$. Note that since all of physical space is covered by object parts or cavities, any configuration of the moving object falls within some c-region. The universe is therefore a simply connected region.

Because of the convexity of object parts, it is possible to show:

Theorem 4. *Every c-region formed by two convex pieces or cavities A and B is a convex region.*

Any configuration of the moving object can be represented qualitatively by the combination of region overlaps which is present in the configuration, expressed as an environment of obstacles and bubbles. Figure 7 shows examples of configurations and their representation by environments. An environment E is *legal* if P is feasible and contains only bubbles (overlaps with open space). Thus, the condition of non-overlap between moving object and obstacles can be modelled as feasibility of environments.

Feasible Skeleton. Figure 8 shows the legal part of the feasible skeleton for the example. It qualitatively represents all legal motions of the moving object. A qualitative solution to the piano movers problem can be given by first mapping the initial and final positions to the minimal places to which they belong, and then finding the qualitative path between initial and final place by searching in the graph. For example, a path between configurations A and B (Figure 7) can be found as follows. First, I map to the minimal feasible places: $A = \{x/8, y/8, y/9\}$ is already minimal, $B = \{x/10, y/10, y/11\}$ is mapped to

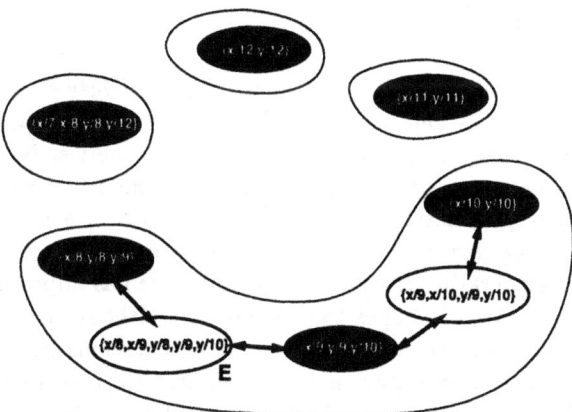

Fig. 8. *The legal part of the feasible skeleton for the example. All places containing overlaps (obstacles) have been omitted. The numbering of the regions refers to Figure 6. Minimal places are shown in black, join places in white.*

$\{x/10, y/10\}$. Next, I find a path in the graph between the two minimal environments, in this case the path $\{x/8, y/8, y/9\} \Rightarrow \{x/8, y/8, x/9, y/9, y/10\} \Rightarrow \{x/9, y/9, y/10\} \Rightarrow \{x/9, x/10, y/9, y/10\} \Rightarrow \{x/10, y/10\}$. Using more complete (d+1)-relations (as described in Section 2), for this example it would be possible to construct a complete graph containing *all* feasible environments (not only minimal ones). This graph also contains environment B so that no mapping to a minimal environment would be required.

I have implemented a prototype which demonstrates the topological reasoning techniques on the piano movers problem for two-dimensional objects. The input to the program is given in the form of three collections of convex bitmaps, representing the parts of the fixed objects, the moving object, and the cavities. A preprocessor uses these bitmaps to determine all possible simultaneous overlaps of 3 pairs of parts. A first version computes this by systematic search of all possibilities, which is rather slow. A much faster solution consists of detecting simultaneous overlaps in randomly generated positions of the moving object. The implementation shows that it is possible to solve such complex planning problems without *any* analytical representations.

This example shows that the minimal skeleton is powerful enough to solve problems of real-world complexity. In this example, the qualitative representation has the advantage that arbitrary object shapes can be dealt with at no penalty. Currently known techniques for solving such problems with curved shapes require approximations with algebraic curves, and extremely complex and brittle computation to determine the possible paths.

Many other applications, such as control system design ([9]) require similar forms of reasoning. It is likely that topological inference can provide similar advantages there as well.

5 Conclusions

In this paper, I have presented the concept of *feasibility* of a set of qualitative spatial propositions and given an algorithm for computing a *minimal feasible skeleton* sufficient to answer many important queries. Note that the set of feasible minimal environments is equivalent to the set of maximal environments of negated propositions: if $E_{min} = p_1 \wedge p_2 \wedge ... \wedge p_i$ is feasible, then $N_{max} = \bigwedge_{j, j \notin 1..i} \neg p_j$ is a maximal consistent environment.

Using topological rather than geometric inference has important advantages. First, regions with curved shapes can be handled without any increase in complexity, as long as they are convex. Object representations as unions of convex parts have long been postulated in vision research ([5]), so this may not be a severe restriction. Second, the methods are simpler and more robust than geometric computation. This is because (i) they do not require computing the presence of precise points, but only the presence of regions with many points, and (ii) topological inconsistencies which may arise in geometric computation cannot occur here.

I am currently working on extending the robot motion planning application to also include rotation. Another interesting direction for further work in the context of the planning application would be to consider the use of abstractions to speed up the computation. I am also considering the applying the approach to other spatial reasoning problems.

References

1. **V. Chvátal:** "Linear Programming," W. H. Freeman, 1983
2. **D.A. Randell, A.G. Cohn, Z. Cui:** "Naive topology: Modellling the force pump," in P. Struss, B. Faltings (eds.): *Recent Advances in Qualitative Physics*, MIT Press, 1992
3. **Z. Cui, A.G. Cohn, D.A. Randell:** "Qualitative Simulation Based on a Logical Formalism of Space and Time," *Proceedings of the 10th National Conference of the AAAI*, AAAI Press, 1992
4. **M.J. Egenhofer:** "Reasoning about binary topological relations," in O. Gunther, H.J. Schek(eds.): *Advances in Spatial Databases*, pp. 143-160, Springer-Verlag, 1991
5. **D.D. Hoffman, W.A. Richards:** "Parts of Recognition," *Cognition* 18, 1985
6. **T. Lozano-Perez, M. Wesley:** "An Algorithm for Planning Collision-Free Paths Among Polyhedral Obstacles," Comm. of the ACM, **22**, 1979,
7. **E. Spanier:** "Algebraic Topology", Mc. Graw Hill, 1966
8. **J.T. Schwartz, C.K. Yap:** *Advances in Robotics, Vol. 1: Algorithmic and Geometric Aspects of Robotics*, Erlbaum, Hillsdale, N.J., 1987
9. **F. Zhao:** "Phase Space Navigator: Towards Automating Control Synthesis in Phase Spaces for Nonlinear Control Systems," *Proceedings of the 3rd IFAC International Workshop on Artificial Intelligence in Real Time Control*, Pergamon Press, 1991

Proximity Operators for Qualitative Spatial Reasoning

Mark Gahegan

Department of Geographic Information Systems, Curtin University,
PO BOX U 1987, Perth 6001, WA, Australia.
tel: +619 351 3309, fax: +619 351 2819, email: mark@cs.curtin.edu.au

ABSTRACT

One way to increase the power of Qualitative Spatial Reasoning is to introduce proximity operators (such as *close* and *far*) that are surrogates for distance measures. These operators appear to be semi-quantitative in nature as opposed to purely qualitative. In the light of observations drawn from psychometric testing of perceived proximity, this paper discusses how a model to support proximal reasoning could be constructed. The relationships between the model and the raw data are described. Fuzzy set membership is used to reason about the degree of *closeness*. The formulation of queries involving proximity is presented, with the meaning of linguistic variables being instantiated within a given *context* at execution time.

1   Introduction

Qualitative Spatial Reasoning is concerned primarily with the abstracted relationships between objects in space, as opposed to the underlying geometry defined by the raw spatial data. Consequently, it is centred around topological relationships [Egenhofer and Franzosa, 1991], [Smith and Park, 1992], [Molenaar, 1994]. Qualitative spatial reasoning offers a means of transforming all relationships in the data to give a view of the universe of discourse (U) that is centred around a single object, here called the reference object R.

In the discussion that follows, relationships are defined between geographic features of interest (here termed objects). For simplicity, each object is considered to be defined in two-dimensional space by a single point $o_{(i,j)}$. The reference object is similarly defined, $R_{(h,k)}$. This 'egocentric' view of the world can be very useful as a means of constructing queries concerning relationships between a particular object of interest and other objects. These relationships can be expressed as:

"Which objects are East-of R?"   or   "Which objects overlap with R"

An example query might be:

"Find a petrol station East of current location"
(where 'current location' is set to be R).

A key reason for the importance of qualitative spatial reasoning is that it allows a user to express queries in a way that is much more intuitive; that is, relationships inherent within the data are used, without the user becoming swamped in the actual values (co-ordinates and angles) that comprise the raw data.

To expect a GIS user to be familiar enough with the raw data to reason using only geometry would seem to be asking too much in all but a few cases. Qualitative spatial reasoning is offered as an alternative, supplementary, means of working with the data. Some precision is lost by moving away from absolute values, but this may be compensated by the increase in *accessibility* afforded to the user. As discussed by Frank, [1992], there is evidence that qualitative reasoning follows along similar lines to the thought processes of humans.

Furthermore, as pointed out by Sharma et al. [1994], qualitative spatial reasoning separates the absolute numeric properties of the data from the "determination of magnitudes and events, which may be assessed differently depending on the *context*". The implication is that if a model for *context* can be created, then the qualitative relationships can be defined in the light of this model, whereas quantitative values must remain fixed. The spatial reasoning is then to some extent adaptive to the context of the current task, and hence may appear more intuitive to the user. The support of *context* for *proximity* forms the main subject of this paper.

1.1   Proximity and Context

The idea that one object has a qualitative relationship by distance to another is an important concept, since it can be used as a predicate on which to relate together objects in a more natural way than resorting to geometry. The example given above can then be further qualified:

"Find a petrol station *South of* and *close to* current location"

Indeed, it has been argued that the directional operators alone are insufficiently powerful, and need to be supplemented by a qualitative proximity measure [Gapp, 1994]. This allows more of the underlying data to be utilised in analysis and therefore increases the qualifying power of the reasoning.

Most topological operators are strictly qualitative, and their definition, once given, is fixed. However, the concept of proximity does not appear to behave in this fashion, but is instead highly dependent on the context within which it is applied. It will be argued in this paper that proximity depends on many factors, including the spatial distribution of the data under consideration, and various aspects concerning the current task. In the light of this dependency, it might be more accurate to refer to proximity as *semi-quantitative* [Roberts and Gahegan, 1991], [Gahegan, 1994]. The definition of a semi-quantitative operator allows adaptation to the data under consideration.

1.2 Recent Research on Proximity

In order to use proximity as a qualifier in spatial reasoning, distances between objects must be described by linguistic variables such as *close* and *far*. These variables correspond to some type of distance metric. Two alternative approaches are given by Frank [1992], and Dutta, [1990]. In the former, a tolerance space and relation are used, with distances being ranked into a number of intermediate steps from 0 to $n - 1$. In the latter, a fuzzy membership function is constructed, $\mu_F : U \rightarrow$ [0, 1], where $\mu_F (u)$ denotes the degree of membership (between zero and one inclusive) of u in the fuzzy set F. In both of the above approaches, all objects can be graded according to their distance from R.

Proximity is described here by a fuzzy membership function, since there are well defined methods for reasoning with data in this form [Zadeh, 1975]. One possible mapping to linguistic variables is:

$0.0 \rightarrow 0.1$ *very far*
$0.1 \rightarrow 0.3$ *far*
$0.7 \rightarrow 0.9$ *near*
$0.9 \rightarrow 1.0$ *very near*

Note that a value of 1.0 signifies co-incidence, and a value of 0.0 signifies maximal separation. Two other important linguistic variables are used to represent the extrema; namely *nearest* and *farthest*. These are analogous to the SQL functions MIN and MAX.

An alternative approach to that developed here is taken by Robinson [1990], in which a fuzzy spatial relation is learnt via interaction with a computer in a question and answer style dialogue. The notion of proximity is constructed implicitly, but the results are equally valid.

Proximity appears to be a richer concept than is currently modelled, and there is a danger that the problem has been over-simplified to some extent. Proximity measures that are included in spatial reasoning must behave in a way that follows a human perception of proximity. Failure to achieve this goal will result in counter-intuitive interaction with the GIS, with a consequence that unreliable results will be produced.

There are some issues that remain to be addressed, namely:

- How do humans reason about proximity?

- How does scale affect proximity?

- Does the nature of the task to be undertaken have any effect?

1.3 Psychometric Testing of Proximity

Various psychometric studies have been conducted to assess how humans make subjective judgements regarding distances across a wide range of different data domains, including geographical space [Lundberg and Ekman, 1973], [Guttman, 1968].

In order to gain further understanding as to how humans judge proximity, a simple test was conducted on fifty subjects, each of whom was asked to comment on the *closeness* of a series of objects to a reference point. Further details of the test can be found in Appendix A. Whilst it is not claimed that these tests are conclusive in any way, some interesting points were observed, namely:

(1) In the absence of other objects, subjects reasoned about proximity in a geometric fashion, and furthermore, the relationship between distance and proximity can be approximated by a simple linear relationship.

(2) When other objects of the same type are introduced, proximity is judged in part by *relative* distance, that is, the distance between an object and R is modified according to the distances from other objects to R.

(3) Distance is affected by the size of the area being considered, or alternatively, it has some relationship to perceived scale.

The above observations are discussed in the section that follows and their consequences are formulated into a contextual model for proximity.

2 A Model Supporting Context in Proximal Relationships

There are several issues that need to be addressed before measures of proximity can be defined at all. Not least of these is the formulation of a metric that can be used to measure proximity given a set of objects.

2.1 Absolute Distance Metrics

Observation (1) suggests that, in the simple case, proximity is directly proportional to distance and that proximity can be modelled adequately with a linear Euclidean distance metric. So, for two dimensional data:

$$\rho\left(o_{(i,j)}, R_{(h,k)}\right) \propto \sqrt{(i-h)^2 + (j-k)^2}$$

To give an actual numeric value to proximity, it is necessary to compare the distance from o to R with some sort of maximal distance, M. This could be measured in a variety of ways, for example as the distance to (i) the furthest object, (ii) the furthest corner of a bounding rectangle, or (iii) the distance given by the diagonal of a bounding rectangle. Figure 1 shows these options as 1, 2 and 3 respectively, with the reference point denoted by ($\bullet$). From the test results obtained, the most appropriate measure would seem to be 3, the diagonal of the bounding rectangle. This represents the maximum possible separation between *o* and *R*.

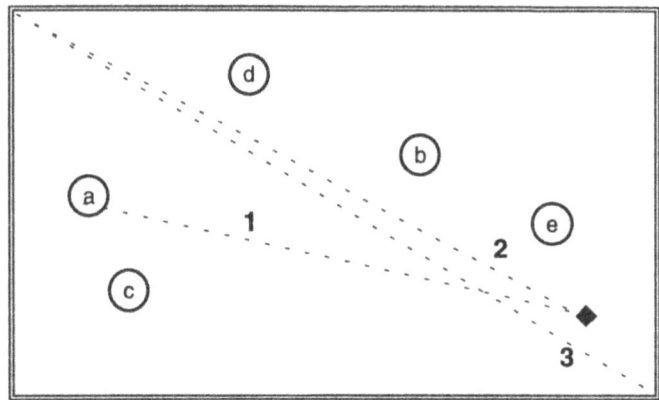

Figure 1. Options for computing maximal distance (*M*)

If the size of the bounding rectangle is described by (0 ,0 : *x, y*), the proximity (ρ) of object *o* to *R* is given by:

$$\rho\left(o_{(i,j)}, R_{(h,k)}\right) = 1 - \frac{\sqrt{(i-h)^2 + (j-k)^2}}{\sqrt{(0-x)^2 + (0-y)^2}}$$

When using an absolute approach, whilst there is a continuum for proximity, (*very close > close > far > very far*) this cannot be extended to include *closest* and *farthest*. This is because it is entirely possible for a point to be the *closest* but not *close*, or *farthest* but not *far*. As a consequence, there is a loss of qualifying power. The fact that a point is *close* says nothing about the distribution of points in general; that is, it is not *close* compared to other points.

A distribution function for geometric proximity is shown in Figure 2. Based on the observed results, the function is modelled simply as a linear relationship, where the diagonal distance *M* defines the extent of the X axis.

This approach has the advantage from a computational perspective that proximity has a fixed relationship to the distance between *o* and *R* only, and so the insertion and deletion of objects will not alter the quality of relationships at all. However, from the point of view of the user, this can be somewhat counter-intuitive.

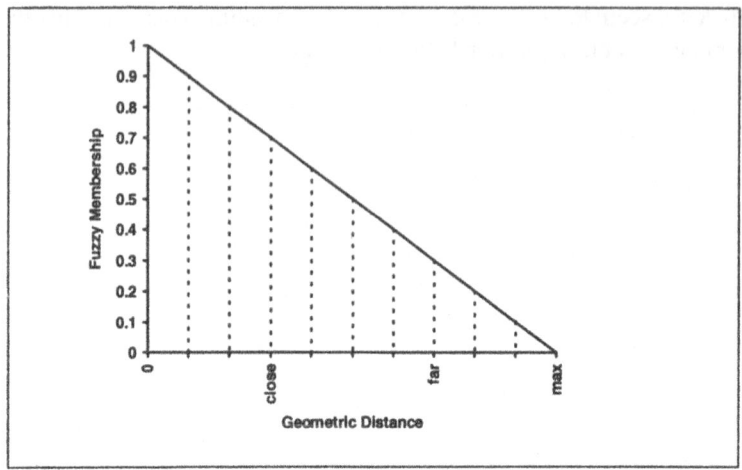

Figure 2. A possible fuzzy distribution function for absolute distance

2.2 Relative Distance Metrics

In the presence of other objects, observation (2) implies that modifications to proximity measures must be made to account for the patterns of distribution of these objects. There are several possible methods for calculating relative proximity, ranging from simple ordinal techniques to point distribution functions from spatial statistics.

The simplest measure is an ordinal approach. The n objects under consideration are simply ranked according to their distance from R, with the most proximal objects being first (R is assigned the zeroth ranking). Using the mapping to linguistic variables given in Section 1.2, an object (o) is considered *close* to R according to:

very close (o, R): ranking (o, R) $\leq$ ($n * 0.1$)

close (o, R):   ranking (o, R) > ($n * 0.1$) $\wedge$ ranking (o, R) $\leq$ ($n * 0.3$)

far (o, R):   ranking (o, R) $\geq$ ($n * 0.7$) $\wedge$ ranking (o, R) < ($n * 0.9$)

very far (o, R):   ranking (o, R) $\geq$ ($n * 0.9$)

closest (o, R):   ranking (o, R) = 1

farthest (o, R):   ranking (o, R) = n

This measure works well if the distribution of objects is fairly even, and provided the number of objects is not too small ($n < 10$). One obvious disadvantage however, is that objects can be separated from R by large distances, but still be regarded as *close*, if there are enough objects even further from R[1].

The fuzzy distribution function used is shown in Figure 3. For the present it is kept deliberately simple as a linear relationship, and is similar to Figure 2, but in this case, the extent of the X axis is defined by n.

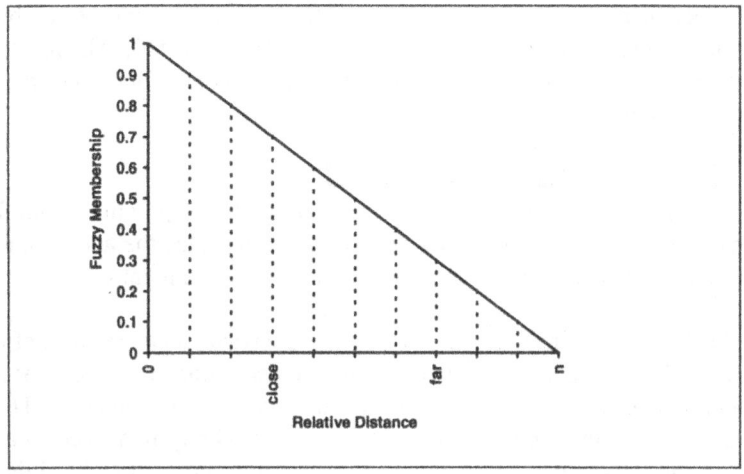

Figure 3. A distribution function for relative distance

2.3   A Combined Proximity Measure

The above two approaches represent extremes in the way in which proximity can be measured; the first relies entirely on absolute distance, the second entirely on relative distance. A more intuitive approach makes use of both approaches together. A number of methods for combining fuzzy values exist, for example the fuzzy AND, fuzzy OR, and fuzzy Algebraic Product. By choosing such a method, the data can be examined for points that are both *close* in a geometric sense and/or *close* in a relative sense.

The membership function for the sets absolute proximity (P_a) and relative proximity (P_r) are applied to all objects $\{o\}$ with respect to the current reference point. With

---

[1]A more sophisticated approach is to use the statistical distribution between points to calculate proximity. Nearest Neighbour distribution function such as those described by Cressie [1991] are perhaps most suitable, where the probability of finding a neighbour in a circle centred on R, of radius r is modelled. An example would be:

$$G(r) = 1 - \exp(-\lambda \pi r^2), \qquad r \geq 0$$

where $G(r)$ is the probability that the distance from R to the nearest object is less than or equal to r, and λ is the Poisson intensity function.

the distribution functions described above, the fuzzy union operator produces useful results.

$$P_a \cup P_r \equiv \left\{ \left(o, \max\left(\mu_a(o), \mu_r(o) \right) \right) \right\}$$

That is to say, an object is considered to be *close* if it is either geometrically close *OR* relatively close. However, as the model for proximity is developed, it will probably become necessary to adopt a more subtle method for evidence combination, such as the Gamma Operator (Zimmermann & Zysno, 1980), which allows all contributing evidence to effect the result, rather than being entirely dictated by the most extreme values.

2.4   The Dynamic Nature of Spatial Data

It follows from Observation (2) that the *relative* nature of proximity may lead to inconsistencies if the data under consideration changes, since the addition, deletion or displacement of objects will change the distribution of the point set.

From a database perspective, the GIS is a dynamic environment. Features of interest may change shape, relocate, and they may not be persistent over time; that is, they may begin their existence at some point in time, and end it at another. The point being made here is that any pre-defined notion of proximity may need to adapt to changes in the data under investigation. It is therefore not sufficient to define an initial proximity zone and to expect its validity to remain constant. Rather it is more appropriate to take the opposite approach; that is to calculate proximity only at execution time, when all influencing factors are known, and can be taken into account. Obviously the calculations involved place some additional burden on the computer, but the compensation is that the proximity measures are guaranteed to be as up-to-date and appropriate as the current raw data allows.

2.5   The Effect of Scale

Observation (3) indicates that proximity may be interpreted as a function of scale. What might be considered *close* at one scale may also be considered *far* at another. For example, there is no logical inconsistency in the following:

> *London (UK) is close to Paris (France)*
> *London (UK) is far from Perth (Australia)*

and

> *London (UK) is close to Perth (Australia)*
> *London (UK) is far from Sea of Tranquillity (Moon)*

A good deal of research has been conducted into the problems associated with scale changes, and specifically into providing seamless and continuous views of space as scale is changed [Roberts et al., 1991], [Richardson, 1994]. Indeed, these notions have become established in some commercial systems (for example Smallworld GIS). It is also desirable to include the same flexibility in qualitative spatial reasoning.

Scale in isolation only defines a ratio of displayed data to on-the-ground distances. Before proximity measures can be calculated, it is also necessary to know a frame of reference against which all distances can be compared. A useful default is to adopt the extents of the current viewing window onto the underlying data; that is, the area that the user is currently viewing on the screen. Assuming that this area is rectangular, the theoretical maximum distance between two points is given by the diagonal extent of the area, and used as described in Section 2.1. The area selected defines the objects under consideration, in this case via the use of a simple point in polygon intersection algorithm. All included objects form a subset which is then used to calculate P_a and P_r, with respect to R.

2.6   The "Attractiveness" of Objects

Although scale would appear to have the major effect, it is also true that a user's perception of what might be considered *close* is not consistent amongst all pairs of object types, and would appear to change according a perceived attractiveness. For example, *"close to the shops"* may be 1 km or less, whereas *"close to the Toxic Waste Dump"* may be 10 km or less, at the same scale. One way of coping with this is to allow the object type to affect the steepness of the absolute fuzzy proximity distribution. A steeper gradient effectively reduces the distance that is considered to be *close*, and a shallower gradient increases it. Before this concept may be used effectively, a much greater level of understanding is required regarding a user's perceptions of the real world. For now, this aspect of proximity is noted but not implemented.

2.7   The Effect of Reachability

Proximity may not simply imply closeness in a geometric sense, but also closeness topologically. For example, it does not matter that a is close to b, if one is trying to get to b from a by train and there is no track. In other words, there may be an imposed transportation network that constrains movement between objects. Distances must then be calculated along this network.

From an implementation perspective, reachability causes little difficulty. Instead of calculating a straight-line distance between points, a network distance is computed, using any imposed restrictions on accessibility, flow, and direction of links. In some static applications, it may be worth storing the pre-calculated distances in a matrix for re-use, since their computation can be expensive.

3   A Definition of Proximity with Context

To summarise, the concept of proximity does not appear to be fixed, but rather it is defined by:
- The (sub) set of objects under consideration.
- The path of connection or route between objects.
- The spatial distributions inherent in the actual data (at a specific time).

- The scale at which the data is being viewed.
- The attractiveness of objects.

The objects under consideration form a set {o} of type O. This set may be restricted using any valid qualifying clauses (for example, see Worboys et al., [1991]), so that only objects with certain properties are considered in calculating proximity. The relative positions and number of objects in {o} define relative proximity. The bounding rectangle is supplied either as part of the query or as a default from the current viewing area. The bounding box and the geometric positions of the objects define absolute proximity. The distance metric is defined in the query as "Euclidean", or "Network" (although others are possible). Objects with a positive or negative attraction should be noted as such in the object definition, although no allowance for this is currently made.

To paraphrase, the context is taken to be "Is object o of type O *close* to R, using the absolute and relative distances from o to R, where distance is calculated using a specified method, within the current region of interest".

As a simple example, consider the query:

"Which petrol stations are *close to R* and *South of R* by road?"

The query is executed in two stages. The concept of proximity as modelled here is semi-quantitative and hence driven from the data. So, before the query can be executed, it is necessary to instantiate the CloseTo relation from the data using the current context as given by:

$$\text{close}:O = \text{CloseTo} \ (o{:}O, \{o\}, R, \{x_1, y_1, x_2, y_2\}, \textit{DistanceMethod})$$

The resulting set of objects is denoted as *close*, and is also of type O. The final stage is to join {close} with the other query qualifiers via:

$$\left\{ \ p{:}\ \text{petrol station} \ \middle| \ \exists \, c{:}\ \text{close} \ \left(\textit{southof}(p, R) \wedge c = p \right) \right\}$$

Concepts such as '*South of*' can also be expressed in fuzzy terms, allowing a straightforward combination of set membership to be applied when formulating the result [Dutta, 1990]. In the above example, the expression can be optimised so that only the objects in {close} are tested for also being *South*. Note however, that the entire set of objects was used to calculate the context for {close}. A different result may be obtained if {close} was instantiated after the qualifier '*South of*' had been applied to {p}, since the context then becomes *close* with respect to petrol stations South of R. Exactly how, or indeed if, this additional flexibility should be presented to the user is a perplexing issue that requires further investigation.

4 Conclusions

There is an established need to support proximity measures as a part of qualitative spatial reasoning. Psychometric testing has implied that humans do not judge proximity in a purely Euclidean manner, with the consequence that measuring and modelling proximity is not a straightforward process. There would seem to be strong justification for adopting a semi-quantitative approach where relationships are only calculated when the context of the problem is fully understood, i.e. at query execution time. A model for such a context has been developed and presented. The model attempts to account for absolute and relative methods of judging distances, and the effects of scale, data and task. It is based on the results of preliminary tests and so may need altering and/or refining in the light of further investigation.

The tests conducted thus far are only exploratory in nature, and additional, more rigorous tests must be constructed before any definite conclusions should be drawn concerning human perception of proximity. Specifically, further testing will focus on the distribution functions described in Sections 2.1 and 2.2, and also on the combination of evidence in Section 2.3. It may also be appropriate for an automated learning strategy to be applied, or to make use of expert judgement about what might be considered *close* at a particular scale.

5 References

Cressie N A C (1993), Statistics for Spatial Data. John Wiley and Sons, USA, Ch. 8.

Dutta S (1990), Qualitative Spatial Reasoning: A Semi-Quantitative Approach Using Fuzzy Logic. In: Design and Implementation of Large Spatial Databases (Eds. Buchmann A, Günther O, Smith T R and Wang Y-F) Springer-Verlag, New York, pp 345-364.

Egenhofer M J and Franzosa R D (1991), Point-Set Topological Spatial Relations. Int. J. Geographical Information Systems, Vol. 5, No. 2, pp 161-174.

Frank A U (1992), Qualitative Spatial Reasoning about Distances and Directions in Geographic Space. Journal of Visual Languages and Computing, Vol. 3, pp 343-371.

Gahegan M N (1994), Support for the Contextual Interpretation of Data Within an Object-Oriented GIS. Proc. Spatial Data Handling '94, (Ed. Waugh T C and Healey R G), pp 988-1001, Edinburgh, Scotland.

Gapp K-P (1994), A Computational Model of the Basic Meanings of Graded Composite Spatial Relations in 3D Space. Proc. Int. Workshop on Advanced Geographic Data Modelling (AGDM '94), (Ed. M Molenaar and S De Hoop), pp 66-79, Delft, Netherlands.

Guttman L (1968), A General Nonmetric Technique for Finding the Smallest Coordinate Space for a Configuration of Points. Psychometrika, Vol. 33, No. 4, pp 469-506.

Lundberg U and Ekman G (1973), Subjective Geographic Distance: A Multidimensional Comparison. Psychometrika, Vol. 38, No. 1, pp 113-122.

Molenaar M (1994), A Syntactic Approach for Handling the Semantics of Fuzzy Spatial Objects. Proc. European Science Foundation, GISDATA, Baden, Austria.

Richardson D E (1994), Generalization of Spatial and Thematic Data Using Inheritance and Classification and Aggregation Hierarchies. Proc. Spatial Data Handling '94, (Ed. Waugh T C and Healey R G), pp 957-972, Edinburgh, Scotland.

Roberts S A, Gahegan M N, Hogg J and Hoyle B S, (1991), Application of Object-Oriented Databases to Geographic Information Systems. Information and Software Technology, Vol. 33, No. 1, pp 38-46.

Roberts S A and Gahegan M N, (1991), Supporting the Notion of Context Within a Database Environment for Intelligent Reporting and Query Optimisation. European Journal of Information Systems, vol. 1, no 1, pp 13-22.

Robinson V B (1990), Interactive Machine Acquisition of a Fuzzy Spatial Relation. Computers and Geosciences, Vol. 16, No. 6, pp 857-872.

Sharma J, Flewelling D M and Egenhofer M J (1994), A Qualitative Spatial Reasoner. Proc. Spatial Data Handling '94, (Ed. Waugh T C and Healey R G), pp 665-681, Edinburgh, Scotland.

Smith T R and Park K K (1992), Algebraic Approach to Spatial Reasoning. Int. J. Geographical Information Systems, Vol. 6, No. 3, pp 177-192.

Worboys M F, Hearnshaw H M, Maguire D J (1991), Object-Oriented Data and Query Modelling for Geographical Information Systems. Proc. 4th International Symposium on Spatial Data Handling, (Ed. Brassel K & Kishimoto H), Dept. of Geography, University of Zurich, Switzerland, Vol. 2, pp 679-688.

Zadeh L A (1975), The Concept of a Linguistic Variable and its Application to Approximate Reasoning. Information Sciences, Vol. 8 pp 199-249 (part I), Vol. 8 pp 301-357 (part II), Vol. 9 pp 43-80 (part III).

Zimmermann H J and Zysno P (1980), Latent Connectives in Human Decision Making. Fuzzy Sets and Systems, Vol. 4, pp 37-51.

Appendix A: Details of the Psychometric Testing

A group of 50 subjects took part in a test designed to investigate how humans reason about proximity. The subjects all had some practical exposure to the use of GIS.

About half of those tested were using GIS in relation to their employment, the other half were full-time students. There did not appear to be any significant differences between the two groups based on employment, but there were some differences observable based on experience with handling geographic data. Specifically, the more experienced subjects tended to respond with more weight given to relative proximity than their less experienced colleagues.

Subjects were asked to rate diagrams similar to that shown in Figure 4, according to how *close* they judged point (*e*) was to *R*, shown as (♦). The distance to *e* was varied (linearly), and averaged responses produced are shown graphically in Figure 6. For the most part, the relationship appears to be linear, (the rightmost value in the graph may be giving a lower result due to proximity with other objects and the edge of the diagram).

The response for the distance between *e* and *R* in Figure 4 averaged 0.67. This can be contrasted with Figure 5, where the actual distance between *e* and *R* is the same but additional, closer points have been inserted. The response in this case was significantly lower at 0.50. This, and similar observations, indicate that a relative measure for proximity is being used, at least in part.

The effect of changing the size of the study region was also examined, but the results were problematic. A modified Figure 4 was shown, with the only difference being that the bounding box itself was enlarged, whilst the pattern of objects was kept constant. Responses fell into two categories, some judging point *e* to be closer, and others further away. A possible explanation is that some subjects judged the distance from *e* to *R* in terms of the boundary, and others judged it in terms of relative distance to other objects. Further investigation is needed to fully explore these trends.

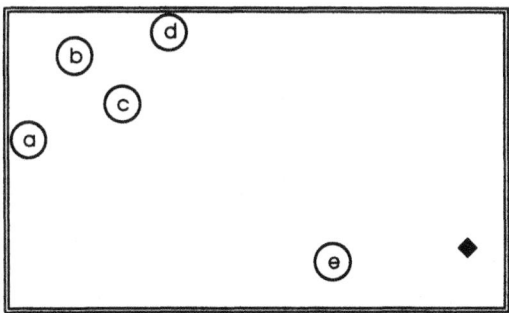

Figure 4. Sample test diagram, ρ (*e*, *R*) = 0.67

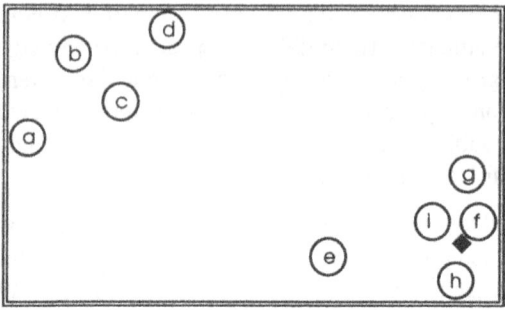

Figure 5. Extra points added, ρ $(e, R) = 0.50$

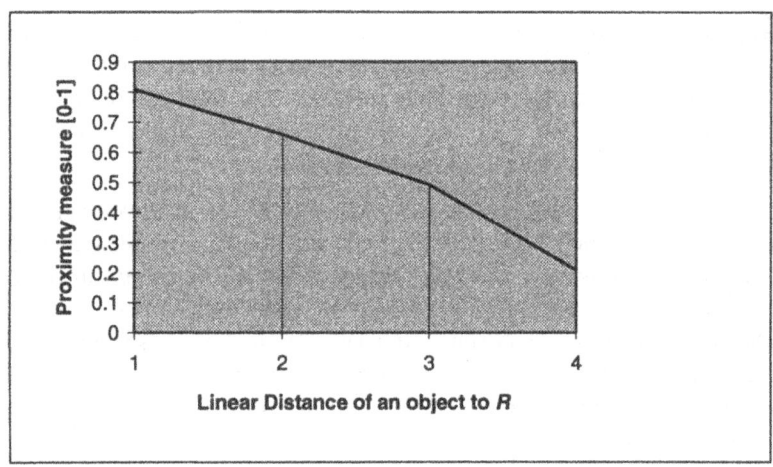

Figure 6. Results of assessing proximity to R from Figure 4.

Qualitative Distances

Daniel Hernández[1], Eliseo Clementini[2], Paolino Di Felice[2]

[1] Fakultät für Informatik
Technische Universität München
80290 Munich, Germany
e-mail: danher@informatik.tu-muenchen.de

[2] Dip. di Ing. Elettrica
Università di L'Aquila
67040 Poggio di Roio, Italy
e-mail: eliseo@ing.univaq.it

Abstract. A framework for the representation of qualitative distances is developed inspired by previous work on qualitative orientation. It is based on the concept of "distance systems" consisting of a list of distance relations and a set of structure relations that describe how the distance relations in turn relate to each other. The framework is characterized by making the role of the "frame of reference" explicit, which captures contextual information essential for the representation of distances. The composition of distance relations as main inference mechanism to reason about distances within a given frame of reference is explained, in particular under "homogeneous structural restrictions". Finally, we introduce articulation rules as a way to mediate between different frames of reference.

1   Introduction

The qualitative approach to the representation of spatial knowledge has gained considerable popularity in recent years. Qualitative representations are characterized by making only as many distinctions in the domain of discourse as necessary in a given context.

In previous work, the qualitative description of space has been mostly restricted to topological relations (Randell and Cohn 1989; Egenhofer and Herring 1991; Egenhofer and Franzosa 1991; Clementini, Di Felice, and van Oosterom 1993; Clementini and Di Felice 1995) and orientation relations (Hernández 1991; Freksa 1992; Latecki and Röhrig 1993; Zimmermann 1993). The combination of topological and orientation relations provides a restricted form of positional information that is mainly useful in small-scale environments such as "the objects in a room" (Hernández 1994).

In this paper we develop a model for the qualitative representation of distances as they are needed in the context of geographic space. People's concepts of space and therefore distance are dependent upon both culture and experience (Lowe and Moryadas 1975). What it means for A to be near B depends not only on their absolute positions (and the metric distance between them), but also on their relative sizes and shapes, the position of other objects, the frame of reference, and "what it takes to go from A to B". With other words, distance concepts are context dependent. Quantitative approaches try to avoid contextual issues by reducing all distance information to an absolute metric scale. However, this is not always feasible or desirable. Qualitative approaches must deal with contextual dependencies since, by definition, only those distinctions relevant in a given context are made. In our approach, we capture contextual information by using frames of reference to qualify distance relations.

We will restrict our attention here to two-dimensional space, which is commonly used as a projection of the three-dimensional physical space. Thus, we can consider a scene to be made up of geometric objects (points, lines, and areas) variously arranged in the plane and possibly overlapping. Furthermore, we assume an isotropic space, in which the effort to move is the same in all directions, and thus the isolines connecting all points at the same distance are concentric circles. Most anisotropic spaces can be translated into isotropic ones by appropriate transformations.

In what follows we will put our contribution in the context of existing literature by briefly reviewing some related work. We will then introduce various levels of distance distinctions and their domain structure (section 3) as well as the composition of distance relations as main inference mechanism (section 4). While those sections assume for simplicity a uniform reference frame, we discuss in section 5 how to mediate between different frames of reference.

2 Related Work

There is a considerable amount of recent work in the area of qualitative spatial reasoning to which the model presented in this paper relates. We shall focus here only on the most closely related papers and refer to Hernández (1994) and Freksa and Röhrig (1993) for a more general review of that literature.[3]

Most related work has concentrated on the qualitative description of size, which being a linear quantity has some similarity to distance. In particular, Allen (1983) briefly describes an extension of his temporal interval reasoner to handle duration, the temporal equivalent of size. Mukerjee and Joe (1990) who extend Allen's approach to multi-dimensional spaces (essentially by maintaining tuples of 1-dimensional relations), base their representation of relative size on the "flush translation operator ϕ". The idea is to observe the relations between two intervals as they move along what the authors call "relation continuum" and deduce the relative size from them. Zimmermann (1991) develops a representation for object sizes based on differences and a partial ordering. The relation $A(>, d_1)B$ denotes the fact that "A is higher/larger than B by the amount d_1", since $|A| = |B| + |d_1|$. In Zimmermann (1993) this "delta calculus" is combined with orientations. However, only a restricted set of distance distinctions is possible in that model. While representations based on direct comparisons

---

[3] There is also a large amount of relevant cognitive and linguistic work which we are excluding here for brevity.

can handle moderately different sizes, other calculi concentrate on differences in the order of magnitude (Raiman 1986; Mavrovouniotis and Stephanopoulus 1988). We shall use order-of-magnitude relations below to express the structure of distance systems.

The most closely related work that explicitly describes a method for qualitative reasoning about distances (*far*, *close*) and cardinal directions (N, E, S, and W) in geographic space is Frank (1992). It is based on an algebra of paths on which the two operations of *inversion* and *composition* are defined. Frank discusses two direction systems, one based on triangular areas and one based on projections, and presents alternatives for the combination of distance and direction, some of which produce only 'Euclidean approximate' results. Our qualitative distance model is superior to Frank's in that not only equally spaced distance intervals but also regions of varying sizes are dealt with.

3   Modeling Distances Qualitatively

We propose a qualitative framework where three elements are needed to establish a distance relation: the *primary object* (PO), the *reference object* (RO), and the *frame of reference* (FofR). The distance between the reference object A and the primary object B is expressed by $d_{AB} = d(A, B)$. We will defer discussion of frames of reference to section 5, while this section and Section 4 treat aspects that are independent or the reference frame.

A distinction has to be made between *comparing* the magnitudes of distances and *naming* distances. For comparing distances, the obvious set of predicates is $<$, $=$, $>$, which characterize the result of direct comparison. With respect to naming, the types of objects involved and the context in which they are embedded are decisive factors for establishing the set of relations to be used. The first level of granularity that comes to mind distinguishes between *close* and *far*. Those two relations subdivide the plane into two regions centered around the reference object, where the outer region goes to infinity. Characteristic of the semantic of qualitative distance relations is that they partition the physical space into regions of different sizes (where the difference can be even in the order of magnitude).

Cognitive considerations suggest the need for systems of distance relations organized along various levels of granularity, for example: a level with three distinctions *close*, *medium*, and *far*, a level with four distinctions *very close*, *close*, *far*, and *very far*, a level with five distinctions *very close*, *close*, *commensurate*, *far*, and *very far*, and so on. Notice that the names given to relations are arbitrary, since we do not discuss linguistic reasons to associate a meaning to a given term. The relations partition the plane in circular regions (see Figure 1).

The qualitative approach deals implicitly with uncertainty in that the next coarser level of distinctions is chosen whenever no decision can be made about the appropriate relation at a finer level. Most of the time this is better than coming up with fuzzy membership numbers, which can be quite arbitrary. However, the

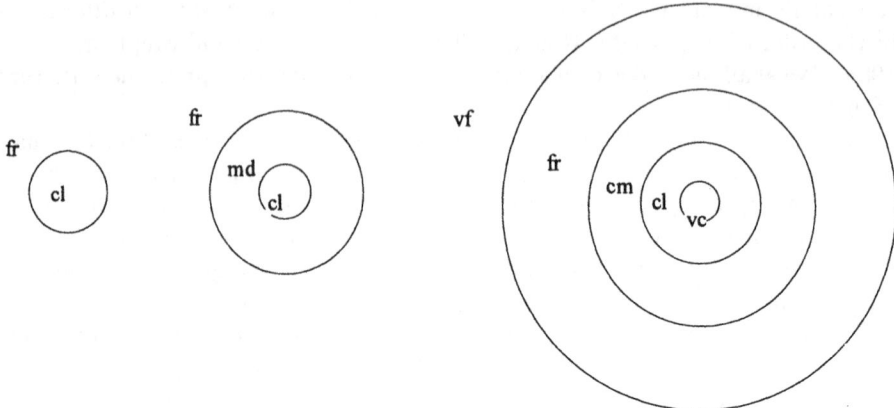

Fig. 1. Various levels of distance distinctions

general framework presented here is independent of the kind of boundary (sharp, fuzzy, overlapping) between the regions.

In general, at a given granularity level space surrounding a reference object is partitioned according to a number of totally ordered distance distinctions $Q = \{q_0, q_1, q_2, \ldots, q_n\}$, where q_0 is the distance closest to the reference object and q_n is the one farthest away (going to infinity). Distance relations are organized in *distance systems* (D) consisting of:

- a list of *distance relations* (i.e., the set of qualitative distinctions being made and their increasing distance order);
- a set of *structure relations* describing how the distance relations in turn relate to each other (e.g., order-of-magnitude relations between the various named distance ranges).

In this paper we will only consider *homogeneous* distance systems, in which all distance relations are related to each other by the same type of property. In order to describe those properties, we distinguish between δ_i being the "distance range i", and Δ_i being the "distance range from the origin up to and including the distance range δ_i" (Figure 2). (Note that the distance symbol q_i labels all distances starting from the origin and falling in the range Δ_i.)

A very common restriction is that of *monotonically* increasing ranges:

$$\delta_0 \le \delta_1 \le \delta_2 \le \ldots \le \delta_n \tag{1}$$

An additional useful *range restriction* is that a given distance range be bigger than the sum of the previous ones:

$$\delta_i \ge \Delta_{i-1}, \forall i > 0 \tag{2}$$

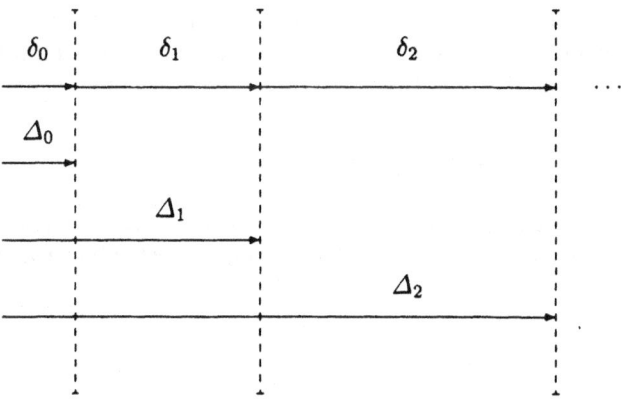

Fig. 2. Distance ranges vs. distance from origin

Finally, if a distance range δ_j is much bigger than a previous one δ_i ($\delta_j \gg \delta_i$), then δ_j will absorb δ_i in the composition:

$$\delta_j \pm \delta_i \simeq \delta_j \tag{3}$$

These restrictions constrain the resulting sets in the composition of relations as we will see in the next section. The restrictions imposed on homogeneous distance systems correspond to the most common types of distance concepts used. The general case of heterogeneous distance systems is dealt with in (Clementini, Di Felice, and Hernández 1995).

4 Composition of Distance Relations

Given the distance $d_{AB} = d(A, B)$ between the reference object A and the primary object B and the distance $d_{BC} = d(B, C)$ between B and C, the composition of distances gives us the distance d_{AC} between A and C. This resulting distance will in general be a range of possible distances, for which we will find a lower and an upper bound.

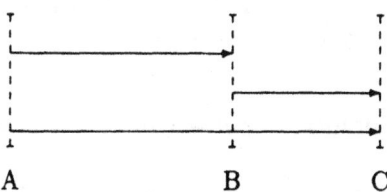

Fig. 3. Composition of distances for same orientation

In the case in which the orientation of B with respect to A is the same as the orientation of C with respect to B (see Figure 3), the composition amounts to adding two "positive quantities", so that the lower bound cannot be less than the bigger of the two distances:

$$\text{LB}(d_{AC}) = d_{AB} \oplus d_{BC} = \max(d_{AB}, d_{BC}). \tag{4}$$

Without structural restrictions, the upper bound for the composition is q_n. Assuming monotonically increasing distance ranges (restriction 1 above), however, we obtain:[4]

$$\text{UB}(d_{AC}) = \text{ord}^{-1}(\text{ord}(d_{AB}) + \text{ord}(d_{BC})). \tag{5}$$

Considering five possible distance symbols, the resulting composition table is

$\oplus$	q_0	q_1	q_2	q_3	q_4
q_0	q_0,q_1	q_1,q_2	q_2,q_3	q_3,q_4	q_4
q_1	q_1,q_2	q_1,q_2,q_3	q_2,q_3,q_4	q_3,q_4	q_4
q_2	q_2,q_3	q_2,q_3,q_4	q_2,q_3,q_4	q_3,q_4	q_4
q_3	q_3,q_4	q_3,q_4	q_3,q_4	q_3,q_4	q_4
q_4	q_4	q_4	q_4	q_4	q_4

Table 1. Composition of distances for same orientation (monotonicity)

given in Table 1.

Some of the entries in Table 1 are possible only in the case of equally spaced distance ranges. By considering the range restriction (2), the composition of two distances can at most be one step bigger than the maximum of the two distances. The upper bound becomes:[5]

$$\text{UB}(d_{AC}) = \text{succ}(\max(d_{AB}, d_{BC})). \tag{6}$$

Therefore, some of the resulting distances in Table 1 can be excluded obtaining the composition table in Table 2.

The upper bound can be further lowered with the absorption rule (restriction 3), which allows us to disregard the effect of the smaller relation (for example, for a difference between distances of two steps, i.e., $p = 2$, we obtain the results in Table 3):

$$|\text{ord}(d_{AB}) - \text{ord}(d_{BC})| \geq p \Rightarrow \text{UB}(d_{AC}) = \max(d_{AB}, d_{BC}). \tag{7}$$

---

[4] The function *ordinal* is defined as: $\text{ord} : Q \to \{1 \ldots n+1\}$, such that $\text{ord}(q_i) = i+1$. Note that $\text{ord}^{-1}(i) = q_n$ for $i > n$.

[5] The function *successor* gives the next symbol in the list, that is: $\text{succ}(q_i) = q_{i+1}$ for each $i < n$ and $\text{succ}(q_n) = q_n$.

$\oplus$	q_0	q_1	q_2	q_3	q_4
q_0	q_0,q_1	q_1,q_2	q_2,q_3	q_3,q_4	q_4
q_1	q_1,q_2	q_1,q_2	q_2,q_3	q_3,q_4	q_4
q_2	q_2,q_3	q_2,q_3	q_2,q_3	q_3,q_4	q_4
q_3	q_3,q_4	q_3,q_4	q_3,q_4	q_3,q_4	q_4
q_4	q_4	q_4	q_4	q_4	q_4

Table 2. Composition of distances for same orientation (range restriction)

$\oplus$	q_0	q_1	q_2	q_3	q_4
q_0	q_0,q_1	q_1,q_2	q_2	q_3	q_4
q_1	q_1,q_2	q_1,q_2	q_2,q_3	q_3	q_4
q_2	q_2	q_2,q_3	q_2,q_3	q_3,q_4	q_4
q_3	q_3	q_3	q_3,q_4	q_3,q_4	q_4
q_4	q_4	q_4	q_4	q_4	q_4

Table 3. Composition of distances for same orientation (absorption rule)

In the case of the composition of distances with opposite orientations—see Figure 4—the upper bound is given by the maximum of the two distances since this corresponds to the difference between two "positive quantities":

$$\text{UB}(d_{AC}) = d_{AB} \ominus d_{BC} = \max(d_{AB}, d_{BC}). \tag{8}$$

Without restrictions, the lower bound is $\text{LB}(d_{AC}) = q_0$, while, applying a similar strategy as in the 'same orientation' case, we can increasingly restrict the lower bound. Imposing restriction 1, the lower bound becomes:

$$\text{LB}(d_{AC}) = \text{ord}^{-1}(|\text{ord}(d_{AB}) - \text{ord}(d_{BC})|). \tag{9}$$

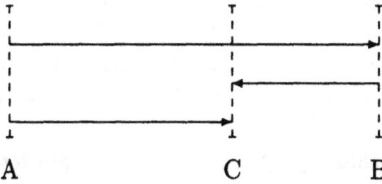

$$A \qquad\qquad C \qquad\qquad B$$

Fig. 4. Composition of distances for opposite orientation

$\ominus$	q_0	q_1	q_2	q_3	q_4
q_0	q_0	q_0,q_1	q_1,q_2	q_2,q_3	q_3,q_4
q_1	q_0,q_1	q_0,q_1	q_0,q_1,q_2	q_1,q_2,q_3	q_2,q_3,q_4
q_2	q_1,q_2	q_0,q_1,q_2	q_0,q_1,q_2	q_0,q_1,q_2,q_3	q_1,q_2,q_3,q_4
q_3	q_2,q_3	q_1,q_2,q_3	q_0,q_1,q_2,q_3	q_0,q_1,q_2,q_3	q_0,q_1,q_2,q_3,q_4
q_4	q_3,q_4	q_2,q_3,q_4	q_1,q_2,q_3,q_4	q_0,q_1,q_2,q_3,q_4	q_0,q_1,q_2,q_3,q_4

Table 4. Composition of distances for opposite orientation (monotonicity)

$\ominus$	q_0	q_1	q_2	q_3	q_4
q_0	q_0	q_0,q_1	q_1,q_2	q_2,q_3	q_3,q_4
q_1	q_0,q_1	q_0,q_1	q_0,q_1,q_2	q_2,q_3	q_3,q_4
q_2	q_1,q_2	q_0,q_1,q_2	q_0,q_1,q_2	q_0,q_1,q_2,q_3	q_3,q_4
q_3	q_2,q_3	q_2,q_3	q_0,q_1,q_2,q_3	q_0,q_1,q_2,q_3	q_0,q_1,q_2,q_3,q_4
q_4	q_3,q_4	q_3,q_4	q_3,q_4	q_0,q_1,q_2,q_3,q_4	q_0,q_1,q_2,q_3,q_4

Table 5. Composition of distances for opposite orientation (range restriction)

The results for the five distance symbols are shown in Table 4. In Table 5 the results that may happen for equally spaced distance ranges (restriction 2) are removed. The results affected by this rule are those for which the difference between the distances is at least two steps, that is:[6]

$$|\mathrm{ord}(d_{AB}) - \mathrm{ord}(d_{BC})| \geq 2 \Rightarrow \mathrm{LB}(d_{AC}) = \mathrm{pred}(\max(d_{AB}, d_{BC})). \quad (10)$$

Eventually, by applying restriction 3, the lower bound becomes:

$$|\mathrm{ord}(d_{AB}) - \mathrm{ord}(d_{BC})| \geq p \Rightarrow \mathrm{LB}(d_{AC}) = \max(d_{AB}, d_{BC}). \quad (11)$$

With a difference $p = 2$, we have the results in Table 6.

In the general case, the composition of two distance relations must take into account any of the possible orientations and not just the two cases of same and opposite orientation (see Clementini et al. 1995). However, these two cases correspond to the two extremes of the range of resulting distances when an arbitrary orientation is taken into consideration. The case of opposite orientations gives the lower bound and the case of same orientation gives the upper bound of the range.

In what has been said up to now, we have ignored the role of frames of reference, which will be addressed in the following section.

---

[6] The function *predecessor* gives the previous symbol in the list, that is: $\mathrm{pred}(q_i) = q_{i-1}$ for each $i > 0$ and $\mathrm{pred}(q_0) = q_0$.

$\ominus$	q_0	q_1	q_2	q_3	q_4
q_0	q_0	q_0,q_1	q_2	q_3	q_4
q_1	q_0,q_1	q_0,q_1	q_0,q_1,q_2	q_3	q_4
q_2	q_2	q_0,q_1,q_2	q_0,q_1,q_2	q_0,q_1,q_2,q_3	q_4
q_3	q_3	q_3	q_0,q_1,q_2,q_3	q_0,q_1,q_2,q_3	q_0,q_1,q_2,q_3,q_4
q_4	q_4	q_4	q_4	q_0,q_1,q_2,q_3,q_4	q_0,q_1,q_2,q_3,q_4

Table 6. Composition of distances for opposite orientation (absorption rule)

5 Frames of Reference and Articulation Rules

Thus far, we have implicitly assumed that the scale in which the distance distinctions apply is known. The scale of a distance system is determined by the context in which the distinctions are made. Using an analogy to the qualitative representation of orientation, we distinguish among three different types of contexts or frames of reference (Hernández 1994):

- **Intrinsic frame of reference**
  The distance is determined by some inherent characteristics of the reference object, like its topology, size or shape. An object like a house, for example can implicitly determine what is *close* and *far* with respect to itself, without the need of any external factors.
- **Extrinsic frame of reference**
  The distance is determined by some external factor, like the arrangement of objects, the traveling time, or the costs involved.
- **Deictic frame of reference**
  The distance is determined by an external point of view. The most immediate case is the one of objects that are visually perceived from an observer standing at the point of view. Deictic frames of reference include also cases in which the point of view is used figuratively, i.e., not in the sense of sight. Often the point of view is related to how an individual builds a mental map of space.

Thus, a frame of reference has to take into account all contextual information. For a given type, we have to establish the criteria fixing the scale and choose an appropriate distance system. Therefore, three components make up a frame of reference:

$$\text{FofR} = (T, S, D) \tag{12}$$

The type T is either intrinsic, extrinsic or deictic. The scale S is, depending on the type: a function of inherent characteristics of the reference object, e.g., $f(\text{size}(\text{RO}))$ (for the intrinsic type), a reference unit given by external factors (for the extrinsic type), or a function of the distance between the point of view and the reference object $f(d(\text{PV}, \text{RO}))$ (for the deictic type). The distance system D is a structure as defined in Section 3 (made up of distance relations and structure relations).

In the general case, the qualitative description of distance among a set of objects is reasonably assumed to be given according to different frames of reference. A "basic" type of qualitative reasoning is therefore to relate the distances to each other and be able to infer new information. Ideally, we would like to transform all distance descriptions to the same ("canonical") frame of reference. However, different distance frames of reference refer to different granularities or scales, thus making a transformation into an implicit frame difficult. We rather must restrict ourselves to giving articulation rules (cf. Hobbs 1985) that state how two particular frames of reference compare. This comparative information, which consists mainly of order information between reference magnitudes, must be explicitly maintained in the knowledge base. It can further constrain the relations maintained in the constraint network, and suggests deferring naming to those cases where it can be done in a disambiguating context. From the definition of frames of reference given above it follows that articulation rules must relate the scales and distance systems of the frames involved. The frame type does not need to be explicitly related, since it already determines the scale factor and is thus contained in it. The distance systems must be compared as to the sets of relations involved and their structure (i.e., the order-of-magnitude relations between the distances). In general, only similar distance systems might be successfully related to each other, and some might be incomparable to each other.

6 Conclusion

In spite the fact of distance being an important cognitive spatial concept, and the increased research activity in the area of qualitative spatial reasoning, no satisfying qualitative model for distances had been developed up to now. This paper's major contribution is providing such a model with the characteristic advantages of the qualitative approach: a flexible set of distance distinctions at various levels of granularity, an implicit way of handling uncertainty, and the corresponding reasoning mechanisms.

This, however, is only the beginning of a longer term collaborative research effort. Several of the explicitly stated assumptions and restrictions in the paper point to further research directions (some of which already have been pursued as reported elsewhere):

- Distance is only one component (the other one being orientation) of positional information. Further work will have to deal with the combination of distance and orientation, which we expect to constrain each other in a way that actually simplifies the reasoning process.
- Here we have only considered the case of homogeneous distance systems, in which all distance relations are related to each other by the same type of

property. The more general case of heterogeneous distance ranges, which is likely to correspond to cognitive distance concepts without *a priori* restrictions, still needs to be investigated.

- We have ignored the case of extended objects: The extension of objects, however, influences the concepts of distance. As a first classification, we will consider three different scenarios. If the distances involved at a given scale are such that the extension of the objects can be disregarded, we use the point abstraction as in this paper. If the extension of the objects is of the same order of magnitude of the distances among them, then the distance will be computed between the boundaries. If the objects connect, the distances will be computed between the centroids of the objects.

- The articulation rules mechanism sketched in the previous section needs further study. This will be done in the context of an application in the domain of vehicle navigation systems. In that context, information at various scales and granularities needs to be dealt with to guide vehicles at the single road level (small-scale environment), at the city level (urban scale), and at the region level (geographic scale).

Acknowledgements

The work of Daniel Hernández reported here has been partially funded by the German Ministry for Research and Technology (BMFT) under FKZ ITN9102B. The work of Eliseo Clementini and Paolino Di Felice has been supported by the Italian MURST project "Basi di dati evolute: modelli, metodi e sistemi" and CNR project no. 95.00460.CT12 "Modelli e sistemi per il trattamento di dati ambientali e territoriali". We would like to thank the anonymous reviewers for their insightful remarks and improvement suggestions.

References

Allen, J. F. (1983). Maintaining knowledge about temporal intervals. *Communications of the ACM*, *26*(11), 832–843.

Clementini, E. and Di Felice, P. (1995). A comparison of methods for representing topological relationships. *Information Sciences*. To appear.

Clementini, E., Di Felice, P., and Hernández, D. (1995). Qualitative representation of positional information. In preparation.

Clementini, E., Di Felice, P., and van Oosterom, P. (1993). A small set of formal topological relationships suitable for end-user interaction. In Abel, D. and Ooi, B. C., editors, *Third International Symposium on Large Spatial Databases, SSD '93*, Volume 692 of *Lecture Notes in Computer Science*, pages 277–295. Springer, Berlin.

Egenhofer, M. J. and Franzosa, R. (1991). Point-set topological spatial relations. *International Journal of Geographical Information Systems*, *5*(2), 161–174.

Egenhofer, M. J. and Herring, J. (1991). Categorizing binary topological relationships between regions, lines, and points in geographic databases. Technical report, University of Maine, Department of Surveying Engineering.

Frank, A. U. (1992). Qualitative spatial reasoning with cardinal directions. *Journal of Visual Languages and Computing, 3,* 343–371.

Freksa, C. (1992). Using orientation information for qualitative spatial reasoning. In Frank, A. U., Campari, I., and Formentini, U., editors, *Theories and Methods of Spatio-Temporal Reasoning in Geographic Space. Intl. Conf. GIS—From Space to Territory,* Pisa, Volume 639 of *Lecture Notes in Computer Science,* pages 162–178. Springer, Berlin.

Freksa, C. and Röhrig, R. (1993). Dimensions of qualitative spatial reasoning. In Piera Carreté, N. and Singh, M. G., editors, *Proceedings of the III IMACS International Workshop on Qualitative Reasoning and Decision Technologies—QUARDET'93—,* Barcelona, pages 483–492. CIMNE, Barcelona.

Hernández, D. (1991). Relative representation of spatial knowledge: The 2-D case. In Mark, D. M. and Frank, A. U., editors, *Cognitive and Linguistic Aspects of Geographic Space,* NATO Advanced Studies Institute, pages 373–385. Kluwer, Dordrecht.

Hernández, D. (1994). *Qualitative Representation of Spatial Knowledge,* Volume 804 of *Lecture Notes in Artificial Intelligence.* Springer, Berlin.

Hobbs, J. R. (1985). Granularity. In Joshi, A., editor, *Proceedings of the Ninth International Joint Conference on Artificial Intelligence,* Los Angeles, CA, pages 432–435. International Joint Conferences on Artificial Intelligence, Inc., Morgan Kaufmann, San Mateo, CA.

Latecki, L. and Röhrig, R. (1993). Orientation and qualitative angle for spatial reasoning. In Bajcsy, R., editor, *Proceedings of the Thirteenth International Joint Conference on Artificial Intelligence,* Chambéry, France, pages 1544–1549. International Joint Conferences on Artificial Intelligence, Inc., Morgan Kaufmann, San Mateo, CA.

Lowe, J. C. and Moryadas, S. (1975). *The Geography of Movement.* Houghton Mifflin, Boston.

Mavrovouniotis, M. L. and Stephanopoulus, G. (1988). Formal order-of-magnitude reasoning in process engineering. *Computer Chemical Engineering, 12,* 867–880.

Mukerjee, A. and Joe, G. (1990). A qualitative model for space. In *Proceedings of the Eighth National Conference on Artificial Intelligence,* pages 721–727. American Association for Artificial Intelligence, AAAI Press/The MIT Press, Menlo Park/Cambridge.

Raiman, O. (1986). Order of magnitude reasoning. In *Proceedings of the Fifth National Conference on Artificial Intelligence,* pages 100–104. American Association for Artificial Intelligence, AAAI Press, Menlo Park.

Randell, D. A. and Cohn, A. G. (1989). Modelling topological and metrical properties of physical processes. In Brachman, R., Levesque, H., and Reiter, R., editors, *Proceedings 1st International Conference on the Principles of Knowledge Representation and Reasoning*, pages 55–66. Morgan Kaufmann, San Mateo, CA.

Zimmermann, K. (1991). SEqO: Ein System zur Erforschung qualitativer Objektrepräsentationen. Forschungsberichte Künstliche Intelligenz FKI-154-91, Institut für Informatik, Technische Universität München.

Zimmermann, K. (1993). Enhancing qualitative spatial reasoning—combining orientation and distance. In Frank, A. U. and Campari, I., editors, *Spatial Information Theory. A Theoretical Basis for GIS. European Conference, COSIT'93*, Marciana Marina, Italy, Volume 716 of *Lecture Notes in Computer Science*, pages 69–76. Springer, Berlin.

Measuring without Measures
The Δ-Calculus*

Kai Zimmermann

Department of Computer Science, University of Hamburg**,
Vogt-Kölln-Str. 30, D-22527 Hamburg
Phone: +49-40-54715-368, Fax: +49-40-54715-385
Email: zimmerma@informatik.uni-hamburg.de

Abstract: In recent years qualitative reasoning approaches have become increasingly popular and are preferred over quantitative numeric approaches for applications in the field of AI. This is due to several factors. The most striking argument in the field of temporal and spatial reasoning is that humans are not able to give precise numeric estimates of their environment, e.g., if asked to estimate temporal duration or object size. Nevertheless we are capable of dealing with our surrounding world in a very efficient manner and are able to produce qualitative descriptions of it. In the field of point like measures, such as object dimensions or the duration of intervals, only a few new qualitative approaches have been developed, such as, Order of Magnitude for technical domains, and thus researchers tend to stick with numeric approaches. In this paper we present a new approach based on cognitive considerations of how humans *perceive* spatial dimensions and how they *reason* with this spatial knowledge. We then describe how reasoning is performed within the new calculus and how it can be adopted for representing not only one-dimensional measures, but also areas, volumes and proportions.

1 Introduction

In recent years qualitative approaches in representation and reasoning formalisms have become increasingly popular. They are usually as efficient as their quantitative counterparts in terms of runtime complexity of algorithms and they seem to model a more human like way of reasoning in spatial and temporal domains, particularly. Several qualitative approaches have been developed in the past, but most of them fail to handle point like measures, such as durations and object dimensions, qualitatively. Some methods even transform qualitative information into a quantitative representation before reasoning. In the following we describe a new calculus, developed with psychological and linguistic considerations in mind, that tries to overcome the deficiencies of former approaches: The Δ-calculus.

---

*   The contents of this paper has not been published yet other than as an informal report.

**   Part of this report has been supported by the German Minister for Research and Technology under grant ITN9102B, while the author was working at the Technische Universität München. The author is solely responsible for the content of this publication.

2 Existing Approaches

For reasoning with temporal intervals [Allen 83] introduces a qualitative method for describing the relative positions of intervals using 13 disjoint relations. This approach is adopted and extended by [Knight & Jixin 92] who develop a set of axioms allowing the extension of Allen's approach with time points and durations. However both Allen and Knight&Jixin stick to numeric representations for the durations. Allen even transforms expressions like "A took longer than B", i.e., Duration(A) > Duration(B), into Duration(A) = f*Duration(B), f ∈]1; ∞[. This is done to handle the qualitative relation *longer* analogous to *longer by some factor*, i.e., "A took between two and three times as long" which is represented as Duration(A) = f*Duration(B), f ∈ [2; 3]. Knight and Jixin introduce axioms that extend the durational reasoning component. For example, from the interval A *meets* B, i.e., the end point of A equals the start point of B, it follows that the two intervals can be viewed as one composed interval having as duration the sum of the durations of A and B.

In the field of technical domains several systems have been developed to deal with point like measures. Early constraint propagation approaches dealing only with numeric data, e.g. [Steele & Sussman 78], can be extended and combined with qualitative reasoning to derive better results. One such system is the Quantity Lattice described in [Simmons 86]. It integrates quantitative knowledge with qualitative knowledge, i.e., relations and algebraic expressions. It is capable of performing qualitative reasoning, e.g., the combination of relations, and numeric evaluations. Inferences may be drawn as well by combining this knowledge. For example, one can derive new numeric constraints from relations which then may lead to the evaluation of algebraic expressions.

To overcome restrictions of a simple relational calculus, several systems use sign information instead, e.g. [Williams 88], and perform qualitative calculation of differential equations. In this formalism measures are characterized by the sign, +, 0, and -, respectively, and reasoning is performed by qualitative evaluation of algebraic equations. Another widely used approach is Order of Magnitude reasoning, O(M). This formalism extends the set of relations that may hold between two points to more than simply <,=, and >, via the introduction of relations like *slightly smaller* and *much smaller*. These new relations are defined in terms of the factor between two measures and a tolerance parameter, e, chosen from]0; 0.4656[. For example, a is *slightly smaller* than b, is defined by: ratio a/b lies in the interval]1/(1+e); 1[, see [Raiman 88] and [Mavrovouniotis & Stephanopoulus 88] for details.

3 Demands on a More Cognitively Plausible Representation

Although the approaches introduced above enable us to deal with point like measures in technical systems, the selected qualitative abstractions are either computationally too weak or unlikely to be used in human reasoning, and the performed operations on these values are not very plausible from a cognitive point of view. We now discuss their deficiencies and try to formulate demands on a cognitively more plausible representation.

First of all humans do not perceive any negative measures directly. There is no such thing as having a negative length or volume, for example. Even a hole in the ground does not have a negative height, but rather we perceive its vertical extent as a *positive* depth. Our natural language does not provide any words to describe negative spatial or temporal measures and we are not able to imagine such objects: Negation is a mathematical abstraction that does *not* correspond to a perception in the real world.

Although it may be very convenient to use negative values in mathematics and the mathematical modeling of physical processes, negative measures are not needed in a calculus based on human perception. Therefore we will use only positive measures and the special symbol Ø, denoting zero, in our calculus. The depth of a hole will be represented as a positive measure, but keeping in mind the fact that the direction of this measure is opposed to height. Mathematically one can consider this as representing only the absolute amounts of measures directly, keeping the sign information separate. We will see that this is not a limitation, but rather that it adds to the capabilities of the formalism.

The formalism is based on relational knowledge between point like measures. As with most of the familiar approaches, this allows to represent the measures without the necessity of having an absolute scale, such as meters, to which all measures are compared and referenced. This is more plausible because humans do not measure their environment permanently, and precise measures of, e.g., length, only occur in specialized technical domains and situations. However, if precise measures are available, they can be represented in the formalism and function as reference points for estimates, such as "between 2 and 3 centimeters".

Existing relational approaches have several deficiencies. The most striking problem is their poor resolution and their lack of granularity. Although the O(M) approach provides seven relations instead of three, one can not choose the granularity of the representation. Sometimes it is sufficient to only represent the fact that a > b. But sometimes one may want to specify by how much it is greater. Since one must choose the parameter that divides the ratio space once, and it is not possible to change it according to the problem within the same reasoning task, one can not provide a mechanism that changes the granularity of the representation depending on the context. Δ-Calculus provides mechanisms to adapt the granularity whenever needed.

Moreover, we do not only criticize the fixed granularity and the necessity of choosing a parameter for multiplication in O(M), but we consider the task of multiplication to be a cognitively implausible operation. Direct multiplication of two measures does not occur in normal situations. If asked, e.g., to multiply the height of a chair with the height of a table, for the typical person there is no way to do so naturally. Instead, he would have to measure the height, e.g., in meters, and then perform this task mathematically, as he might have learned in school.

From psychological experiments it is known that the time to estimate the ratio between the length of two presented lines is proportional to the ratio, [Hartley 77] and [Hartley 81]. This suggests that we do not perceive ratios directly or derive them by multiplication, which would take constant time, but rather that they get evaluated during the estimation task by adding the shorter line to itself until it supercedes the longer.

Thus, multiplication of a measure with a numeric factor should be modeled as a summing process in a cognitively plausible representation. Therefore, the Δ-calculus does not provide any means to multiply measures directly and multiplication with numeric factors is restricted to natural number and ratios composed of natural numbers.

4 The Δ-Calculus

In the following sections the Δ-calculus is introduced. It has been developed according to the above stated demands and overcomes several deficiencies of other approaches. An earlier version of the formalism is described in [Zimmermann 91].

4.1 The Basic Representation

The basic idea behind the Δ-calculus is that whenever we compare two point like measures and perceive that they are unequal, we know that there exists a difference. Although this fact may seem trivial at first, because we are so used it, we now introduce a difference measure so that we can say, "a is greater than b by some x". This is represented in the formalism by defining a new three valued relation a (>, x) b, meaning: a > b and a = b+x, where a, b, and x are restricted to positive values. Fig. 1 shows the graphical representation. The relation is symmetrical with respect to b and x, that is: a (>, x) b ⟺ a = b+x ⟺ a = x+b ⟺ a (>, b) x.

Fig. 1: The graphical representation exploiting the symmetry
of the smaller measure and the difference measure.

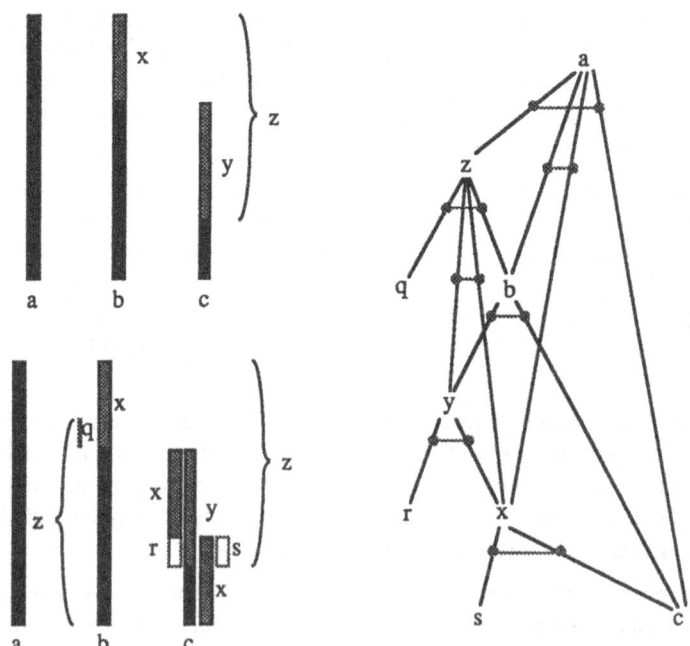

Fig. 2: A simple example modeled with the Δ-calculus.
Not all derivable relations are represented.

If we were to only introduce these difference measures and no new relation we would not gain much, since the formalism could make no use of the fact that x is representing the exact difference between a and b. Fig. 2 shows an example scene modeled with the Δ-calculus. As can be seen easily this is a much more informative representation than we would get if we only knew a > b > c.

The special symbol Δ is used as a generic symbol for these differences whenever no further use is made of the difference measures. It just represents the fact that there exists a difference and prevents the system from generating unused measures. Thus when we only want to represent "a is greater than b" we denote a (>, Δ) b. If we now perceive the additional fact that the resulting difference is, say, smaller than b, we transform the generic difference into a difference measure and denote a (>, x) b and x (<, Δ) b.

Since b and x may be the same, we are also able to model products with natural numbers as a sum. Thus a (>, b) b means a is twice b. To allow for greater products we introduce a muti-set (bag) for the measures, «x», and can now represent, e.g., "more than three boxes maybe stacked into the closet" as: c (>, «b, b, b») x. To represent that just three boxes maybe stacked into the closet we need only add: x (<, Δ) b. Note that one should make limited use of factors. Only factors in a restricted range are useful, since the result of such a multiplication usually would not be comparable to other measures. If there is an object d just as high as three boxes we can use sets of the multisets and denote: c (>, {«b, b, b», «d»}) x.

4.2 Reasoning with the Δ-Calculus

Reasoning with the Δ-calculus is performed through composition and addition of relations. This can be accomplished through propagation in a manner similar to ordinary relational reasoning. From a (>, x) b and b (>, y) c we derive: a (>, «x, y») c, as can be seen easily. But now, since we may have additional knowledge about the differences, we can make use of this knowledge. Whereas the composition of a > b and b < c only yields: a ? c, we can now look at the relation between the differences. If a (>, x) b and b (<, y) c and x (<, z) y, we conclude that a (<, z) c. Fig. 3 illustrates this situation. Due to this fact we can get much better results through propagation than with existing approaches.

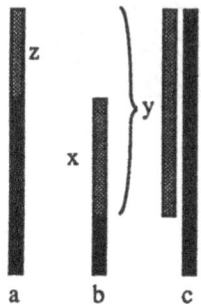

Fig. 3: Knowledge about the relations between the differences
allows for better results in combination.

This can also be done with more than one possible relation, e.g., a (>, x) b or a (=, Ø) b combined with b (>, y) c yields a (>, z) c and z (>, x) y or z (=, Ø) y. Thus, the uncertainty about the relation between the input measures is reflected by the uncertainty of the relation between the differences in the result.

But there is an important case to take into account to avoid running up against a combinatorial explosion. Consider the case of a (>, x) b or a (=, Ø) b combined with b (>, y) c or b (=, Ø) c. Now four possible relations may hold between a and c: a (=, Ø) c, a (>, x) c, a (>, y) c, or a (>, «x, y») c. If we were to represent all these possible relations explicitly, in the case of a ? b combined with b ? c, ordinary combination would yield 13 possible relations between a and c. These combinations would then have to be fed into the propagation algorithm resulting in a combinatorial explosion, without adding any information. To avoid this, we use the Δ-construct in the result, i.e., we denote in the above example: a (=, Ø) c or a (>, Δ) c, and in the case where nothing is known about the relation, we get a ? c as result. This problem does not occur in ordinary relational reasoning, because the different cases are distinguished.

If the retrieval algorithm is built such that it holds all possible decompositions of measures into smaller measures encountering in path analysis, we can deduce even more. Consider that we know that into closet c1 exactly two boxes may be stacked and into another closet c2 more than two boxes may be stacked: c1 (>, «b, b») Ø and c2 (>, «b, b») x. Since x is restricted to positive values greater than Ø we can conclude that c2 (>, x) c1, without requiring a direct path in the relation graph from c2 to c1.

5   Representing Areas, Volumes, and Proportions

Up to now only point like measures, such as the height of boxes and closets, and reasoning over relations between them has been considered. But of course, we can group measures that belong to one object, and thus describe its volume and proportion. Fig. 4 and 5 show examples.

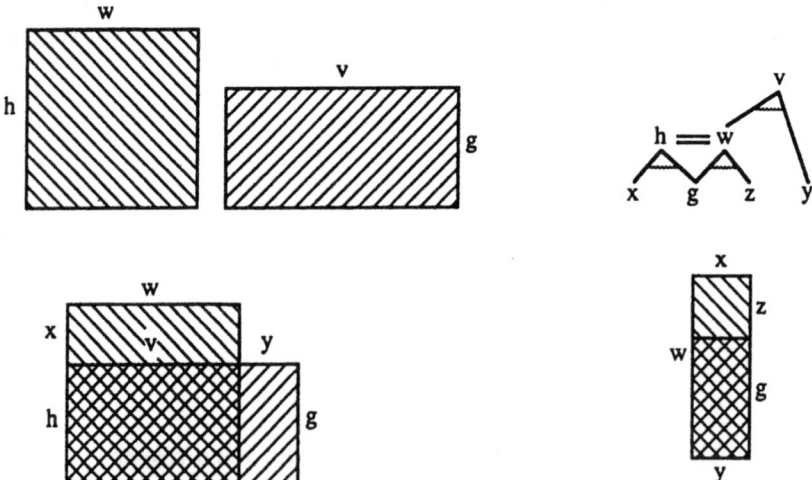

Fig. 4: Areas may be related by extracting the overlapping area and recursively relating
the remaining areas. Only little knowledge about the relation between the
side lengths is needed to evaluate the relation between the areas.

In the case of reasoning over areas and volumes respectively, we can make use of the simple fact that when in any orientation both sides of area A are longer than the sides of an area B, A is greater than B. Thus if in the original configuration this does not hold, we can try to rotate one of the areas by 90° by simply exchanging the measures for the sides of the object. If there is no configuration such that both sides of one object are longer than the sides of the other, start with a configuration where one of the sides is longer. We can then extract the overlapping area and try to relate the remaining areas recursively, see Fig. 4. It should be noted that according to psychological experiments there is evidence that humans do not estimate areas correctly. For example, square areas are consistently underestimated when compared to rectangles [Mates et al. 92]. It is not clear to the author whether this stems from some mapping performed in the visual system shortly after the retina as Mates suggests or wether it is due to the fact that humans estimate areas by the length of the surrounding border line, as other authors claim. Both reasons for the observed behavior might be modeled using the Δ-calculus.

Instead of representing the areas and volumes of objects one can also represent the proportions of objects qualitatively. Fig. 5 shows an example.

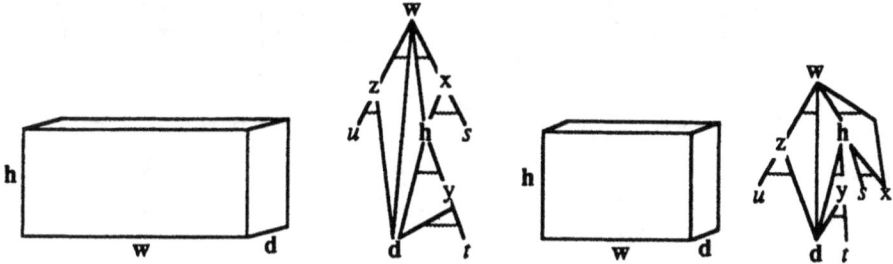

Fig. 5: Two objects with different proportions
and their representation using the Δ-calculus.

When the sides of objects are related to each other they form patterns which are characteristic of classes of objects with similar proportions. The derived patterns describe not only exact ratios, but ranges of acceptable ratios for a class. Consider the following example, Fig. 6.

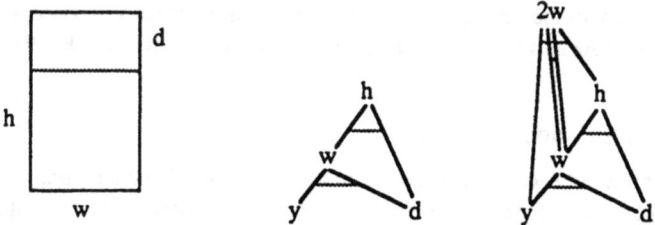

Fig. 6: Extracting ranges for the proportion ratios.

The height of the block is greater than the width, but the remaining difference is less than the width, i.e., $h(>, d)w$ and $d(<, y)w$. Since $h=w+d$ from $d(<, y)w$ follows that $d<1/2h<w$, i.e., $w \in]1/2h, h[$. On the other hand, from $h=w+d=d+y+d$ we can derive that $2w= d+y+d+y =h+y$, i.e., $2w (>, y) h$ and therefore $h \in]w, 2w[$.

Thus, we have an efficient way of defining estimates of proportions based on simple percep-tions, in this case the comparison of the two sides and the remaining difference. Both, upper and lower limits, can be derived by Δ-calculus, if needed. The derived classes may than be divided into subclasses, providing levels of increasing granularity, see Fig. 7. Note that not only ratios of the power of two occur, i.e., four, eight, and so on, but that the next subdivi-sions are ratios of three, four, five, and so on. This is achieved in Δ-calculus without having to introduce a *division* operation explicitly.

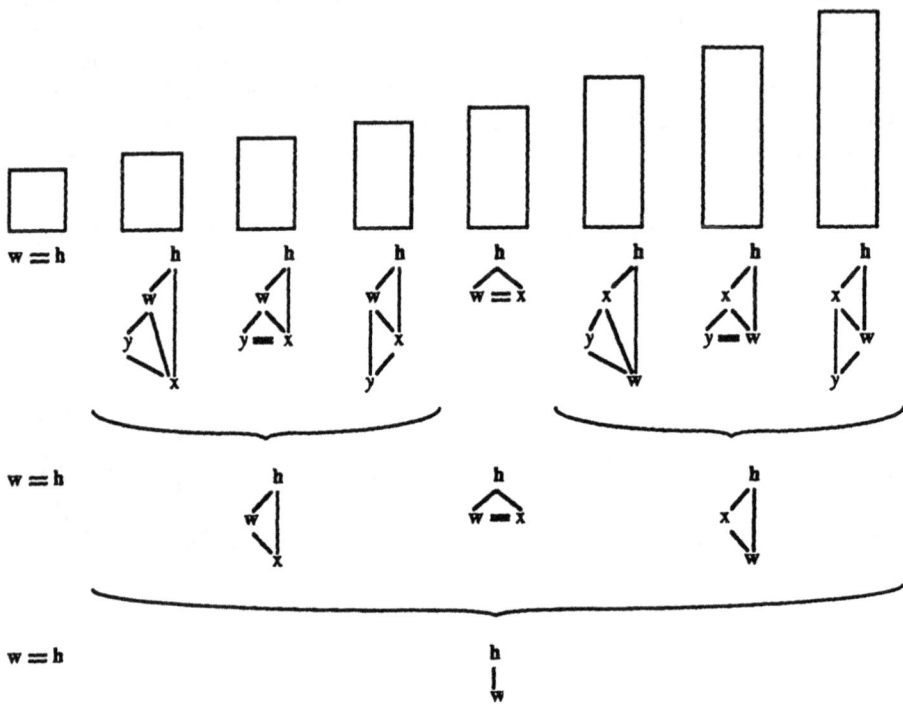

Fig. 7: Proportion classes at different levels of granularity. The w stands for the width and the h for the height of the correspondings rectangles. The classes form conceptual neighbor-hoods such as those introduced for temporal relations in [Freksa 92].

6   Conclusion and Outlook

The above introduced Δ-calculus allows us to represent spatial and temporal point like mea-sures in a more cognitively plausible manner. It provides a means for representing point like measures qualitatively with increased expressability on different levels of granularity and to derive better results for the combination of relations, than with existing approaches. It fulfills all the given psychological and linguistic demands outlined previously. Although it has been developed to model commonsense spatio-temporal reasoning, it may be also used in technical systems, modeling the human understanding of physical processes, instead of evaluating differential equations using qualitative values, as most systems do. The Δ-calculus has been used as an extension to a temporal reasoning system performing duration reasoning [Zimmermann 91] and for modeling spatial scenes, see [Zimmermann 93].

Acknowledgements

I thank my advisors Christian Freksa and Wilfried Brauer and my colleague Daniel Hernández.

References

Allen, J.F. (1983). Maintaining Knowledge about Temporal Intervals. *Communications of the ACM* 26(11): 832–843.

Freksa, C. (1992). Temporal Reasoning Based on Semi-Intervals. *Artificial Intelligence* 54: 199–227.

Hartley, A.A. (1977). Mental Measurement in the Magnitude Estimation of Length. *Journal of Experimental Psychology: Human Perception and Performance* 3: 622–628.

Hartley, A.A. (1981). Mental Measurement of Line Length: The Role of the Standard. *Journal of Experimental Psychology: Human Perception and Performance* 7: 309–317.

Knight B. and Jixin, M. (1992). A General Temporal Model Supporting Duration Reasoning. *AI Communications* 5(2): 75–84.

Mates, J., Maio, V.D. and Lánsky, P. (1992). A Model of the Perception of Area. *Spatial Vision* 6(2): 101–116.

Mavrovouniotis, M.L. and Stephanopoulus, G. (1988). Formal Order of Magnitude Reasoning in Process Engineering. *Computer Chemical Engineering* 12: 867–880.

Raiman, O. (1988). Order of Magnitude Reasoning. *Proc. of AAAI-88*, 100–104.

Steele, G.L. and Sussman, G.J. (1978). *Constraints*. MIT AI-Memo 502 (11), Cambridge.

Simmons, R. (1986). "Commonsense" Arithmetic Reasoning. *Proc. of the AAAI-86*, 118–124.

Williams, B.C. (1988). MINIMA: A Symbolic Approach to Qualitative Algebraic Reasoning. *Proc. of the AAAI-88*, 254–269.

Zimmermann, K. (1991). *SEqO: Ein System zur Erforschung qualitativer Objektrepräsentationen*. Report FKI-154-91, Technische Universität München.

Zimmermann, K. (1993). Enhancing Qualitative Spatial Reasoning — Combining Orientation and Distance. A. U. Frank and I. Campari (Eds.): *Proc. of the International Conference on Spatial Information Theory. A Theoretical Basis for GIS*, Elba, Italy, Springer: 69-76.

The Millennium Project:
Constructing a Dynamic 3+D Virtual Environment for Exploring Geographically, Temporally and Categorically Organized Historical Information

Earl Rennison and Lisa Strausfeld

Visible Language Workshop
MIT Media Lab
20 Ames St.
Cambridge, MA 02139
E-mail: *{rennison,straus}@media.mit.edu*

Abstract

This paper describes a conceptual and computational approach for enabling understanding of a large, multidimensional set of information. The goal of the Millennium Project is to provide a knowledge seeker the ability to move through virtual time and space to explore and discover the connections between artifacts of philosophy, painting, music, literature, science, and political events of a pivotal time in world history: the years from 1906 to 1918. This virtual space continually constructs and reconstructs itself based on the knowledge seeker's movements through and within it, much like the process of moving through the conceptual spaces of our minds as we construct meaning.

The conceptual framework for this research is based on linguistic metaphor theory and cognitive science. In this paper we show how our concepts of *embodied cognitive models* and *visual discourse* assist us in designing and building a computational environment that enables people to understand large bodies of information.

1. Introduction

In 1912 the S.S. "Titanic" sank on it's maiden voyage, Woodrow Wilson won the U.S. presidential election, Sun Yat-sen founded Kuomintang (the Chinese National Party), C.G. Jung published "The Theory of Psychoanalysis," Edwin Bradenburger invented a process for manufacturing cellophane, and Marcel Duchamp painted "Nude descending a Staircase." How, if at all, do these events relate to one another? Where, when and what were the confluences of ideas and people that influenced the outcome of these events? How do we acquire the knowledge to understand the complex associations between people and ideas, across time and place, based on the artifacts and events they created?

Perhaps the most resourceful approach to understanding a large corpus of information is to seek the help of a person who has carefully studied the artifacts for a long time: an expert. We could meet with this person and ask questions. If we were lucky, the expert would have all the original artifacts with her and would show them to us as she described the relationships between them. We could direct the conversation to areas that we understand or are most interested in, and the expert would be able to fill in more and more detail. This would be one of the most expedient ways of understanding the information.

Imagine, further, that our expert, like a computer, has perfect memory. Imagine that our expert has studied the corpus of information in intimate detail and has built up a representation of how each piece of information relates to all the other pieces of information. She has classified the information into categories, organized events by when and where they occurred, and sorted events by cause and effect relationships. Further, imagine that during the course of a conversation our expert is able to show us visually what she is thinking and how the information is organized, as though we are seeing things through the mind's eye of the expert. Imagine that the expert is able to construct a virtual *space*, representing a mental space, to illustrate the relationships between the original artifacts, and that we are able to move through this mental space to examine the relationships more closely. As we move through the space, the expert responds to our movements by dynamically restructuring the space to show more detail or more abstraction/generalization depending upon how we move.

Simulating the expert visual interlocutor we described above would clearly fulfill our objective for the Millennium Project. In order to create this expert we need to do three things:

1. Build up a computational representation of how each information object in our database relates to all other pieces of information

2. Project the representation onto the virtual space of a computer display that resembles a mental space

3. Enable interaction by reading knowledge seekers' interests in the information via their movements in the space, and respond by dynamically restructuring the virtual information space.

1.1 The Database

Our database consists of a set of files that contain information objects that describe events, artifacts, people and ideas pertaining to the years 1906-1918. They are displayed as 3D text objects that sometimes include images, video clips, or sounds. Each of these information objects contain annotations that describe the properties of the information objects. These basic properties include:

1. date
2. location
3. associations (term, object)
4. cause-effect relationships
5. size measurements

1.2 The Objective

Our objective is to enable understanding of historical information by allowing knowledge seekers to explore and examine information objects in dynamic 3D spaces. These dynamic 3D spaces should adapt to the dynamic interests of the knowledge seekers, following the typically nonlinear path of mental understanding.

In the sections that follow we discuss our computational approach to these three tasks. Before we explain this approach, however, we need to describe the conceptual framework that structures our computational approach, and its relation to linguistic metaphor theory and cognitive science.

2. Conceptual Framework: Pathways to Understanding

2.1 Embodied Cognitive Models

How do we understand information? This is the subject of the next five sections: context, scale relations, experience, perception, and analysis. We will discuss each one in relation to the problem of understanding, in relation to our research on embodied cognitive models [Strausfeld, 95b], and in relation to the Millennium Project.

First, it is important to understand understanding as a process, like a conversation, or a journey. Understanding comes from our own experience. We learn about the physical world by experiencing it: we look, we touch, we move around, and we relate things to our bodies.[Johnson, 92]

Information, however, is typically abstract. We cannot literally see it, touch it, move though it, or relate it to our bodies. Yet we are able to somehow structure abstract ideas in such a way that we understand them and can store them in our minds. Linguistic metaphor theory claims that the way we experience the physical world through our bodies makes its way into the cognitive models we use to structure abstract ideas. Understanding is *embodied* in that our bodily experience directly influences the way we structure thought. [Johnson, 92].

Metaphor is how we map the concrete cognitive structures or models in our minds to an abstract domain, like information. The problem of representing the abstract with the concrete is the problem of language. We encounter this problem every time we attempt to express ideas outside the realm of the physical world. In *Metaphors We Live By*, Lakoff and Johnson expose the way language allows us to implicitly (and often subconsciously) reference our physical and cultural experience in the world to express or understand abstract concepts or ideas. Lakoff and Johnson show that language is based on a conceptual system that is metaphorical in nature. They write: *"The essence of metaphor is understanding and experiencing one kind of thing in terms of another"* [Lakoff, 80].

In our research we focus on embodied cognitive models for the following reasons:

1. Language provides evidence that understanding is structured by our bodily experience in the physical world

2. Body-related metaphors, because they refer to our ongoing experience in the physical world, can be extended and combined. Other metaphors (like the desktop metaphor) have limiting structures that inhibit extension and combination.

3. Body-related metaphors are inherently spatial because they relate to the way we interact in physical space. Our daily experience in physical space provides us with good intuitions about virtual space interaction via movement and object manipulation.

In the next five subsections we present a number of kinesthetic cognitive models that support our five-part approach to enabling understanding of information. These cognitive models encapsulate ideas put forth on metaphor theory by linguists George Lakoff and Mark Johnson [Lakoff, 80; Lakoff, 87; Johnson, 92]. According to Lakoff, there are four types of cognitive models around which thought is structured.

1. Propositional
2. Image schematic
3. Metaphoric
4. Metonymic

Propositional models specify elements, their properties, and the relations among them (like the *desktop* and its relationship to *file folders*). Image schematic models specify schematic images that operate on a more conceptual level such as *trajectories* or *containers*. Metaphoric models are mappings from a propositional or image-schematic model in a source domain to an analogous structure in a target domain (like the mapping of the desktop model to a graphical user interface). Metonymic models involve the substitution of a part of a one of the above model types for the whole, or for another part of the model (e.g. "The White House refused to comment on the issue.")

The first two cognitive models, propositional and image schematic, characterize concrete thought structures while the second two models characterize mental mappings of these concrete structures onto abstract thoughts.

In Table 1, we have categorized a set of kinesthetic cognitive models. While we attempt, in the next sections, to relate these models as directly as possible to The Millennium Project, we present them here in a more conceptual form for two reasons: 1) These models illustrate and support the claim that understanding (even of abstract ideas) is embodied, and 2) we consider these models to be extremely useful for designers of interactive media spaces.

The first column of this table, labeled "body", identifies the body parts or features that we have associated with a group of cognitive models. Column 2 lists two to three cognitive models, each one in one of the four categories described above. Columns 3 and 4 provide language examples and a visual example to show the relationship between the group of cognitive models and the body.

2.1.1 Context

We experience our bodies both as containers of objects (e.g. internal organs) and as objects in containers (e.g. buildings). We experience our bodies as having centers (internal organs) and peripheries (fingers and toes). We generally view the centers as more important than the peripheries.

The container schema models a fundamental and inescapable logic of 3D space: everything is either inside a container or out of it – P or not P. The process of categorization relies on image schemas that use the body as a reference. Categorization is central to understanding and plays a significant role in the information environment we create in the Millennium Project.

Table 1: Kinesthetic Cognitive Models

Body	Cognitive Models	Language Examples	Visual Examples
Body Location [Context] *see section 2.1.1*	the CONTAINER image schema the CENTER-PERIPHERY image schema the IMPORTANT IS CENTRAL metaphor	"I can figure it *out*." "How did you get *into* computers?" "Let's get to the *heart* of the matter." "What are the *central* points?"	
Body Movement [Experience] *see section 2.1.2*	the SOURCE-PATH-GOAL image schema the MIND IS A BODY MOVING IN SPACE metaphor	"I'm *approaching* an understanding of the situation." "We've got a *long way to go* in our studies." *"Follow the path* of the argument."	
Body Size [Scale Relations] *see section 2.1.3*	the SCALE image schema the IMPORTANT IS BIG metaphor	"I'm feeling *up* today." "The chances are *higher* today that there will be *low* humidity." "This is a *big* opportunity for me."	
Eyes [Perception] *see section 2.1.4*	the SEEING AS UNDERSTANDING metaphor the MIND'S EYE metaphor	"I *see* what you mean." "It's *clear* to me now." "I need to get a new *point of view* on this issue and *focus* on what is essential." "Let's keep things in *perspective*."	
Hands [Analysis] *see section 2.1.5*	the GRASPING AS UNDERSTANDING metaphor the TOUCHING AS THINKING ABOUT metaphor	"I *get* it." "I've *reached* an understanding." "Let's *examine* the issue further." "*Let go* of the past."	

Understanding the different roles of objects in 3D virtual space has influenced our research in both the design of 3D information environments as well as the navigation and object manipulation within these environments.[1] In this project we introduce the idea of *context containers*. A context container encloses a collection of information objects from our database and gives meaning to the spatial relationships of objects with respect to the container, and with respect to other contained objects. An example context container could contain information objects of philosophy, painting, music, and architecture from Europe between 1909 and 1911. The information objects in this context are automatically assigned meaningful locations and sizes based on geographical location, time, and categorical similarity. We can move through this container of philosophy, painting, music, and architecture objects as we might move through a museum gallery. We are able to pick up objects to analyze and examine from different points of view. If we want to explore an object further, we can enter the object as a container of new and more detailed information objects. Conversely, if we move well outside of the original container of objects (from Europe between 1909 and 1911), we can pick up the container itself to examine the distribution of objects within it over time, category, and location.

2.1.2 Experience

Our movement through space takes us from starting points (*sources*) to ending points (*goals*), along *paths*. There may be obstacles along the paths or diversions that may take us on new paths.

Movement between different contexts is characteristic of the understanding process. In the Millennium Project we introduce the idea of *transitional spaces,* spaces that literally transition us from one context container (a source context) to another (a target context). Transitional spaces are like corridors between rooms, or narrow urban streets between piazzas. These pathways reveal structural connections between contexts and avoid the disorienting context shifts of hypertext environments. In addition to maintaining context by smooth transitions, transitional spaces allow us to fork off the path to our target context. Transitional spaces invite us to pursue other paths, doorways, or narrow streets, to other contexts. [Lynch, 60]

2.1.3 Scale Relations

We understand our world in terms of qualitative and quantitative relationships between objects and events (e.g. *more, less*, and *the same*). The scale schema has helped

---

[1.]We have observed, for example, that when we perceive ourselves to be *inside* a object in virtual space (i.e. when the boundaries of the object are no longer completely in sight), it is most natural for the navigation controls to control our movements in the space (i.e. "flying" mode). When we perceive ourselves to be *outside* of an object, however (i.e. when the boundaries of the object are completely in sight), it is most natural for the navigation controls to control the movements of the object in the space, much like the idea of "direct manipulation" (i.e. "examining" mode). We are currently experimenting with a gesture input device that automatically transitions between these two modes based on a continuous evaluation of our position relative to other objects in the virtual space.[Allport, 95]

us make design decisions about the relative sizes and locations of information objects in our virtual spaces.

We understand scale most readily when we relate things to our bodies. It is much easier for us to comprehend the scale difference between X and Y as roughly equivalent to the height difference between a 12-story building and a paper clip, than as a roughly 2000-to-1 ratio. Most of the objects in our physical environment are familiar to us and have identifiable human scales that help us understand new and unfamiliar objects that we encounter. We know (or at least think we know) the size of a chair, for example, from any distance but do not know the size of a cube from any distance. Further, we can deduce the distance of a chair because we know its size.

Few of the scale cues of physical space exist in abstract virtual spaces of text and images because they lack *embodiment*. In the Millennium Project, we attempt to embody our dynamic spaces by introducing a sense of human scale. This approach better enables us to understand how information objects relate to one another.

2.1.4 Perception

Vision, our primary source of information about the world, plays a crucial role in our acquisition of knowledge. Mental attention is typically connected to the gaze of the viewer. [Lakoff, 87]

In the Millennium Project, we make use of transparency, perspective, and 3D point of view. Spatial perception, of course, is integrally tied to movement. [Arnheim, 54] As we move around a space, or within a container, we see objects from different perspectives. When we pick up an object, we rotate it in different directions in order to view all of its sides. Both activities, which involve movement of ourselves within a space, or movement of an object, enable us to gain a more thorough understanding of our environment and the objects within it. Information objects in the Millennium Project are designed to be viewed from several points of view and against several context container backgrounds. The use of 3D point of view in our previous work is described in [Strausfeld, 95a].

2.1.5 Analysis

We use our hands to examine physical objects, to turn them around in order to see all of their sides. We allow that our eyes may sometimes fool us, but if we can touch things, we feel confident in our ability to understand them.

In addition to flying through information environments, we are allowing for direct manipulation of information objects as well as context containers in the Millennium Project. Movement through space alone is not sufficient for understanding spatialized information. In our context containers, we are able to pick up objects to examine as well as alter the container itself in order to redistribute the objects contained within it. For example, if we are inside a context container of "philosophy in Vienna between 1909 and 1911", we can pick up a Wittgenstein object to analyze, or we can stretch the timeline wall of the container to broaden the time span we are examining.

2.2 Visual Discourse

The above embodied cognitive models illustrate how we understand concepts relative to our bodily experience. The dynamic component to these models emerge when we consider the process of discourse used in natural language. One important cognitive model for describing aspects of meaning construction in natural language is the concept of "mental spaces" as developed by Fauconnier [Fauconnier, 94]. From models of natural language discourse, we have correlated and extrapolated a visual and interactive process for exploring information. We call this process "Visual Discourse" and describe in subsequent sections the computational environment that supports this process [Rennison, 95].

2.2.1 Mental Spaces and Discourse Representation

The concept of visual discourse is precipitated from Fauconnier's theories on mental spaces. Mental spaces are characterized by the following properties [Lakoff, 87]:
- Spaces may contain mental entities
- Spaces may be structured by cognitive models, such as
  - Categories
  - Hierarchical Structure
  - Relational Structure
  - Radial structure
  - Foreground-background structure
  - Linear quantity scales
- Spaces may be related to other spaces by "connectors"
- An entity in one space may be related to entities in other spaces by connectors
- Spaces are extendable, in that additional entities and idealized cognitive models may be added to them in the course of cognitive processing

Fauconnier maintains that we communicate meaning through a process of accessing mental spaces through conceptual connections [Fauconnier, 94]. Examples of this include mappings between source and target domains in conventional metaphor [Fauconnier and Turner, 93; Lakoff, 87; Lakoff and Johnson, 80; Lakoff and Turner, 89]; and discourse involving time, viewpoint and reference [Seuren, 84]. Fauconnier represents discourse as a process where 1) categories are used to set up mental spaces, 2) temporary connections to other spaces are established, and 3) new frames are created dynamically as the discourse unfolds. In this process, participants must keep track of the maze of spaces and connections that are built, and this happens through the use of point-of-view and point-of-view-shifts that are grammatically encoded by means of tenses, moods, space builders, anaphors and other cognitive operators. [Fauconnier, 94]

2.2.2 The Visual Discourse Analog to Natural Language Discourse

Mental spaces and discourse suggest a compelling model for the communication of meaning. As illustrated in Figure 1, natural language is used as a medium for building mental spaces as discourse unfolds. In this process, the expert traverses preestablished cognitive models, and the knowledge seeker, listening to the expert, constructs a mental space in his or her mind based on the grammatical structures of natural language elicited by the expert. In this way, the utility of the conversation is measured by the ability of both parties to construct similar mental spaces.

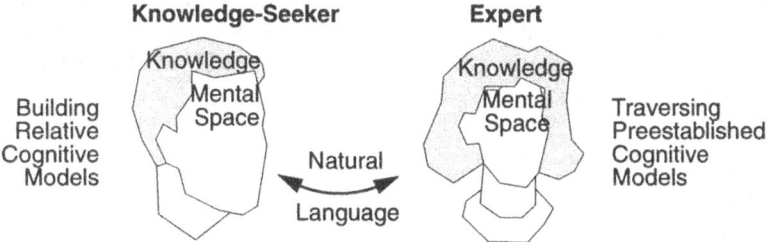

Figure 1. Representation of discourse using natural language as the exchange medium.

Our goal is to find the analog of this process, where natural language is replaced with a virtual visual information space and a users movement and interactions in the space form the basis for communication. In this way, visual discourse addresses the communication of meaning through a process of

- Visualizing relationships between information elements, where the relationships are illustrated by their relative position in space that has been phrased by contextual information [Fauconnier, 1994; Strausfeld, 1995][2]
- Dynamic point-of-view shifts, where the information elements are held constant and the types of relationships are changed, and
- Dynamic context shifts, where a new set of information elements are dynamically established based on the constraints of the new context.

In the next section, we describe our computational approach to enabling visual discourse.

3. A Computational Process for Enabling Understanding of Information

In this section we describe a computational process we have developed to enable dynamic exploration if information based on the conceptual framework described in the previous section, and on the approach explored with the Galaxy of News system [Rennison, 94]. There are two primary objectives of our computational process: 1) to automatically compute conceptual structures that describe information, and 2) to dynamically present the conceptual structure to the users to aid them in understanding the information.

In the Millennium project, we have developed a system that automatically analyzes a corpus of information to derive conceptual structures that aid us in understanding the relationships between information objects that represent concepts spanning both space and time. These conceptual structures include categorical structures, hierarchical

---

[2.]An analogical example of this is a 2D graph where the data points, or in our case information elements, are positioned in a space that is phrased by the labels on the x- and y-axis.

structures, relational structures, radial structures, linear quantity scales, and foreground-background structures. Each of these structures help us understand the relationships between information elements. Table 2 presents an overview of a conceptual and computational framework we use to project information organized in conceptual structures into virtual information spaces. This table shows how conceptual structures relate to the five information organization structures proposed by Richard Saul Wurman [Wurman, 89]. These include

- Location
- Alphabet position
- Time
- Category
- Hierarchy[3].

In addition, Table 2 also shows the correspondence between conceptual structures and image schemas (as introduced in Section 2.1), and between conceptual structures and metaphorical mappings. This table presents the computational structures we use to represent the conceptual structures.

We have also developed methods for dynamically presenting the information to the user through "visual discourse", a process that interactively unfolds over time [Rennison, 95]. There are two important aspects of visual discourse: 1) how the conceptual structures are mapped to virtual space such that they convey meaning, and 2) how the computer interprets user interaction.

Our computational process consists of the following steps:

1. Analyzing the information-base to construct a representation of the relationships between the information objects, namely analyzing the underlying *structure* of the information-base
2. Presenting the information relationships in a 3D virtual space that provides a particular contextual view on the information, and
3. Interpreting user movements and actions in the 3D virtual space to dynamically query for additional information and dynamically reconstruct the virtual space to show the relationships between the objects returned from the query.

The relationships between these steps and important subprocesses are illustrated in Figure 2. Each of these steps is discussed in detail in the following sections.

3.1 Structure

In this section we describe our approach to structuring our information-base to enable people to understand the complex relationships among information objects. We have defined a process for analyzing information objects and deriving structures that convey relationships between the information elements. Information elements in this definition include the original information objects as well as features that are extracted

---

[3.]Wurman refers to hierarchy as the relative size and position of things. This differs from categorical hierarchies.

Table 2: Projection from Conceptual Structures into Virtual Information Spaces

Conceptual Structure	Information Organization Structure	Computational Structure	Corresponding Image Schema	Metaphorical Mapping	Virtual Space Mapping
Categorical Structure	Categorical Temporal (i.e. periods)	ARN/Graph	CONTAINER	CONTEXTUAL as INSIDE	Graphical objects as containers that either define space or occupy space
Hierarchical Structure	Hierarchical	Acyclic directed graph	PART-WHOLE UP-DOWN SCALE	GENERAL as WHOLE ABSTRACT as HIGHER IMPORTANT as BIG	Graphical objects as containers inside other containers Graphical objects scaled relative to importance
Relational Structure	Categorical Temporal (cause-effect) Location (geographical)	ARN TARN LARN	LINK	RELATED as CONNECTED SIMILAR as CLOSE	Graphical objects attached Graphical objects positioned relative to one another (as using MDS)
Radial Structure	Categorical (fuzzy)	Fuzzy cluster graphs	CENTER-PERIPHERY	IMPORTANT as CENTRAL	Spherical, axial and hyperbolic spaces
Linear Quantity Scales	Hierarchical Alphabetical	Sorted list	UP-DOWN LINEAR ORDER	MORE as UP	Graphical objects viewed sequentially
Foreground-background Structure	Temporal Alphabetical	Sorted list	FRONT-BACK	FUTURE as IN FRONT	Graphical objects viewed sequentially

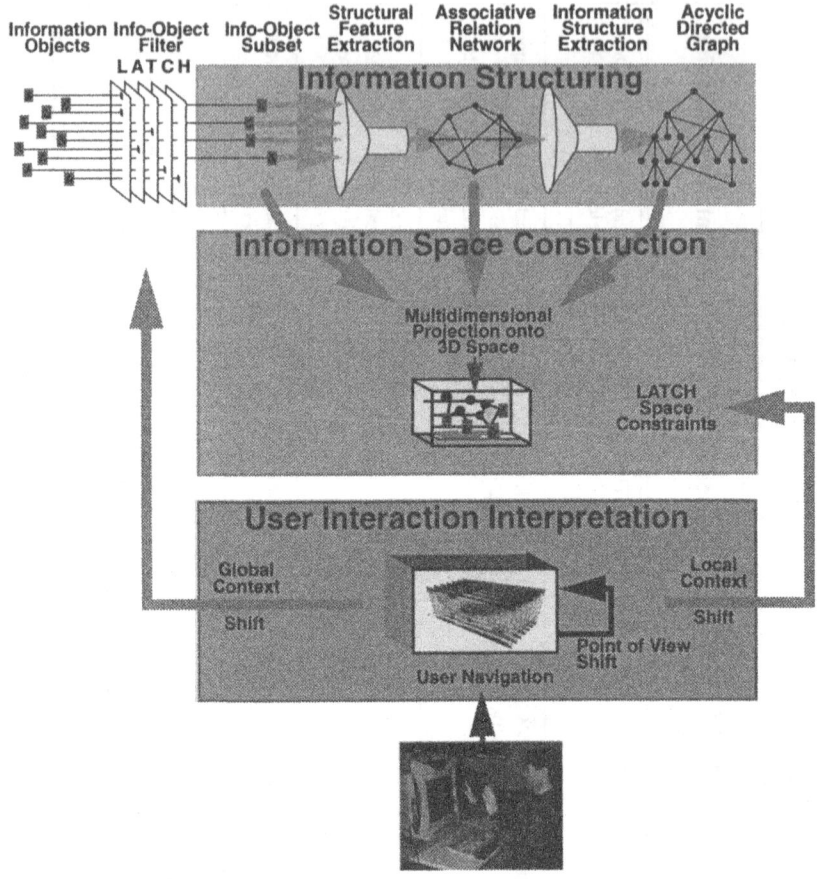

Figure 2. The Computational Process of Visual Discourse.

from the information objects (such as keywords). We use the extracted features to analyze the structure of the information objects (we describe this process below). The structure analysis process yields structures that correlate to cognitive structures such as categories, hierarchical structures, relational structures, and radial structures [Lakoff, 87]. These structures will in turn be used in the process of mapping the structural relations onto a visual space that is presented to the user (as described in Section 3.2). In addition, these conceptual structures aid the user in navigating through the virtual information spaces, as well as aid in understanding the relationships between information objects describing events and artifacts that span place and time.

As illustrated in Figure 2, the conceptual structures are derived through the following process:

1. Filtering the original set of information objects to a reduced subset (via a LATCH filter, optional)

2. Extracting key features from the reduced set of information objects (e.g. keywords)

3. Constructing a computational representation that captures the structural relationships between extracted features and the underlying information objects (e.g. Associative Relation Network)

4. Processing the structural relationship representation to extract computational structures that correspond to conceptual structures (e.g. Acyclic Directed Graph).

We describe each of these steps in the following subsections.

3.1.1 Information Object Filtering

The first step of the structure analysis process is to filter the original set of objects to a reduced set. This essentially establishes the initial or global context for a discourse. This filtering process is based on an initial condition specified by the user. For example, "Let's start will information that pertains to the geographical location of Vienna, Austria, during the period from 1911 to 1912, that fall into the categories of painting and abstraction." This sentence formulates a query or filter that screens information objects to derive a subset of objects. Queries for information objects are either made explicitly, via a text entry mechanism such as a dialog box, or through implicit interaction within an information space. Implicit information queries, which are based on users' movements in the information space, are describe in more detail in Section 3.3.

In the Millennium Project, we have defined a filtering process based on Richard Wurman's five methods for organizing information, as described above. We call our initial filter a "LATCH Filter." In this initial stage, objects are passed through a LATCH filter to establish the initial set of information objects.

It is also important to note that this filtering stage is optional. If the user does not specify an initial condition, the entire database is used as an initial context and the following process continues from there.

3.1.2 Extracting Key Features

The second stage of the analysis process is to extract key features from the information objects. These features include such information as the dates/duration that an event occurred, location an event occurred, and sets of symbols that describe the information object (refer to Figure 3). The symbols in this case refer to elements such as nouns, noun phrases, verbs, and verb phrases that describe the subjects, actions, and objects of the information context. They may also include constructs such as Universal Record Locators (URLs) and names.

In the Millennium Project, we allow for three levels of feature definition: 1) features extracted from the content or body of the information object, 2) features defined by an object annotator, and 3) features associated with the object by the end-user, or knowledge seeker. Each of these features are treated separately and the user has control over how the system applies them in constructing the information spaces.

The features fall into two categories: general properties and structural relations. General properties include information such as size, date/time, location, and so forth. The general properties of the information objects vary according to the type of object.

```
<ODFile 0.9>
<ObjType people>
<ObjName `alma-mahler.html'>
<Annotator (Lisa Strausfeld, Earl Rennison)>
<Author `'>
<Location (`Vienna, Austria', `New York, New York, USA')>
<Date (`Aug. 31, 1879', `Dec. 11, 1964')>
<Source `Britannica Online'>
<!-- Association Sets that describe this object -->
<AssociationSet Subjects ((music, art, piano, writer),
    (woman, marriage, wife, divorce, relationships, affairs, love),
    (Mahler Symphony No. 6, Mahler Symphony No. 8, The Tempest
    Wozzeck,And the Bridge Is Love),
    (Gustav Mahler, Oskar Kokoschka, Gustav Klimt, Walter Gropius,
    Franz Werfel, Arnold Schoenberg, Gerhart Hauptmann,
    Enrico Caruso, Alban Berg)) >
<AssociationSet Influenced (Gustav Mahler, Oskar Kokoschka,
    Gustav Klimt, Walter Gropius, Franz Werfel) >
<TITLE>Alma Mahler</TITLE>
<H1> Alma Mahler </H1>
(b. Aug. 31, 1879, Vienna, Austria-Hungary--d. Dec. 11, 1964, New
York, N.Y.,U.S.) <p>
Alma Mahler (also known as Alma Maria Schindler, Alma Gropius,
and Alma Werfel) was wife of Gustav Mahler, known for her
relationships with celebrated men. <p>
The daughter of the painter Emil Schindler, Alma grew up
surrounded by art and artists. She studied art and became friends
with the painter Gustav Klimt, who made several portraits of her.
Her primary interest, however, was in music: she was a gifted
pianist and studied musical composition with Alexander von
Zemlinsky. <p>
In 1902 she married Gustav Mahler, who at first discouraged her
from composing; he is said to have changed his mind after hearing
her songs. Mahler left a musical portrait of her in the first
movement of his Symphony No. 6, and he dedicated Symphony No. 8
to her. After his death in 1911 Alma had an affair with Oskar
Kokoschka, who painted her many times, most notably in "The
Tempest" (1914; "Die Windsbraut"). In 1915 she married the
architect Walter Gropius; they were divorced after World War I.
She married the writer Franz Werfel in 1929. In the late 1930s
the Werfels left Nazi Germany, eventually settling in the United
States. <p>
During her lifetime Alma Mahler became friends with numerous
celebrated artists, including the composer Arnold Schoenberg, the
writer Gerhart Hauptmann, and the singer Enrico Caruso. The
composer Alban Berg dedicated his opera Wozzeck (1921) to her. <p>
Alma Mahler published two collections of Gustav Mahler's letters
as well as her memoirs, And the Bridge Is Love (1958). She also
published a number of songs. <p>
```

Figure 3. Example Information Object File.

For example, information objects that pertain to artifacts may contain a size of the artifact, date produced, location produced, and who created it. Information objects that pertain to events would not include a size (unless some conceptual size can be specified), the date may be specified as a duration, the location may be specified as a region that may change over time, etc. Structural information consists of sets of symbols that indirectly bind an information object to other related objects.

We extract key symbols and symbol sets from the contents of textual information via one of three techniques. First, we provide a mark-up language that allows authors or annotators to explicitly embed specifications of AssociationSets[4] in the body of an information object description file (as illustrated in Figure 3). These AssociationSets can have a hierarchical structure such as the "Subject" AssociationSet illustrated in Figure 3. This hierarchical structure is similar to the sentence-paragraph-section-chapter-book type structures that bind words together, but operates on the principles of association as opposed to grammatical structures[5]. Second, we can use automatic text indexing techniques based on symbol frequencies to extract keywords from a text document. And, third, we can use a part-of-speech tagger [Brill, 92] to identify the nouns, noun phrases, verbs, and so forth.

3.1.3 Constructing Relationship Representation

Once we have extracted important features from the documents, we use these features to construct a representation that captures the *emergent relationships* between the information objects. A key element of our research is to find emergent structural properties that are not globally or explicitly defined, but rather emerge from the amalgamated properties of the individual objects. Hence, we do not impose a global structure on the information spaces; they are derived automatically from the contents of the information-bases through this bottom-up structuring process.

In the Millennium Project, we specifically use associative relations that define co-occurrences of symbols as the basis for our structural representation [Rennison, 94]. In addition, we also use temporal-causal relationships, and geographical and absolute temporal parameters (as specified by the authors of the information objects) to build a representation of the underlying structure. Figure 4 illustrates our core representation of the information structures. As described above, each information object can contain a set of dates, a set of locations, and associated sets of symbols (AssociationSets). When these sets of symbols, dates, and locations are inserted into the core representation they strengthen weights between the symbols, dates and locations.

The main element of our representation is an Associative Relation Network (ARN) [Rennison, 94]. An ARN captures the relationships between symbols contained within information objects. The relationships between symbols contained in an ARN define

---

[4.]AssociationSets are sets of symbols (such as keywords and URLs) that co-occur and are bound together for some structural or grammatical reason, such as a sentence. The symbols in the list can also have weights and counts.

[5.]In the future, we plan to analyze the emergent properties of amalgamated grammatical structures, such as the subject-action-object relationships.

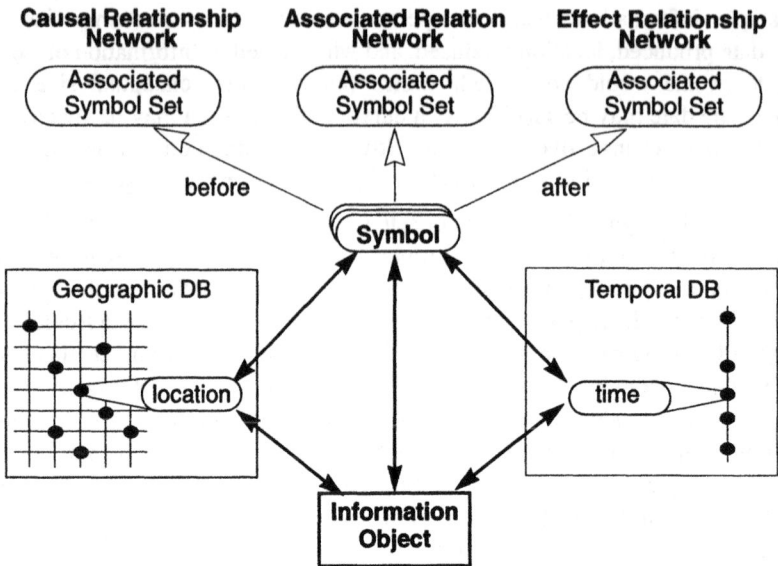

Figure 4. Information Relationship Representation

the relationships between information objects. An ARN maintains weighted relationships between symbols contained in the network, as well as the relationships between symbols and the information objects with which they are associated.

An ARN defines an N-dimensional space that contains N^2-N terms. The basis vectors of the space are defined by symbols extracted from the information objects. Associated with each basis vector (i.e. symbol) is a vector that defines the relationship between itself and all the other basis vectors (i.e. symbols). With an ARN, the information objects reinforce the associative weights between symbols that represent the relationships between information objects. A symbol also forms a link between objects. However, the link that the ARN forms between information objects is not a simple index between information objects. It contains structural information that determines the strength of the relationship between the objects.

The ARN described above is also used to capture relative temporal relationships between information objects, and implicitly the cause and effect relationships between information objects. Our current information object mark-up language allows authors and annotators to specify sets of symbols that the subject of the information object was *influenced by*, and a set of symbols that the subject of information object *influenced* (as shown in Figure 4). Each of the *influenced by* symbols are associated with each of the symbols that describe the information object, and these relationships are maintained in a separate ARN that also maintains the temporal distance between associated symbols. We call this extended ARN a Temporal ARN, or TARN. Likewise, each of the *influenced* symbols are associated with each of the symbols that describe the information object. This relationship is maintained in a separate TARN.

In addition, each symbol in the representation has a reference to all the locations and times that the symbol occurred as defined by an information object. Likewise, each location and time has a reference to associated symbols, and back to the information objects that contain the location or time. The locations are also stored in a geographic database that facilitates quick filtering and searching of either symbols or information objects. Times are stored in a temporal database that facilitates quick filtering and searching for related symbols and information objects.

The primary utility of this representation is the ability to compute probability, similarity, and distance measures between symbols and information objects. These measures are used in computing categorical classifications, fuzzy clusters, hierarchical structures and sorted lists as described in Table 2. The complex representations described above are dynamically processed to extract these structural relationships that are implicitly maintained by the representation. This process is discussed in the next section.

3.1.4 Computing Conceptual Structures

The most important step of the structuring process is deriving computational structures that correspond to conceptual structures and implicitly define structural relationships between information elements. We specifically compute the following computational structures:

- *graph* where each node in the graph corresponds to a category[6] and linked nodes correspond to related symbolic categories
- *acyclic directed graphs* where each node in the graph corresponds to a symbolic category and linked nodes correspond to symbolic sub-categories
- *fuzzy cluster graphs* where each node in the graph corresponds to a symbolic category and linked nodes correspond to related symbolic categories such that the node is the central theme (as in a conceptual radial structure)
- *sorted lists* where each node represents a place in some linearly ordered sequence or scale.

A brief description on how these are computed is provided below.

A graph is generated from an ARN. Essentially an ARN represents a graph structure; however, since this structure has a very high dimension, it can be pruned by applying a *similarity threshold*. This process simply removes nodes from the ARN that fall below the similarity threshold.

We use several techniques to compute acyclic directed graphs [Rennison, 95]. These techniques fall into two categories: clustering and probabilistic sorting. Within these two categories we use two primary techniques: top-down and bottom-up. The clustering algorithms use similarity and distance measures calculated from an ARN. The probabilistic sorting techniques use probabilities measures computed from an ARN [Rennison, 94]. The following recursive process describes one of the techniques we use to compute acyclic directed graphs:

1. Search through the ARN and find all the statistically independent symbols

---

[6]·Note that in some cases the nodes may correspond to times, locations or the information objects depending upon the type of conceptual or information structure we are generating.

2. For each independent symbol, find all the symbols statistically dependent on the independent symbol

3. For each set of dependent symbols, find the independent symbols

4. Repeat steps 2 and 3 until all the dependent symbols are independent of one another.

The information hierarchy resulting from this process is used to aid the user in navigating through information structures. This process essentially defines a technique for abstracting and generalizing. As the philosopher William James noted "we acquire knowledge through a process of differentiating characteristics. This process of differentiation is based on finding dissociations between elements" [Arnheim, 69]. This process captures the essence of this objective.

Currently, we compute a fuzzy cluster graph by first computing an acyclic directed graph using a top-down probabilistic approach. Then, we apply a clustering algorithm using each node in the graph as a centroid and searching for all symbols that fall within the range of the symbol, where the range is defined as the farthest distance from the node symbol to a child symbol.

The result of the computational processes described above is a set of computational structures that map to conceptual structures. In the next section, we describe how these computational structures are used to construct spaces that reflect the underlying conceptual meaning.

3.2 Space Building

The presentation aspect of the Visual Discourse process consists of projecting the multi-dimensional structural model into a three dimensional visualization. Because of the high dimensionality of the underlying space (a direct correlation to the number of features extracted from the information objects), it is not possible, or at least not meaningfully intelligible, to project the entire underlying space into a 3D representation directly. The construction of the projection, therefore, must be carefully considered. The projection should be a direct representation of the cognitive structures derived from the information objects. Our objective is to generate dynamic virtual spaces that correspond to the mental spaces we continually construct during natural language exchanges.

We have defined a model and process for projecting the structural information into a 3D space. The process is dependent upon the type of view, or the conceptual viewpoint, on the information for a given space. Currently, we have parameterized the types of spaces that can be generated according to location, alphabetical position (though the use of this constraint is limited), time (*absolute*, e.g. at time T, and *relative*, e.g. before, after), category, and hierarchy, or as Wurman terms LATCH.[Wurman, 89] These parameters may be specified individually, or by combinations. For example, a space can be generated to illustrate the temporal relationships between information elements (which may include combinations of the original information objects, and features extracted from the information objects). Or, a temporal relationship may be combined with a geographical relationship. Specification of these parameters essentially define the *context* in which the information elements are positioned in space. Some particularly meaningful contexts include the following:

- Categorical[7]
- Categorical-Temporal (absolute)
- Categorical-Temporal (relative)[8]
- Categorical-Geographical
- Categorical-Geographical-Temporal

3.2.1 Constructing 3D Information Spaces

The multidimensional structural representation of our information-base allows our system to dynamically generate meaningful sets of information objects that adapt to our continuous queries, as expressed by our continuous movements in the information space. In order to dynamically explore and interact with these information sets, we have to display them in such a way that invites investigation and allows for intuitive interaction. To this end, a 3D space builder automatically constructs information contexts from a list of information objects and a list of extracted features (such as keywords) which are also displayed as graphical objects. An information context is displayed as an enclosure that contains the set of information and feature objects. Figure 5 shows a sample container with France on the ground plane (the xz-plane) and time on the vertical axis (the y-axis).

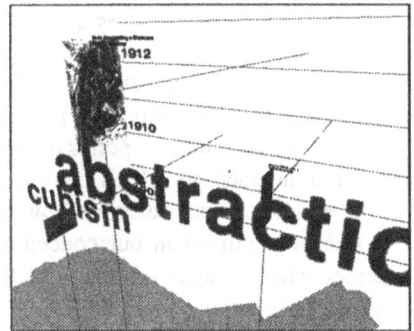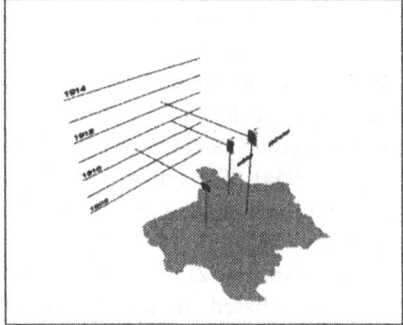

Figure 5. Screen captures of a simple 3D virtual information space

All information objects contained in a 3D information context are assigned a context-specific XYZ location, XYZ axial rotation, scale, color, and transparency based on a mapping of each one of these display attributes to an appropriate information content attribute. In addition, each information object displays different representations of itself relative to our position and orientation in space.

Another task of our 3D space builder is the generation of *transitional spaces*. Transitional spaces are connectors from one context container to another. A transition between a context that is contained inside another (i.e. the information object list of the

---

[7.]Categorical spaces correspond directly to conceptual spaces and conceptual structures.
[8.]This in effect shows causal relationships.

new context is a subset of the old information object list) is experienced like a power-of-ten shift or an infinite zoom. [Morrison, 94]

3.3 Interaction Interpretation

An important aspect of meaning communication, and hence understanding information-bases, is the dynamic process of shifting point-of-view and shifting context. As Fauconnier clearly delineates, a central theme in meaning construction is access through conceptual connections that define mappings between source and target domains.[Fauconnier, 94] In our computational environment we define a context to be a set of information objects and the relationships between them. A context is represented and presented to the knowledge seeker as a container and a set of contained objects, where the container defines the relationship between the objects. We define a context shift to be either global or local. Establishing a global context implies *filtering* or refiltering the original information objects into a working subset of information objects. For example, we may wish to establish a global context to be all objects "in the geographic area of 'France' during the period of 1911 to 1912." Local context shifts imply a change in *conceptual viewpoint* on the subset of information objects and *illustrate* a new set of relationships between the context of objects. For example, we can shift between a categorical view, to a categorical-temporal view, to a geographical-temporal view. Key questions that arise from this process are: How does the user indicate these context shifts? How are these context shifts executed?

Table 2, *Interpretation of User Interaction*, outlines our approach to these questions. It lists the possible user interactions and their effect on the display of objects and contexts as well as the underlying information representation. An information object will display more detailed information up close than it will from far away, for example, and will foreground and background different information from different points of view [Strausfeld, 95a]. The left column lists possible user interactions which consist of movement of self and manipulation of objects. The middle column describes how we interpret user actions based on the cognitive models we outlined in our conceptual framework (Section 2). The right column describes what changes are made to the current context based on the our interpretation of the user's actions.

4. Conclusion

In this paper we have described the Millennium Project, a research effort to construct a dynamic 3+D virtual environment for exploring and discovering connections between artifacts of philosophy, painting, music, literature, science, and political events of world history from the years 1906 to 1918. We have shown how *embodied cognitive models* and *visual discourse*, our conceptual framework based on linguistics and cognitive science, supports our approach to enabling understanding of information by:

- building structural representations of sets of information objects in our information-base that correspond to appropriate conceptual structures,
- automatically displaying meaningful 3D information contexts with spatial structures that correspond to these conceptual structures, and

Table 3: Interpretation of User Interaction

User Interaction		Action Interpretation	Computational Operation
Movement of Self	WITHIN a container	User wants to explore current context	Different views generated within current context
	INTO an object	User wants the object to establish a new context container	NEW SUB-CONTEXT New space constructed Transition performed between old and new spaces
	TOWARDS object(s)	User wants more detail about the object	NEW OBJECT REPRESENTA-TION by adding detail
	AWAY (BACK-ING-UP) from object(s)	User wants less detail/more abstraction	NEW OBJECT REPRESENTA-TION by removing detail When outside container, NEW CONTEXT, original objects replaced with abstracted representations
	THROUGH an object	User wants next in a sequence	NEW SUB-CONTEXT with next in sequence computed
	OVER, UNDER, AROUND object(s)	User wants to see object(s) from different points of view	Different views generated within current context
Manipulation of Object(s)	Translate: MOVE, PUSH, PULL or DRAG	User wants to see object(s) in different relation to other objects in space	Currently, no operation. (Future work may use object manipulation to generate a new context that user "builds" interactively.)
	Rotate: TURN	User wants to see object(s) from different points of view	CHANGING OBJECT REPRE-SENTATION based on view angle with respect to context container
	Scale: STRETCH or COMPRESS	If scaling container, user wants to extend or contract context constraints (e.g. time)	NEW CONTEXT generated by extending or contracting constraints mapped to the XY, or Z axes

Key:
NEW CONTEXT: Context generated by establishing new filtering constraints and refiltering the original set of information objects
NEW SUB-CONTEXT: Context generated by adding additional constraints and refiltering the information objects in the current context to generate a new set of information objects
NEW OBJECT REPRESENTATION: Space is restructured by adding or removing related- and/or substructures that correspond to the object

- interpreting user interaction to enable continuous querying and processing of the information-base through spatial movement and object manipulation rather than mouse clicking.

Lastly, this project has been motivated by a desire to create a unique virtual experience not possible in the physical world. We have aimed to create virtual spaces as rich, dynamic, and compelling as the spaces of our minds as we understand, create, and dream.

5. Acknowledgments

The authors would like to acknowledge the continued support, advice and direction provided by William Mitchell and Ron MacNeil. A special thanks also goes to David Allport, Suguru Ishizaki, Robin Kullberg, Ishantha Lokuge, Dave Small, Louis Weitzman, Yin Yin Wong, and Xiaoyang Yang of the Visible Language Workshop for providing many critiques and suggestions. This work was sponsored by ARPA, JNIDS, NYNEX and Alenia.

6. References

[Allport, 95]   Allport, D., E. Rennison, L. Strausfeld, Issues of Gestural Navigation in Abstract Information Spaces, *Proceedings of CHI 95, Conference Companion*.Denver, Colorado.

[Arnheim, 54]   Arnheim, Rudolf. *Art and Visual Thinking*. London: University of California Press, 1954.

[Arnheim, 69]   Arnheim, Rudolf. *Visual Thinking*. London: University of California Press, 1969.

[Brill, 92]   Brill, Eric. A Simple Rule-Based Part of Speech Tagger. *Proceedings of Third Conference on Applied Natural Language Processing*. 1992. Trento, Italy: ACL.

[Fauconnier, 94] Fauconnier, Gilles. *Mental Spaces: Aspects of Meaning Construction in Natural Language*. Cambridge, UK: Cambridge University Press, 1994.

[Jackendoff, 83] Jackendoff, R. *Semantics and Cognition*. Cambridge, MA: MIT Press, 1983.

[Johnson, 92]   Johnson, Mark. *The Body in the Mind*. Chicago: The University of Chicago Press, 1992.

[Lakoff, 80]   Lakoff, G., and M. Johnson. *Metaphors We Live By*. Chicago: University of Chicago Press, 1980.

[Lakoff, 87]   Lakoff, G. *Woman, Fire, and Dangerous Things: What Categories Reveal about the Mind*. Chicago: University of Chicago Press, 1987.

[Lynch, 60]   Lynch, Kevin. *The Image Of The City*. Cambridge, MA: MIT Press,

1960.

[Morrison, 94]   Morrison, Philip & Phyllis, and the Office of Charles and Ray Eames. *Powers of Ten: About the Relative Size of Things in the Universe*. New York: Scientific American Library, 1994.

[Ortony, 93]   Ortony, Andrew (Ed.). *Metaphor and Thought*. Cambridge, UK: Cambridge University Press, 1993.

[Rennison, 94]   Rennison, E. Galaxy of News: An Approach to Visualizing and Understanding Expansive News Landscapes. *Proceedings of UIST*. 1994. Marina Del Ray, California.

[Rennison, 95]   Rennison, E. *The Mind's Eye: An Approach to Understanding Large Complex Information-Bases through Visual Discourse*. MS Thesis. Massachusetts Institute of Technology. Cambridge, MA. August, 1995.

[Seuren, 84]   Seuren, P. *Discourse Semantics*. Oxford, UK: Oxford University Press, 1984.

[Strausfeld, 95a]   Strausfeld, L. Financial Viewpoints. *Proceedings of CHI 95, Conference Companion*. Denver, Colorado.

[Strausfeld, 95b]   Strausfeld, L. *Embodying Virtual Space to Enable Understanding of Information*. MS Thesis. Massachusetts Institute of Technology. Cambridge, MA. August, 1995.

[Wurman, 89]   Wurman, Richard Saul. *Information Anxiety*. New York: Bantam Books, 1989.

Providing Spatial Navigation for the World Wide Web

Andreas Dieberger
Georgia Institute of Technology
School of Literature, Communication, and Culture
Atlanta, GA 30332-0165
Tel.: (404) 894-2730, Fax.: (404) 853-0373
mail: andreas.dieberger@lcc.gatech.edu
home: http://www.gatech.edu/lcc/idt/Faculty/andreas_dieberger/

Abstract

The World Wide Web (WWW) is a rapidly growing distributed hypertext on the Internet. This paper presents a way to enable users to navigate the WWW spatially by providing a spatial user interface metaphor in a textual virtual environment. This may help users to orient themselves in the masses of information available. The term "spatial navigation" stresses the fact that the structure of the information space is made explicit in this type of navigation whereas hypertext itself, and especially the WWW tend to hide this structure. We review common navigational strategies on the WWW and point out how these strategies indirectly make use of the underlying structure of the information space. Whereas hypertexts hide their structure, virtual environments explicitly show it. We outline navigational differences between hypertexts and (textual) virtual environments and describe a way to combine the advantages of both. In our system, presently implemented at the Georgia Institute of Technology, a textual virtual environment is combined with the WWW to create a mirror space to a part of the WWW. Navigating this virtual environment users also navigate the WWW, but in a spatial way. The system supports interaction between users and therefore collaborative navigation. This allows conducting guided tours and describing paths to information vaguely, like in real environments.

1. Introduction

The World Wide Web (WWW) is a distributed hypertext system based on the Internet. Like in ordinary hypertext a WWW page consists of information and (hyper) links, objects containing a reference to another WWW page. When activating the link the corresponding page is fetched and displayed. As the WWW is a distributed hypertext the fetched page does not have to be available locally; instead it can be located on any WWW server worldwide. The linking process is transparent and the user does not need to know what server the information comes from. This transparency, the relative ease of use of most WWW browser programs and the rapidity of the access is the main reason for the popularity of the WWW.

The World Wide Web - sometimes also simply called "The Web" - presently is the fastest growing man-made construct. The larger the Web gets, the more obvious navigational problems become. Many of these problems are typical hypertext problems, others are typical to the Web.

In the hypertext field it is generally assumed that navigation in hypertext is not difficult per se but that it can be easy when the structure of the information space is made explicit [SHUM90]. See also [KiHi94] for a study that transfers way-finding knowledge to the hypertext field.

Ironically the most popular feature of the WWW -- the transparency of the linking process -- works against making the structure of the information space explicit. It is easy and pleasurable to browse or "surf" the Web, which means to navigate it in an undirected fashion. Locating information *quickly* is more or less impossible however.

Besides this impossibility of goal-directed navigation there are other navigational problems which will not be covered here.

In section 2 of this paper we describe typical navigation strategies on the Web and show how users often try to supply navigable structures for the Web themselves. In section 3 we describe textual virtual environments (MOOs), their space concept and how they are navigated. Section 4 presents how the MOO and the WWW can be combined to create a spatialized Web and how navigation in this system works. Section 5 describes a particular example of a spatial navigation metaphor for the Web -- a URL shipping mall metaphor. A few other possibilities are outlined in section 6. The paper concludes with a summary of navigational problems of the Web and how we think the MOO/WWW combination can solve some of them.

2. Navigation in the WWW

Hypertext is nonlinear information, which means that the author does not have complete control in which order users access it. Instead the author provides a set of nodes that are linked by hypertext links (pathways). Reading hyperdocuments is a task of navigation through an information space defined by the linking structure.

2.1. Navigation in hypertext

Parunak compared real world navigation to hypertext navigation and found 5 main navigation strategies in hypertext [PARU89]:

- The identifier strategy permits the searcher to identify the target upon encountering it.

- The path strategy uses a procedural description to get to the target.

- The direction strategy depends on two characteristics of the space navigated: texture and comparability. Texture is the existence of a distinguished point relative to which directions can be established, whereas comparability is the existence of a relation between two points of the space.

- The distance strategy tries to reduce the distance between searcher and the target.

- The address-strategy requires knowledge on how to resolve an address and then navigates directly to the address of the target.

The usefulness and availability of these strategies for hypertext navigation relies on the topology of the hypertext (see also section 2.2.). Hypertext essentially defines a topological information space where the hypertext pages are nodes connected by links. The presence or absence of links between pages defines a connectedness or distance relation. Pure hypertext defines no other concept of distance between nodes.

It was mentioned that hypertext navigation is assumed to be easier in an explicit structure. Such a structure would define concepts of location, distance and direction. "A possible way to localize a user in a hypertext is to impose a structure on the hypertext and to identify the user's location within that structure" [RiBO94] (p.88) (see also [NiWe80].

Such a structure helps users to develop a structural understanding of the information space based on spatial metaphors most users use anyway: "(...) Users tend to make heavy use of spatial metaphors: they report feeling "lost", speak of going "up" and "down" between levels or going "in" and "out" of situations. Users often spontaneously construct spatial mental models or mental "maps" in order to move easily from one context to another. This has obvious implications for design. Reducing the memory load on the user is one benefit of making these mental "maps" explicit." [SeNi90] (p. 150)

2.2. Navigational strategies used on the WWW

Most hypertexts, and especially the WWW hide concepts of location and structure from the user. This lack of navigational structure in the Web lead to an interesting social behavior: Users freely provide the necessary structure for themselves and for other users. We see this behavior as a signal that a more explicit structure is needed so badly on the Web, that users start to generate this structure themselves.

Many navigational strategies rely mainly on this structure provided by users and therefore they are significantly different from strategies in single-user hypertexts. Of the five strategies described by Parunak only the address strategy is commonly used on the Web whenever a node is referred to directly by its address. Such an address is called Uniform Resource Locator or URL. The distance, direction, and path strategies are nonexistent on the Web as the necessary structural concepts are not available.

The only way to make use of the identifier strategy is by employing "Webcrawlers", programs navigating the Web and collecting keyword-information in databases. These services blur the structure of the information space even more as all contextual information is lost. Both the address and the identifier strategy ignore all structure on the Web. All other strategies (except for free browsing which is no "strategy") rely on the Web's hidden structure.

2.2.1. Navigating using other people's knowledge

Navigation strategies on the Web often rely on the expertise of other users. As it is difficult to *re-find* information users collect lists of addresses (URLs) of interesting pages. These lists are called hotlists. Many users invest much time and effort to create well structured Web pages from these lists and make them available to other users by linking them to their home-pages[1]. When looking for information it is therefore a useful strategy to look at home pages of people with related interests [ERIC95].

Another possibility is to consult a web directory, most of which are again maintained for free by web users. An example is the Yahoo List[2]. The WWW has reached a state where the important issue for information providers is not to have information on the Web but to have a pointer to this information in a Web directory or many hotlists.

2.2.2. Navigation using an existing structure on the Web

The Web has also a geographical structure, even if the browser programs try to hide this fact from the user. This geographical structure allows employing a navigational strategy that is tied to the location of the institution running the Web server -- this strategy is particularly useful when looking for university or company related information. When looking for specialized information it is a good start to look at the server of a university strong in that field, or at the company selling the product interested in. These servers often carry also pointer lists to related information providers. Whereas university servers are geographically localized this is not necessarily true for large companies. So this strategy makes use of a geographical structure in the former and a more abstract structure in the latter case.

---

[1] Home-pages are Web pages associated with individuals or institutions.
[2] The Yahoo list is at http://www.yahoo.com/

While these strategies use an existing structure of the WWW without making it explicit, there are URL-lists that are organized only according to geographical structure. Using such a list makes the geographical structure visible to the user.

Fig. 1. The Virtual Tourist map for Europe.

An example is the Virtual Tourist page, that allows people to navigate from a global view to continents (see Fig. 1) and further to countries (see Fig. 2), regions and cities[3]. Navigation is geographical till the user reaches a list of servers in a selected region or city. This list often contains a summary of the server contents. The Virtual Tourist page therefore combines a geographical strategy with using "somebody else's knowledge".

A similar approach is realized in the City.Net pages [4]. Again the user can navigate geographically but City.Net present countries and regions as sorted lists and therefore might be better suited for geographically less proficient users.

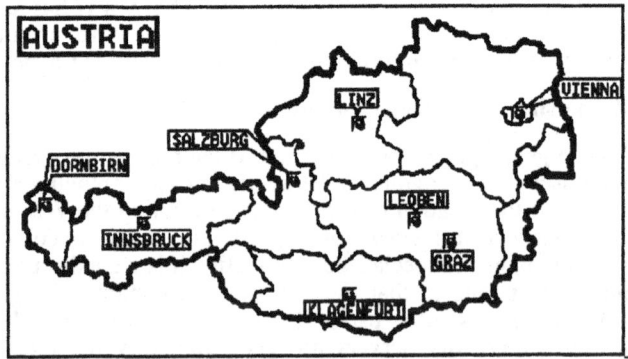

Fig. 2. The Virtual Tourist map for Austria.

---

[3] The Virtual tourist can be found at: http://wings.buffalo.edu/world/

[4] The City.Net can be found at: http://www.city.net/

2.2.3. (Re-) defining structures on the Web

Instead of relying on existing structures it is possible to create virtual geographies for the Web [DiBo95]. An example was the now defunct WebWorld system, which defined a virtual landscape (see Fig. 3).

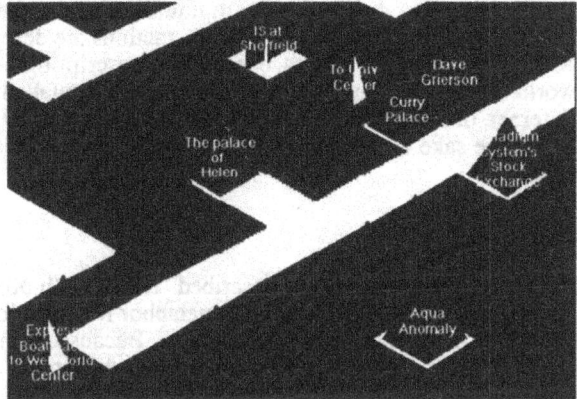

Fig. 3. A screen shot of the now defunct WebWorld system. Spatial grouping in this system was done by the users and often signified topical closeness.

In this landscape users could place objects that represented links to WWW pages or transitions to other WebWorlds. Users with related interests or work areas often gathered their objects in an area. Closeness therefore indicated relatedness to a group topic even if the object names didn't directly hint at a relationship.

Another approach to handling hypertextual information spatially is realized in the VIKI system. VIKI detects structures in the arrangement of objects manipulated by the user. VIKI has been used to organize WWW hotlists but it is not available on the WWW [MaSC94], [MaSh95].

2.3. Give the user what she needs

Web pages essentially are user interfaces for information access. The designer of these interfaces (the authors) have to anticipate navigational needs to provide useful information pathways. In this respect Web pages are similar to real life spaces, where the "author" (city planner or architect) also has to anticipate the navigational needs of the users (inhabitants) to provide useful pathways.

To support navigation well it is not only necessary to define a useful, visible structure but also to give additional information about locations (link destinations). Most Web browsers show if a user has visited a link already [5]. Examples of useful additional information are the size of the object reached through a link, the speed of the network connection (often related to geographical distance) and so forth. The Audible Web system, for example, is a WWW browser, that tries to communicate such additional information using sound cues [AlBe95].

The discussion so far ignored approaches of visualizing hypertext structures as overview maps or fish-eye views. These approaches rely on a discernible structure in the hypertext and it is this lack of obvious structure we are concerned with. The entire

---

[5] Revisiting nodes is sometimes considered a sign of disorientation in hypertexts. However it is also a necessary prerequisite for learning paths and landmarks in an environment. See also [LYNC60], [DIEB94] and [KiHi94].

Web also is far to big to be visualized this way. For information on these approaches see for instance [MuFH94], [MuFH95], or [SaBr94].

3. Navigation in textual virtual environments

Textual virtual environments differ from graphical ones in their text-based representation. Locations, objects, users, and their interactions are described textually [CuNi93], [ERIC93]. The textual interface provides descriptions of a rich and detailed virtual world. Typical examples of such systems are textual adventure games played over the Internet, often called MUDs (Multi-User Dungeon) or MOOs (MUD Object Oriented). For the sake of simplicity we refer to all types of these systems as MOOs in this paper.

3.1. How do MOOs work?

In a MOO the user is located in a textually described "room" with other objects and other players. The concept of the MOO room is a metaphor for "location" or "mode". Exits from a room act as links between these locations. Because of the similarity to the hypertext node and link model MOOs are sometimes considered a special case of hypertext.

MOO users may communicate by talking using the say command. A line of text "said" is displayed to all other players in the same MOO room. There are also other forms of communication in the MOO. Players can also interact by giving objects to each other, by manipulating objects and by moving through the MOO space.

Navigational command like "go north" or simply "north" move a player character through the north exit of the MOO room to the next room. Typically MOO rooms provide rectangular exits like up, down, north, west, east, south but other directional exits like southwest occur as well. Non-directional exits use the name of a location or a direction in a different reference frame -- for example "shop" or "out". While directional exits are associated with directions in the environment the direction of these non-directional exits has to be inferred by the user from knowledge about the environment. Exits can also be realized using special commands like "climb rope" or "enter tramway", which are also non-directional exits. It is up to the designer of the virtual space if these exits follow the same conventions as in a real environment (see section 3.3.2.).

Here is a short example of a MOO session that shows how users can interact with objects and other users. User input is shown in bold letters. The liana described in one rooms is a non-obvious room exit and can be used with the command "climb liana".

```
You are in the local pub.
(...)
> west
A small town yard surrounded by houses.
To the west you can see a small break in the buildings.
North street starts here.
         Obvious exits: north, south, west and east.
> south
A long road going east through the village. The road
narrows to a track to the west. There is an alley to
the north and the south.
You see a liana hanging down from the sky.
         Obvious exits: north, south, up, west and east.
   Juggler is here.
>say hello Juggler!
You say, "hello Juggler"
```

```
>emote smiles
Andreas smiles
Juggler gives a book to you.
>look at book
This is a heavily used address book. It contains pointers into
the WWW. The book belongs to Juggler.
```

Because of the rich interaction possibilities in the MOO the space of the MOO is also a social "place". Many MOOs on the Internet evolved to virtual communities and in these communities similar social behavior as in real environments can be observed. For a discussion of how such spatial interfaces evolve to "interplaces" see [ERIC93]. The rooms in these systems can also be interpreted as sites or modes of a spatial user interface as it was described in [NiWe90]. Navigation between such sites and modes is possible along paths or trails which again stresses the spatiality of the system.

3.2. The space concept of the MOO

Directions in the MOO are often given directly in the naming of exits. which does not define a location for a MOO room. Locations are inferred from the naming of exits, from the room contents, and from how the transitions between room are described. MOO rooms possess neither size nor form and the player character does not inhabit a certain location inside the room. They are described as if the player would perceive the room as a whole. Users therefore assume the rooms to have a size and shape that "fits" their mental model of the area described. This allows to create spatial structures that do not occur in real environments (see section 3.3.2.) and that allow novel types of spatial navigation. Most users are surprisingly flexible in coping with unusual spatial structures [DIEB94].

3.3. Navigation in MOOs

Most MOO areas are designed to represent virtual cities or landscapes and navigation in these areas is similar to navigation in an unknown city. Users always log in at the same location and explore from there. They first find a few interesting places nearby and learn how to get there and back, which comprises basic route knowledge. From these few known locations they explore further till they gain overview knowledge of a limited area of the MOO. Almost nobody ever learns the entire environment because it is far to large. Many MOO systems contain over 10.000 rooms.

As in real environments navigation in the MOO relies on learning landmarks, that is places with a "distinct look" or "functionality". In a set of informal interviews (described in [DIEB94]) people described rooms as distinct, when they contain many objects or people, are described as having a special layout, are often visited, have many exits, are near to another important place and so forth. In this respect MOO rooms function similar to locations in real environments, see also [LYNC60], [DIEB94], [TrDi95]).

One of the problems in MOO navigation is how to communicate this distinct "look" to the user. It may seem a good idea to describe MOO locations in much detail but large textual descriptions tend to look similar on first glance. Also most users will not bother to read a long description. Drawing on genre knowledge a lot of information can be given in a few words however, as in the following description:

```
You discover a secret chamber in the pyramid. Wherever your
direct your torch you see gold gleaming.
        Obvious exits: out.
```

Even this short description manages to evoke a strong mental image which would be difficult to evoke using graphics alone. Whereas most users are able to perceive a picture in one glance MOO users have to *read* the whole text. Very few users perceive the formatting of a room description as a "picture" [DIEB94]. Also when several rooms are described using similar text they cannot easily be perceived as different. Despite all these problems navigation in MOO environments works surprisingly well after a short period of getting used to the system.

3.3.1. Navigation strategies in MOOs

Most MOO users learns only a small part of the MOO but they are able to navigate this part quite effectively. They often reach a location by executing a series of walk commands like "s, e ,e ,open door ,down ,n ,e ,e" without even looking at the screen. Such a series of commands acts as a relative address for a location. For destinations outside the well-known set of rooms users refer to other people by asking them for directions.

People refer to rooms by their names, their functionality, keywords (objects) in the description or by relative location ("it is s,e,e from the shop"). Like in a real environment it is possible to describe a path to a location using incomplete and even slightly incorrect information.

MOO navigation often is a collaborative task. Players are not restricted to communication within the same room. Instead they can communicate also with players in other MOO rooms. It is therefore possible to guide players by giving instructions remotely. Should a player get lost in the environment there often are also teleport features that transport them to a landmark [DIEB94].

3.3.2. Peculiarities of navigation in virtual space.

If a MOO is designed with an overall consistent spatial concept any deviation from this concept is perceived as something magic and called a magic feature. When magic features are not well designed and when their effects are not described well they may be more hindrance than help. Especially in MOO systems designed to access information well designed magic features often make navigation much more efficient.

Most MOO designers try to create a more or less realistic environment and they even design areas on maps before coding them. Spatial discrepancies in the environment like overlapping rooms or exits that tunnel through the MOO space therefore occur by intent. These features often make the environment easier to navigate. Several magic features have been described in more detail in [DIEB94] and [TrDi95].

4. Combining the advantages of the WWW and the virtual environment

In a system we currently build at Georgia Tech we combine the navigational facilities of a MOO with the WWW. Such combinations have been realized for gaming purposes already - however, as far as we know, all these systems used Web pages to navigate the MOO (for example the htMUD system[6]), whereas we use the MOO to spatially navigate the Web.

We provide a MOO environment that represents a spatialized version of parts of the WWW. The spatial navigation in the MOO does not replace the pure hypertextual navigation in the Web, but enhances it. The MOO actually serves as a spatialized hotlist containing landmarks in the WWW and as such a spatial environment in which the user can reach important entry points into the Web by navigating spatially.

---

6 Information on htMUD is available at http://www.elf.com/~phi/htmud.html

The discussion so far seems to state that the spatialization of hypertext essentially will help in navigation. Although there is evidence that making the structure of information explicit helps people using it there is little evidence that all spatial structures are equally helpful. Therefore we based the design of our system on navigational research on MOO spaces as reported in [DIEB94], [TrDi95].

4.1. The MOO client - combining the MOO and the WWW

In our MOO environment we associate URLs with rooms, objects and activities. Activity in the MOO (for example entering a room) retrieves the associated URL and causes the corresponding Web page to be displayed in a Web client. Navigation in the MOO-space therefore results in navigation in the Web-space. For the user this creates the impression of moving through *one* information space on which she has two different views[7]. This illusion is particularly strong when MOO objects contain also a textual description of the Web page. Especially when Web pages are large (like in the case of video files) or when the connection to the Web browser is temporarily disabled this double occurrence of information makes the system more usable.

Well-designed information spaces require the designer to separate structure information describing space from content information. In a textual virtual environment this separation is a difficult design problem as structure and content both are described using text. It is possible to convey both types of information in a MOO room description but this approach requires a strong enough visual separation, for example separating lines [DIEB94].

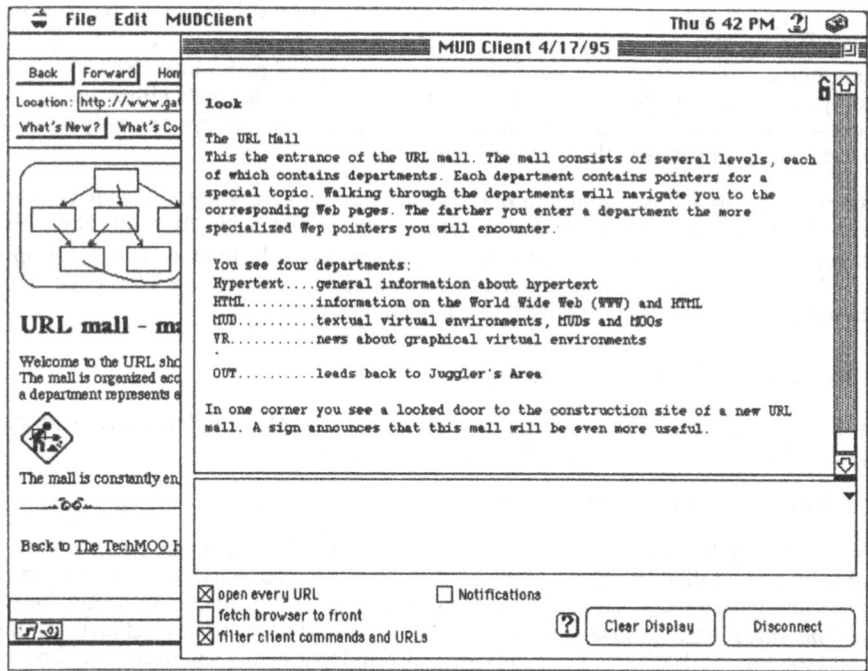

Fig. 4. The combination of MOO and Web client shows two views of a single space.

---

[7] The association of information in the WWW to objects in MOO space realizes a mnemonic space like it was used in mnemotechnics by Greek rhetoricians as described in [YATE66].

For the sake of simplicity our system uses two separate windows. This solution is less than optimal, but it allowed us to create a usable prototype of our MOO client within a few days. The client is implemented in HyperCard 2.2., a rapid prototyping tool for the Macintosh, and we use the Netscape WWW Browser as the Web client. The two clients act as two windows into separate but related information spaces (see Fig. 4).

Presently we explore various possibilities of spatial navigation in the Web. In the summer '95 quarter we will try to hold a telecourse in technical writing using this system. We hope to gain valuable insights in how people make use of the spatial environment to organize and exchange information.

4.2. Navigating the WWW by navigating the MOO

What are the advantages of the MOO/WWW combination over using a Web client alone? One way to see this combination is as a spatialized WWW hotlist that also allows to interact with other users. The spatialization can be done either by an administrator or by the user community. The structures we represent in our system are only small subsets of the Web. Representing the whole Web or even large parts thereof is impracticable as it changes its contents and its structure too rapidly. A spatial environment that changes its spatial structure that fast would be impossible to navigate. Therefore we use mainly quite stable Web pages or Web pages that are likely to exist for a longer time. These pages represent landmarks in the WWW that can be found easily using spatial navigation in the MOO. The spatial structure in the MOO mirrors a logical structure of the material on the Web.

Our system is also a useful tool to design fixed paths through the Web. As MOOs allow guiding other users it is also very easy to conduct a guided tour through the WWW using this system. According to Zellweger "Users are less likely to feel disoriented or lost when they are following a pre-defined path rather than browsing freely, and the cognitive overhead is reduced because the path either makes or narrows their choices." [ZELL89] Such a type of path structure can be easily defined in the MOO as a pathway through the MOO/Web.

The focus on spatial navigation in our system does not hinder users to switch to the Web client and to use hypertext-style navigation. While using the Web client the player stays stationary in the MOO. In case she should get lost in the Web she simply switches back to the MOO client and updates to the corresponding Web page. This feature is similar to teleporting back to a known landmark.

5. Example - a shopping mall metaphor

In this section we describe a simple shopping mall metaphor implemented using our system. The mall is used to spatially navigate a hotlist by topic.

5.1. How to navigate structured lists

The "URL Mall" is arranged according to subject and to level of detail inside a subject. The mall itself uses a spatial metaphor but the subsections of the mall (departments) presently make use of a more abstract space concept by using "next" and "previous" exits. An elevator (not visible in the log) connects several levels of the mall (see Fig. 5). Each level contains specialized departments and each level houses related topics. The rooms in the departments are arranged so that the user navigates to more specific information the more she enters the department. Note how the problem of separating structure and content re-arises in this area as each room is designed to contain a short description of the corresponding Web page (see section 4.1.).

> **look**
The URL Mall
This the entrance of the URL mall. The mall consists of several
levels, each of which contains departments. Each department
contains pointers for a special topic. Walking through the
departments will navigate you to the corresponding Web pages.
The farther you enter a department the more specialized Web
pointers you will encounter.

You see four departments:
Hypertext....general information about hypertext
HTML.........information on the World Wide Web (WWW) and HTML
(...)

> **hyper**
The Hypertext 1
You are in the hypertext department. The link [mall] leads back
to the mall entrance, whereas the links [next] and [prev] lead
to the next or previous room in the sequence respectively.
[first] leads to the first (this) room in the sequence.

This room contains the URL of the "World Wide Web FAQ". This
FAQ (Frequently Asked Questions) answers most typical beginner
questions.

> **next**
The Hypertext 2
(...)
This room links you to introductory information about how to
create you own Web pages. The rooms further ahead provide links
to tools that make this task easier.

> **first**
The Hypertext 1
You are in the hypertext department. The link [mall] leads back
to the mall entrance,
(...)

Other conceptual directions are easy to realize inside departments. An example is the
use of "up" and "down" exits for linking more or less detailed information. We plan
to create several departments that contain the same Web pointers but differ in the use
of directions to learn how users react to these differences.

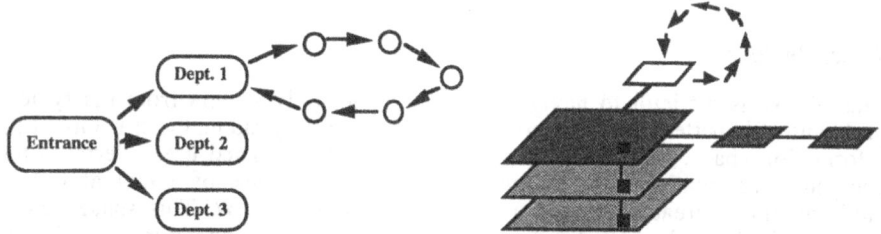

Fig. 5. The conceptual space of the mall metaphor. The left figure shows the essentially
hierarchical information space. The right sketch shows that these structures can easily be
connected using additional pathways leading up or down or in any other fixed direction to
create pathways between spatially separated but related rooms.

6. Further adventures in spatial Web navigation

Even if the mall itself is a useful spatial structure the full potential of the MOO/WWW combination comes from the rich interaction possibilities with the spatial environment and other users. MOO essentially are social places [ERIC93] and the combination of the MOO and the Web therefore can such a social place as well.

Typical interactions in MOOs include (almost) real-time or asynchronous conversation between users and direct interaction between users or between users and the environment. Complex types of Web navigation therefore can be realized easily. Examples are automatic presentations of Web page collections, address books that provide sorting and updating of URLs, or robots that wander through the MOO and monitor changes or provide navigational services.

More interesting is direct support for spatial navigation. As users can give directions in the MOO so can virtual users (programmed characters or robots). An example for a useful virtual user is a tour guide who leads novice users though the MOO giving a guided tour and pointing out major landmarks in the MOO (and the Web).

Other navigational tools are transport systems connecting locations in the MOO (which essentially means "locations in the combined MOO/Web space") either by navigating along regular paths or by teleporting there. Possible metaphors for these two types of navigation are a tramway system in the first and a subway system in the latter case. For a detailed description of these metaphors see [DIEB94].

Presently the connection between the two clients is a one-way direction. Movement in the MOO reflects in the Web client but not the other way round. In section 4.2. this was described as a positive feature. A bi-directional connection allows more flexible and less textual navigational features however, like a map of the shopping mall, which the user can click on to navigate to a certain department.

Another interesting possibility is to use the MOO system itself as a Web server. Output from the MOO can then be displayed as Web document in the Web-client. Examples are objects collecting information about the use of the MOO system, or MOO areas with a dynamic layout.

We are only beginning to understand the capabilities of the MOO/Web combination. One of our planned projects is to use a modified version of the client to navigate an interactive movie or to create a navigable auditory space as an art installation. However our main objective in the near future is to study spatial navigation in the Web in more detail.

7. Conclusions

The WWW is difficult to navigate because it not only suffers from the typical problems of hypertext but makes these problems worse by hiding the structure of its information space. Many WWW pages act as badly designed user interfaces for accessing information. The system we present makes use of a textual virtual environment to create a mirror space to parts of the WWW. This space makes structures in the Web explicit and provides spatial navigation, and user interaction. It can be structured in a way to define new structures on the WWW space that are tailored to the users needs and can be enriched with additional information to help in navigation. The rich interaction possibilities in the MOO allow to give lectures, to create automated presentations, and to guide users through the information space. Locations in the MOO are associated with Web pages and act as landmarks in the Web space. Using the combination of MOO and WWW the same types of interaction can be performed in the Web. This even includes the possibility to describe paths in the Web using incomplete information. We see many possible applications for such a system including further research of navigation in spatialized hypertexts.

Acknowledgments

This work is made possible by a research grant from the Austrian "Fonds zur Förderung der wissenschaftlichen Forschung" (Dr. Erwin Schrödinger Stipendium), Grant J01021-MAT for which we are very grateful. We further wish to thank Prof. Jay D. Bolter and Dawn Chapelle Jones for their continuing support and Sabine Timpf, Werner Kuhn, and Christian Unfried for valuable comments on drafts of this paper.

References

[AlBe95]   Albers M., Bergman E.: "The Audible Web: Auditory Enhancements for Mosaic", CHI'95 Conference Companion, pp. 318-319

[CuNi93]   Curtis P., Nichols D.A.: "MUDs Grow Up: Social Virtual Reality in the Real World", electronic publication, 1993, ftp://parcftp.xerox.com/pub/MOO/papers/MUDsGrowUp.ps

[DiBo95]   Dieberger A., Bolter J.D.: "On the design of hyper"spaces"", sidebar in Comm. of the ACM, Special Issue on Hypertext, to appear in August 1995

[DIEB94]   Dieberger A.: "Navigation in Textual Virtual Environments using a City Metaphor", PhD Thesis at the Vienna University of Technology, November 1994

[DiTr93]   Dieberger A., Tromp J.G.: "The Information City Project - A virtual reality user interface for navigation in information spaces", presented at Virtual Reality Vienna '93, December 1993, available at http://www.gatech.edu/lcc/idt/Faculty/andreas_dieberger/VRV.html

[ERIC93]   Erickson T.: "From Interface to Interplace: The Spatial Environment as a Medium for Interaction", Proc. of COSIT'93, Springer LNCS 716, Springer 1993, pp. 391-405

[ERIC95]   Erickson T., personal communication, March 1995

[KiHi94]   Kim H., Hirtle S.C.: "Spatial Metaphors and disorientation in hypertext browsing", to appear in Behaviour and Information Technology, in Press

[LYNC60]   Lynch K.: " The image of the city", MIT Press, 1960

[MaOs93]   Masinter L., Ostrom E.: "Collaborative Information Retrieval: Gopher from MOO", presented at INET'93, ftp://parcftp.xerox.com/pub/MOO/papers/MOOGopher.ps

[MaSC94]   Marshall C.C., Shipman F.M., Coombs J.H.: "VIKI: Spatial Hypertext Supporting Emergent Structure", Proc. of ACM European Conference on Hypertext '94, Edinburgh, September 1994, pp. 13-23

[MaSh95]   Marshall C.C., Shipman III F.M.: "Spatial Hypertext: Designing for Change", Comm. of the ACM, Special Issue on Hypertext, to appear in August 1995

[MuFH94] Mukherjea S., Foley J.D., Hudson S.E.: "Interactive Clustering for Navigating in Hypermedia Systems", Proc. of ECHT'94, pp. 136-145

[MuFh95] Mukherjea S., Foley J.D., Hudson S.E.: "Visualizing Complex Hypermedia Networks through Multiple Hierarchical Views", Proc. of CHI'95, pp. 331-337

[NiWe80] Nivergelt J., Weydert J.: "Sites, Modes, and Trails: Telling the user of an interactive system where he is, what he can do, and how to get to places", in: Guedj et al. (Eds.): "Methodology of Interaction", North Holland, 1980

[PARU89] Parunak H. VanDyke: "Hypermedia Topologies and User Navigation", Proc. of ACM Hypertext '89, pp. 43-50

[RiBO94] Rivlin E., Botafogo R., Shneiderman B.: "Navigating in Hyperspace: Designing a structure-based toolbox", Comm. of the ACM, 37(2), February 1994, pp. 87-96

[SaBr94] Sarkar M., Brown M.H.: "Graphical Fisheye Views", Comm. of the ACM, 37(12), December 1994, pp. 73-84

[SeNi90] Sellen A., Nicol A.: "Building User-centered On-Line Help", in Laurel B. (Ed.): "The Art of Human Computer Interface Design", Addison-Wesley 1990, pp. 143-153

[SHUM90] Shum S.B.: "Real and Virtual Spaces: Mapping from spatial cognition to Hypertext", Hypermedia, 2(2), 1990, pp. 133-158

[TrDi95] Tromp J.G., Dieberger A.: "MUDs as text-based spatial user interfaces and research tools", Journal of Intelligent Systems, to appear in 1995

[YATE66] Yates F.A.: "The Art of Memory", Chicago University Press, 1966

[ZELL89] Zellweger P.T.: "Scripted documents: A hypermedia path mechanism", Proc. of ACM Hypertext '89, ACM Press 1989, pp. 1-14

Structural Analysis of Geographic Information and GIS Operations from a User's Perspective

May Yuan* and Jochen Albrecht**
* Department of Geography, University of Oklahoma
** Institute for Spatial Analysis and Planning in Areas of Intensive Agriculture,
University of Vechta

Abstract

Geographic information and GIS operations constitute the kernel of a geographic information system. However, most GIS fail to structure geographic information and operations with direct mappings to users' conceptual schemata and analytic needs. As a result, GIS data and operations tend to be system-dependent and switch from one system to another is not trivial. Since the conceptual schemata for structuring both declarative and procedure knowledge are system-independent, this paper suggests frameworks for structuring geographic information and operations from users' perspectives, hereby making GIS data and operations interoperable. Four related user conceptual models are identified. Location snapshots and mosaics represent a location-centered conceptualization, whereas entity and entity snapshot models suggest an entity-centered view of reality. Four levels of GIS functions include the task level, the semantic level, the syntactic level, and the interaction level. This paper concludes that the four conceptual models can provide direct mappings from users' concepts to data objects, whereas functions at the task and semantical levels appear to fit well into users' models for procedure knowledge. Therefore, structuring geographic information compatible to the four conceptual models and designing GIS functions at the task and semantical levels will advance to data models and GIS function independent of hardware and software.

1   Introduction

A GIS is a system which facilitates query and analysis of geographic information. The kernel, spatial databases and GIS functions, determines what and how much geographic information can be derived from the system. Ideally, spatial databases and GIS functions should be independent of hardware and software so to enable the support for interoperability of spatial data and universal GIS operations. Unfortunately, a strong emphasis on technical aspects in the design of most GISs results a significant drawback of application-specific data and system-confined operations.

A portable database requires a generic data model to organize data that facilitates a wide variety of inquired geographic information. The relational data model (Codd, 1970) is an example for such a generic data model in non-spatial domains but the development of GIS generic data models is still in its infancy. What is special about spatial data that impedes the achievement of a good spatial data model? Goodchild (1992a) epitomized that spatial, or rather, geographic data are unique in four ways: (1) the need for dual keys access to spatial database by either attributes or by locations, (2) the continuous two-dimensional nature of the spatial key, (3) spatial dependence that adjacent locations tend to possess similar attributes, and (4) the curved surface of the globe, over which geographic data are distributed. The emphasis on the spatial key and spatial property signifies the importance of locational concerns in spatial data handling. Furthermore, geographic data are not static since the world is a dynamic system; geographers and other scientists are not only interested in the current state of reality, but also its past, its future, and many processes involved in these transitions. If the spatial considerations have complicated problems in spatial data modeling, the temporal aspects certainly further perplex the task.

Interoperability is probably the GIS buzzword of the year. Yet, with the exception of the long-term OGIS project (OGF, 1993; Buehler, 1994), there are no concepts in sight that address the often described needs for true connectivity. A "live-link" among a few vendor products surely is insufficient. Anybody who tried to run an Arc/Info™ AML program within Idrisi™ has an idea of the difficulties involved. On the other hand, there exists only a limited set of spatial analytical operations across all the different products, platforms, and data models. If these could be united so that each operation that essentially does the same also bears the same name, then a lot of confusion among the user community could be avoided. Even better though, would be a shell that performs all data model-specific operations invisibly, confronting the user only with a handful of elementary analytic operations that look the same no matter what system is actually used.

Based on the belief that users' conceptual models are crucial to the development of a generic GIS data model and universal GIS functions, this paper starts with user interviews from data and functional perspectives. The data oriented approach investigates users' conceptual models used to solve geographic problems. Wildfire is used as an example in the survey of required components of geographic data for various wildfire studies. Wildfire studies need to consider great variations in both space and time, and therefore the results are applicable to studies of other geographic phenomena with spatial and temporal characteristics. The functional oriented approach inspects the tasks that a GIS user wants to solve and suggests a taxonomy of elementary operations that are universal, yet can be used to create context-dependent processing plans. This paper analyzes users' views of geographic information and GIS operations and suggests basic frameworks for structuring a generic GIS data model and universal GIS functions. By suggesting the possibility of dissociating data and functions from hardware and software, this paper intends to signify the importance of users' perspectives to the design of an effective and efficient GIS.

2 User-Centered Views of Geographic Information Modeling

Structurally, a GIS could be said to consist of five components: hardware, software, data, procedures *and people*. If the design and operation of a GIS is to be optimized, all five components need to be considered. A key aspect of this approach is the inclusion of the *user* as a part of the total system. People are part of GIS and their requirements and behavior are crucial elements of any study of how to make GIS more efficient and effective.

People, on the other hand, are quite difficult to assess. Empirical Social Research is a discipline that specializes on formalizing techniques that should be applied in interviews. However, these are arduous to apply here. The pitfall lies in the nature of the knowledge that the user possesses. Those ignorant of GIS have trouble to imagine possible applications that extend beyond the range of operations that they perform manually. Experienced users on the other hand, are usually preoccupied by the philosophy of the system that they experienced. This is reflected by the proverbial forester who describes a stand of beautiful old trees as a magnificent polygon. Experts, who have used a variety of GIS intensively, and therefore could be expected to be able to abstract their knowledge are rare. A formal questionnaire to identify the general user requirements therefore runs the danger of being too conservative when users have a poor understanding of current technological developments and/or where GIS functionality has the potential to markedly alter products and procedures. It also has limited potential to explore user's mental models or differences between individuals in the user population.

GIS data models are representations of reality. GIS users do not have direct access to reality. Instead, they interact with the systems through the data models based on which geographic data are organized. A data model defines data objects, operations, and integrity rules (Codd, 1981). As described by Medyckyj-Scott and Blades (1990):

> "For the GIS designer the geometric data model is a formalized abstract set of spatial objects classes, operations on them and the relationship between them, and the rules that govern the operations and relationships (Herring, 1990). From the user's point of view the geometric data model defines what he sees at the user interface and how he can act with what he see; in other words it controls the view of the world, as portrayed by the GIS, that the GIS user has."
> (p. 13)

As a result, GIS data models had better reflect the mental models of both the system designer and the user in order to best facilitate the communication between GIS and the user. Cognitive studies suggest that people use schemata (or object schemata) as a formalism for representing concepts or knowledge, and use scripts (or event schemata) as a framework for structuring a sequence of actions (Anderson, 1990). Based on this cognitive proposal, this study aims to find primitive objects and concepts that GIS users employ to conceptualize reality and build up their geographic knowledge, and to disclose elementary operations that GIS users apply as building blocks for problem solving. To wit, this approach is taken based on two beliefs: (1) data objects modeled in the systems should have direct mappings to the concepts that the user applies to acquire and/or analyze reality, and (2) GIS operations should have direct correspondence to the user's problem solving scripts.

This paper emphasizes the human component of a GIS and stresses that users' requirements and behavior are crucial elements of any study of how to improve efficiency and effectiveness of a GIS. As a result, user interviews are the main techniques used to reveal both primitive objects of their schemata and elementary functions of their scripts in the search for components of geographic information. Results from interviews of wildfire experts at twelve wildfire research sites across the United States and Canada are analyzed according to semantical, spatial, and temporal concepts, since geographic phenomena, in general, can be observed in terms of attributes, spatial, and temporal characteristics (Sinton, 1978). The tapes of interviews were reviewed. Except for technical terms, the meanings and concepts of the conversations were analyzed rather than the exact wordings. Analysis was first focused on the semantics and relationships of concepts and then on conceptual differences in spatial and temporal data requirements. The acquired knowledge of wildfire was then refined and used as a basis for reconstructing user's mental models of wildfire and disclosing their object schemata.

As for the survey of GIS operations, both questionnaires and lineage diagrams were employed to solicit the use of actual GIS functions. The questionnaires, based on a list of 144 GIS operations described in Burrough (1992), Goodchild (1992b); de Man (1988); Rhind and Green (1988); and Unwin (1990) were distributed at a number of international conferences in 1993 and 1994 (Albrecht, 1995). This group is certainly not representative for the GIS community as a whole, i.e. the far majority of them were Arc/Info™ users with Erdas™ and Idrisi™ ranking second and third. Most of them did not have experience with more than one GIS.

However, considering the constraints, there are a number of results in this informal survey that can be regarded as applicable to the GIS community.

3   Characterization of Geographic Data and Functions

A classification should reflect the purposes or objectives of the activity for which the classification is created. Calkins and Obermeyer (1991) give one of the very few process-oriented flow charts representing GIS-supported information winning (Fig. 1). It depicts an optimistic view assuming that the information content is enhanced, the amount of data decreased, and that value is added at each stage of transformation.

As implied in Calkins and Obermeyer (1991), GIS-supported information winning starts with data-centered operations and then advances to function-centered operations. Data-centered operations include data collection, data organization, and data transformation, whereas function-oriented operations focus on input-output mechanisms and interdependency among functions for completing certain tasks. This study follows the data-function framework in search for components of geographic information through data-oriented interviews of wildfire experts and function-centered interviews of GIS users. Outcomes from interviews on both wildfire concepts and analytic operations reflect sets of primitive abstracts and primary procedures in conceptual models, which are users' basic frameworks used for structuring knowledge and solving problems. These results categorize both user views of spatio-temporal wildfire views and of geographic analytic operations.

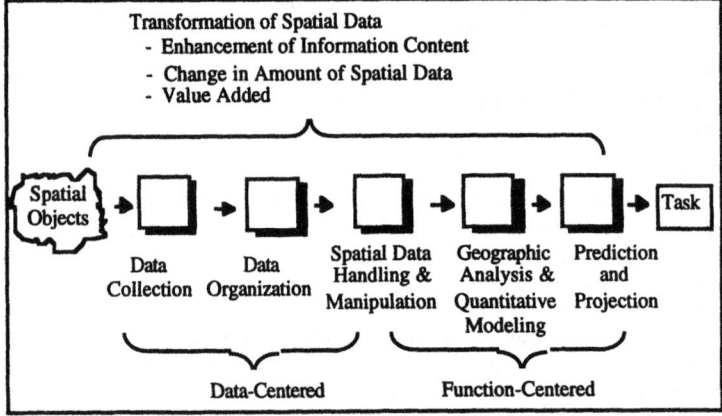

Fig. 1. Process-oriented flow charts representing of GIS-supported information winning (modified from Calkins and Obermeyer, 1991)

3.1   User Views of Space and Time in Wildfire

According to conceptualization of spatio-temporal wildfire data the survey suggests three categories of user views and further concludes four conceptual models. The three user views of spatio-temporal data include the potential of having fires, individual fires, and multiple fires (Table 1). Each of them is mapped to a conceptual data model except for the view of individual fires. The view of individual fires includes two distinct conceptualization modes: one describes a fire itself, and the other, fire's effects. Therefore, the three views result in four conceptual data models for wildfire.

The four conceptual models embrace fire mosaics, locational snapshots, fire entities, and entity snapshots (Fig. 2).

User views of fire potential and multiple fires map to the conceptual models of locational snapshots and fire mosaics, respectively. As mentioned above, the user view of individual fires can be further categorized into two classes of research foci: fire processes and fire impacts. Studies of fire processes stress the development and progress of a fire event, which is conceptualized by a fire entity model. In contrast, fire impact research emphasizes changes in an environment before and after a fire event and the entity snapshot model is the framework users use to acquire and analyze data. This section elaborates the three user views of spatio-temporal wildfire data, whereas section 4.1 will detail the four conceptual models.

Examples of user views on the potential of having fires at locations include studies and applications of fire forecasting and fire simulation. Parameters considered in fire potential modeling include fuel moisture, atmospheric circulation, and occurrences of lightning strikes. These parameters frame indices for fire risk at locations and these indices are updated on a regular basis, such as every day or every 12 hours. As such, views on fire potentials first specify spatial units, measure parameters and calculate an index of fire risk for every spatial unit, and determine time steps for updating the index. Both space and time are conceptualized as artificially discrete and pre-defined units, as cells and chronons. Attributes can not be determined without specifying locations and instants, and the value of any attribute is assumed to be a constant for all locations in a cell and through the duration of a chronon.

Table 1: A taxonomy for spatio-temporal wildfire views

	Snapshot for fire potential	Entities for fire individuals	Mosaics for fire populations
Fires in study	no fire	one fire or interacting fires	multiple fires
Spatial resolution (m^2)	10^6-10^8	1-10^6	10^2-10^8
Temporal resolution	1-24 hour	1 second - 1 hour	1 year
Spatial scale	regional-national	local-regional	regional
Temporal scale	day	hour-year	decade-millennium
Areal fire data		burning areas burned areas areal ignition	the most recent burns
Linear fire data		linear ignition	
Point fire data	lightning	point ignition	fire scars tree rings
Measure	weather	fires	landscape patterns
Record	fire potential index	fire paths fire's rate of spread fire intensity	fire regimes
Analysis		fire progress	fire regimes
Information Processing	update fire potential	track fire progress	update and overlay fire mosaics

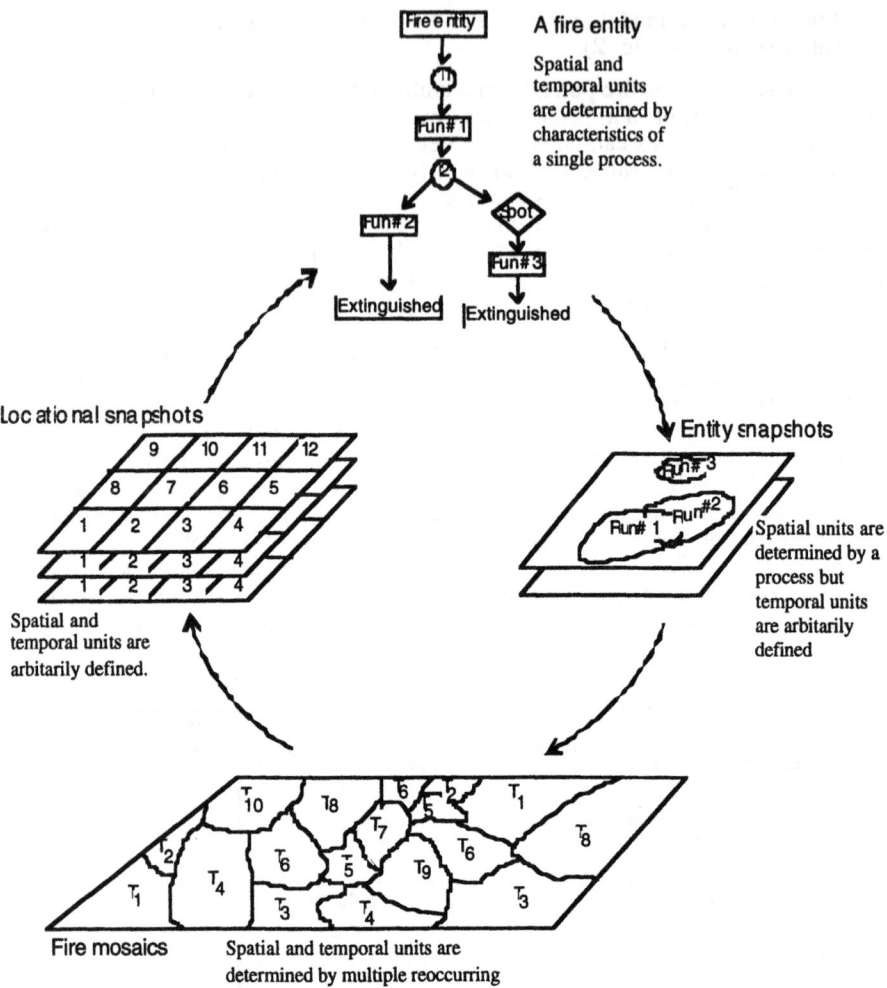

Fig. 2. Four GIS conceptual models

Studies of individual fires include modeling of fire behavior and fire growth, analysis of fire phenomena, and fire effect assessment. Two major approaches appear: one describes the burning status of a fire, and the other analyzes ecological, environmental, or socio-economic impacts of a fire. As mentioned above, studies of fire effects apply a conceptual model of entity snapshots to depict the impacts of a fire on the environment, whereas the other studies use the fire entity model to describe the progress of a fire. Fire behavior modeling aims at describing fire processes and therefore emphasizes descriptions of fire spread and its interactions with weather, vegetation, and terrain. Fire growth modelers conceptualize fires as separate entities spreading across a landscape through time, especially fire's rate of spread and intensity. Fire effect assessment looks for impacts of a fire by a comparison of pre- and post-fire conditions. Spatio-temporal data are collected so as to describe what has happened inside a fire or a burned patch, i.e. a fire run. Spatial and temporal resolutions of their measurements depend on the burning/burned areas of a fire and the fire's rate of spread, ranging from seconds and minutes to hours. Therefore in both approaches, concepts about space and time are, in fact, functions of the fires under study. Measures of attributes, space, and time refer to the properties of fires on focus, rather than properties of locations.

Views of fire populations conceptualize space as a result of multiple fire events through the time being. In this respect, space is a function of processes and time. Such studies include fire history modeling and fire management, which conceptualize space as mosaics of burns through a course of time. One of the main tasks in these studies is to define a fire population and fit the population into a good distribution for statistical descriptions of spatio-temporal distributions of fire's occurrences in a defined space and time frame. Determination of a study area and time is crucial since it decides the fire population under study. However, time exists only at instants when fires have occurred. Their spatial units are spatial patterns of burns from past fires.

3.2   Hierarchical Organization of GIS Functionality

A GIS is intended to provide the user with information for problem solving and decision making (Densham and Goodchild, 1992). To achieve these goals, the user has to process a sequence of tasks. Users have mental models about the task they want to accomplish with a system, and the way the system lets them accomplish those tasks. These models are defined by the user's prior experience, existing knowledge, and preconceptions about tasks (Albrecht, 1994).

GIS operations can be differentiated into functions and tasks. 'Task', as it is used here, shall describe all actions that require human input, i.e. the knowledge about context; whereas the notion 'function' is used for singular actions or sequences that can be automated. Tasks are usually composed of functions. GIS operations (comprising of either one, tasks or functions - the definition often depends on the domain) occur at a variety of hierarchical levels, prohibiting reduction to a simple goal-task-function hierarchy. A parallel phenomena is that a number of operations are essential for some applications while superfluous for others. This reflects the heterogeneity of the user community and leads to the conclusion that it is not possible to define a single universal GIS task taxonomy.

4   Conceptual Models for Geographic Data and Operations

4.1   Four Conceptual Models for Spatio-Temporal Knowledge

The wildfire interviews suggest three user views from the perspective of data collection and requirements. The user views on fire potential and multiple fires have direct mappings to user's conceptual models for encoding wildfire knowledge. However, the view on individual fires needs to be further divided into emphases on fire processes and on fire impacts, and the two subdivided views correspond to the fire entity model and the entity snapshot model, respectively. As a result of a closer look at the four conceptual models, two main approaches appear to the acquisition and organization of fire knowledge: location-based and entity-based conceptual models of structuring wildfire or, in general, geographic knowledge. The location-based approach conceptualizes reality as a set of snapshots (locational snapshots) or mosaics (fire mosaics), whereas, the entity-based approach, as a course or process of an entity's behavior (the fire entity model) or the entity's imprints (entity snapshots).

Fig. 2 structures the four conceptual models and suggests an information cycle of mutually supportive data input/output flows among the four conceptual models. For example, data output from the locational snapshot model represent the distribution of fire potential indices, which is crucial to determine fire's rate of spread and intensity in fire behavior modeling. The model of fire entity is then used to encode the course of fire's spread and/or to compare the prediction of a simulated fire with a real fire event to calibrate parameters in fire behavior models.

In the location-based approach, geographic knowledge accumulates per spatial unit. Spatial units can be delineated arbitrarily by a matrix of cells as in the view of fire potential snapshots or empirically by a pattern resulted from a process as in the view of fire mosaics. Attribute information is acquired and described in accordance with associated spatial units. Temporal information is either adjoining to all spatial units in a form of snapshots on a universal time frame (a snapshots model), or attached to individual spatial units in a form of temporal strings on an event time scheme (a mosaics model). No variations exist within any spatial unit. The two location-based schemata match the raster-cell (regular tessellations) and vector-polygon (irregular tessellations) data models in GIS. Location-based schemata require reconstruction of geographic knowledge ascribable to any reconfiguration of spatial units.

The entity-based approach conceptualizes geographic knowledge as attributes of individual entities. Entities may be solid objects (watersheds), events (wildfires), or concepts (counties). Geographic knowledge describes semantical characteristics, spatial properties and temporal behavior associated with identified entities. The entity-based views of reality describe either the current status of an entity (such as wildfire behavior modeling), or entity's vestige (such as fire effects analysis). Both views require identification of entities first, and then build geographic knowledge upon these entities.

4.2 Conceptualization of Tasks

Figures 3 and 4 show lineage diagrams for typical geographic modeling tasks using Arc/Info™ (Fig. 3) and Erdas™ (Fig. 4). Project 'A' was a one-year project of GIS students at the University of Osnabrück, Germany. Their task was to evaluate all city parcels with respect to their qualification for conversation measures. The actual lineage diagram is a lot more complicated than what is shown in Fig. 3 and represents rather the result of an expert operator who optimized the necessary queries to the Arc/Info™ system used.

Project 'B' is the introductionary example of a GIS class. This time, the image processing and GIS software Erdas™ is used for a location finding problem. The raster-based system shows some strength in the overlay and reclassification routines, however, as Fig. 4 reveals strikingly, the user needs to spend much time with the delineation of objects. The majority of operations exists only because the data model or the development history of the particular GIS requires them. These chores are left to the user although because the GIS developers have not done their job. Section 6 will depict the same task based on a radically reduced, yet complete set of elementary analytical GIS operations.

The lineage diagrams, together with a content analysis of vendor-supplied manuals and the questionnaires revealed a dominance of data-centered views. Most people interviewed were even surprised when they realized that parts of our studies concentrated on procedures rather than data. They learned to conceptualize their GIS use in terms of data which is reflected by the sheer amount of data-related functions offered by GIS as opposed to those that could be attributed to analysis. This is easy to understand as the majority of these users spend most of their time with data capturing, transformation and cleaning (error pruning).

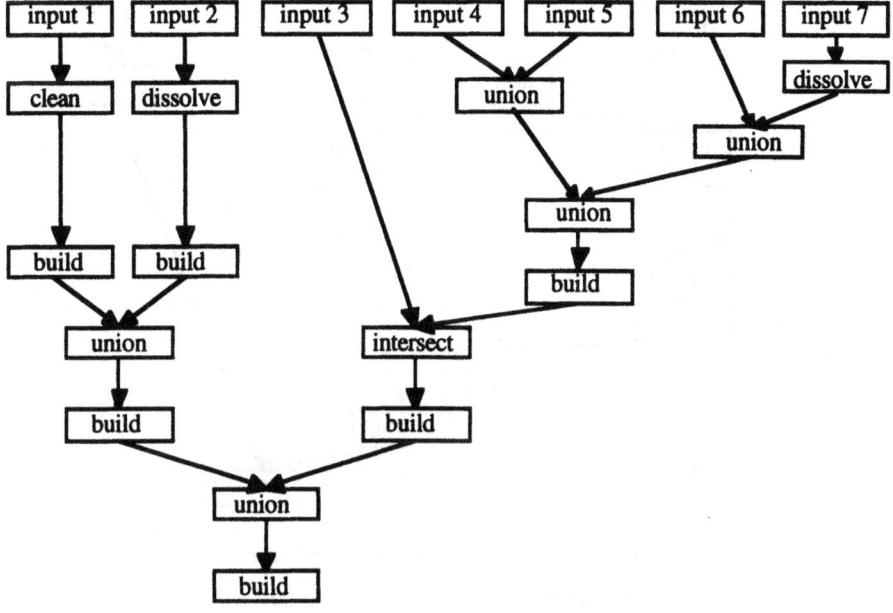

Fig. 3. Lineage of project 'A' using Arc/Info™

One of the major problems in current GIS is their evolution from either raster or vector origins. This data structure distinction has dictated differences in analysis functionality (e.g. planar enforcement vs. clump/labeling operations). The available set of analysis procedures and the names used for them differ between individual GIS products. The workflow analysis (Figures 3 and 4) showed that some the far majority of all operations during the analytical stage are due to the data structure and hence are a nuisance to the domain specialist who needs to concentrate on the analysis itself.

For the interviews, users were asked to rank 144 operations within a goal-task-function hierarchy (Huxhold, 1989). The list of operations was drawn from an analysis of GIS literature and software manuals. The responses assert the diversity of the GIS community and rendered an attempt impossible to define a list of elementary GIS tasks that is universally applicable (Albrecht, 1995).

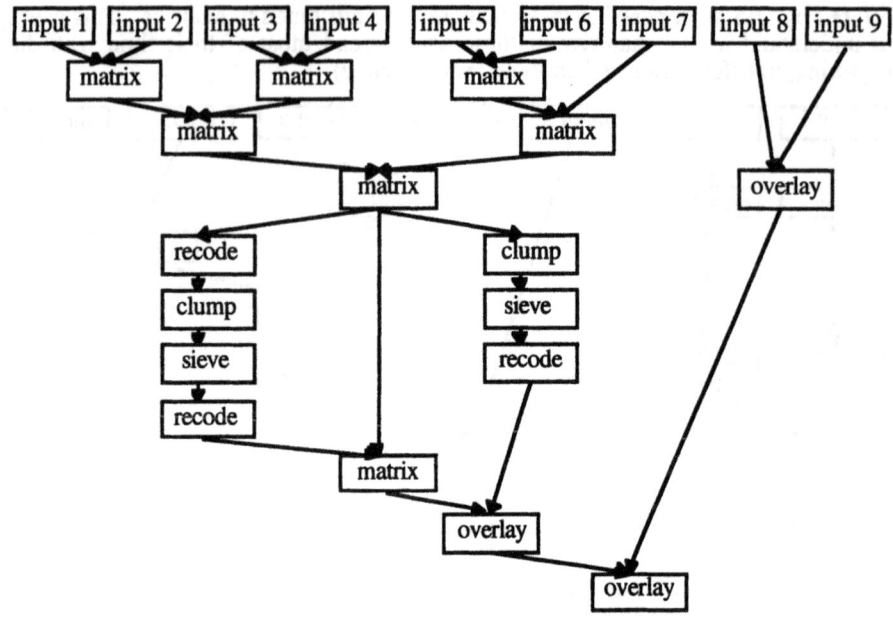

Fig. 4. Lineage of project 'B' using Erdas

Conscious conceptualization of tasks seems to be restricted to expert users who formed their own strategies for the aggregation of operations into work units that they deal with as if they were single units. These experienced users refer to tasks in a form that reflects implicit knowledge. This is why users work data-centered; they can't concentrate on their tasks unless they are experts who perform all the data-related routines by heart.

5   Organizing Geographic Information

Based on the three spatio-temporal views of wildfire data and the four inferred wildfire conceptual models, there are two main approaches to acquire and organize spatio-temporal knowledge, location-centered and entity-centered. The two approaches can be applied to other domains of geographic knowledge because wildfire represents a good variety of spatio-temporal variations and dynamic characteristics common to general geographic phenomena (Yuan, 1994). The location-centered approach models attributes and temporal characteristics according to spatial units; whereas the entity-centered approach depicts spatial and temporal characteristics based on entities with a set of attributes. As a result, delineation of spatial units and identifications of geographic entities are key issues to extract components of geographic information from a data's perspective.

5.1 Modeling Geographic Information from a Location-Centered Approach

The determination of spatial units is fundamental to the location-centered approach in modeling geographic information. Spatial units are locations or sizes of resolution, in which attributes are assumed to be spatially homogeneous and temporally uniform.

Locations can be points or lines, which indicate positions but no areal extents. Areal spatial units need to examine the issues of spatial resolutions. If the resolution is a constant, spatial units are equivalent to cells in a raster layer, whereas if the resolution varies according to the spatial distribution of a phenomenon or a geographic process, spatial units are resels in a vector layer (Tobler, 1995). Many factors are important to the determination of spatial units, such as spatial variations of a phenomenon of interests, sampling techniques, and research/operation goals. Usually, cell sizes are pre-determined by sampling technology but resels are likely to be delineated by partitioning space into a closest category of measured attributes. Geographic knowledge accumulates through an increment of descriptions of individual spatial units. Apparently, the size of spatial units determines the amount of geographic information derivable from a spatial data set and the shape of spatial units influences spatial drawn from the spatial data set.

The configuration of spatial units outlines a basic spatial framework to aggregate and generalize geographic data at a snapshot or through a period of time. A snapshot approach samples and measures attributes from all spatial units at an instant in time, or at least handles the data as if they are all collected at once. In contrast, attributes can be sampled individually in each spatial unit at different instants in time or through various chronons, and consequently these spatial units become space-time composites as described in Langran (1992).

5.2 Modeling Geographic Information from an Entity-Centered Approach

An entity-centered approach to the modeling of geographic information requires first to identify an entity of interests and then describe its thematic, spatial, and temporal characteristics. However, thematic properties are in accordance with user's perspectives rather than neutral to interpretations. 'Semantic' is better used to imply the relative meanings to users of different interests. For example, fire fighters see wildfire as a set of fire runs but fire effect analysts conceptualize wildfire as overlapping burned patches. Semantics is important to the identification of entities and is subject to classification, generalization, aggregation, and association. A fire consists of multiple runs at different points in time from a fire fighter's perspective but, in contrast, it is composed of patches of homogenous burning severity to fire effect analysts. Geographic entity types defined in the Spatial Data Transfer Standards (SDTS, 1992) as well as many taxonomies of phenomena or processes can be directly employed to set the definition schemes for semantical entities. Definitions of geographic entities appear to have overlaps semantically, especially in different languages (Mark, 1993). However, their definitions may be independent of space and time. For example, a grass fire is any fire occurring in a grassland. There is no specifications with regards to the shape, size, or locations of the burned areas or to the limitations of a particular season.

Once the entity of interest is identified, spatial and temporal characteristics are to acquire and describe in order to learn and represent a good geographic knowledge about the entity. Obviously, this conceptualization defines space and time by these spatial and temporal characteristics of the identified entity. In other words, this conceptualization breaks the integrity rule of space exhaustion, which enforces there is one and only one spatial object (a digital representation of an entity in reality) in any given space and time, and there is no empty space. Obviously, two events (entities) can occur at the same place coincidentally and certain events do not occur in certain areas. The entity model describes spatial characteristics at each point in time and therefore, geographic knowledge is developed by observing and measuring the changes of spatial characteristics of the entity through time.

Significant points in time appear when mutations of the entity occur and spatial and temporal characteristics are therefore depends on the stability of the entity. As a result, there is no uniform spatial units nor chronons in the entity model. A unique concept embedded in the entity model depicts the continuity in space and time unless mutations of the entity or interruptions by other events occur.

The entity snapshot model also requires to define an entity of interest first and then examines the entity's spatial and temporal characteristics. Space is therefore confined according to the spatial extent of the entity and spatial units are determined by the spatial distribution of the entity's semantic attributes, such as severity of burns in fire aftermaths. However, chronons are arbitrarily defined, for instance, right after a fire been extinguished or a month after a fire event. Similar to the locational snapshot model, all semantical and spatial data collected are assumed to be measured synchronically at a point in time. Comparison of semantical attributes is made between two time points within a spatial unit. Since the entity snapshot model assumes spatial homogeneity within each spatial unit, spatial configuration has to be re-structured if any changes in spatial properties occur, such as shape, size, and geometry.

6   Organizing GIS Functions

In section 4.2 we showed how the domain scientist is forced to work on the data-centered side of Fig. 1. Unless they became GIS experts themselves, those scientists are reduced to tedious button-pressing operators who have no chance to concentrate on their assignment.

Card et al. (1983) identified four levels of operations, namely the task level, the semantic level, the syntactic level and the interaction level. The *task level* corresponds to the overall GIS project management. Here, the user's ideas of the task are analyzed and structured. The result is a description of a task domain and the basic entities in the user's universe of discourse. The task level is domain-specific, which is why an off-the-shelf GIS always needs to restrict itself to the next-lower *semantic level*. Within current GIS use the semantic level could be compared to the GIS specialist who writes batches or macros. The system's functional capability (from a user's point of view) is represented by conceptual objects and the operations on them. It is at the semantic level that tasks are specified in terms of the conceptual objects and operations. This paper concentrates on the analysis of the semantic level.

The further we get down in Card's hierarchy, the further the user gets away from his *own* conceptions. The *syntactic level* is somewhat similar to the way that GIS are described in user manuals, where the abstract syntax and semantics of the conceptual objects (which scale down to pixel or point, line, area objects) are defined. The *interaction level*, finally, describes the most common level of interaction with GIS. Here the syntax is specified in all its detail including the sequence of key strokes required to specify commands and their arguments. From a user's point of view this is at such an abstract level that any relationship with the task level is indiscernible — which is why so many users remain at the level of keystroke sequences without ever to understand why and what it is that they are doing.

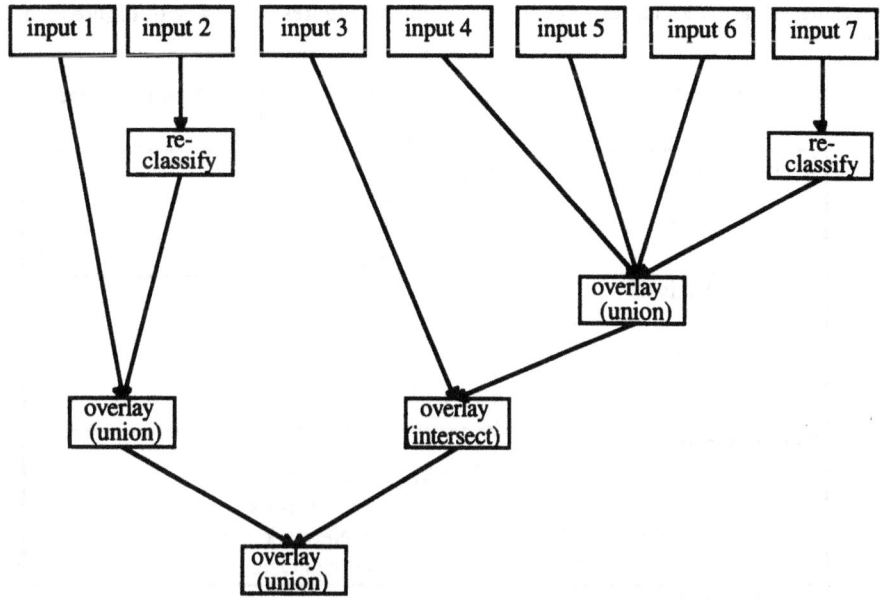

Fig. 5. Lineage of project 'A' using VGIS

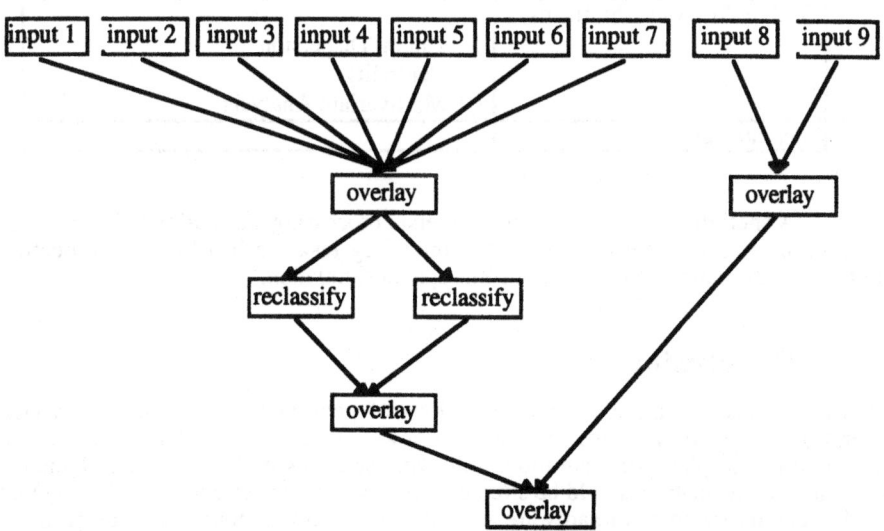

Fig. 6. Lineage of project 'B' using VGIS

Table 2 lists all the functional groups that could be employed to build a truly data model-independent task-based GIS user interface. This list contains the full analytic functionality of the major current vendor-provided GIS, yet it saves the user from some 70% or 80% of the operations that are currently necessary to perform the same task. Processing scripts based on this limited, yet almost all-perceivable tasks covering set, are a lot easier to deal with (compare Figures 3 and 4 with Figures 5 and 6).

Table 2. Domain- and data model- independent GIS processing steps as building blocks for any GIS task

Search / (Re-)classification	
	⟨ Thematic Search
	⟨ Search by Region
	⟨ (Re-) Classification
Location Analysis	
	⟨ Buffer
	⟨ Corridor
	⟨ Overlay
	⟨ Thiessen / Voronoi
Terrain Analysis	
	⟨ Slope and Aspect
	⟨ Catchment / Basins
	⟨ Drainage Network
	⟨ Viewshed Analysis
Distribution / Neighborhood	
	⟨ (cumulative) Costs, Diffusion / Spread
	⟨ Proximity
	⟨ Nearest Neighbor
Spatial Analysis / Statistics	
	⟨ Pattern/Dispersion
	⟨ Centrality
	⟨ Multivariate Analysis
Measurements	

A shell that is based on these universal processing steps allows the analysis-oriented user to concentrate the spatial modeling task, be it a location/allocation problem, path finding, districting, or some spatial simulation.

7 Conclusion

This paper has discussed conceptualizations of geographic information and GIS operations based on a series of user interviews. Users conceptualize geographic information and plan GIS operations following schemes in their problem domains. Geographic information should be structured in a way compatible to users' conceptual models and independent of hardware and software. GIS functions should be properly designed and categorized so to enable users to concentrate on problem solving and decision making tasks at the semantic level rather than to be distressed by command syntax and sequences. Since users' conceptual models are system-independent, geographic information and GIS operations structured according to these models are therefore universally operational.

From the data's perspective, three views of spatio-temporal data are identified: distributions of an attribute (such as potential of having fires), descriptions of a single event or process (such as individual fires), and results from multiple events or processes (such as multiple fires). People view reality from different perspectives and require various representations to structure geographic information in an optimal way so to obtain a direct support for reasoning in a problem domain. As a result, four conceptual models are further revealed from detailing spatio-temporal knowledge encoding and reasoning schemes applied to the three views. The locational snapshots and mosaics model reality from a location-centered perspective: everything is conceptualized as attributes of locations. The difference is the locational snapshots define both spatial and temporal units arbitrarily in a discrete framework, whereas the mosaics model configures space according to the results of a set of processes but describes time as an attribute of spatial units. The entity model and entity snapshots, on the other hand, represent reality by descriptions of individual entities. Only the spatial and temporal extents associated with the entity of interest are the main concerns in these models; in other words, they allow empty space and time. While the entity model defines space and time by the progress of an event or a process, the entity snapshots represent space by spatial units of homogeneity at arbitrarily defined points in time.

From the function's perspective, we have to conclude that current GIS do not allow a task-centered approach. All major GISs force the user to deal with data-related issues which thwart the job that the system is supposed to support. The proposed list of universal, elementary GIS processing steps (Table 2) is intended as a ground for further discussions. It represents the current layout of the screen menu that is used for some task building software developed at the University of Vechta. The elegance of the user interface has a price though: all the expertise that usually only comes with many years of GIS use now needs to be gathered in the pre-processing files briefly described in section 6. Unfortunately space constraints do not allow this to be elaborated any further here. From the perspective of the data models discussed above, Table 2 suits the locational approach better than the entity-based one. While the first treats time as an attribute, the second requires operations that support movement and change of entities.

8 References

Albrecht, J., 1994. Universal elementary GIS tasks: beyond low-level commands, *Proceedings Sixth International Symposium on Spatial Data Handling*, pp. 209-222. Edinburgh.

Albrecht, J., 1995. Semantic net of universal elementary GIS functions. In *ACSM/ASPRS Annual Convention and Exposition, Technical Papers, Vol. 4, (Auto-Carto 12)*, pp. 235-244.

Anderson, J. R., 1990, *Cognitive Psychology and its Implications*, 3rd ed., W. H. Freeman and Company, New York.

Buehler, K., 1994. OGIS Project Document 94-016. *OGIS Reference Model*. The Open GIS Foundation, Cambridge.

Burrough, P., 1992. Development of intelligent Geographical Information Systems. *International Journal of Geographic Information Systems*, VI (1): 1-11.

Calkins, H. and N. Obermeyer, 1991. Taxonomy for Surveying the Use and Value of Geographical Information. *International Journal of Geographical Information Systems*, 5(3):341-352.

Card, S., Moran, T. and A. Newell, 1983. *The Psychology of Human-Computer Interaction*. Lawrence Erlbaum Associates, Hillsdale.

Codd, E., 1970. A Relational Model of Data for Large Shared Data Banks. *Communications of the ACM,* 13(6).

Codd, E., 1981. Data Models in Database Management. *ACM SIGMOD Record* 11(2).

Densham, P. and M. Goodchild, 1992. Spatial Decision Support Systems: scientific report for the specialist meeting. NCGIA Technical Report 90-5, NCGIA, Santa Barbara.

Goodchild, M. F., 1992a, Geographical information science. *International Journal of Geographical Information Systems,* 6(1):31-45.

Goodchild, M. F., 1992b. *Spatial Analysis Using GIS: a seminar workbook.* NCGIA, Santa Barbara.

Huxhold, W., 1989, *An Introduction to Urban Information Systems.* Oxford University Press, New York.

Langran, G., 1992, *Time in Geographic Information Systems.* Taylor & Francis, New York.

Mark, David M., 1993, Toward a theoretical framework for geographic entity types. In Frank, A. U. and Campari, I.(Eds.): *Spatial Information Theory: A Theoretical Basis for GIS,* pp. 270-283, Springer-Verlag.

de Man, E., 1988. Establishing a Geographic Information System in Relation to Its use. *International Journal of Geographical Information Systems,* 2(3):257.

Medyckyj-Scott, D. and M. Blades, 1990. User's Cognitive Representations of Space: relevance to the design and use of GIS. *Report 13, Midlands Regional Research Laboratory,* University of Leicester.

OGF, 1993. *The Open Geodata Interoperability Specification,* Version 1.0, Preliminary Draft, November 15, 1993. Open GIS Foundation, Cambridge, MA.

Rhind, D. and N. Green, 1988. Design of Geographical Information System for a Heterogeneous Scientific Community. *International Journal of Geographical Information Systems,* 2(1):23-28.

Sinton, D., 1978, The inherent structure of information as a constraint to analysis: mapped thematic data as a case study. *Harvard Papers on GIS,* Volume 7, edited by G. Dutton, Addison-Wesley Publishing Company, Inc., Reading, MA.

SDTS, 1992. *Spatial Data Transfer Standard.* Federal information processing standards publication 173. Computer Systems Laboratory, National Institute of Standards and technology, Gaithersburg, MD.

Tobler, W., 1995, The resel-based GIS, *International Journal of Geographical Information Systems,* 9(1):95-100.

Unwin D., 1990. A syllabus for teaching Geographical Information Systems. *International Journal of Geographical Information Systems,* 4(4):461-462.

Yuan, M., 1994, *Representation of Wildfire in Geographic Information,* unpublished Ph.D. dissertation. State University of New York at Buffalo.

A Loosely Coupled Interface to an Object-Oriented Geographic Database*

Jean-Paul Peloux[1,2] and Philippe Rigaux[2]

[1] Fleximage, 43 rue de la Brèche aux Loups, F-75012 Paris, France
[2] Cedric/CNAM, 292 rue St Martin, F-75141 Paris Cedex 03, France
{peloux,rigaux}@cnam.fr

Abstract. Communication between end-users and spatial databases raises important and specific issues. Unlike classical systems, it requires strong cooperation between two major tools: spatial query language and graphical interface. This paper addresses the problem of building an interactive querying system for spatial databases with an open architecture. We focus on the following issues: (i) design of a database-independent Graphical User Interface (GUI) providing query expression support and advanced user-database interaction functionnalities (ii) software architecture for dealing with exchanges between the database logical representation and a graphical representation at the interface level (iii) implementation of spatial data and functionnalities within a DBMS without requiring changes in the logical data model or query language syntax. We present a prototype coupling the O_2 DBMS, its query language O_2Sql, and C++/X-Window-based toolkits as graphical components for the interface implementation.

1 Introduction

One of the more challenging issues raised by Geographic Information Systems (GIS) is the design of tools allowing the end-user to easily and efficiently access to spatial information. Unlike classical databases where information can be directly displayed to any end-user without changing its structure or abstraction level, spatial databases store large amounts of complex-structured data that need to be converted toward graphical representation.

Our first objective was to allow the user to access heterogeneous GIS databases through a unified interface. This implied to define an open architecture for dealing with exchanges between the database logical representation and a graphical representation managed by an external interface module. A related issue was to verify the effectiveness of an implementation of spatial data structures and functionnalities on top of the OODBMS O_2 [BDK91] without requiring changes to its logical model or query language syntax.

---

* Work partially supported by the French *CNRS GDR Cassini*, the ESPRIT BRA *AMUSING* and *Fleximage Inc.*

The result of this work is a prototype built upon O_2 and using X11/Motif interface generators such as XFaceMaker [NSL90] and Ilog Views [Ilo94]. The current O_2 geographic database uses for a spatial model GeO$_2$ [DRSM93, RDS95]. This experiment represents a first step toward open GIS and the access to spatial data servers through information networks.

The paper is organized as follows. Section 2 is devoted to user interface design issues. Section 3 describes the software architecture. We give in section 4 a description of the current implementation of the prototype and illustrate it by a simple querying session.

2 User interface design

There are two basic requirements for an interface devoted to spatial databases: (i) displaying geographic information within a spatial context according to visual communication efficiency and (ii) supporting the end-user query expression. We briefly develop these considerations below.

2.1 Managing graphical representation

The interface must provide a graphical representation of spatial data retrieved from the database. This involves at least the following functionnalities (see [Ege94] for a comprehensive study):

1. *Graphical combination of query results*: a *map*[3] is obtained by overlaying several layers, each layer being a repository for a set of spatial objects obtained from the database by a selection or computation operation. It should be possible to add or remove a layer, or to change the overlay order [Voi91, VvO92].

2. *Graphical attributes*: while the spatial part of geographic objects is easily rendered with graphic objects, the descriptive part must be represented through *graphic attributes* (color, pattern, ...). This involves (i) partitionning the query results into *graphic classes*, (ii) assigning a *graphic style* to each *graphic class* (i. e. defining a *legend*), (iii) designing tools for allowing the end-user control of these operations.

3. *Multi-map representation*: as a consequence of previous considerations, an object might be displayed with different graphic attributes, and within different contexts. This leads to a *multi-map* representation, each map managing its specific layers set given specific display parameters (map size, overlay order, graphic styles).

---

[3] We are aware thap the map metaphor has some drawbacks: because of space limitations, we refer the interested readers to [Kuh91, Mar89]. This is true as well for the important spatial concept of *scale* which will not be discussed here. See [RS94].

2.2 Querying support

It is the user interface responsability to help as much as possible the end-user in his querying task. We distinguish three kinds of query according to the part played by the graphical interface.

1. *Alphanumeric queries* only address the descriptive part of the geographic object.

2. *Spatial queries* involve spatial data by means (i) of selection criteria using geometric parameters, or (ii) spatial functions applied to the spatial part of the object. The end-user cannot express geometric parameters (the end-user does not know for instance rectangle coordinates when expressing a window query). The following support is thus required: (i) tools for drawing any geometric primitive the query language could deal with, (ii) creation of a logical representation of these parameters to be included within the query language syntax, (iii) conversion of spatial parameters into an appropriate numeric representation for the query interpreter.

3. *Visual queries*: because graphical syntactic elements (color, pattern, line style and so on) are not rich enough to represent the complexity of objects semantic description within the database, it is a common situation that the user asks for a further description of an already displayed object. We will denote as *Visual queries* those queries which deal with displayed objects. The GUI should be able to satisfy visual queries with a minimal syntactic user action (usually mouse-pointing).

Therefore it appears that graphical interface advanced tools are necessary as a complement to the query language retrieval functionnalities. The architecture presented below is intended to satisfy these requirements.

3 Architecture

There are many ways of designing a user interface for GIS databases. We could (i) integrate display and other interactive requests to the query language [EF88, Ege91, Ege94], (ii) create an external module specifically devoted to visualisation and interaction with database extracted informations [Voi91]. We chose the latter approach in order to build an open interactive system that would be able to access many heterogeneous DBMS (performance is another argument in favor of separating the interface level from the database level: see [RSV93] for a detailed argumentation). We describe below an architecture which is the result of an initial work presented in [Voi91], and whose first implementation was described in [RSV93].

3.1 Overall description

The architecture is composed of three levels. The first one is that of the database, storing spatial objects according to the DBMS's logical model. There are not particular requirements for spatial database design. We give in section 4 the actual implementation which is based upon O_2, but we could have used another DBMS as well. We will only assume in the sequel that spatial data may be retrieved using a spatial query language.

The second level is the *Abstract Representation Model* (or *ARM*) which stands as an intermediate between database and maps. It provides a logical level independent from the database, and it acts therefore as a support for implementing interactive functionnalities.

The third level is the *User Representation Model* (denoted *URM* in the sequel). It consists in *maps* displaying extracted objects with a user-friendly and customized representation.

ARM level

Basic informations managed by the *ARM* level are *Graphic objects* defined as follows[4]:

```
Class GraphObj
        Type    tuple [db-link : DbObj,
                       geom  : Spatial,
                       graphic-type : string]
```

Each graphic object *graphobj* is the result of a conversion process applied to a database object *dbobj*: *graphobj = convert(dbobj)*. A graphic object keeps a link toward its associated database object (*db-link* attribute in the above class) in order to fetch this object's description when needed. Nevertheless, we store at the *ARM* level some informations which will avoid to go back to the database each time a function is triggered:

- The *graphic-type* attribute is intended to partition the query results as described in section 2.2. As a matter of fact, it is computed from some descriptive attribute of the associated database object.

- The geometric description attribute (*geom* in *GraphObj* class) is obtained from the query result as a conversion from the database spatial description to type *Spatial* belonging to the *ARM* logical structures.

We need a geometric description at this level to support the *display* operation, as well as many end-user functionnalities such as *zooming, scrolling, mouse-selection* that do not need querying the database again. A simple, spaghetti-like,

---

[4] We use a syntax similar to that of O_2.

spatial representation should therefore be sufficient. In particular, topological information, intended for tasks such as spatial analysis, query optimization and database consistency, is not useful at the interface level. Note however that topology is easily and naturally communicated through graphical representation.

Each *Spatial* value may be any heterogeneous combination of points, lines and polygons (with holes).

```
Class Spatial
     Type tuple [points : set(Point),
                 lines  : set(Line),
                 polys  : set (tuple [poly : Polygon,
                                      holes : set(Polygon))]
                ]
```

The last important logical structure of the *ARM* level is the *Layer* class. A *layer* is a set of graphic objects:

```
Class Layer
     Type set(GraphObj)
```

The reader is referred to [Rig95] for a comprehensive study of *ARM* structures and functionnalities.

URM level

URM logical structures mainly consist in *Map* objects.

```
Class Map
     Type tuple [mbr    : Rectangle,
                 legend : set(Style),
                 layers : list(Layer),
                 size   : Rectangle]
```

A *map* is a graphical representation created by overlaying a set of layers *layers* within a graphical support whose size is *size*. The overlay order is defined by the *list* constructor. *mbr* is the minimal bounding rectangle of the map. This implies that a *clipping* operation is applied to each layer so that only graphic objects that fit into *mbr* are displayed. Together with *size*, *mbr* determines the map scale. The *legend* attribute is composed of *Style* objects:

```
Class Style
     Type tuple [graphic-type   : string,
                 color          : Color,
                 line-style     : LineStyle,
                 pattern-style  : PatternStyle,
                 icon           : Bitmap]
```

Note that the *icon* attribute is used for displaying *Point* components of *Spatial* objects.

A graphical representation in a map *m* is obtained by a natural join between styles of *m.legend* and graphic objects of *m.layers*. Each map is therefore a specific representation of graphic objects stored at the *ARM* level.

3.2 Operations

This architecture is intended to offer a user-friendly environment for querying a spatial database and efficiently manipulating the result of queries. We present now the two important operations discussed in section 2: display and query.

Display

Assuming an initial layer *l*, an initial map *m*, and a query *Q*, displaying the result of *Q* involves the following steps:

```
// Database access
1.  Add (1,convert (DbQuery(Q)));
// Join with graphic attributes
2.  Layer 1' = Join(1,m.legend);
// Homothetic and translation operations
3.  Layer 1" = ChangeScale (Translation(1',mbr), m.size/m.mbr));
// Visualization in m
4.  Display (1");
```

DbQuery returns the set of geographical objects answering *Q*. Once converted, this set is added into layer *l*. From the second step to the last one, we are only dealing with the last two levels of architecture: (i) a join with appropriate styles assigns to graphic objects the attributes needed for graphical representation, (ii) a translation is applied to adjust the lower-left corner of the map window with the lower-left corner of *m.mbr*, followed by a change of scale relative to the respective sizes of *m* and *mbr*. Finally l is displayed within the map. The following points are noteworthy:

-   The end-user can create layers, and use each layer as a target for a given query. The result of the query is then simply added to the layer content.

-   New maps can also be created, with specific styles and specific size. If the map is created from already displayed objects, then (i) it only involves steps 2 to 4 of the above algorithm, and (ii) no duplication of graphic objects at the *ARM* level should be needed.

Many functions can be defined at the *ARM* level to control graphical representation parameters independently from database: *legend* display and update, changing the overlay order of a set of layers, hiding a layer from a particular map, etc. Therefore, a consistent set of interactive operations may be given to an end-user to customize the spatial objects representation according to his needs (see [Rig95]).

Querying

Our three level architecture design principle is best illustrated with *visual queries*. A visual query can be split into two basic operations: (i) find the graphic object selected with the mouse, (ii) access, from this graphic object, to the associated database object.

The first operation can be processed at the *ARM* level, because it only requires an intersection algorithm, taking the coordinates of the point as an input parameter, and based on the spatial representation of graphic objects stored in layers. Once the object has been identified, it can be accessed in the database. Such a functionnality exhibits the cache role played by the *ARM* level. All end-user operations are transmitted to this level, and from here can possibly be redirected to database if needed.

4 The prototype

Our experiment consists of two parts: (i) a database prototype written on top of the O_2 DBMS, (ii) a GUI developed in C++ with the help of the X11 interface generators *XFaceMaker* and *Ilog Views*. We present each component, and give a description of the global architecture. The last subsection presents a short interactive session, both from user and software point of view.

4.1 Database prototype

The geographic database prototype was developed for Fleximage's GQL project [Rap93]. The prototype integrates into the Object Oriented DBMS O_2 a geometric toolbox developed in C [Rap93] and the GeO_2 component [DRSM93, RDS95] which models spatial basic primitives. Spatial queries are expressed by means of O_2Sql [BCD89], the object query language of O_2. We shortly describe below the main features of the GQL spatial database prototype.

In our implementation, we follows the now classical relational approach, extended with abstract data types for modelling spatial data [Güt88, ODD89, Ege91]. It provides the user a uniform approach for spatial and non spatial attributes of geographical objects. A single *Geometry* data type supports any spatial property. For closure under spatial operators, an object of *Geometry* type is any combination of points, lines and areas. CARTECH [ANT94] or GéoSabrina [LPV93] are examples of implementations which follow a similar approach with a relational DBMS. For a user writing queries, the *Geometry* O_2 class with its methods is the only visible part of the spatial model.

A description of the *Geometry* class is done in figure 1. the *geom* attribute defines the geometric components of the object. It is a set of objects of the *Primitive class*, a virtual superclass of *Point, Line* and *Area* elementary classes[5]. Three levels of topological structuration (Spaghetti, Network and Topology) are available. The GeO$_2$ implementation is the basic repository for this multi-level spatial data model. Figure 1 shows a simplified view of GeO$_2$ classes.

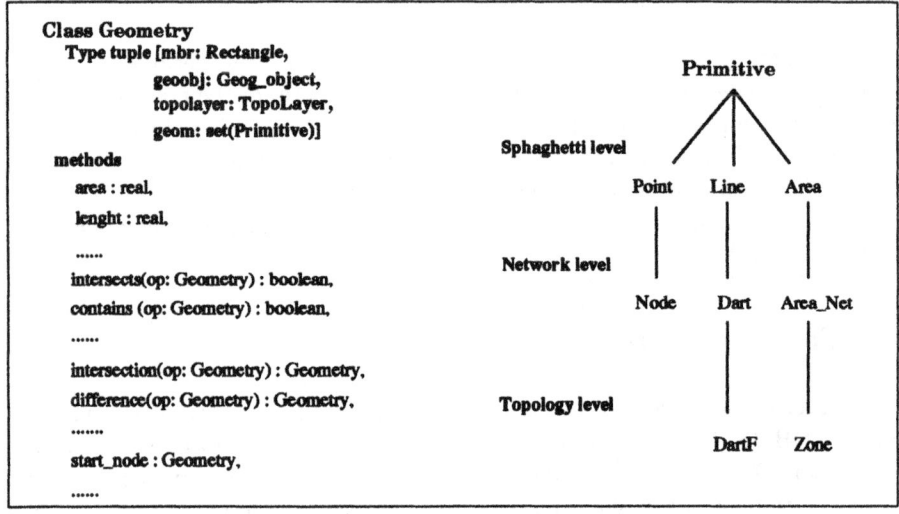

Fig. 1. : The *Geometry* class and the GeO$_2$ inheritance graph

The manipulation of geometric objects is done by means of a large set of methods of the generic *Geometry* class. These spatial operators include predicates (*is_empty, contains, intersects,...*), as well as operators with scalar result (*length, area,...*), or operators with geometric result (*intersection, difference, union, boundary,...*). Some agregates like *union* or *section of terrain elevation according to a given line*, operate on a set of objects of the *Geometry* class.

Some operations are only significant for objects with a specific shape (such as objects reduced to a unique component), or/and for objects belonging to a given topological level. The value returned by these operators will be significant only when applied to those specific objects, otherwise this value will be null.

---

[5] *mbr, geoobj* and *topolayer* attributes of *Geometry* are respectively the minimum bounding rectangle of the geometric object, a backward reference to the geographical entity supporting this object and a reference to the topologic layer, repository of the coordinates system and, in case of need, basic unit for topological relationships.

Primitives for navigation through the topology are examples of such operators.

The geometric toolbox is used for implementation of most of the operators, except those mentionned above. It operates on a Spaghetti representation of data, and when geometric values are returned, they are themselves stored into Spaghetti structure. Howewer, some topological operations, such as adjacency, are computed with higher performance using topological links, if stored. Such topological operations are redefined at the topological level.

Spatial queries are easily expressed with O_2Sql. We use the powerful functionnalities given by a query language for object oriented database, and particularly, the ability of applying operations to objects in any query expression. *Geometry* class methods are thus used for the spatial attributes of geographical objects. Simple query examples are given in subsection 4.4.

4.2 GUI

A GUI interface following the approach described in section 3 was developed, independently from the database prototype, in C++ and X11/Motif. The graphical interface was built with the following toolkits.

- *XFaceMaker* [NSL90] is an advanced *interface generator* which provides interactive building of Motif interfaces. Its main features are: (i) *interactive design of presentations* by selecting Motif widget classes and inserting them within a widget composition hierarchy, (ii) *interface behavior description* by association of *callback routines* with Motif components (buttons, menu item, ...)[6]. A Motif module built with XFaceMaker is accessible through a C function.

- *Ilog Views* [Ilo94] C++ graphic library was used to implement the *ARM* level. The library can be used at different levels, among which the following were of particular interest for us: (i) *Graphic Objects* are pre-defined graphic classes, each of them implementing a specific geometric shape and behavior, (ii) *Managers* allow to group *Graphic Objects* within *layers* and control composition of many layers to create *views* which define final representations onto the screen. Many high-level interactive functions such as *select-by-pointing* are also provided at the *Manager* level.

The database and GUI prototypes were then connected as described in the following section.

4.3 Global achitecture

Figure 2 summarizes the whole software architecture. Integration was realized in C++ within a central module *Control*. The three levels of architecture are: (i) the

---

[6] At the interface level, only the signature of a routine is required: the body belongs to the underlying application.

O_2 DBMS[7], (ii) *ARM* (Ilog Views/C++ modules), and (iii) *URM* implemented by an association of Ilog Views and XFaceMaker components. *URM* is itself split into a *MAP* module for displaying map and styles, and an *ENV* module which proposes all the graphical devices (buttons, menus) to bring access to software functionnalities.

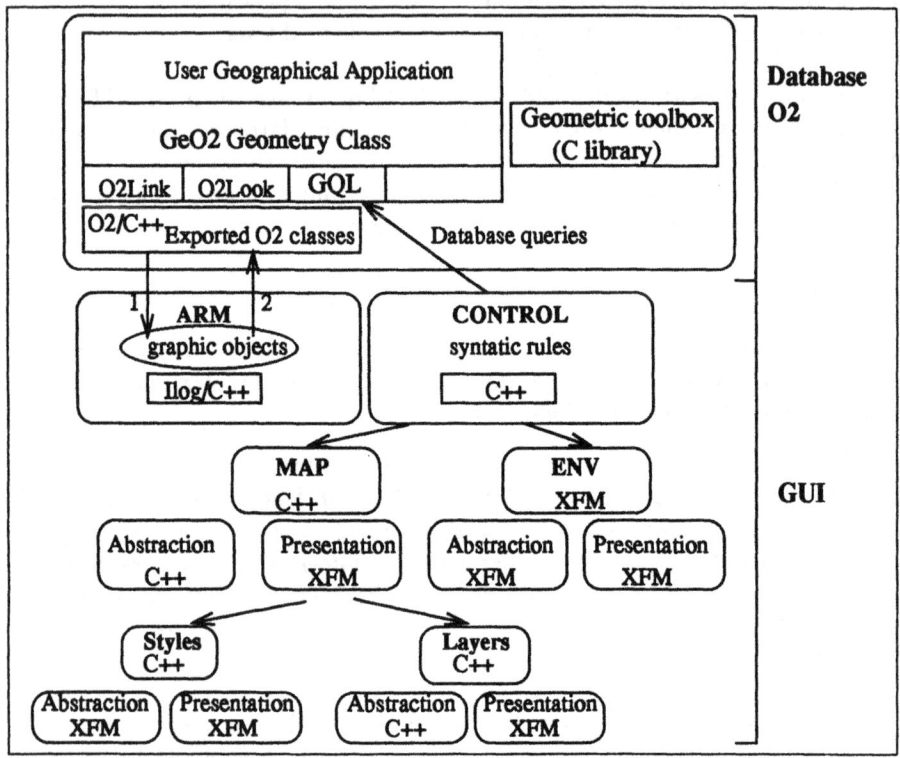

Fig. 2. : Architecture

4.4 A simple session example

Figure 3 shows the main components of the interface: a *query window* and a *map window*. The query window simply consists in a menu bar, some buttons and a text area to type queries. The O_2Sql query stands for displaying the land use map (t is an object in the *LUSE* named value [BDK91]). We construct a tuple containing the following items: (i) *LEGEND* gives the graphic type of each

---

[7] An O_2 application can communicate with a C++ environment through an utilitarian program called O_2Link. The C++ application sends messages to O_2 objects or directly access to the values of these objects.

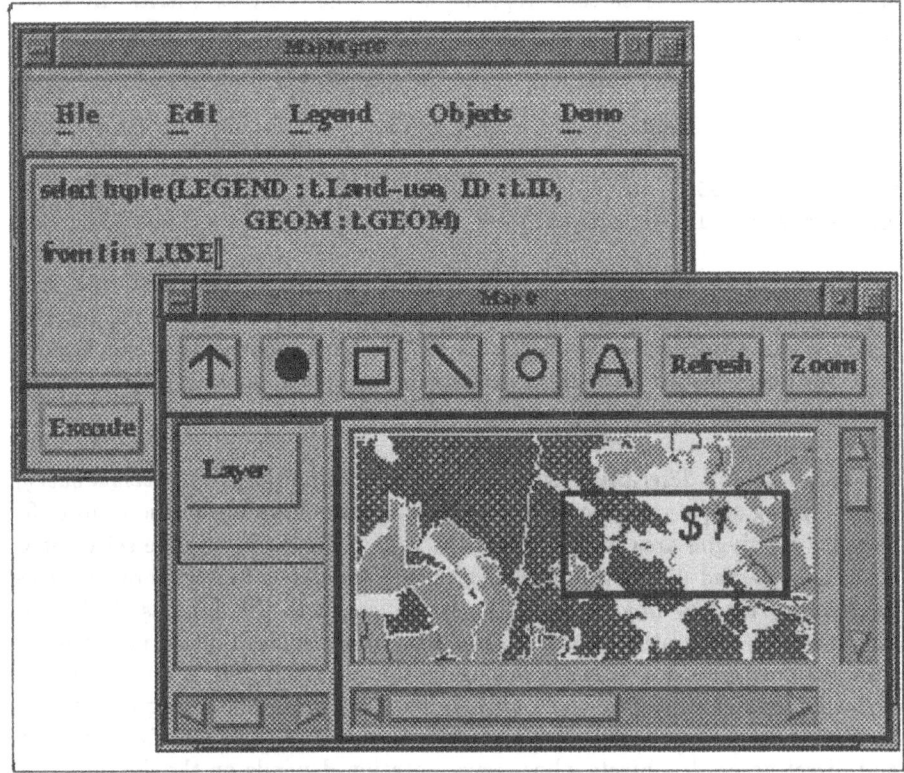

Fig. 3. : Query and map windows

selected object[8], (ii) *ID* transmit the object identifier in the database which will be used for further access[9], (iii) *GEOM* denotes the spatial database attributes.

The result is stored in a first layer after a conversion (see section 3.2) applied to the spatial attribute. Note that this conversion function is the only part that would have to be rewritten for another spatial data model in the database or for another DBMS connection. A graphical representation of this layer is finally displayed (figure 3).

A map comes with its own tools: selection (arrow), geometric parameters drawing tools, zoom, legend display and update, layers overlay customizing. Let

---

[8] We apply in this case a simple projection on the *Land-use* attribute, but more complex computation could be made to customize the number of *graphic classes* depending on the scale: see [RS95].

[9] Of course this is true only for persistent database objects. We can as well compute new spatial objects using geometric functions. In this case, the *ID* attribute is not useful.

us consider some of these tools together with more complex queries. Here is a *spatial query* that selects roads (r is a road in the map of roads, ROADS) whose geometry intersects a rectangle R, and prints as a result only the part inside R:

```
select tuple(LEGEND: r.TYPE, ID : r.ID,
             GEOM : r.GEOM->Intersection($1))
from r in ROADS
where r.GEOM->Intersects($1)
```

The $1 parameter (figure 3) is the formal parameter associated with rectangle R drawn by the user. The result is shown in figure 4 (the roads selected inside R are emphasized by darker lines). We have created a second layer to store the query result: we could also have displayed it in another map, or chosen interactively any combination of these operations.

To get the full description of an object displayed on the screen (*visual query*), the following steps are processed: (i) the end-user clicks inside the object (for instance a polygon area on the left of the map in figure 4), (ii) the select operation is processed at the interface level, using geometric functions of the *ARM* level as provided by Ilog Views, (iii) a query is sent to the database[10], (iv) the alphanumeric description of the selected object is displayed into a new window. This window does not actually belong to the interface, but is created by the $O_2 Look$ graphical interface tool. This shows that, while presenting a graphical, highly interactive, and spatially-oriented representation, this interface offers full and direct access to objects whose representation depends on the database.

5 Conclusion

We described in this paper an interactive GUI system loosely coupled to the O_2 OODBMS. This work shows that coupling of spatial languages with graphical interfaces involves at least the following features: (i) spatial data exchanges between interface and database should be hidden to the end-user[11], (ii) once an object has been fetched from the database and displayed onto the screen, its graphical representation is an anchor to its database representation through *visual queries*, (iii) complexity of spatial data representations implies enriching query languages with tools to manipulate and customize parameters like scale, styles and composition of maps.

The GUI is currently used as a GIS kernel for experiments on spatial data management with O_2 (namely multiple representation modelling and querying [RS94, RS95] and spatial indices implementation [PdSMS95]). A future work related to end users/databases interaction will fully exploit the independence of

---

[10] We could have applied any other function to this object, such as *delete, update, ...*

[11] This is true not only for the graphical representation of database spatial attributes, but also for the symbolic representation of geometric parameters.

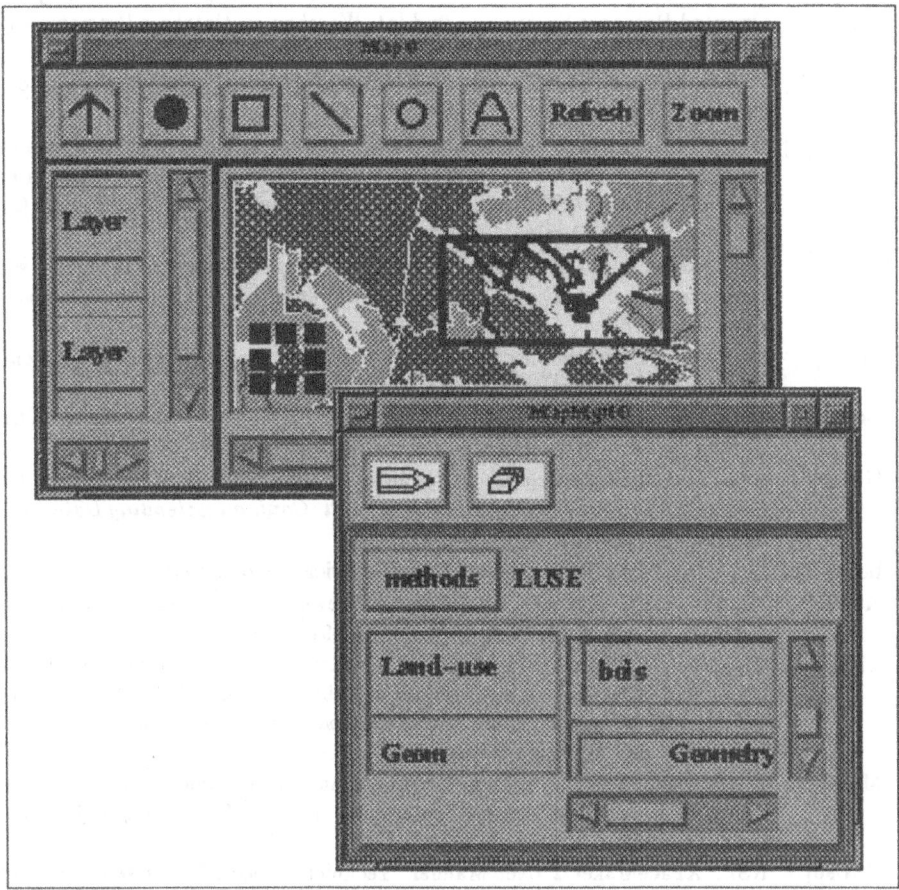

Fig. 4. : Visual query

the GUI: we plan to experiment connection of ou interface through Internet to any O_2 spatial database or more generally to any spatial server.

Acknowlegments. We are grateful to Michel Scholl for his encouragement and support which helped much to improve the quality of this paper. We also thank Véronique Mansart and Joël Thorner for their careful readings.

References

[ANT94] F. Arcieri, E. Nardelli, and M. Talamo. CARTECH: a Prototype of Geographical Information System. In H.J. Scheck J. Nievergelt, T. Roos and P. Widmayer, editors, *Int. Workshop on Advanced Research in GIS (IGIS)*, pages 178–191, Ascona, Switzerland, 1994. LNCS no 884, Springer Verlag.

[BCD89]   F. Bancilhon, S. Cluet, and C. Delobel. A Query Language for an Object-Oriented Database System. In *2nd Int. Worshop on Database Programming Languages (DBPL)*, pages 301–322, 1989.

[BDK91]   F. Bancilhon, C. Delobel, and P. Kannelakis, editors. *The O₂ Book*. Morgan Kaufmann, 1991.

[DRSM93]   B. David, L. Raynal, G. Schorter, and V. Mansart. GeO2: Why Objects in a Geographical DBMS ? In D. Abel and B.-C. Ooi, editors, *Advances in Spatial Databases (SSD'93)*, pages 264–276, Singapore, June 1993. LNCS No. 692, Springer-Verlag.

[EF88]   M. Egenhofer and A. Frank. Towards a Spatial Query Language: User Interface Considerations. In *Int. Conference on Very Large Database (VLDB)*, 1988.

[Ege91]   M. Egenhofer. Extending SQL for Cartographic Displays. *Cartography and Geographic Information Systems*, 18, 1991.

[Ege94]   M. Egenhofer. Spatial SQL: A Query and Presentation Language. *IEEE Transactions on Knowledge and Data Engineering*, 6:86–95, 1994.

[Güt88]   R. H. Güting. Geo-Relational Algebra : A Model and Query Language for Geometric Database Systems. In *Proc. Intl. Conf. on Extending Data Base Technology (EDBT)*, pages 506–527, 1988.

[Ilo94]   Ilog. Ilog Views Reference Manual. Technical report, 1994.

[Kuh91]   Werner Kuhn. Are Displays Maps or Views ? In *Proceedings of AUTO-CARTO 10*, pages 261–274, Baltimore, 1991.

[LPV93]   T. Larue, D. Pastre, and Y. Viémont. Strong Integration of Spatial Domains and Operators in a Relational Database System. In D. Abel and B.-C.Ooi, editors, *Advances in Spatial Databases (SSD'93)*, Singapore, 1993. LNCS No. 692, Springer-Verlag.

[Mar89]   D. M. Mark. Cognitive Image-Schemata and Geographic Information: Relation to User Views and GIS Interfaces. In *GIS/LIS'89*, pages 551–560, Orlando, 1989.

[NSL90]   NSL. XFaceMaker 2 User Manual. Technical report, NSL (Non Standard Logics), 1990.

[ODD89]   B. C. Ooi, Ron Sack Davis, and K. J. Mc Donell. Extending a DBMS for Geographic Applications. In *5th Int. Conf. on Data Engineering*, 1989.

[PdSMS95] J.P. Peloux, G. Reynal de St Michel, and M. Scholl. Evaluation of Spatial Indices Implemented with the O2 DBMS. *Ingénierie des systèmes d'information*, 3(4), 1995. To appear.

[Rap93]   Rapport des spécifications GQL. Technical report, Société Fleximage, Paris, September 1993. In French.

[RDS95]   Laurent Raynal, Benoit David, and Guylaine Schorter. Building an OOGIS Prototype: Some Experiment with GEO2. In *AUTO CARTO 12*, 1995. Charlotte, North Caroline, To appear.

[Rig95]   P. Rigaux. Interfaces Visuelles et Représentation Multiple dans les Bases de Données Spatiales. Thèse de doctorat CNAM, 1995.

[RS94]   P. Rigaux and M. Scholl. Multiple Representation Modelling and Querying. In *Int. Workshop on GIS (IGIS'94)*, pages 59–69, Ascona, Switzerland, 1994. LNCS No. 884, Springer-Verlag.

[RS95]   P. Rigaux and M. Scholl. Multi-scale Partitions: Application to Spatial and Statistical databases. In *Int. Conference on Large Spatial Databases*, Portland, Maine, USA, 1995. To appear.

[RSV93]   P. Rigaux, M. Scholl, and A. Voisard. A Map Editing Kernel Implementa-
          tion: Application to Multiple Scale Display. In *Spatial Information Theory
          (COSIT)*, pages 341–365, Elba, Italy, 1993. LNCS No. 716, Springer Verlag.
[Voi91]   A. Voisard. Towards a Toolbox for Geographic User Interfaces. In *Design
          and Implementation of Large Spatial Databases (SSD)*, Zurich, 1991. LNCS
          No 525, Springer Verlag.
[VvO92]   T. Vijlbrief and P. van Oosterom. The GEO++ System: an Extensi-
          ble GIS. In *Int. Symposium on Spatial Data Handling*, pages 40–50,
          Charleston, South Carolina, 1992.

Overcoming the Knowledge Acquisition Bottleneck in Map Generalization: The Role of Interactive Systems and Computational Intelligence

Robert Weibel, Stefan Keller and Tumasch Reichenbacher

Department of Geography
University of Zurich
Winterthurerstrasse 190
8057 Zurich (Switzerland)
Phone: +41-1-257 5152 / Fax: +41-1-362 5227
E-mail: {weibel,keller,tumasch}@gis.geogr.unizh.ch

Abstract

Past research in cartographic generalization has shown that algorithmic methods are well suited to handle narrow tasks, but appear to have limited potential so solve the entire generalization process comprehensively. Attempts to use systems based on explicit knowledge representation (e.g., rule-based or expert systems) also had relatively little success. The major limiting factor to explicit knowledge systems in generalization is the scarcity of formalized knowledge available. That is, knowledge acquisition (KA) forms the major bottleneck to progress of knowledge-based techniques.

In this paper, we discuss what options are available for cartographic KA, assess their potential, and propose alternatives which are based on the integration of techniques of computational intelligence (CI) with interactive environments. CI methods have the advantage of avoiding explicit knowledge formulation. Interactive systems allow to keep the human expert in the loop and thus augment his/her productivity as well as the potential for KA. Two examples are presented for this approach. The first example focuses on knowledge acquisition by process tracing in interactive systems. A comprehensive KA methodology supported by inductive learning algorithms is proposed and its technical feasibility assessed by an experiment. In the second example, genetic algorithms are used as an optimization method for interactive control of parameter settings for line generalization operators in an approach termed 'generalization by example'.

1 Introduction

In recent years, applications of geographic information systems (GIS) have matured, and databases of considerable size have been built. Users are now beginning to realize the lack of generalization functionality with respect to the development of value-added products from the initial database and the update of existing databases, particularly when multiple scales are involved. Since the majority of results of GIS-based modeling activities are still communicated to the end user in graphical form, functions are needed for automated *cartographic generalization*. It must be possible to derive display products from a basic database at arbitrary scale or symbolization, and maintain good readability. In a digital environment the requirements of generalization also extend beyond the original focus on graphics to include functions for *model generalization* (Muller et al. 1995). This objective mainly relates to data reduction and filtering and is thus not directly oriented towards graphical output.

Of the two basic purposes of generalization, this paper focuses on cartographic generalization. Cartographic generalization (or short, generalization) is prototypical of many map design processes: it represents a semi-structured problem, whose goals and outcomes may vary and are often ill-defined (Armstrong 1991). Algorithmic methods for generalization, which formed the core of research until the mid-1980s, offered only isolated solutions to partial problems (Weibel 1991, Ruas and Lagrange 1995). Thus, the focus of research subsequently shifted towards methods that would better relate to 'processing based on understanding' (Brassel and Weibel 1988), that is, towards methods based on knowledge and insight rather than deterministic and mechanistic views of the problem domain. At first, the attention was on expert systems (ES), most often building on production rules. Very few successful examples of implementations exist from that period, however, with some notable exceptions such as Nickerson (1988).

The lack of success of ES in cartographic generalization was mainly due to two problems. Firstly, and most importantly, any knowledge-based system, no matter what knowledge representation scheme it employs, can only be as powerful as the knowledge it contains. It soon became clear that for most map design and generalization problems, sufficiently detailed and accurate knowledge did not exist, and was also hard to formalize in the knowledge acquisition (KA) process. The scarcity of knowledge formed the limiting factor to progress of automation in generalization and thus created a situation which may be termed the *knowledge acquisition bottleneck*. The second deficiency of ES is their need for *explicit knowledge representation* (e.g., as production rules). Explicit knowledge formulation is problematic in the cartographic context for a variety of reasons, including the problem of having to deal with mixed topological and semantic notions occurring in generalization, the need for numerous exception rules (and the resulting conflicts between opposing rules), the difficulty to extend the knowledge base, and others (Ruas and Lagrange 1995, Keller 1994).

Any approach intended to improve the situation of knowledge-based methods in generalization would thus need to offer a solution to both problems outlined above. In this paper, we are discussing the use of techniques of computational intelligence (CI) — also termed machine learning or machine intelligence — as a possible path to achieve this goal. As will be shown below, CI offers options to either help generating explicit knowledge from implicit sources automatically, or even avoid the specification of explicit knowledge completely, by virtue of its *computational* approach to *learning* from implicitly represented knowledge (e.g., sets of positive or negative examples) rather than following the trails of explicitly stated knowledge. While the application of CI methods forms the first design element of our research, the use of interactive systems is the second design element. Generalization, as a design process, is inherently intuitive and can be understood as a decision making process (Weibel 1991, Armstrong 1991). Thus, we believe that the human should be kept in the loop, leading to a strategy which we termed *'amplified intelligence'* (Weibel 1991). The cartographic system or GIS visualization module should provide decision support to the user for his/her cartographic design work. This holds true for both the objectives of map production and knowledge acquisition.

The major objective of this paper is thus to show that the combination of the two key design elements — computational intelligence imbedded in interactive environments — provides a powerful strategy for developing cartographic visualization modules which can adequately handle complex tasks such as generalization. To that end, we will briefly discuss options for cartographic knowledge acquisition, including CI methods. The core of the paper is formed by a presentation of two applications of CI techniques: 1) the use of inductive learning procedures to interpret process tracing output generated from interactive systems, and 2) the application of genetic algorithms for interactive control and optimization of parameters of generalization algorithms.

2 Sources for Knowledge Acquisition

2.1 Types of Knowledge in Generalization

According to Muller (1991) and Armstrong (1991) cartographic knowledge takes three different forms. *Geometrical knowledge* describes the geometry (locations) and distribution of cartographic features. *Structural knowledge* represents the structure of cartographic features in terms of their geomorphological, economic or cultural meaning, and thus relates to the term 'semantic knowledge' used in the KA literature (McGraw and Harbison-Briggs 1989). Apart from general descriptive attributes, measures may be extracted describing the shape and topological relations of features. Finally, *procedural knowledge* is used to select the appropriate generalization operators, algorithms, and parameter settings required to perform a generalization task. It is the knowledge that is needed to control the flow of operations.

2.2 Sources of Cartographic Knowledge

The ultimate source of cartographic knowledge always is the human expert, that is, the cartographer. The paths to this fundamental source may differ, however, and the manifestations of cartographic knowledge may have been sufficiently transformed by intermediate technical steps to justify the distinction of several indirect knowledge sources. Cartographic knowledge can thus be derived from four different sources: 1) human experts, 2) text documents, 3) maps, and 4) process tracing in interactive systems.

Human Experts. Eliciting knowledge directly from human experts (cartographers) would seem the most logical choice for knowledge acquisition, with the least danger of introducing interpretation errors. The classical KA methods used in domains outside cartography include interviewing experts, learning by being told, and learning by observation (McGraw and Harbison-Briggs 1989). Several supporting techniques have been developed to make knowledge elicitation more effective: structured interviews, repertory grids, critical incidents, artificial problems, and querying by ES shells. A recurring problem with knowledge acquisition in cartography is that cartographic knowledge is different. Unlike the primary application domains of knowledge-based systems such as medical diagnosis, systems configuration, taxonomy, or fault diagnosis, which are based on complex, yet rather well-documented knowledge, cartographic design involves a great deal of intuition (particularly with respect to procedural knowledge). What makes cartographic knowledge most special, however, is that it is essentially encoded graphically and thus hard to describe by words. Given the specific situation of generalization, it thus seems natural that the conventional KA methods must be refined and extended for the purposes of generalization.

Text Documents. The analysis of text documents represents another option for knowledge acquisition. While cartographic textbooks offer the general principles, compilation and production guidelines in use at mapping institutions provide a rather extensive potential source of semi-formal knowledge, especially of procedural knowledge. However, the descriptions contained in such documents are often rather vague, incomplete, and fall particularly short of explaining the difficult aspects of cartographic operations. In some cases, they are even kept in a predominantly graphical form, showing illustrations of favorable and unfavorable examples (e.g., SGK 1975). Another problem is that the information is usually declarative or even normative, and may thus not correspond to actual practice.

Maps. As an alternative to analyzing text documents, graphical documents – maps – may be studied. This approach attempts to extract knowledge by identifying the type and degree of modifications that occur to the individual map elements across the scales of a map series (Buttenfield et al. 1991). The strategy has also been dubbed 'reverse engineering' (Weibel 1995), since the process starts with the end product, and attempts to identify the operations that led to this result. The major problem with this knowledge source is the fact that the final map is usually the product of a series of complex and convoluted design operations. Thus, apart from technical problems involved with measuring and tracing the effects of different generalization operators, it is frequently not possible to reliably identify the operations that shaped the end product, and determine their sequence and relation.

Process Tracing in Interactive Systems. Due to the problems involved with other knowledge sources — human experts as the direct source on the one hand, and text documents and maps as indirect sources on the other — several authors have proposed the use of interactive systems for knowledge acquisition (Weibel 1991, McMaster and Mark 1991). Originally, the approach of 'amplified intelligence' was designed as a software strategy for building operational systems for map production (Weibel 1991). However, the basic model can be further exploited by integrating an interaction logging mechanism which allows to trace the interactive generalization process. Thus, this method forms a special instance of process tracing as it occurs in traditional knowledge acquisition (McGraw and Harbison-Briggs 1989). Interactive process tracing was expected to alleviate the deficiencies of both direct knowledge elicitation and indirect analysis of text documents and maps. In contrast to direct elicitation, the interactive system has the potential to act as an 'interpreter' that enables (or even forces) the expert user to express his/her design operations in the 'language' of the digital system (i.e., by selecting appropriate generalization operators and parameter settings). Intuition is thus cast into structured decisions. In comparison to using indirect knowledge sources, the advantage that would accrue is that the expert can always provide further comments and guidance due to the fact that he/she is kept in the loop, and the transformation of knowledge is not completely detached from the expert.

Although this strategy is appealing as a concept, it has only been partially developed so far. McMaster and Mark (1991), for instance, discussed design options for user interfaces for interactive knowledge acquisition. In the next section, we will propose a comprehensive methodology for process tracing by interaction logging and present preliminary experimental results.

2.3 The Potential of Computational Intelligence

According to Carbonell's (1990) widely accepted classification, four major paradigms of computational learning can be distinguished: *inductive learning* (e.g., acquiring knowledge by induction from large sets of positive and negative examples), *analytic learning* (e.g., case-based reasoning by deduction), *genetic algorithms* (e.g., classifier systems), and *connectionist learning methods* (e.g., artificial neural networks). Since the application of CI methods to cartographic tasks is a relatively new research topic, only few publications exist that present a theoretical assessment of the potential of these techniques, and even fewer ones report on actual implementations and experiments.

The theoretical potential of inductive learning as an auxiliary technique for knowledge acquisition in generalization is assessed in Weibel (1995). Keller (1994) reviews the use of case-based reasoning (CBR) to solve generalization problems, and defines the prerequisites and a computational framework for CBR applications. Armstrong (1991, 1993) discusses the possibility of using genetic algorithms (GA) to automatically generate and test generalization alternatives in order to empirically arrive at optimal generalizations. Muller (1992) and Werschlein and Weibel (1994) examine the application of neural networks (NNs) to solve generalization tasks which can be stated as classification problems.

The general conclusion which may be drawn from the study of the available literature is that the greatest potential for CI techniques may accrue if they are applied to rather specific tasks which relate well to their strengths, and if they are used as auxiliary techniques for knowledge acquisition and representation in conjunction with more traditional software paradigms (e.g., algorithms or explicit knowledge systems). The following sections will present two examples of inductive learning and genetic algorithms, respectively, applied to generalization problems.

3 Interaction Generates Knowledge: KA by Process Tracing

3.1 A Methodology for KA by Interactive Process Tracing

A methodology for cartographic knowledge acquisition was designed and implemented which integrates different, to date isolated methods into a comprehensive framework (Reichenbacher, in print). The underlying idea is to log the interactions between an expert user and a generalization system during a working session and later interpret the logs using inductive learning algorithms. In solving the given generalization problem with the help of the operators available in the interactive system, the user is capable of contributing his/her subjective (procedural) knowledge about the generalization process (Figure 1). Shape measures are used to achieve a description of the structure of the features contained in the cartographic database (providing structural knowledge). Thus, procedural knowledge can be traced and related to structural knowledge, allowing one to answer questions such as which generalization operators are selected and which parameter values are specified for a particular operator in relation to scale, feature class, and line complexity.

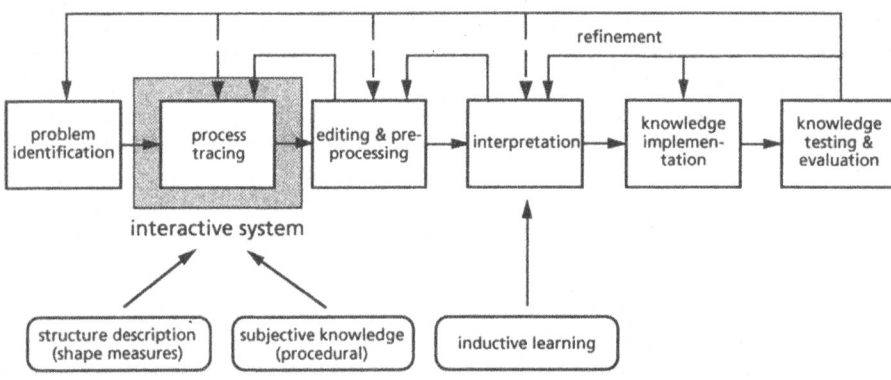

Figure 1: A methodology of knowledge acquisition using process tracing by interaction logging and inductive learning.

The proposed methodology represents an extension to the general model of knowledge acquisition by Buchanan et al. (1983) and consists of six modules (Figure 1). *Problem identification* involves epistemological and knowledge level analysis aiming at structuring the problem domain and defining different types and levels of generalization knowledge on which appropriate KA techniques can be applied. *Process tracing* involves logging of user interactions (e.g., the selection of algorithms and corresponding parameter settings), as well as descriptive measures characterizing the structure of the features processed. *Editing and preprocessing* encompasses tasks such as format conversion, noise removal, and data classification in order to prepare the logged data for the subsequent interpretation step. *Interpretation* makes use of inductive learning algorithms with the objective of deriving decision trees or rule sets. In *knowledge implementation*, the prototype rules derived in the previous step are implemented within the experimental interactive system (or possibly in a different system). Finally, *knowledge testing and evaluation* assesses the quality of the derived knowledge, attempting to identify inconsistencies and rule conflicts. Obviously, this KA methodology does not represent a strict sequence of steps; various possibilities for feedback loops exist (Figure 1).

3.2 An experiment

An experimental study which had the following objectives was carried out: 1) empirically assess the feasibility of the proposed methodology, 2) extract prototype rules, and 3) obtain experience in the application of the techniques involved.

The experiment made use of an interactive generalization system developed in a previous project (Schlegel and Weibel, in print), which in turn employed a commercial GIS as a carrier system. The generalization system was extended by a KA interface to control the interactive session using GUI elements (menus, check boxes, sliders, etc.). The main menu of the interface allows the user to invoke a set of dialogs to define global control parameters (source and target scales, map purpose, feature classes), calculate and display structure measures, browse through feature attribute tables, perform generalization, and turn process tracing on/off (Reichenbacher, in print).

Test data from the French National Mapping Agency (Institut Géographique National – IGN) were used in the experiment. Two data sets were available for the study area located in the region of Valence in the Rhone valley. The first data set represents an extraction from IGN's BD Carto product at a scale of 1:100,000. To keep the experiment at a manageable level of complexity, only the road network was selected. The second data set contains the road network for the same area which was manually generalized to a scale of 1:250,000 and later digitized. In this experiment the manual result served as the target generalization.

Problem Identification. Since the emphasis of this initial study was on a technical feasibility assessment rather than knowledge acquisition, the experimental setup was simple, reducing the need for problem identification. A number of constraining assumptions were made. Firstly, the source data set was presumed to be cartographically correct. Secondly, it had to be possible to perform a generalization using the simplification and smoothing algorithms available in the prototype system (Schlegel and Weibel, in print). Thirdly, as mentioned above, only a single feature class (road network) was involved. Finally, it was assumed that the user generalizes the source data using the manual result as a template (visual backdrop); the task is to try to match this solution applying the tools available in the system. This latter constraint was introduced to simplify the KA process, but would certainly need to be relaxed in future experiments.

Process Tracing. A logging format was implemented to enable process tracing. An interaction is defined as any operation the user performs invoking or responding to a system action. Examples of interactions include the selection of features, choosing a generalization operator and algorithm, specifying parameters values for an algorithm, or defining feature symbology. As Table 1 shows, information is logged to two outputs which are both tables of the relational database system used by the carrier system. Each interaction is written to a record of the 'interaction table', while for each cartographic feature, structural information is stored by adding further attributes to the 'feature attribute table' which already exists as a table of the carrier system's database. The shape measures used in this experiment are taken from Jasinski (1990), who built in turn on the work by McMaster (1986) and Buttenfield (1991).

Interaction Table (stored for each **interaction**)	Feature Attribute Table (stored for each **feature**)
interaction_type (algorithm)	feature_id
user_id	map_purpose
map_id	source_scale
feature_id	target_scale
date	length (line length)
time	coor (number of coordinates)
parameters	anchor (anchor line length)
	seg (segmentation)
	band (bandwidth)
	ev (error variance)
	cv_ev (coefficient of variance)
	lratio (length ratio)
	asl (average segment length)
	cv_asl (coefficient of variance)
	aa (average angularity)

Table 1: Data logged for interactions (relating to procedural knowledge) and features (structural knowledge)

The source data set contains 161 cartographic lines. 52 lines were generalized using the interactive system. Initially, only the Lang algorithm (Lang 1969) was chosen since it uses two parameters (instead of just one like most other algorithms) — a distance tolerance and a look-ahead number — and preserves the character of cartographic lines well (McMaster 1987). For each line, parameters were selected using the GUI sliders so as to best approximate the manual generalization displayed as a visual backdrop. Since the interactive simplification of a line usually represents a trial-and-error process, with several tries rejected until the optimal parameter setting is found, an 'undo' mechanism was implemented. For this purpose an 'interaction cycle' was defined to include all interactions between the selection of a feature to its deselection. Interactions rejected by the undo operator are tagged as invalid in the corresponding record of the interaction table and removed in the subsequent preprocessing step.

Editing and Preprocessing. Before the logged data can be passed to inductive learning algorithms further data editing and reformatting is necessary. First, the interaction table is parsed and the invalid interactions are deleted. The result is a table containing one record for each line (52 in our case). Using the feature_id as a key, the interaction table is then joined with the feature attribute table in order to be able to also access the structural information calculated before (see Table 1). In our experiment, the following attributes were subsequently written to a text file (Table 2).

t	n	coor	length	anchor	lratio	band
seg	ev	cv_ev	asl	cv_asl	aa	road_class

Table 2: Attributes for the interpretation with machine learning. (t = distance tolerance; n = number of points to look ahead). Road_class represents the road classification as shown in the topographic map.

Since only a single algorithm was used for a single map data set, the attributes interaction, map_purpose, source_scale and target_scale were not included as they remain constant. The inductive learning algorithms used in this experiment require input in the form of a list of symbolic descriptions of an example. The structural data, however, were measured on a metric scale. The scale of measurement was therefore transformed by cluster analysis, resulting in two or three classes per attribute with symbolic labels such as high, low, medium, few, etc. The file was then formatted to the final list structure in a text processor.

Interpretation. This step was performed using public domain inductive learning tools. These tools running under Macintosh Common LISP represent implementations of ID3 (Quinlan 1986), AQ15 (Hong et al. 1986), and PRISM (Cendrowska 1987), which have the advantage of producing easy-to-read production rules. The goal of this step is to organize the knowledge contained in the unstructured interactions. Considering the fact that the number of interactions logged in a session quickly reaches several hundreds, an automated approach is certainly justified.

Our experiments initially focused on deriving prototype rules for the selection of the distance parameter t of the Lang algorithm. The conditional part of the rules then contains structural attributes. Two examples out of the 10 rules generated by the AQ15 algorithm are shown here.

Rule 1:

```
(IF
   ((COOR FEW) and (LRATIO MEDIUM) and (CV_EV LOW))
or
   ((ASL LONG) and (CV_ASL HIGH) and (CV_EV
MEDIUM))
   THEN (CLASS T8))
```

Rule 2:

```
(IF
   ((LRATIO BIG)) or
   ((COOR MEDIUM) and (ANCHOR MEDIUM) and (CV_EV
MEDIUM)) or
   ((SEG MEDIUM) and (EV LOW) and (LRATIO MEDIUM))
   THEN (CLASS T16))
```

The target CLASS into which the algorithm tries to classify the data represents the value for the distance tolerance of the Lang algorithm (denoted by T#). Rule 1 states that the distance parameter is equal to class 8 if the line has only few coordinates, the length ratio (the ratio of the line length and the anchor line length) is medium, and the coefficient of variation of the error variance is low. Furthermore, the same distance tolerance applies if the average segment length is long, the coefficient of variation of the average segment length is high, and the coefficient of variation of the error variance is medium. Rule 2 can be read analogously. It should also be noted that the symbolic values (high, medium, long, etc.) which may seem somewhat fuzzy could easily be constrained by numeric upper and lower bounds of the corresponding class and range checked by the inference engine of the knowledge-based system which serves for implementation of the prototype rules. Likewise, the number of classes for each attribute could be increased to achieve greater resolution.

Knowledge Implementation, Knowledge Testing and Evaluation. The two final steps of the proposed methodology — knowledge implementation, and knowledge testing and evaluation — were not carried out yet in this experiment. Thus, the prototype rules have neither been verified nor falsified. It should be noted, however, that the prototype knowledge extracted by inductive learning does not necessarily need to be implemented in the form of production rules. In the context of systems such as the experimental platform used here, an algorithmic implementation via adaptive defaults to be retrieved from a lookup table according to the structural measures of cartographic features may be more appropriate.

3.3 Discussion

The prototype rules generated in this simple experiment are certainly not yet very sophisticated. Yet, the experiment has allowed to clearly show the technical feasibility of this novel approach. In order to be useful for the purposes of knowledge acquisition, further research must be carried out with the dual objective of achieving technical improvement as well as running more sophisticated KA experiments.

In technical terms, the major critical factors constraining the methodology are the interactive generalization system and the methods for achieving the structural descriptions of map features. The generalization system used here is still rather simple. It would need to be extended by further generalization operators based on more powerful data models. Additionally, more research on user interfaces for generalization is required. With respect to structural feature description, better shape measures must be developed and implemented. For instance, the measures by Buttenfield (1991) used in this study appear to be biased by the fact that they are based on the 'anchor line' (the line connecting the endnodes of a line), which represents a poor approximation of the general trend of the cartographic line. The measures developed by Plazanet (1995) seem much more promising. Finally, structure description should also be enriched by contextual information (e.g., topology, neighboring features, feature clustering, etc.).

In terms of KA experiments, future research needs to relax the above constraining assumptions and increase the range of generalization operators and feature classes under study. Great care must be taken to design meaningful experiments in collaboration with expert cartographers. Additionally, future KA studies must involve significant numbers of cartographers. Under such terms, it will eventually be feasible to address the issues of knowledge implementation as well as testing and evaluation.

4 CI Supports Interaction: 'Generalization by Example' for Interactive Parameter Control

In an interactive cartographic system, there are usually various generalization operators with different characteristics available. Given this choice, it is difficult even for the expert user to select the right algorithms and the right parameter settings to solve a given generalization problem. Current systems offer little help to the user in terms of good defaults. Further limitations may be caused by the available user interaction techniques. If the parameter settings have to be entered purely numerically, it is difficult for the user to translate numbers to graphical results. Yet, even if parameter control via sliders with 'real-time' generalization is available, the result can only be definitely assessed if displayed at target scale.

The objective here was therefore to design a new user interaction technique that is supported by CI methods and allows the user to more intuitively control line generalization algorithms and associated parameters. We termed this user interaction technique 'generalization by example'. The user interactively draws, for a representative stretch of linework, a generalized line at target scale, called the 'sample line' (Figure 2). This manual generalization is then used to derive the parameters for a previously selected algorithm which would best match the sample line. For this parameter optimization, genetic algorithms (GA) are used. Once the control parameters for the generalization operator have been established by the GA and accepted by the user, they can be expanded to the entire window or layer. As a side effect, KA functions can be hooked into this process to trace the actions that occur (e.g., the critical points that are selected on a line) and log the data in light of further interpretation, thus complementing the KA methodology discussed in the previous section.

4.1 Implementation

The proposed interaction technique is implemented in five steps. User interaction is involved in the first step (drawing the sample line) and the last step (accept/reject result). In this early phase of the project, it is further assumed that the user has previously selected an algorithm, thus restricting the task to finding appropriate parameters for this algorithm. In the example shown in Figure 2 the algorithm by Douglas and Peucker (1973) was used. Following is an outline of 'generalization by example' for line generalization:

1) A sample line, which should be a small but representative part of a selected cartographic line, is drawn interactively by the user to indicate the implicit geometric properties such as the sinuosity of the intended generalization output (Fig. 2).

2) The part on the original line between the endnodes of the sample line is taken as input for the line generalization algorithm.

3) The parameter value(s) which, using the given generalization algorithm, would need to be applied to generate the sample line from the original line are determined by an optimization procedure using GA (see below).

4) Using the parameter setting determined in step 3, the entire selected original line is generalized and displayed at target scale.

5) The output is being presented to the user who can make his/her decision: accept, modify, or reject and backtrack. If the parameter setting is not rejected, it is expanded to a larger selection: the entire feature class, the window, or an arbitrary set of selected lines.

Genetic algorithms were chosen for several reasons as a method for searching the parameter space for a suitable solution. Firstly, they are well documented and particularly suited for parameter optimization (Goldberg 1989). GA are a particular instance of the broader class of genetic or evolutionary strategies. Common to evolutionary strategies is the simulation of evolution of individuals by processes of selection, mutation and reproduction. Individuals are evaluated by their performance (or fitness) in the collective environment. GA apply a parallel search using general heuristics for exploration in a population of individuals. Thus, GA lend themselves to parallelization. They provide near-optimal solutions which are close to the global optimum at a very high probability. Unlike many other optimization techniques, GA are not based on constraining assumptions such as continuity or derivability of the function to be optimized. They are also capable of scaling up to larger problems.

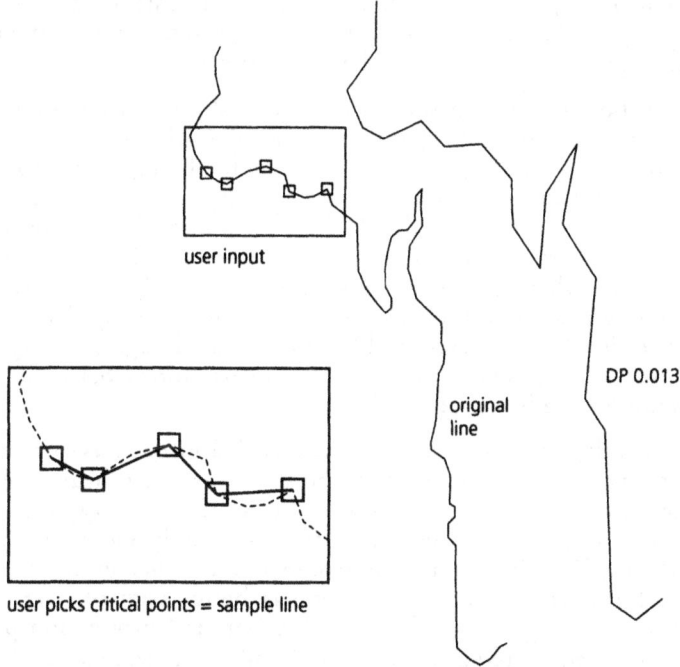

Figure 2: 'Generalization by example'. The inset shows how the user draws the 'sample line' (squares). Optimization by GA finds the parameter value (0.013 map units) which is appropriate to generate the sample line from the original line using the Douglas-Peucker algorithm. The entire selected line is then generalized using the determined parameter setting.

In our case, 'individuals' passed to the GA are equivalent to alternative parameter values for the selected generalization algorithm (e.g., the perpendicular distance tolerance for the Douglas-Peucker algorithm in Fig. 2). An initial set of individuals can be generated randomly or taken from a lookup table if previous knowledge exists. The 'fitness' of these parameters is evaluated as the degree of displacement between the sample line and the line that would result when applying the generalization algorithm with the given parameter. This displacement should be minimized in order to achieve a good match between the two lines. It is expressed by the vector and area displacement measures of McMaster (1987). Figure 2 illustrates the interaction technique and shows a sample result for a line taken from the IGN road data base also used in the experiments discussed in the previous section.

4.2 Discussion

The above study represents an attempt at exploring the options for user interaction in cartographic systems and finding adequate levels of interaction. On the one hand, the objective is to investigate the design and use of systems which support users at their level of understanding. On the other hand, it is hoped that at the same time more insights can be gathered about the cognitive processes involved in the interactions.

The proposed approach was termed 'generalization by example' in analogy to 'learning by example' (which is part of CBR); it also relates to the classification in psychology called 'learning by being told'. Expert users often have a good understanding of the necessary 'coarseness' and degree of generalization for the target scale. With the proposed interaction technique, users give the system a 'clue' of their expectations by sketching a prototypical sample line which in turn helps to establish adequate settings. This approach implements key HCI elements like 'direct manipulation', or 'user access and feedback', and helps establishing a 'shared cognitive responsibility' between user and system (Turk 1990): humans thinks better in graphical terms, computers are better at numbers.

First results indicate that the proposed approach is promising. An obvious technical limitation of GA, however, is their heavy demand on computing resources leading to slow response times on typical platforms. Hardware performance, on the other hand, is steadily increasing and at the same time parallel computing environments are becoming increasingly affordable. Furthermore, since the user has better control over parameter specification, time is saved in the overall design and production process by minimizing or even avoiding the usual trial-and-error sequences.

Several improvements to the method as shown here are possible. The performance increase that could be achieved by parallelization should be studied. The approach could also be extended to propose more than one parameter setting for one algorithm, for instance, if the user enters multiple sample lines. Related to this point is the problem of handling heterogeneous data sets (e.g., road data for both mountainous and lowland areas) where multiple algorithms and parameter settings would seem a necessity. Finally, the approach could be extended by allowing more than one algorithm, with the objective of determining an optimal combination of algorithm *and* associated parameters. For the current as well as for future extended versions of the interaction technique, more systematic experiments would also need to be run in order to reach a more conclusive assessment of their potential.

As a long-term perspective, it should be attempted to extend the current 'generalization by example' approach towards a 'generate-and-test' strategy. In other words, the potential of GA to generate alternatives and assess their fitness with respect to a given evaluation function should be more fully exploited. The system would then attempt to find an appropriate solution automatically by an evolutionary process rather than trying to match the input provided by the user. Such a strategy, however, would increase the demands on criteria, measures, and functions for evaluation. While in the current approach, the user informally shows what a good result looks like, this information would need to be formalized and parameterized in order to be built into the evaluation function used by the GA. As was discussed in the previous section, however, research on structural measures for cartographic features and the evaluation of cartographic quality still has a long way to go (Muller et al. 1995, Weibel 1995, Plazanet 1995).

5   Conclusions

We have presented an overview of options to overcome the current bottleneck of knowledge acquisition and representation in generalization, with a focus on two concrete experimental studies. The novel elements of the work reported here include a comprehensive methodology of KA by process tracing in interactive systems, as well as the use of genetic algorithms for control and optimization of parameters of generalization algorithms. On the conceptual level, we have provided strategies for integrating CI methods with interactive systems.

The experiments carried out in this work may be quite simple, and the results that could be achieved thus far are rather preliminary. However, we are convinced that the research presented here shows sufficiently well how CI methods can greatly contribute to improving the present situation of carto-graphic generalization. In general, we see the best potential for CI techniques if they are imbedded in interactive systems, and designed to solve rather specific tasks in concert with other software techniques, most notably algo-rithmic tools and explicit rule bases. In such a context, CI methods provide possibilities to generate prototype explicit knowledge, support the human decision making process by offering new options for interaction, or perhaps even replace algorithmic generalization operators.

Methods of computational intelligence provide a technological source which has largely remained untapped in GIS, and particularly in digital car-tography. Future research therefore should address a variety of issues. Firstly, a range of initial experiments such as the above must be carried out to test the ground for the application of CI techniques, followed by more pro-found studies designed to extend and reinforce proposed methodologies such as the model for KA in interactive systems. Also, the issue of how precisely CI-based tools are best imbedded in the HCI loop will require further attention.

Probably the most important research problem, and thus the key to further progress in the use of CI techniques in generalization is, however, the development of effective measures and methods for evaluating the quality of generalization results. Such procedures would be needed, for instance, as functions to evaluate the fitness of alternatives generated in GA, or associate descriptive information with generalization operations recorded in interactive process tracing. An underlying problem which must be solved in order to develop appropriate evaluation methods is the definition of accurate criteria of what constitutes a good map and/or generalization. As was discussed above, the recent evolution of generalization research indicates a renewed strong interest in this topic as a prerequisite to advancing the field. We are therefore confident that a more solid theoretical and technical foundation for the application of CI methods in generalization will soon be developed.

Acknowledgments

We would like to thank the team of the Laboratoire COGIT at the Institut Géographique National, Saint-Mandé (France) for providing the data used in our experiments as well as for fruitful discussions with the first author during his stay at the lab in April 1994 and repeatedly thereafter. We are also indebted to Regula Ehrliholzer for her assistance in preparing the figures, and to Caroline Westort for reading an earlier version of the manuscript.

References

Armstrong, M.P. (1991): Knowledge Classification and Organization. In: Buttenfield, B.P., and McMaster, R.B. (eds.): *Map Generalization — Making Rules for Knowledge Representation*. London: Longman, 86-102.

Armstrong, M.P. (1993): A Coarse-Grained Asynchronous Parallel Approach to the Generation and Evaluation of Map Generalization Alternatives. *NCGIA Specialist Meeting on Formalizing Cartographic Knowledge*, Buffalo, 23.-27.10. 1993, 37-43.

Brassel, K.E., and Weibel, R. (1988). A Review and Framework of Automated Map Generalization. *Int. Journal of Geographical Information Systems*, 2(3): 229-44.

Buchanan, B., Barstow, D., Bechtal, R., Bennett, J., Clancey, W., Kulikowski, C., Mitchell, T., and Waterman, D. (1983): Constructing an Expert System. In: Hayes-Roth, F., Waterman, D., and Lenat, D. (eds.): *Building Expert Systems*. Reading, MA: Addison-Wesley.

Buttenfield, B.P. (1991): A Rule for Describing Line Feature Geometry. In: Buttenfield, B.P. and R.B. McMaster (1991): *Map Generalization - Making Rules for Knowledge Representation*. London: Longman UK: 1150-171.

Buttenfield, B.P., Weber, C.R., Leitner, M., et al. (1991): How Does a Cartographic Object Behave? Computer inventory of topographic maps. *GIS/LIS '91*, 2: 15-104.

Carbonell, J.G. (1990): Introduction: Paradigms for Machine Learning. In: Carbonell, J. (ed.): *Machine Learning: Paradigms and Methods*. Cambridge, MA: MIT Press.

Cendrowska, J. (1987): PRISM: An Algorithm for Inducing Modular Rules. *International Journal of Man-Machine Studies*, 27(2,3,4).

Douglas, D.H., and Peucker, Th.K. (1973): Algorithms for the Reduction of the Number of Points Required to Represent a Digitized Line or its Caricature. *The Canadian Cartographer*, 10(2): 112-122.

Goldberg, D.E. (1989): *Genetic Algorithms in Search, Optimization, and Machine Learning*. Reading, MA: Addison-Wesley.

Hong, J., Michalski, R.S., and Mozetic, I. (1986): AQ15: Incremental Learning of Attribute-Based Descriptions from Examples, the Method and User's Guide. *Reports of the Intelligent Systems Group*, Department of Computer Science, University of Illinois at Urbana-Champaign, ISG-86-5, 57 pgs.

Jasinski, M.J. (1990): The Comparision of Complexity Measures for Cartographic Lines. *National Center for Geographic Information and Analysis, Technical Paper*, 90-1, 73 pgs.

Keller, S. (1994): On the Use of Case-Based Reasoning in Generalization. *Sixth International Symposium on Spatial Data Handling*, Edinburgh, 5.-9.9. 1994, 2: 1118-1132.

Lang, T. (1969): Rules for the Robot Draughtsmen. *The Geographical Magazine*, 42(1): 50-51.

McGraw, K.L., and Harbison-Briggs, K. (1989): *Knowledge Acquisition: Principles and Guidelines*. Englewood Cliffs, NJ: Prentice Hall.

McMaster, R.B., and Mark, D.M. (1991): The Design of a Graphical User Interface for Knowledge Acquisition in Cartographic Generalization. *GIS/LIS '91*, 1: 311-320.

McMaster, R.B. (1986): A Statistical Analysis of Mathematical Measures of Linear Simplification. *The American Cartographer*, 13(2): 103-116.

McMaster, R.B. (1987): The Geometric Properties of Numerical Generalization. *Geographical Analysis*, 19(4): 330-346.

Muller, J.-C. (1991): Generalization of Spatial Databases. In: Maguire, D.J., Goodchild, M.F., and Rhind, D.W. (eds.): *Geographical Information Systems: Principles and Applications*. London: Longman, 1: 457-475.

Muller, J.-C. (1992): Parallel Distributed Processing: An Application to Geographic Feature Selection. *Proc. Fifth International Symposium on Spatial Data Handling*, Charleston, SC, 1: 230-240.

Muller, J.-C., Weibel, R., Lagrange, J.-P., and Salgé, F. (1995): Generalisation: State of the Art and Issues. In: Muller, J-C., Lagrange, J.-P., and Weibel,

R. (eds.): *GIS and Generalization: Methodological and Practical Issues.* London: Taylor & Francis.

Nickerson, B.G. (1988): Automated Cartographic Generalization for Linear Features. *Cartographica,* 25(3), 15-66.

Plazanet, C. (1995): Measurements, Characterization, and Classification for Automated Line Feature Generalization. *Auto-Carto 12,* Charlotte, NC, 27.2.-1.3. 1995, 59-58.

Quinlan, R. (1986): Induction of Decision Trees. *Machine Learning,* 1(1): 81-106.

Reichenbacher, T. (in print): Knowledge Acquisition in Map Generalization Using Interactive Systems and Machine Learning. *17th International Cartographic Congress of the ICA,* Barcelona, 3.-9.9. 1995.

Ruas, A., and Lagrange, J.-P. (1995): Data and Knowledge Modelling. In: Muller, J-C., Lagrange, J.-P., and Weibel, R. (eds.): *GIS and Generalization: Methodological and Practical Issues.* London: Taylor & Francis.

Schlegel, A., and Weibel, R. (in print): Extending a General-Purpose GIS for Computer-Assisted Generalization. *17th International Cartographic Congress of the ICA,* Barcelona, 3.-9.9. 1995.

SGK (Schweizerische Gesellschaft für Kartographie)(1975): Kartographische Generalisierung — Topographische Karten. *Kartographische Schriftenreihe,* Schweizerische Gesellschaft für Kartographie, 1.

Turk, A.G. (1990): Towards an Understanding of Human-Computer Interaction Aspects of Geographic Information Systems. *Cartography,* 19(1): 31-60.

Weibel, R. (1991): Amplified Intelligence and Rule-Based Systems. In: Buttenfield, B.P., and McMaster, R.B. (eds.): *Map Generalization — Making Rules for Knowledge Representation.* London: Longman, 172-186.

Weibel, R. (1995): Three Essential Building Blocks for Automated Generalisation. In: Muller, J-C., Lagrange, J.-P., and Weibel, R. (eds.): *GIS and Generalization: Methodological and Practical Issues.* London: Taylor & Francis.

Werschlein, Th., and Weibel, R. (1994): Use of Neural Networks in Line Generalization. *EGIS '94,* Paris, 30.3.-1.4. 1994, 76-85.

Spectral Representations of Linear Features
for Generalisation

Emmanuel Fritsch and Jean Philippe Lagrange

Laboratoire Cogit
Institut de Géographie Nationale
2 av Pasteur - 94160 St. Mandé - France
fritsch@cogit.ign.fr

Abstract: In this paper we propose the use of new representations of linear features in order to make up for the weakness of classical generalisation algorithms. Such representations are developed from spectral tools: Fourier series and wavelet decomposition. The theoretical possibilities of representations are discussed from a generalisation point of view. Algorithms are built on these representations. The experiments are presented and discussed, particularly the encouraging results leading to caricature.

1 Introduction

Generalisation is the complex process that consists in deriving a 'smaller' and 'simplified' from a set of spatial data, with regard to some specifications, such as a new semantic resolution and a possibly legibility and acale constraints [Weger, 93], [Lagrange & Ruas, 94]. This definition does not distinguish cartographic from GIS-related generalisation. In this paper, we focus on cartographic generalisation of linear features.

The observation of manual generalisation shows that it may be decomposed into basic operations [McMaster, 89]. For linear feature generalisation, we distinguish filtering, smoothing, selective elimination and enhancement. Displacement is not considered here as it may be seen as being a subsequent operation.

The current algorithms used for linear generalisation, for example [Douglas & Peucker, 73], [Opheim, 82], provide more of a filtering than a smoothing or an enlargement, and no selective elimination. They produce some undesirable results, especially for high-scale reduction ratios. The drawbacks may consist in: an unaesthetic angularity, a loss of important details and, more importantly, topology degradation (see fig. 1).

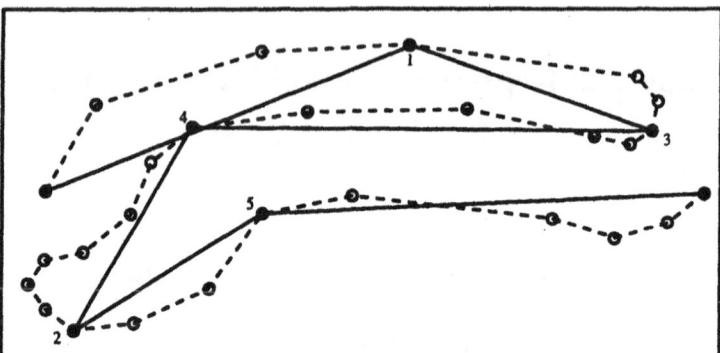

Fig. 1. An example of ill performance of the Douglas & Peucker algorithm. The dashed line is the original line, while the thicker line is the result of the Douglas & Peucker algorithm, with a 25% compression rate.

The main cause of such poor results can be found in the use of lists of coordinate pairs which provide a segment-by-segment representation of a line. Such a representation is not really helpful in generalisation because relevant information needed for generalisation is global and lists of coordinate pairs do not render it.

Other representations have been developed [Freeman, 74], [Hough, 62] but few with the purpose of generalisation [Affholder 93]. In order to achieve an effective generalisation process, we should use a representation which is better at revealing the abstracted information of the curve than a set of coordinates of vertices. Indeed, in order to generalise a given detail by enlargement or smoothing, we need a representation which does account for this detail. So, alternative suitable representations have to perform an abstraction of information. The curvature constitutes a basic representation of the curve in which the bends are well detected. In this way, it performs the first abstraction of information. Other mathematical tools that can be used for generalisation are spectral decompositions. Fourier series have already been used in cartography [Clarke et al, 93], but often more for modelling than for generalisation purposes.

Therefore we want to devise new representations, which fit in better with the specific goals of generalisation. Then, on those representations, we want to develop algorithms which will generalise the information as revealed by the new representation. This approach leads to the generalisation process illustrated in fig 2.

The first section of this paper is devoted to the quality criteria that should be met by every new representation. These criteria are developed in accordance with the specific purposes of generalisation. We will then show how spectral representations of the curvature should meet the quality criteria. Finally, we will examine two spectral representations, Fourier series and wavelet decomposition, applied on the curvature of a line and, the associated algorithms.

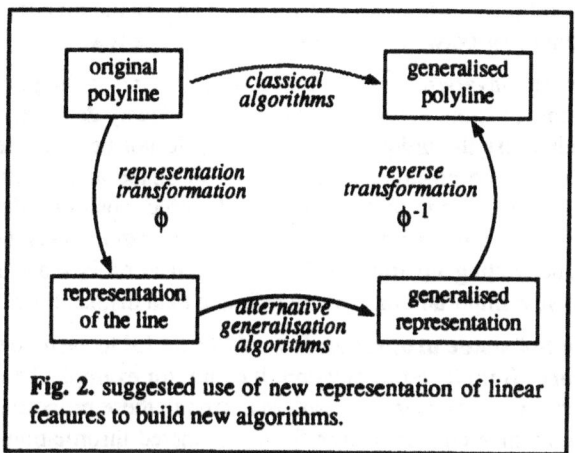

Fig. 2. suggested use of new representation of linear features to build new algorithms.

2 Quality Criteria

Some shape representation criteria have been listed and discussed in [Mackworth & Mokhtarian, 92]: Invariance, uniqueness, stability, efficiency, ease of implementation and computation of shape properties. Although these criteria were defined for computational vision, but not for generalisation, they are relevant to our purpose. Nevertheless, a seventh criterion has to be defined to take into account the specificity of generalisation: reversibility. These criteria need to be specified more precisely:

- **Uniqueness:** It is an intrinsic property of a representation. We want the representation transformation to realise a bijection between the planar curves space and the representation space.

- **Reversibility:** Contrary to computational vision, which aims only at providing information on a curve, the goal of generalisation is to build a new curve deduced from the original one. We therefore need a reverse transformation in order to go back to the "curve domain" once the desired transformation has been applied. We know this reverse transformation exists, once the uniqueness criterion is verified, and we merely want it to be computable.

- **Stability:** From a mathematical point of view, stability is a continuity property of transformation. In fact, we have to consider the continuities of both transformations: The direct and the reverse one. Indeed, since we are working in the hilbertian space of planar curves, whose dimension is infinite, the continuity of direct transformation does not imply the continuity of the reverse one. In fact, the most important continuity for generalisation should be the continuity of reverse transformation, which ensures similar representations to provide similar curves. Without this continuity, there is no guarantee that our generalised curve resembles the target curve. The direct property does not seem so important, but is useful in reducing the complexity of the representation. For generalisation purposes, we have to keep the end points of curves unchanged to ensure the maintenance of possible existing connexions betwenn curves, for example in a road network.

- **Invariance**: The invariance to rotation, translation and scaling is a basic criterion in computational vision. A representation, as previously defined, cannot be totally invariant, otherwise the uniqueness criteria would not be respected. But a simili-invariance should be ensured, by separation of shape and position information. Translations, rotations and scaling should only affect position information, and let shape information remain invariant. We must distinguish between displacement invariance and scaling invariance: Since generalisation is scale-dependent, scale invariance is less important or even undesirable than displacement invariance

- **Computation of shape properties**: It would be useful to easily detect some properties of a curve from its representation. It could, for example, be a symmetry or a regular repetition of characteristic patterns. Among these properties, we must consider the ones which constitute significant abstracted information that we have to preserve in the process of generalisation.

- **Efficiency**: This criterion is not crucial, as for the time being we do not have to overcome constraints in real time. Nevertheless, we want a low-order polynomial complexity in memory space, and, if possible, in time.

- **Ease of implementation**: The simpler a representation, the easier the construction of the generalisation algorithm and computation. Moreover, for the sake of stability, we will prefer a low number of complex numerical operations that can introduce approximation errors.

3 Fourier Series

3.1 Fourier series on parametrisation of curve

To determine the processing a detail needs to undergo in the course of generalisation the first criterion is its size; the global shape has to be kept, small details have to be smoothed and intermediate details, whose size is approximately equivalent to the resolution of the targeted map, have to be enlarged to remain readable. This first trivial separation can be done by use of decomposition in Fourier series. Indeed, the Fourier series representation consists in a set of coefficients, each one measuring the extent of a spatial frequency, whose inverse is the wave length of the details you want to detect [Bony, 91], [Hörmander, 83]. We can then apply different operations according to the frequencies: no change of lower frequencies, cuting off higher frequencies and amplification of intermediate frequencies in order to perform some enlargements.

A classical result guarantees that in the hilbertian space I of integrable complex functions of period T, each element f can be decomposed into:

$$f(x) = \sum_{k \in Z} c_k \cdot e_k(x)$$

with e_k the trigonometric functions of period T/k, and:

$$c_k = \left\langle f | e_k \right\rangle = \int_0^T f(t)\, e_k(-t)\, dt$$

The periodicity condition on a linear feature is not actually met, except for closed curves. Nevertheless, Fourier series can be applied by periodisation of the functions: this can apply to every function defined on a limited interval.

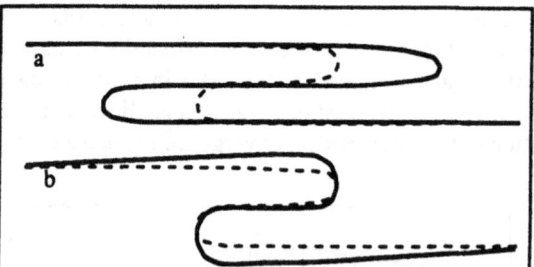

Fig. 3-a: An amplification on the Fourier series of polyline parametrisation. 3-b: The expected result. In both cases the single line is the generalised polyline and the dashed line is the original one. Amplification does not work in the right direction.

At first we decided to apply this method on a parametrisation of polylines [Fritsch 94]. We simply used a uniform parametrisation, stabilised at the extremities by a constant function. The approximation of the curve produced by means of this representation was quite good, but the results on enlargements were very bad. For instance the bends were not enlarged, as we hoped, but elongated in the direction of the bend axis (see fig. 3). This drawback is easy to understand. The increase of Fourier coefficient affects the amplitudes of the sinuosities and not their widths. From a theoretical point of view, the invariance criterion defined above was not ensured. Indeed, the values of the coefficients c_k are changed by rotation and translation of the curve.

3.2 Fourier series on curvature

To overcome this faults, we decide to apply the Fourier series not directly on a parametrisation of the curve but on another function built on this parametrisation. [Boutoura, 89] used such an altenative, with the slope of the curve. For above-explained reasons, we decided to use a function invariant by rotation and translation: The curvature. In fact, it has already been used elsewhere for representing lines [Mackworth & Mokhtarian, 86], [Thapa 88], [Wuescher & Boyer, 91]. The bends and sinuosity appear clearly through curvature as well as fast changes of direction, and invariance to rotation and translation is guaranteed. To ensure the uniqueness of the representation we add the end points of the line, in order to retain locational information.

Then the difficulty lies in the computation of the curvature along a polyline: polylines are not regular enough to define a curvature everywhere since they are not differentiable at their vertices and show a nil curvature between the vertices.

To solve this problem there are some classical techniques, such as Splines or Bezier curves, but the resulting representations are not well-suited for computing Fourier Series so that we decided to use a simpler but adequate method. As a matter of fact, it is possible to define an extension of the notion of curvature, applicable to the polyline. This extension is a distribution defined as being the limit of a sequence of curvature functions which are defined on regular curves tending towards the polyline. This curvature distribution is in fact the sum of the Dirac-functions centred on the curvilinear abscissa of the vertices, and weighted by the value in radian of the direction variation at the respective vertices. Some analytical theorems allow us to build Fourier series on every distribution but the numerical convergence in such a case is very bad.

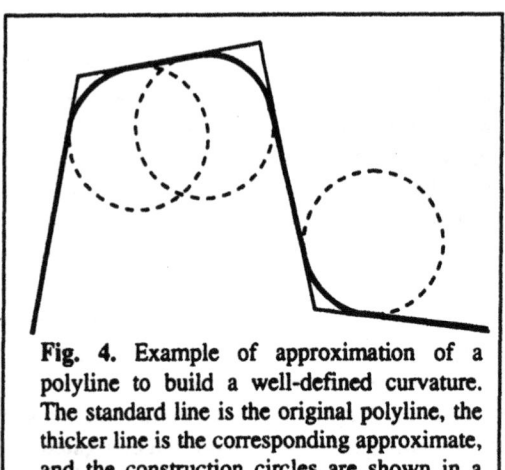

Fig. 4. Example of approximation of a polyline to build a well-defined curvature. The standard line is the original polyline, the thicker line is the corresponding approximate, and the construction circles are shown in a dashed line.

That is why we use a regular approximate of the curve, by smoothing the vertices with circle arcs (see fig. 4). We force the radius of the circles to be smaller than the resolution value in order to guarantee a good approximation and naturally, the contact points on both sides of the vertices are differentiable. In this way, the resulting curvature is a well-defined, piecewise constant function. Thus, the Fourier series coefficients are defined explicit formule.

The reverse transformation is easily computationable. For, the curvature is the derivative of the angular orientation of the curve tangent:

$$\alpha(t) - \alpha(0) = \int_0^t \rho(t)\, dt$$

if c_k are the coefficients of the Fourier series of the curvature, the coefficients of the curve orientation are given by:

$$d_k = \frac{c_k}{ik\pi}$$

For every point, let $\overline{u}(t)$ be the unit tangent vector.

The coordinates of each point of the curve are given by:

$$\overline{M_0 M_t} = \int_0^t \overline{u}(t)dt.\frac{\overline{M_0 M_1}}{\int_0^t \overline{u}(t)dt}$$

This final normalisation introduces locational information and ensures the stability of the end points.

The representation looks quite well; fig. 6 shows that a polyline of 140 points is well approximated with only 200 coefficients. By increasing the number of coefficients the difference between the two curves disappears. So the representation is quite stable. The increase of data volume is neglectable compared to other representations. For instance, the Curvature Scale Space Image (CSSI) used by [Mackworth & Mokhtarian, 86] produces usually 100x200 pixel-representations.

The compliance of this representation by Fourier series with the seven criteria above-defined is actually good:

- **Uniqueness, reversibility and invariance** are inherent to the construction process.

- **Stability** -i.e. the continuity of the reverse transformation- needs to be examined in the direct and reverse case:
  — Continuity of direct transformation:

  This property is false in a general case, but if we take into account only curves of limited curvature, continuity can be reliable upon. In fact, this limitation confirms the pertinence of approximation of vertices by arcs of circles. Since direct continuity is not very important, the demonstration is omitted here for the sake of conciseness.

  — Continuity of reverse transformation:

  It is more important, and we demonstrate it in the following.

let $(\Delta c_n)_{n \in Z}, \Delta\rho, \Delta\alpha, \Delta\overline{u}_t, \Delta\overline{M}_t$ be respectively the variation of $(c_n)_{n \in Z}, \rho, \alpha, \overline{u}_t, M_t$

We know from the Pythagoras theorem that:

$$\| (\Delta c_n)_{n \in Z} \|_2^2 = \sum_{n \in Z} |\Delta c_n|^2 = \int_0^1 |\Delta\rho(t)|^2 dt = \| \Delta\rho \|_2^2$$

We have already seen that: $\forall t \in \Re,\ \alpha(t) - \alpha(0) = \int_0^t \rho(t)\, dt$

therefore: $\forall t \in \Re,\ |\Delta\alpha(t)| = |\int_0^t \Delta\rho(t)\,dt| < \|\Delta\rho\|_2 = \|\Delta(c_n)_{n\in Z}\|_2$

then:

$\forall \epsilon \in \Re_+, \forall t \in \Re,\ \|\Delta(c_n)_{n\in Z}\|_2 < \epsilon \Rightarrow \forall t \in \Re,\ |\Delta\alpha(t)| < \epsilon$

$\Rightarrow \forall t \in \Re,\ \|\Delta\overline{u}(t)\| < \epsilon$

with $\overline{M_0 M_t} = \int_0^t \overline{u(t)}dt. \dfrac{\overline{M_0 M_1}}{\int_0^1 \overline{u(t)}dt}$ and L the length of the curve:

if ϵ is chosen so that $\epsilon < \frac{1}{2}\,min\,(\,1,\,\|\overline{M_0 M_1}\|\,)$

then: $\forall t \in \Re,\ \|\Delta\overline{M_t}\| < |\int_0^1 \Delta\overline{u}(t)dt.\,L| + |\int_0^1 \overline{u}(t)dt.\,L(1 - \dfrac{1}{\int_0^1 \overline{u}(t)dt})|$

$< L(\epsilon + \dfrac{|\int_0^1 \overline{u}(t)dt - 1|}{|\int_0^1 \overline{u}(t)dt|}) < L(\epsilon + \dfrac{\epsilon}{1-\epsilon})$

$< 3L\epsilon$

Finally:

$\forall \epsilon \in \Re_+,\ \|\Delta(c_n)_{n\in Z}\|_2 < \epsilon \Rightarrow \forall t \in \Re,\ \|\Delta M_t\| < 3L\epsilon$

which demonstrates continuity of reverse transformation.

- **Computation of shape properties**: The representation was built to detect the repetition of patterns. Other properties like symmetry could be detected, but nothing has been done yet to turn these properties to good account.

- **Ease of implementation and efficiency**: Since we apply the representation on piecewise constant functions, the corresponding Fourier coefficients are given by an explicit formula, and we do not need an integration algorithm to calculate them; this property allows to strictly monitor the errors. Indeed, "errors" will come only from floating point truncation, and the omission of high frequencies.

3.3 Algorithms

We have envisaged three kinds of algorithms on the representation:

- Filters on frequencies, i.e. transformation of each frequency independently from the others.

- Linear application in the frequency space (which physicists call linear filtering !)

- Non-linear methods, such as neural networks.

The first kind can be tried in an empirical way. For the second and the third one, a great deal of indetermination and complexity oblige us to use heuristical methods, such as machine learning or statistical approaches which require a very large data set.

Moreover, they cannot be implemented without a good knowledge of the representation used in order to simplify the heuristic research.

This is why we implemented so far only the first kind, filters on frequencies. This processing consists in multiplying the frequency series with functions classically called transfer functions.

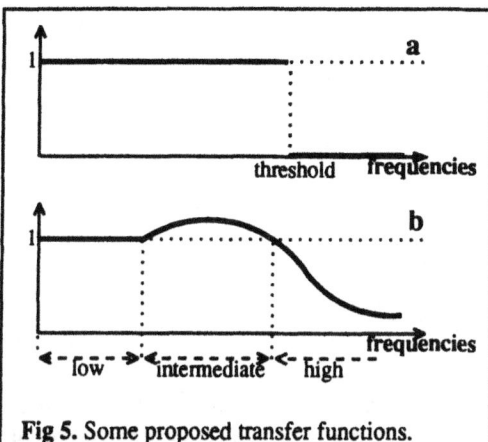

Fig 5. Some proposed transfer functions.
5-a: a truncation of high frequencies. 5-b: a more sophisticated filter, which should fit better with generalisation purposes.

We wanted these filters to keep low frequencies unchanged, to amplify intermediate frequencies, and to erase high frequencies. A transfer function corresponding to this purpose is presented in fig. 5-b.

In fact, no filter goes much beyond the truncation in frequency representation and the amplification of intermediate frequencies does not produce any positive effects. Whatever filter we used, some general trends have been observed (fig 6):

- Small details are well-omitted. This is not surprising, as the volume of information is reduced by the truncation.

- the smoothing of angularity is quite good. Whereas most generalisation algorithms provide too angular results, this one provide well-smoothed curves. In return, there are too many points on the generalised curves, but this fault can be diminished by using a compression algorithm.

- The enlargement of bends is the first step to caricature. This property, which does not depend on any amplification of coefficients, is most welcome. The fewer frequencies kept, the more enlarged the details.

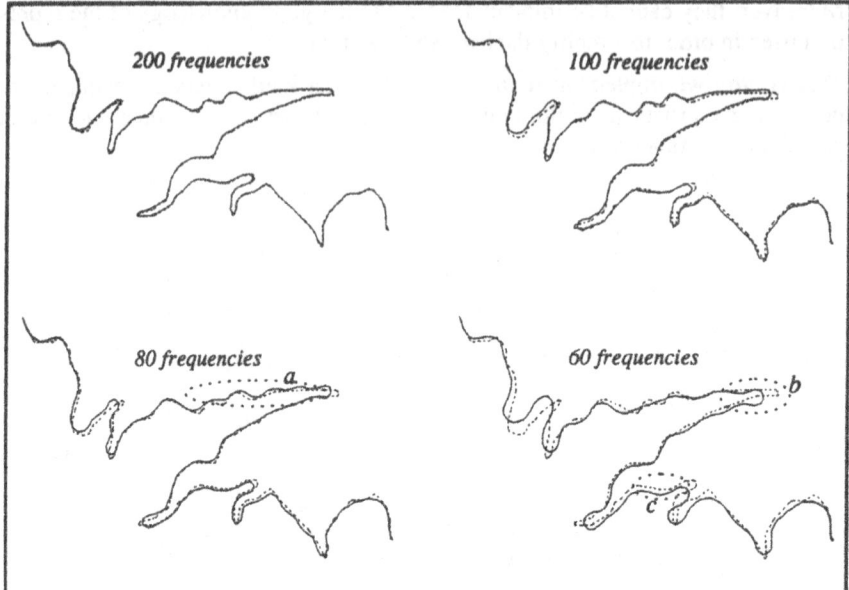

Fig. 6. Truncation of a series of frequencies with 200, 100, 80, and 60 frequencies succesively. The standard line is the generalised line, the dashed line is the original one. Enlargement and shortening of bends appear clearly. An example of interference effects (a) appears on the 80 frequencies curve. The location drift (b) and the resistance of low amplitude large details (c) are shown on the 60 frequencies curve.

But:

- For strong truncations, smoothing is too marked and topological faults may appear.
- Some intermediate details show a strong resistance to smoothing
- a drift in the location of details often appears.
- We observe some resonance effects; a wave length of a detail we want to remove reappears in the neighbourhood of the original location of the detail.

The first theoretical explanation of these drawbacks seems to be a lack of locational information in the representation. Indeed, each coefficient of the Fourier series shows the extent of its frequency on the entire curve.

4 Wavelets

4.1 Wavelet representation

To reduce this lack of locational information we decided to use another spectral method, wavelet decomposition which takes the locational information into account.

The mainspring of the method is the same as for the Fourier series but with another base. The wavelet set $(g_{i,j})$ is constructed by translating and rescaling a function g, called mother wavelet [Gasquet & Witomski, 90]:

$$\forall (i,j) \in Z^2, \quad g_{i,j}(x) = 2^{\frac{-i}{2}} g(2^i x + j)$$

Under some conditions on g, the set $(g_{i,j})$ is a Hilbertian base of I, so that we can write as for the Fourier series:

$$f = \sum_{(i,j) \in Z^2} c_{i,j} \cdot g_{i,j}$$

with:

$$c_{i,j} = \left\langle f | g_{i,j} \right\rangle = \int_{-\infty}^{\infty} f(t) \cdot g_{i,j}(-t) dt$$

but now the integration domain is not limited.

Many mother wavelets have been proposed in the literature, but the first and simplest one, the Haar function (fig 7), is the most relevant one for us. Indeed, this function is generally criticized for its discontinuity but we apply the wavelet decomposition on the previously described approximated curvature. Such a piecewise constant function is well matched by the wavelet base built on the Haar function which is piecewise constant too.

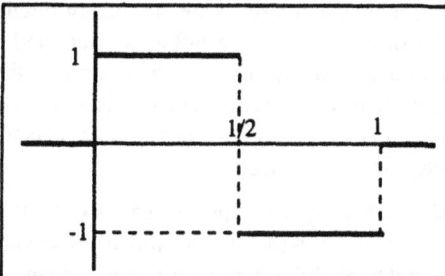

Fig. 7. The simplest wavelet, called Haar wavelet. It is usually not applied because of its irregularity: It leads to slow convergence on regular functions. But our functions are not regular, so that more complicated wavelets are not useful.

Then, for this mother wavelet, and for a function whose support is in [0,1], when aggregating some coefficients, the decomposition reads:

$$f = c + \sum_{i \in N} \sum_{i=0}^{2^i-1} c_{i,j} \cdot g_{i,j}$$

with $c_{i,j}$ defined as above, and c:

$$c = \int_0^1 f(t)dt)$$

Therefore, the choice of the Haar wavelet has some advantages:

- We have summarized in coefficient c all the coefficients $c_{i,j}$ with i<0

- Computation of coefficients c and $c_{i,j}$ is explicit. As mentionned above in the Fourier series, we avoid numerical intégration which could give rise to numerical errors

- Each coefficient corresponds to a closed and limited interval.

The considerations and the demonstrations on uniqueness, reversibility, invariance and stability on Fourier series can be extended to wavelet decomposition. Only the computation of shape properties and efficiency need special commentary. To compensate for the increase in locational information, the detection of pattern repetition is likely to be less efficient. But the assessment of the method must come first before any experimentation.

4.2 Algorithms

The first attempts were very simple and we merely tried an elementary filter: On each frequency, we apply a selective algorithm which keeps the coefficient unchanged if its absolute value is above a threshold which depends on the frequency, and if not, reduce the coefficient to zero. Such an algorithm is particularly expected to erase the low amplitude details among the intermediate frequency details and to enlarge the others. After several adjustments of the thresholds, we obtained moderately satisfying results (fig 8). When compared to the Fourier representation, there is indeed a selective answer on the amplitude of the intermediate details and the resonance effects disappear as expected. These two encouraging tendences are offset by the persistence of the locational drifts which are not reduced.

These contrasted results are easy to explain: The locational information in our wavelet representation only deals with the curvilinear abscissa of details along the curve but not the coordinates of these details in the euclidean plane. That is why interference effects are reduced by wavelet algorithms but not the locational drift, which needs more absolute Euclidean spatial information.

We have to develop the ability of the filter, for instance, with a maximum curvature threshold to help in producing a caricature. But nothing ensures a reduction of the locational drifts: The lack of spatial information implies a modification on the representation. Therefore, additional spatial constraints need to be developed which would force the generalised curve to remain close to the original one. These spatial constraints have to be loose, otherwise the generalised curve will just become a copy of the original one. The smallest drifts have not to be corrected as they are naturally relevant to the normal process of generalisation, whereas the higher the distance between the two curves, the greater the need to correct it. So we have to use a constraint increasing with the distance between the two curves, as a flexible force in order to preserve the previous spectral processing. In this way, information will be treated in two ways: spatial and spectral.

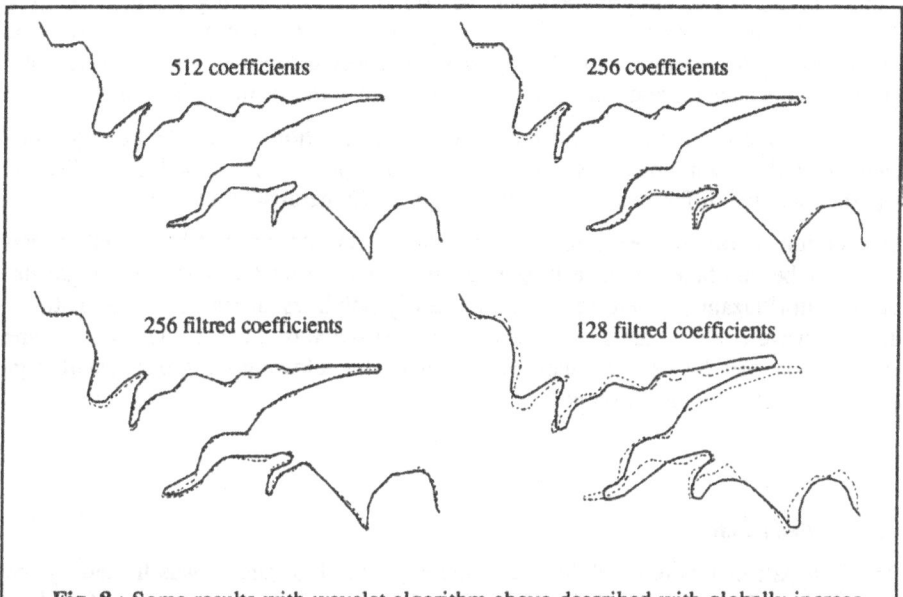

512 coefficients   256 coefficients

256 filtred coefficients   128 filtred coefficients

Fig. 8.: Some results with wavelet algorithm above-described with globally increasing thresholds. The thresholds have been chosen interactively. The comparison with fig. 6. reveals the vanishing of interference effects, and of the resistance to generalisation of large details of low amplitude. But the locational drifts still persist.

Conclusion

We have developed two frequency-based representations of linear features, and shown that from a mathematical point of view, they satisfy the quality criteria defined above. These representations allow to develop new algorithms which can be described as global algorithms [McMaster, 91], and which perform smoothing and enlargment. Although both representations have some limits, their fidelity and the general trend to caricature indicate that the above-explained methods and criteria are quite pertinent to developing generalisation tools. Furthermore, the encouraging results of both representations confirm the relevance of the curvature in the description of a curve for generalisation purposes.

Now we have to develop more sophisticated algorithms on representations. Separately, the improvements resulting from wavelet decomposition on Fourier series and the persistence of locational drifts prompt us to think that additional locational information should still improve the possibility of representation. We are currently elaborating some adjusting constraints opposed to the locational drift in the curve reconstruction process. Such constraints modelled using a readjustment force from the original polyline to the generalised one.

After amplification and smoothing operators we have to develop elimination tools to complete the list of simple operations needed in linear feature generalisation. The typical elimination case consists in omitting some bends in a sequence of close and tight bends. This operation should appear in the spectral representation as a decrease of the frequencies representing a set of bends which could be implemented.

These operations -amplification, smoothing and elimination- could then be used in a general tool, which would segment the line and apply the developed generalisation operators on the different sections, according to the generalisation needs.

In addition preliminary experiments have shown that the wavelet-based representation can be combined with inflection point determination (with the technique described in [Plazanet, 95]) to segment and qualify polylines, at least for road features. Indeed, wavelet coefficients characterise the local shape of a curve and such a shape information can be used to segment polylines by allowing to group small segments.delimited by inflection points.

Acknowledgments

The first part of this research has been started as the first author was hosted by the department of geography of the Univesity of Zürich, under supervision of Robert Weibel. The authors wish to thank Jean Georges Affholder, Corinne Plazanet and Jean François Hangouët for their helpful commentaries, and Ywonna M'Kenzie for correcting the language.

References

[Affholder, 93] J.G. Affholder, 1993. "Road modelling for generalisation". NCGIA Initiative 8. Spec. Meet. Buffalo.

[Bony, 91] J.M. Bony, 1991. "Cours d'Analyse". Ecole Polytechnique.

[Boutoura, 89] C. Boutoura, 1989. "Line generalisation using spectral techniques". Cartographica Vol 26, Nos 3&4, pp 33-48

[Clarke et al, 93] K.C. Clarke, R. Cippoletti and G. Olsen, 1993. "Empirical comparison of two line enhancement methods". Auto-Carto11 Proceedings pp 72-81

[Douglas & Peucker, 73] Douglas and Peucker, 1973. "Algorithms for reduction of the number of points required to represent a digitized line or its caricature". The Canadian Cartographer 10/2.

[Freeman, 74] H. Freeman, 1974. "Computer processing of line drawing images". Comput surveys, vol 6.

[Fritsch, 94] E. Fritsch, 1994. "Recherche d'outils et de representation pour la generalisation". Rapport de stage de DEA. ENSG-IGN.

[Gasquet & Witomski 90] C. Gasquet and P Witomski, 1990. "Analyse de Fourier et applications: Filtrage, calcul numerique et ondelettes". Ed Masson.

[Hörmander, 83] L. Hörmander, 1983. "The Analysis of linear partial differential operators". Springer-Verlag.

[Hough, 62] P.V.C Hough, 1962. "Method and means for recognizing complex pattern". U.S. patent 3069654.

[Lagrange & Ruas, 94] J.P Lagrange and A. Ruas, 1994. "Geographic information modelling: GIS and generalisation". proc of SDH'94 vol2 pp 1099-1117.

[Mackworth & Mokhtarian, 86] F Mokhtarian and A.K. Mackworth, 1986. "Scale-based description and recognition of planar curves and two-dimensional shapes". IEEE Transactions on Pattern Analysis and Machine Intelligence. vol 8 pp 34-43

[Mackworth & Mokhtarian, 92] F Mokhtarian and A.K. Mackworth, 1992. "A theory of multi-scale, curvature based shape representation for planar curves". IEEE Transactions on Pattern Analysis and Machine Intelligence. vol 12 pp 789-605.

[McMaster, 89] R.B. McMaster, 1989. "The integration of simplification and smoothing algorithms in line generalization". Cartographica 26. pp 101-121.

[McMaster, 91] R.B McMaster, 1991. "A conceptutual frameworks for geographic knowledge". Map Generalisation. McMaster & Buttenfield Editors. Longman Sci.entific & Technical.

[Opheim, 82] H Opheim, 1982. "Fast data reduction of a digitized curve". Geo-Processing. pp 33-40

[Plazanet, 95] C. Plazanet, 1995. "Measurement, characterisation and classification for automated line feature generalisation". Auto-carto12, technical papers, vol 4, pp 59-68

[Thapa, 88] K. Thapa, 1988. "Automatic line generalisation using zero-crossing". Photogrammetric engineering and remote sensing. vol 54-4, p 511-517.

[Weger, 93] G.Weger, 1993. "Cours de Cartographie". ENSG-IGN

[Wuescher & Boyer, 91] D.M Wuescher and K.L. Boyer, 1991. "Robust contour Decomposition using a constant curvature criterion". IEEE Transactions on Pattern Analysis and Machine Intelligence. vol 13 pp 41-51.

A Triangulated Spatial Model for Cartographic Generalisation of Areal Objects

J. Mark Ware, Christopher B. Jones and Geraint Ll. Bundy
Department of Computer Studies,
The University of Glamorgan, Pontypridd,
Mid Glamorgan CF37 1DL, UK.
email: jmware@glam.ac.uk

Abstract

Cartographic generalisation involves interaction between individual operators concerned with processes such as object elimination, detail reduction, amalgamation, typification and displacement. Effective automation of these processes requires a means of maintaining knowledge of the spatial relationships between map objects in order to ensure that constraints of topology and of proximity are obeyed in the course of the individual generalisation transformations. Triangulated spatial models, based on the constrained Delaunay triangulation, have proven to be of particular value in representing the proximal and topological relations between map objects and hence in performing many of the essential tasks of fully automated cartographic generalisation. These include the identification of nearby objects; determination of the structure of space between nearby objects; execution of boundary simplification, merge and collapse operations; and the detection and resolution, by displacement, of topological inconsistencies arising from individual operators. In this paper we focus on the use of a triangulated model for operations specific to execution of merge operations between areal objects. The model is exploited to identify the regions of space between nearby objects and to execute merge operations in which the triangulation is used variously to adopt intervening space and to move adjacent rectangular objects to touch each other. Methods for updating the triangulation are described.

1 Introduction

Cartographic generalisation is concerned with the presentation of spatial information at varying levels of abstraction that are constrained by the purpose of communication and by the scale of the required graphic. It is a process that is essential to map making and is one that has to date eluded successful automation. When map data are presented at different scales in a geographical information system the display is in general based on geometric components obtained by digitising existing maps that have been generalised manually. Changes in the map scale are usually achieved by selecting relevant components and plotting them at a geometrically variable, but not cartographically variable, scale.

Cartographic generalisation involves the orchestration of several individual processes that include elimination of map features, line, area and surface detail reduction, amalgamation, typification (or caricature), exaggeration, reduction in dimension, from areal to line or point objects, and displacement (see Shea and McMaster, [1989] for a more detailed typology of operations). Several, but not all, of these individual processes have been automated with various degrees of success [Lagrange, 1993], but despite considerable discussion of, largely, rule-based methods for controlling the processes and the interactions between them [Buttenfield and McMaster 1991], little progress has been made in the very important aspect of map generalisation, referred to by Brassel and Weibel [1988] as process control. The emphasis on automating individual operators, as opposed to overall process control, is reflected in some commercial generalisation software, which is based on the assumption of user interactive control of individual operators [Lee, 1992].

Successful automation of process control will depend upon automating the human visual capacity for recognising contexts and localised situations that require appropriate action to resolve conflicts resulting from the individual operators. Operations such as boundary simplification (line detail reduction), exaggeration and displacement can result in loss of integrity of the map due to change in the original topological relations between map objects. Thus line simplification may result in a line crossing from one side to the other of a point-referenced object; areal object exaggeration may result in overlap with a nearby object; while displacement of one object away from another, in order to retain appropriate graphic separation, may result in the object moving too close to or overlapping another object. In some situations, no amount of shuffling objects around will resolve conflicts, in which the case the solution may be either to eliminate certain objects or to merge nearby objects of the same or similar classification.

In this paper we address the problem of automatically maintaining knowledge of the interaction between areal map objects, for purposes of areal object generalisation, through the use of a triangulated spatial model which is adapted to the storage and computation of data on proximal relations between map objects. The simplicial data structure SDS [Bundy, 1995a,b] is based on a constrained Delaunay triangulation of polygonal objects and of the space in which they are embedded. By the nature of Delaunay triangulation, nearby vertices will be connected to each other and the triangulation can therefore provide an explicit record of such connectivity [Preparata and Shamos, 1988]. In a constrained Delaunay triangulation, nearby vertices will not be connected if there is an intervening edge that is the boundary of an object, in which case the closest vertices to the edge will be connected to one or both of the bounding vertices of the constraining edge.

The SDS maintains explicit data on relationships between objects, triangles, edges and vertices. It has several important benefits for the purpose of map generalisation. By maintaining connectivity between neighbouring connected and disjoint objects it enables rapid determination of the nearest neighbouring objects to any given object and calculations of the distances to each of these neighbours [Jones, 1995a]. This is essential in generalisation in obeying constraints on the minimum separation distances between neighbouring map objects and in identifying candidate objects for amalgamation. If nearby objects are to be merged, the SDS provides immediate access to information on the parts of two or more objects that are adjacent to each other and the regions of space between the objects. Amalgamation of disjoint neighbouring objects may then be performed variously by adopting the intervening region, or regions, or by moving the objects together across the intervening region. When component objects of the SDS are transformed, such as by enlargement, displacement or boundary simplification, resulting topological conflicts between objects can also be readily identified due to the occurrence of singularities in the triangulation, expressed by the presence of (conceptual) folds in the triangulation. Such situations create inversions in individual triangles, in the sense that the edge order becomes reversed. The SDS provides further benefits in resolving such conflicts, as the shape of affected triangles can be used to derive appropriate displacement vectors, if displacement were to be chosen as the appropriate action.

The SDS is exploited by process control software which employs a frame-based reasoning paradigm. Research to date has concentrated particularly on the design and implementation of the SDS in order to create a set of primary operators that will facilitate a wide range of generalisation operations and their control through the frame-based reasoning system. The purpose of this paper is to illustrate the application of the SDS to the specific problem of areal object amalgamation and to describe procedures for maintaining the integrity of the SDS data structure, and hence of the associated model.

In what follows we provide a brief review of the current state of automated cartographic generalisation in section 2, before describing the basic properties of the SDS in section 3. Section 4 describes the application of the SDS to the specific generalisation process of areal object amalgamation, also referred to here as the merge process. We distinguish between several types of merge operation and show how the SDS is applied to each. In section 5 we focus on the issue of maintenance of the SDS consequent upon the execution of individual SDS operations. The paper concludes in section 6.

2 Previous Research in Automated Map Generalisation

Most of the research efforts in automated generalisation have in the past concentrated on the specific problem of line generalisation in which, typically, constituent defining vertices of the line are selectively eliminated [McMaster, 1987]. A distinction has been made between local band processing, in which points are selected or eliminated on the basis of local context, and global band processing, typified by the Douglas-Peucker algorithm in which points are selected in a recursive manner, on the basis of their being extreme or critical points [Douglas and Peucker, 1972]. Thapa [1988] has promoted the use of curvature based techniques, derived from image processing, in which criticality is determined on the basis of maximum change in curvature. Of the relatively limited number of studies of individual areal object detail reduction, notable work by Lichtner [1979] addressed problems specific to buildings, while Weber [1982] initiated a series of studies of the use of expand and contract operators for raster data. Most work on surface detail reduction has focussed on the selection of critical vertices in triangulated irregular networks [Heller, 1990; Lee 1991] and with smoothing gridded surfaces. Weibel's work was notable for attaching importance to structural features of surfaces, as well as performing smoothing operations [Weibel, 1992].

Relatively little research has been conducted on amalgamation or merging of areal features. Some of the earlier techniques were usually raster-based involving expansion and contraction [Weber, 1982; Monmonier, 1983] and were analogous to the mathematical morphology operations in image processing of dilation and erosion. Schylberg [1993] has continued the raster-based approach. A vector mode version of the expansion and contraction method was presented by Muller and Wang [1992], who experimented with maps of groups of islands, combining the merge operation with elimination governed by the criterion of area.

In the process of typification, detail of a particular character is replaced in the generalisation by a larger scale representation which, though lacking locational accuracy, communicates the distinguishing character of the real-world phenomenon. There appears to have been very little research addressed specifically to this task.

Dimensionality reduction, also referred to as collapse, is usually applied in cartographic generalisation to representations of ribbon-shaped phenomena such as rivers and roads, which are to be reduced to a centre line representation. Areal phenomena such as entire cities may be reduced to a single point, which is then used as a locational reference for a symbol. Collapse to a line is equivalent to the medial axis transformation [Duda and Hart, 1973; Kirkpatrick, 1979]. Interest in this transformation in cartography has largely focussed on its use in finding the representative trend of areas for purposes of name placement [Freeman, 1984]. Chithambaram [1991] presented a technique for producing the skeleton, for cartographic purposes, by means of the Delaunay triangulation of a polygon.

Techniques for displacement of buildings relative to adjacent roads were presented by Lichtner [1979 and Leberl [1986]. Notable progress in displacement of representations of natural phenomena such as rivers was made by Nickerson and Freeman [1986] and by Nickerson [1988] who demonstrated techniques involving the derivation of locally calculated displacement vectors that differentially displaced parts of the same object.

3 The Simplicial Data Structure (SDS)

This section provides a description of the SDS, introducing details of its primary components (objects, triangles, edges and vertices), how they relate to each other, and how the SDS is applied to the modelling of a map for the purpose of generalisation. In particular, consideration is given to a number of implicit proximity relationships inherent in the SDS, and the application of these relationships in the derivation of information useful to the task of automated generalisation.

3.1 SDS Definition

The SDS is made up of four primary entities - objects, triangles, edges and vertices - which are combined to form a spatial model M. This model is used to represent a collection of planar polygonal objects O and free space regions F lying between, or contained within, objects. The entire space within M is described by a constrained Delaunay triangulation T, in which the boundary edges of all objects O act as constraints. Constraining edges within T are referred to a real edges, while all other edges (ie. those edges lying within objects or lying in free space) are referred to as virtual edges.

Each object belonging to O and each free space region belonging to F is defined in terms of references to those triangles of T lying within its boundary. In turn, each triangle of T references its three constituent edges and the object or free space region to which it belongs. In addition, each triangle also stores three Boolean values indicating the direction (clockwise or anti-clockwise) in which each of its edges lie relative to the triangle. Triangles belonging to objects are referred to as object triangles, while those belonging to free space regions are referred to as free space triangles. Each edge is described by references to its two defining vertices, plus adjacency information in the form of references to the two triangles to which it belongs. In the case of an edge lying on the boundary of M, one of these adjacent triangle references will be a NULL pointer. Each edge also stores a Boolean flag denoting if it is a real or virtual edge. Vertices are defined by references to coordinate information in the form of x and y values.

3.2 Proximity Definitions

If two objects o1 and o2 share a common edge or vertex then o1 is said to be contiguous with o2, and vice-versa. The extent of contiguity between objects can be expressed by use of the terms edge contiguity and vertex contiguity. The contiguity relationship is also applied to the other SDS primitive entities, and to mixtures of entity types. For example, two triangles sharing a common vertex are termed vertex contiguous triangles, while an object and triangle sharing a common edge are said to be edge contiguous.

If two objects are not contiguous, but there exists a free space path between them, then they are said to be proximal. Furthermore, two objects that are minimally separated by a single triangle are said to be proximal-1 to each other.

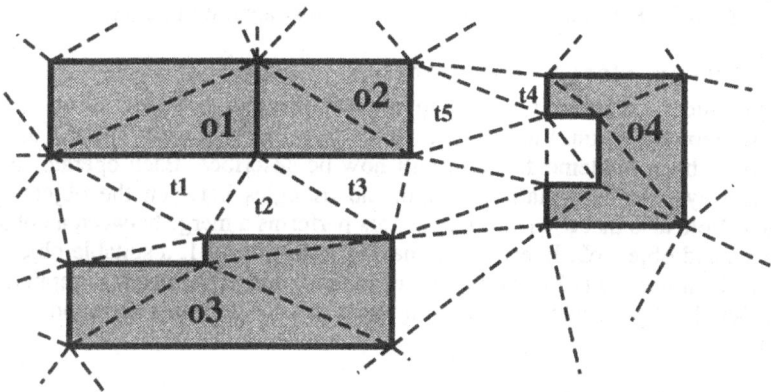

Fig. 1. An example of some SDS proximity relationships. We see that o1 is edge contiguous with o2, o3 is proximal-1 to both o1 and o2, t1 is edge contiguous with o1 and point contiguous with o2 and o3, and t1 and t2 (among others) are edge contiguous triangles. Triangles t1 and t2 connect o1 and o3, while t2 and t3 connect o2 and o3. Triangles t4 and t5 represent the space intervening between o4 and o2.

The set of triangles C connecting two objects o1 and o2 is made up of those triangles that are both edge contiguous with o1 and vertex contiguous with o2, plus those triangles that are both edge contiguous with o2 and vertex contiguous with o1.

We now describe a process for retrieving the list of free space triangles intervening between two objects. Defining the space S lying between two discrete objects is a rather subjective process, as is identifying the free space triangles intervening between o1 and o2. The method described here is therefore not purported to be an absolute solution to the problem, but rather as a minimal solution in that S will always consist of intervening triangles, but will not always include all triangles possibly perceived to be intervening. In order to simplify our explanation we introduce two additional terms. Firstly, the *halo* of an object o_i is defined as the list of triangles that are vertex contiguous with o_i but which lie outside o_i. Secondly, the *enclave* triangles of o_i are triangles that lie in concavities of o_i. These triangles are characterised by having each of their vertices connected to o_i, and by being external to o_i. Collections of edge contiguous enclave triangles are referred to as enclave regions.

The process of finding the region intervening between two objects o1 and o2 begins by finding the halo of each object, and then proceeds by retrieving those triangles which exist in both halo lists. These common connecting triangles are referred to as the intersection list. Certain triangles, characterised by having no real edges and having only one adjacent triangle belonging to the intersection list, are deemed not to be intervening between o1 and o2, and are therefore deleted from the intersection list. However, additional intervening triangles are found by identifying all enclave regions of o1 and o2 that are edge contiguous with the intersection list. The triangles of such regions are added to those belonging to the intersection list, the result being stored in the intervening list.

A number of SDS proximity relationships are illustrated in Figure 1.

4 Prototype Merge Operators

The graphical combination of map objects through merging is an essential generalisation operation. Various types of merge operator, making specific use of the SDS, have been implemented, and will now be described. Each operator uses, in various ways, the information held in the triangles between the objects being merged. Note also that each of the operators performs a merge between an object o1 and a second object o2. In each case, having been given o1, a suitable object o2 is identified automatically by making use of the proximity relationships implicit in the SDS. Details of procedures for deriving these relationships can be found in [Jones, 1995a].

4.1 Append Merge Operator

This is conceptually the most straightforward of the merge operations and consists of amalgamating two edge contiguous objects o1 and o2 to make a new object o3. Merging takes place along the edges common to o1 and o2. The operation begins by identifying these common edges, which are all real. Next, the new object o3 is created (which initially references no triangles), to which the triangle references of o1 and triangles references of o2 are added. Following this, o1 and o2 are deleted from the SDS. This is followed by the re-attribution of any real edges internal to o3 (ie. those edges which were common to o1 and o2) by making them virtual. Finally, redundant data, which now may be present in the form of collinear points, are removed and the internal edges of o3 are optimised (see sections 5.2 and 5.3).

4.2 Direct Merge Operator

Two types of merge operator have been implemented solely for use on rectangular objects such as buildings. The first of these is the direct merge operator, which performs the amalgamation of two proximal-1 objects o1 and o2. The merge, illustrated in Figures 2 and 3, involves the following operations : displacing o1 towards o2 so as to make the two objects meet; rotating o1 to align it with o2; creating a new object o3 made up of the displaced and rotated triangles of o1 and the triangles of o2; and finally deleting o1 and o2 from the SDS.

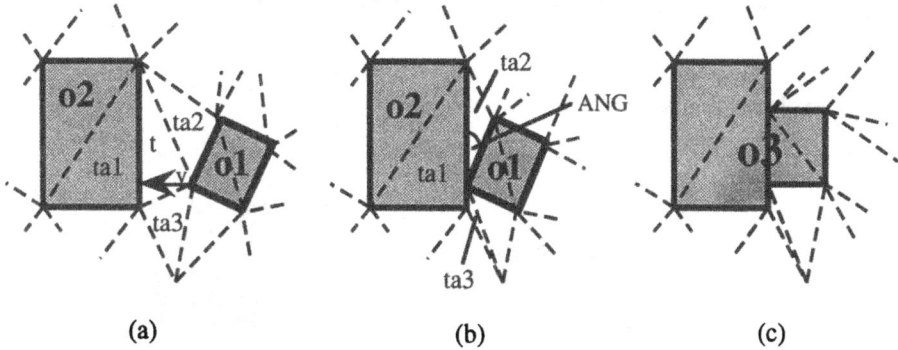

(a)   (b)   (c)

Fig. 2. Displacement of o1 towards o2, followed by the rotation of o1, resulting in a new object, o3.

The displacement of o1 (Figure 2a) is achieved by applying a vector v to the component vertices of o1. This vector is derived from the set of triangles C which connect o1 and o2 (see section 3.2). The derivation involves identifying the triangle t belonging to C with smallest altitude A_t. Altitude is measured with respect to each triangle's real edge. The vector v is then assigned a magnitude equal to A_t and a direction that is perpendicular to t's real edge and positive relative to o1.

Subsequent to the displacement of o1 towards o2 it will sometimes be necessary to align o1 and o2 (Figure 2b). It is noted that t will always have one adjacent triangle ta1 belonging to either o1 or o2, and its two other adjacent triangles ta2 and ta3 belonging to the free space region lying between o1 and o2. After displacing o1, t will have become flat (ie. it will have zero area), and in some cases ta2 and, or, ta3 will also have become flat. If neither ta2 or ta3 are flat then the merge operation continues by rotating o1 in such a way as to make one of them flat, resulting in o1 becoming better aligned with o2. The rotation angle ANG is set equal to the angle that the flattened triangle makes between o1 and o2. The triangle to be flattened is chosen as the triangle having a real edge belonging to either o1 or o2. In the event of both ta2 and ta3 meeting this criterion then the triangle which would result in the least rotation of o1 is chosen.

Having carried out any required rotation of o1, the new object o3 is created and added to the SDS, while o1 and o2 are deleted from the SDS. The object o3 is, at this stage, made of those triangles previously belonging to o1 and o2 (Figure 2c).

The final stage of the direct merge operation involves the re-attribution of certain of o3's edges and update of the SDS in and around o3. This process, which involves re-triangulation, is illustrated in Figure 3. Firstly, all flat triangles are deleted from the SDS (Figure 3a, b & c). This is followed by the deletion of all triangles belonging to o3, as well as all o3's internal edges (Figure 3d). Thirdly, the edges e1 and e2, which have become redundant, are deleted, and the virtual edges e3 and e4 are re-attributed as real (Figure 3d & e). Next, o3 is re-triangulated, using a process similar to that described in section 5.1, resulting in a new list of triangles belonging to o3 (Figure 3f). Finally, the free space region surrounding o3 is checked for overlapping triangles. If any are found then the region is re-triangulated (see section 5.1), otherwise the region is optimised (see section 5.3) (Figure 3g).

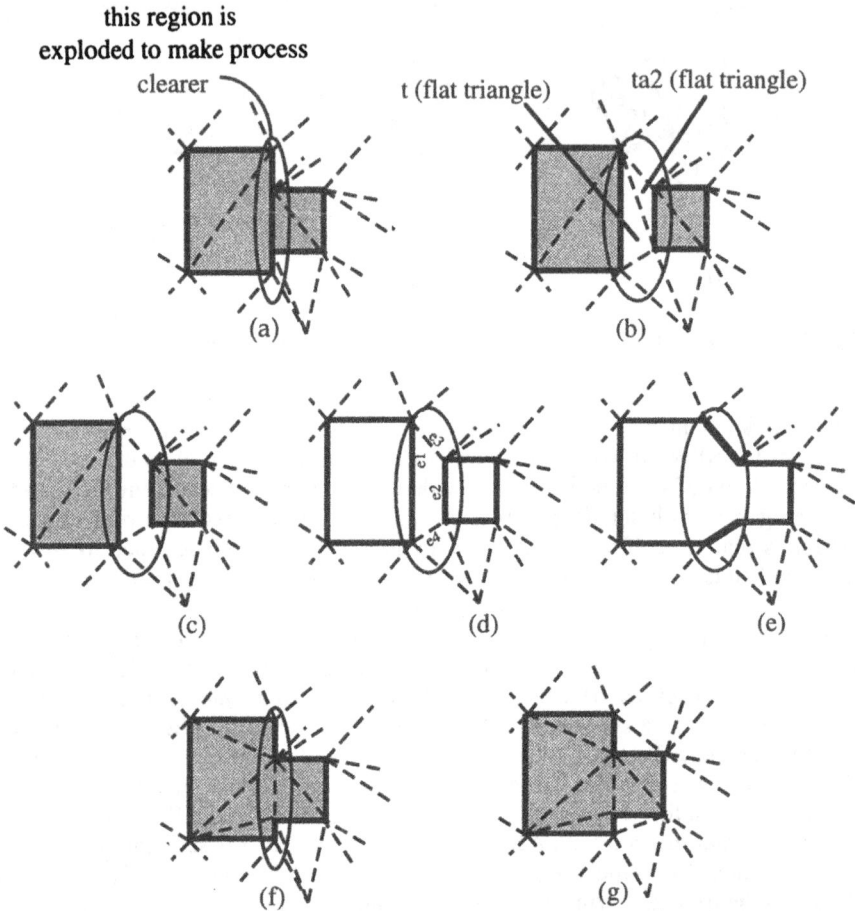

Fig. 3. Re-attribution of edges of o3 and SDS update (see text).

4.3 Snap Merge Operator

The second type of merge operator designed specifically for application to rectangular objects is the snap merge operator. This operator differs only slightly from the direct merge, in that whereas the latter involved initially displacing an object o1 directly towards a second object o2, the former displaces o1 in such a way as to align the nearest vertices of o1 and o2 (Figure 4a & b). This is achieved by aligning the displacement vector with the shortest connecting edge between o1 and o2. The remainder of the snap merge operation follows that of the direct merge (Figure 4c).

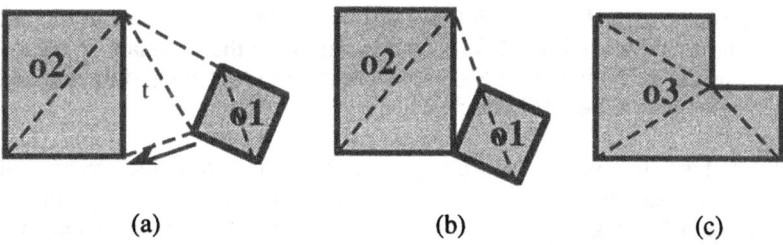

(a)   (b)   (c)

Fig. 4. The snap merge operator.

4.4 Adopt Merge Operator

When merging natural, as opposed to rectangular, objects a freer approach can be taken since there is usually no need to align objects. An operator thought to be particularly suited to merging natural objects is that of adopt merge. This operator is used to amalgamate a pair of proximal-1 objects o1 and o2, forming a new object o3, in such a way that all or some of the triangles belonging to the free space region separating o1 and o2 become part of the new object. The process involves (see Figure 5) : finding all free space triangles S lying between o1 and o2; identifying which of these triangles S_{sub} are to form part of the amalgamated object; combining the triangles belonging to o1, o2 and S_{sub} to form a new object o3 (re-attributing certain edges and deleting o1 and o2 in the process); and finally carrying out any required clean-up of the SDS.

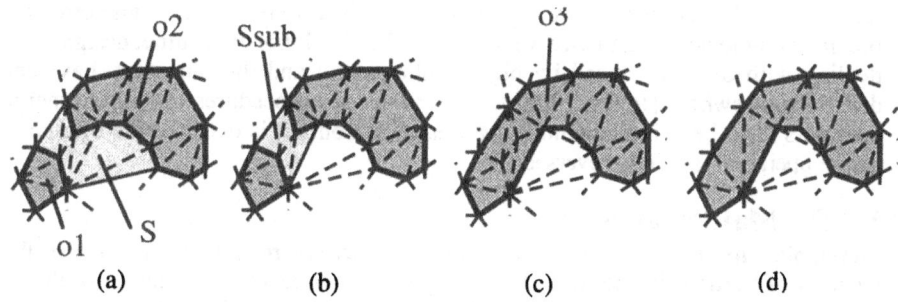

(a)   (b)   (c)   (d)

Fig. 5. The adopt merge operator.

The initial step of the adopt merge operation involves finding all free space triangles S lying between o1 and o2 (see section 3.2) (Figure 5a). In some cases it might be useful to constrain the adopt merge process to take place only in those parts of the intervening region where the distance between o1 and o2 does not exceed some threshold. This is implemented by removing from S any triangles with an edge directly connecting o1 and o2 of length greater that the threshold value. This results in a refined list of intervening triangles S_{sub} (Figure 5b).

The next step in the adopt merge process is the creation of the new object o3, which is assigned the triangles belonging to o1, those belonging to o2 and those belonging to S_{sub}. This is followed by the deletion of o1 and o2 from the SDS. It is now necessary to re-attribute certain edges of S_{sub}. The rule is to re-attribute any real edges of S_{sub} to virtual (since it follows that these edges are internal to o3) and any edges lying on the boundary of S_{sub} (and therefore on the boundary of o3) to real. Edges lying on the boundary of S_{sub} are characterised by having only one adjacent triangle belonging to S_{sub} (Figure 5c).

The final stage involves a clean-up of the SDS. This entails the removal of any vertices which are internal to o3 (see section 5.2), followed by the optimisation of the triangles belonging to o3 and those belonging to the free space regions to which the intervening triangles S belonged (see section 5.3) (Figure 5d).

Example output resulting from the application of each of the described merged operators to large scale Ordnance Survey data is given in Figures 6 to 9.

4.5 Conflict Resolution
It will sometimes be the case that, subsequent to the application of certain generalisation operators, some of the triangles belonging to an object o1 (which was directly involved in the generalisation) will have been forced to overlap some of the triangles belonging to another object o2 (which may or may not have been directly involved in the generalisation). In other words, o1 and o2 will have been forced to overlap, giving rise to spatial conflict. It has been shown [Bundy, 1995a,b] that the SDS can be used to detect and resolve such conflict by searching for, and then correcting, triangle inversions. Inverted triangles (ie. the triangles which have caused the conflict) are identified by virtue of the fact that the order of their component vertices (derived from the relative directions of their component edges) is reversed relative to their order before overlap (ie. vertices which were in a clockwise order are ordered anti-clockwise, and vice-versa). A conflict resolution procedure, designed for dealing with overlap caused by object enlargement and displacement, has been described elsewhere [Jones, 1995b]. As yet, no such procedures relating to merge operators have been implemented. At present, if conflict is detected the offending merge operation is merely reversed.

5 SDS Maintenance
The application of a particular generalisation operator results in a change in the location of vertices in the SDS. Some of this change is valid, in the sense that it does not affect the integrity of the SDS, and is due to transformations in the form or the position of an object or group of objects. However, additional, unwanted changes will sometimes occur, resulting in three types of error, each of which is now described.

5.1 Loss of Topological Integrity
The first type of error is that of a loss of SDS topological integrity, which occurs when free space triangles are forced to overlap other triangles (be they free space triangles or object triangles). The presence of overlapping triangles is detected when one or more triangles become inverted. Inverted triangles are themselves overlapping triangles, as can be non-inverted triangles in their immediate vicinity. Topological errors are particularly serious due to the fact that the successful application of an individual generalisation operator will often be dependent on the SDS retaining topological integrity. Since a particular generalisation operator will rarely be applied in isolation, but rather as part of a sequence of operations, it is important to correct

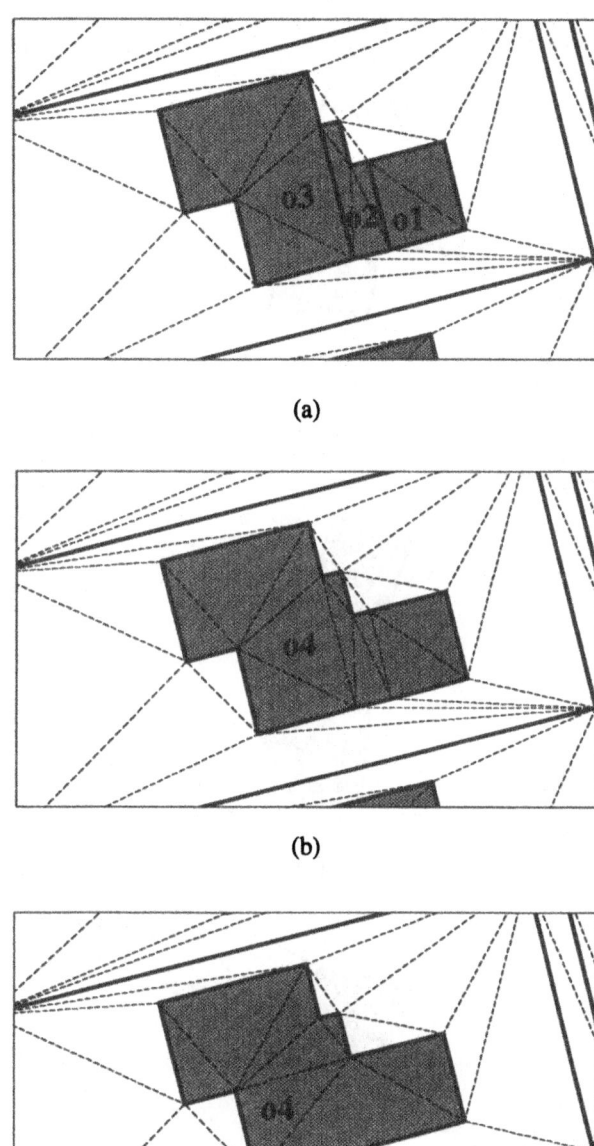

(a)

(b)

(c)

Fig. 6. The append merge operator. (a) Pre-generalisation.
(b) Post-generalisation with errors. (c) Post-clean-up.

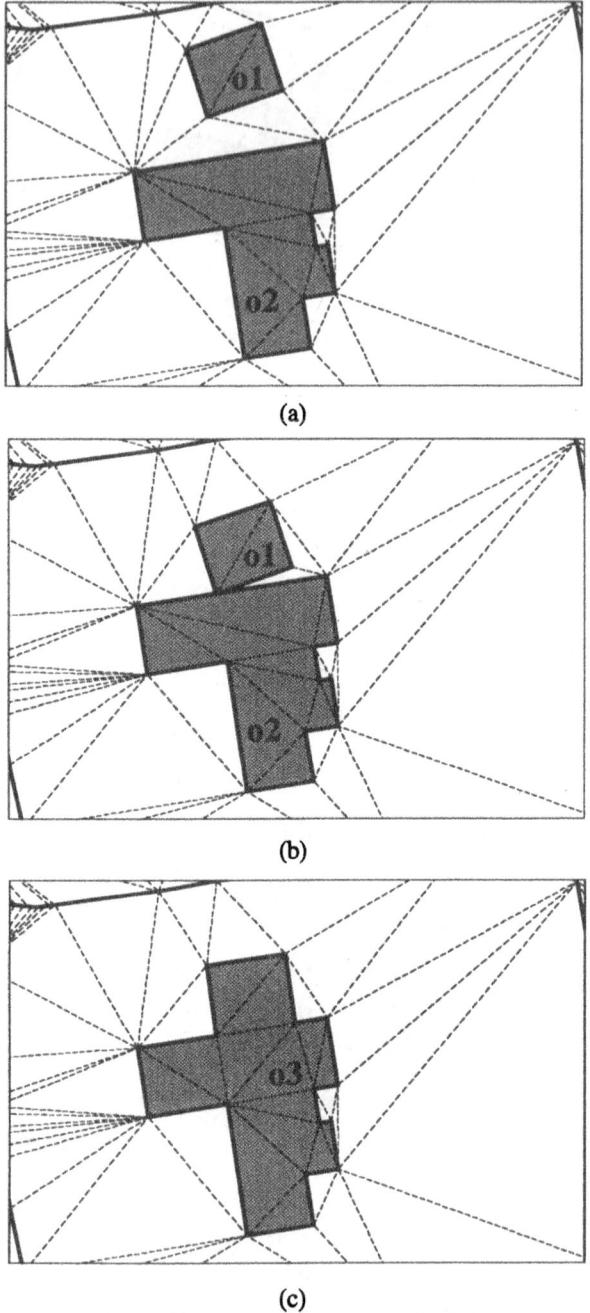

Fig. 7. The direct merge operator. (a) Pre-generalisation.
(b) Prior to object rotation. (c) Post-generalisation.

185

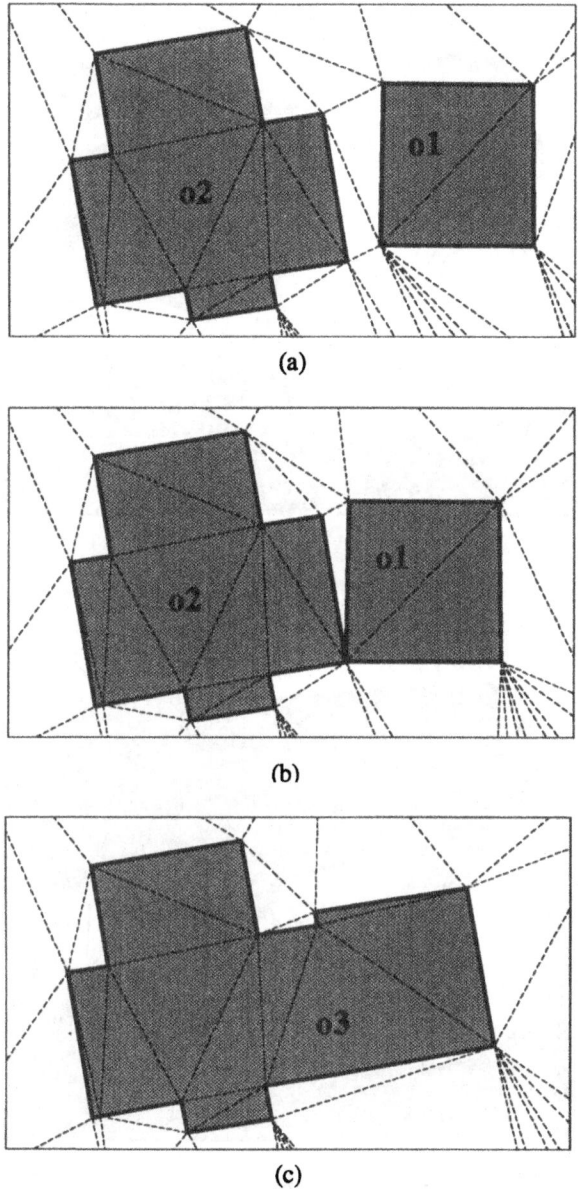

(a)

(b)

(c)

Fig. 8. The snap merge operator. (a) Pre-generalisation.
(b) Prior to object rotation. (c) Post-generalisation.

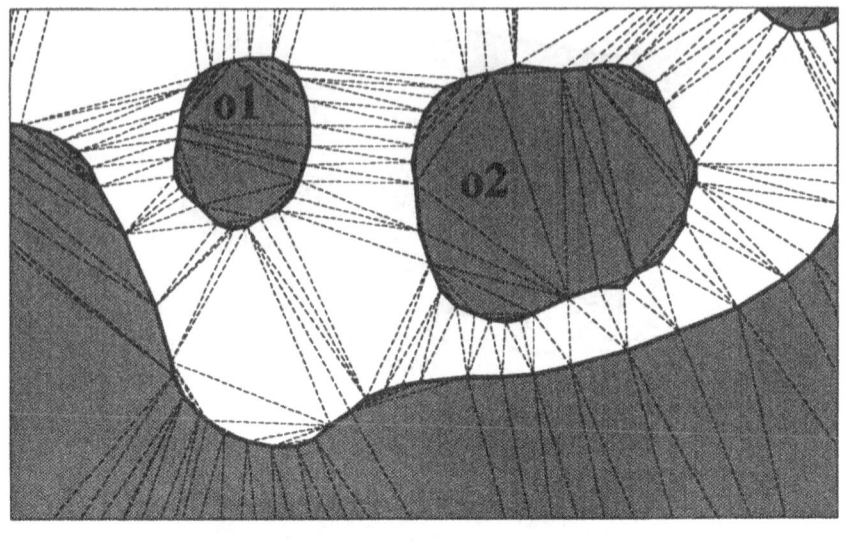

(a)

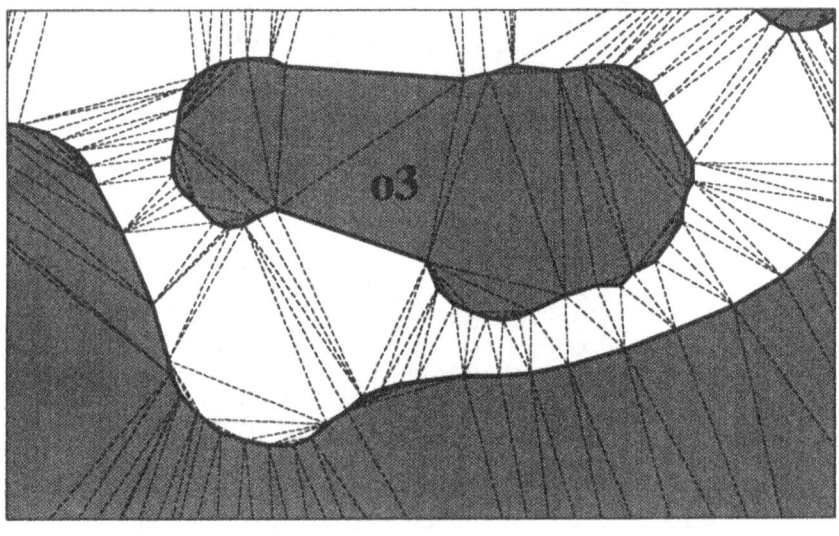

(b)

Fig. 9. The adopt merge operator. (a) Pre-generalisation. (b) Post-generalisation.

all topological errors immediately after application of each operator. This sub-section will discuss a method for performing these corrections. Note that the problem of object triangles overlapping other object triangles (ie. spatial conflict) is discussed in section 4.5.

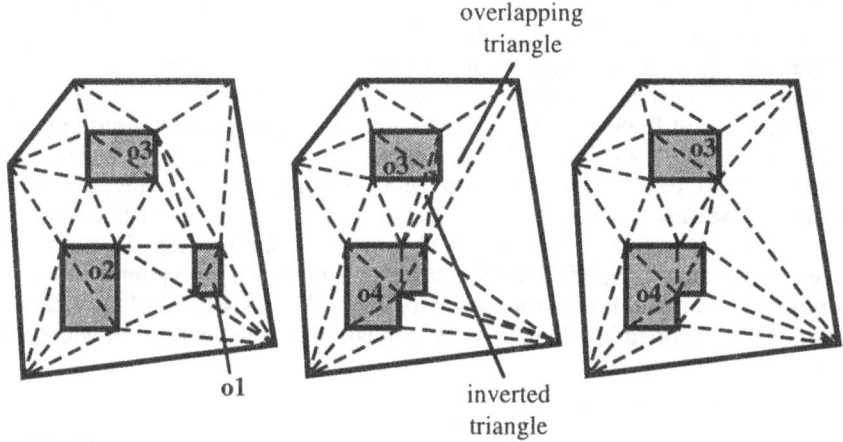

Fig. 10. Resolving problems caused by free space triangles being forced to overlap other triangles.

A general approach to resolving the problems caused by free space triangles which are subject to to overlap is to delete all triangles in the area affected by the overlapping triangles and then to re-triangulate within the empty free space area (Figure 10). It is necessary to define what is meant by the area affected by the overlapping triangles. An optimal definition, that is, one which would result in a minimal affected area, might be regarded as that which only includes free space triangles intersected by the overlapping triangles (including the overlapping triangles themselves). However, in practice, finding these triangles involves computationally expensive line intersection tests, and so our approach is simply to include all triangles belonging to the free space region to which the overlapping triangles belong. The problem of re-triangulating this region equates to that of triangulating a complex polygon (ie. a polygon that may be concave and include holes). Several solutions to the problem of polygon triangulation can be found in the literature (eg. Chazelle, [1983]; De Floriani, [1988]). The method used in our implementation, involves the following : extracting a list of vertices from the region boundary and Delaunay triangulating them; constraining this triangulation using edges obtained from the region boundary; deleting any triangles from the constrained triangulation that do not lie within the free space region; and updating the adjacent triangle pointers of edges lying on the boundary of the region.

5.2 Redundant Data

The second type of error concerns the introduction of redundant data in the form of internal vertices and collinear vertices. Internal vertices will sometimes occur as a result of applying the adopt merge operator, while application of the append merge operator can result in collinear vertices appearing on the new object's boundary. Both these types of vertex are deleted from the SDS, if and when they occur, using the delete vertex operator. This operator, which deletes a vertex v, and is based on the procedure given in Midtbø [1994] (see Figure 11), involves a number of steps. Initially, all triangles in the SDS having v as a vertex are found, forming the enclosing polygon P. It can be seen that any two consecutive edges along the boundary of P, sharing a common vertex b, can be used to form a boundary triangle. The second step is concerned with finding the boundary triangle of P with smallest circumscribing circle. Having done this, the SDS edge between the boundary triangle vertex b and the vertex to be deleted v is swapped, forming a new enclosing polygon P. The process of finding the boundary triangle with smallest circumscribing circle and swapping the appropriate edge is repeated until only three SDS edges remain inside P. These edges are subsequently deleted, along with v. Finally, the triangle formed by the three boundary edges of P is added to the SDS. Note that correct SDS edge adjacency pointers are maintained throughout the process. This algorithm appears to be quite straightforward, but there are a number of exceptional cases, such as when the polygon P is not a convex polygon. Methods for dealing with these situations are reported in Midtbø's paper, and will therefore not be dealt with further here.

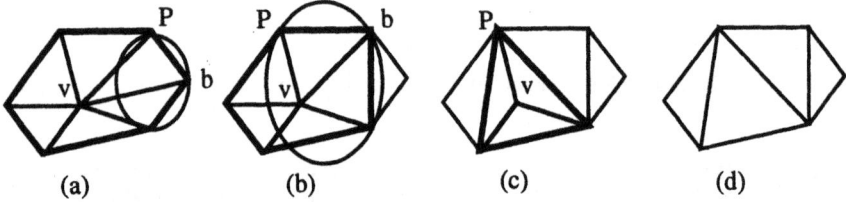

Fig. 11. Deleting a vertex from the SDS. The first edge to be swapped is found in (a), followed by the second in (b). This process continues until P consists of only three triangles (c), which are subsequently deleted from the SDS (d).

5.3 Non-Optimally Shaped Triangles

It will sometimes be the case that following a merge operation some triangles will be non-optimally shaped, ie. the SDS constrained triangulation will not be as close as possible to the Delaunay triangulation. This is due either to triangles becoming stretched or to the re-attribution of edges (from real to virtual and vice-versa). This situation is remedied by applying an optimisation algorithm [De Floriani, 1988] to all triangles involved in the merge. The relevant triangles are those belonging to the object resulting from the merge and any triangles belonging to free space regions affected by the merge. The optimisation algorithm used in our implementation makes use of the max-min angle property of a constrained Delaunay triangulation. This property guarantees that for any pair of adjacent triangles t1 and t2, sharing a virtual edge e and forming a convex quadrilateral Q, then the swapping of edge e with the opposite diagonal of Q does not increase the minimum of the six internal angles of the resulting triangulation of Q. Optimization of the SDS is therefore achieved by swapping the common virtual edge of all pairs of adjacent triangles not meeting the above criterion (Figure 12).

189

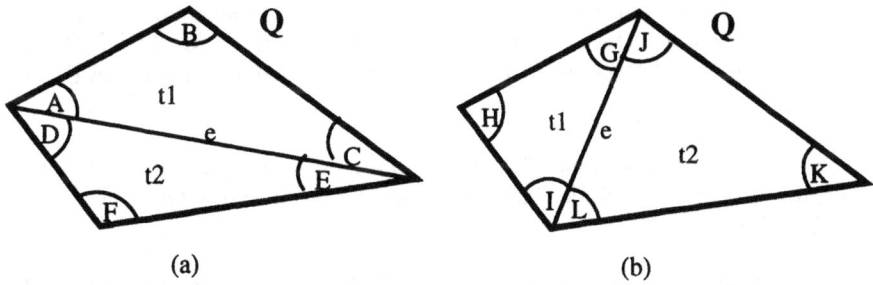

(a)   (b)

Fig. 12. Triangle optimisation making use of the max-min angle property. Here,
triangulation (b) is preferred to triangulation (a) due to the fact that Minimum(G, H, I, J,
K, L) is greater than Minimum(A, B, C, D, E, F).

6 Closing Remarks

This paper has presented details of a spatial data model, termed the SDS, which
offers potential benefit to the future development of automated cartographic
generalisation. In particular, the rich proximity data implicit in the SDS has been
shown to facilitate the rapid derivation of relationships between disjoint map objects
which can subsequently be used to assist in the generalisation process. The paper has
presented details, and actual examples, of how the SDS can be used to perform a
number of fundamental generalisation operations relating specifically to object
amalgamation. It should be noted that the operators described here are regarded as
experimental and that there are clearly options for developing modifications and
enhancements. This is particularly true with regard to aesthetic appearance. For
example, the output produced by the merge operators would probably benefit from
post-processing to perform boundary simplification. However, the main research
objective to date has been the design and implementation of a spatial model which
ensures the maintenance of spatial data integrity during generalisation, particularly
with regard to topological and proximal relations. The SDS achieves this objective
by maintaining knowledge of these relations prior to and during generalisation and
by the application of post-generalisation conflict resolution procedures.

It is intended that future research will fall into two, possibly overlapping, stages.
The first stage will involve the refinement of current generalisation operators and the
development of additional operators. The second stage will be concerned with
bringing these operators under the control of the frame-based reasoning system
referred to in section 1.

Acknowledgements

G.L.Bundy was in receipt of an EPSRC CASE studentship in collaboration with
Ordnance Survey, Southampton. The topographic map data used in the procedures
illustrated in Figures 6,7,8 and 9 was kindly supplied by the Ordnance Survey.

References

Brassel, K. E. and R. Weibel (1988). "A Review and Conceptual Framework of
Automated Map Generalization." *International Journal of Geographical Information
Systems* 2(3), 229 - 244.

Bundy, G. L., C. B. Jones and E. Furse (1995a). Holistic Generalization of Large Scale Cartographic Data. In *GIS and Generalisation Methodology and Practice,* edited by J-C Muller, J-P Lagrange and R. Weibel. Taylor and Francis, 106-119.

Bundy, G. L., C. B. Jones and E. Furse (1995b). A Topological Structure for the Generalization of Large Scale Cartographic Data. *Innovations in GIS 2.* Taylor and Francis, 19-31.

Buttenfield, B. P. (1991). A rule for describing line feature geometry. *Map Generalization: Making Rules for Knowledge representation.* Longman. 150-171.

Chazelle, B. M. and J. Incerpi (1983). Triangulating a polygon by divide-and-conquer. *21st Allerton Conference on Communication and Control Computing,* 447-456.

Chithambaram, R. and K. Beard (1991). Skeletonizing Polygons for Map Generalization. *Technical Papers ACSM-ASPRS Convention,* Baltimore, USA, 44-55.

De Floriani, L. and E. Puppo (1988). Constrained Delaunay triangulation for multiresolution surface description. *Ninth International Conference on Pattern Recognition,* Rome, IEEE, 566-569.

Douglas, D. H. and T. K. Peucker (1973). "Algorithms for the Reduction of the Number of Points Required to Represent a Digitized Line or its Caricature." *The Canadian Cartographer* 10(2): 112-122.

Duda, R. O. and P. E. Hart (1973). *Pattern Classification and Scene Analysis.* Wiley-Interscience, New York.

Freeman, H. and J. Ahn (1984). AUTONAP - an expert system for automatic name placement. *First International Symposium on Spatial Data Handling,* Zurich, International Geographical Union, 544-569.

Heller, M. (1990). Triangulation algorithms for adaptive terrain modeling. *Fourth International Symposium on Spatial Data Handling,* Zurich, International Geographical Union, 163-174.

Jones, C. B., J. M. Ware and G. L. Bundy (1995a). Proximity relations with triangulated spatial models. Manuscript, Department of Computer Studies, University of Glamorgan.

Jones, C. B., G. L. Bundy and J. M. Ware (1995b). Map generalisation with a triangulated data structure. Manuscript, Department of Computer Studies, University of Glamorgan.

Kirkpatrick, D. G. (1979). Efficient computation of continuous skeletons. *Twentieth Annual IEEE Symposium on Foundations of Computer Science,* 18-27.

Lagrange, J. P., A. Ruas and L. Bender (1993). Survey on Generalization. Unpublished (available from Institut Geographiqe National).

Leberl, F. W. and D. Olson (1986). "ASTRA - a system for automated scale transition." *Photogrammetric Engineering and Remote Sensing* 52(2): 251-258.
Lee, D. (1992). Cartographic Generalization. Manuscript, Intergraph Corporation.

Lee, J. (1991). "Comparison of existing methods for building triangular irregular network models for terrain from grid digital elevation models." *International Journal of Geographical Information Systems* 5(3): 267-285.

Lichtner, W. (1979). "Computer-assisted Processes of Cartographic Generalization in Topographic Maps." *Geo-Processing* 1: 183-99.

McMaster, R. B. (1987). "Automated line generalisation." *Cartographica* 24(2): 74-111.

Midtbø, T. (1994). Removing points from a Delaunay triangulation. *Sixth International Symposium on Spatial Data Handling*, Edinburgh, International Geographical Union, 739-750.

Monmonier, M. (1983). "Raster-Mode Area Generalization for Land Use and Land Cover Maps." *Cartographica* 20(4): 65-91.

Muller, J.-C. and Z. Wang (1992). "Area-patch generalization: a competitive approach." *The Cartographic Journal* 29: 137-144.

Nickerson, B. G. (1988). "Automated Cartographic Generalisation For Linear Features." *Cartographica* 25(3): 15-66.

Nickerson, B. G. and H. Freeman (1986). Development of a rule-based system for automatic map generalization. *Proceedings of the Second International Symposium on Spatial Data Handling, Seattle*, Washington, 537-556.

Preparata, F. P. and M. I. Shamos (1988). *Computational Geometry*. Springer-Verlag, New York.

Schylberg, L. (1993). Computational Methods for Generalization of Cartographic Data in a Raster Environment. Royal Institute of Technology, Department of Geodesy and Photogrammetry.

Shea, K. S. and R. B. McMaster (1989). Cartographic Generalization in a Digital Environment: When and How to Generalize. *Proceedings AutoCarto 9*, Baltimore, ACSM/ASPRS, 56-67.

Thapa, K. (1988). "Automatic Line Generalization Using Zero-Crossings." *Photogrammetric Engineering and Remote Sensing* 54(4): 511-517.

Weber, V. W. (1982). "Automationsgestutzte Generalisierung." *Nachrichten aus dem Karten und Vermessungswesen* 88: 77-109.

Weibel, R. (1992). "Models and experiments for adaptive computer-assisted terrain generalization." *Cartography and Geographic Information Systems* 19(2): 133-153.

Object Orientation and Location Updating During Nonvisual Navigation: The Characteristics and Effects of Object- Versus Trajectory-Centered Processing Modes[*]

M-A. AMORIM[1], S. GLASAUER[2], K. CORPINOT[1], and A. BERTHOZ[1]
1 LPPA - Collège de France - CNRS, Paris, France.
2 Department of Neurology, Klinikum Großhadern, University of Munich, Germany.

Abstract: The present study investigates the effect of two distinct processing modes on object location and appearance updating during a guided walk without vision. As a calibration procedure, 12 subjects rotated a head-fixed miniature model until it matched the memorized orientation of the corresponding object, to measure initial (mis)perception of object orientation before the walking task. In the main experiment, observers either continuously kept track of the memorized object appearance during the walk (object-centered task), or they deduced object attributes at a terminal viewing position from continuous trajectory-mapping and knowledge of the object appearance at the initial position (trajectory-centered task), depending on the experimental session. Heading toward the memorized object location and object model rotation supplied information on respectively object location and orientation. Results showed that the two processing modes affected differently spontaneous walk velocity, object orientation updating and retrieval time. Estimation of walked distance and spatial inference processes are the two main sources of errors when updating object location and orientation while walking blind under external guidance.

1 Theoretical Considerations

Updating the memorized location and orientation of a distant object or building while walking in complete darkness may be crucial and performed in two ways. On one hand, we can keep up to date on the changing structure of an internal representation of the previously viewed scene (Rieser and Rider 1991; Loarer and Savoyant 1991; Loomis et al. 1992; 1993) on the basis of self-to-surroundings movement perception (i.e., exproprioception, see Lee 1980). On the other hand, we can update only the walked trajectory from vestibular, kinesthetic, and motor command information, a process called "path integration" (Mittelstaedt and Mittelstaedt 1980; Mittelstaedt and Glasauer 1991), and then reconstruct the whole scene from the new vantage point using spatial inference. This last processing mode is what Huttenlocher and Presson (1973) referred to as a "regenerative strategy" (p.295).

In contrast to previous studies on blind walking (Loomis et al. 1993) which were mainly concerned with updating of the location of single objects or configuration of objects (Rieser and Rider 1991; Amorim et al. 1995), here we

---

[*]This research was supported by a doctoral Grant from the Centre National d'Etudes Spatiales to the first author. We thank Michel Ehrette and Michel Loiron for the technical assistance they provided, respectively, in the stimuli fabrication and HF transmission setting up.

examine updating of *both* previewed *object orientation* and *object location* during guided nonvisual navigation. We were interested in contrasting the effects of the two above-mentioned accounts (i.e., with or without spatial inference) of the changes in the mentally represented perspective structure (Gibson 1979; Rieser 1989) during blind walking. Accordingly, after memorizing an object initial location and orientation, the viewer would either focuse attention on the internally represented object appearance (object-centered task, or OC task) or deduce the object perspective at a final vantage point from cognition of the walked trajectory and recall of the object initial view stored in the episodic memory (trajectory-centered task, or TC task). The cognitive processes involved in these two tasks are sketched in Figure 1 and discussed below.

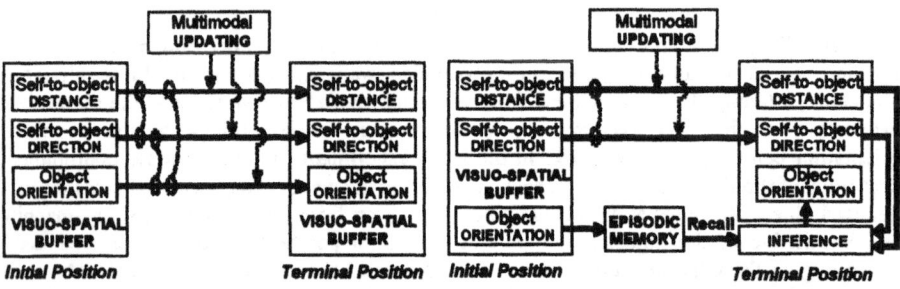

Fig. 1. Cognitive processes involved while updating object location and orientation during nonvisual navigation using either object- or trajectory-centered modes of processing. Arrows and links represent respectively, spatial information transfer and interaction.

According to Kosslyn (1981), in the OC task, the existing object representation would be modified incrementally (shift transformation), whereas in the TC task, once mapped the walked trajectory (Golledge et al. 1994) and retrieved the initial object appearance, the existing representation would be replaced with an altered one in a discrete step (blink transformation). During TC task blind walk, two information channels (self-to-object distance and direction) would be updated, whereas in the OC task, the additional to-be-updated channel (object orientation) would slow down the walking velocity. As a consequence, at the terminal position, in the OC task, the object appearance would be already depicted in the visuo-spatial buffer (short-term memory), whereas in the TC task, the inferential process would increase the overall processing time. At the end of the blind walk, knowledge of *object location* was retrieved by asking the observer to face the (non-visible) object, and *object orientation* information was supplied once the subject rotated a head-fixed miniature model of the object to its estimated bearing from the current vantage point.

2 Experimental Evidence

Material. Two wooden objects were built in the form of an uppercase letter F and both had their borders (2.5 cm thick) painted with a phosphorescent substance to render them visible in complete darkness (see Figure 2a). The first object ('large F') was 45 cm high, whereas the size of the second was 1/5 of the first one and placed inside a clear PVC box which could be held by the subject. The orientation of this miniature 'F' was controlled by a small rotating knob under the box and indexed via a protractor (on the top of the box) rotating with the model (see Figure 2b). Even for an observation or a response taking up to 30 sec to complete, the final object borders luminance of 0.02 cd/m2 remained above photopic threshold (0.01 cd/m2).

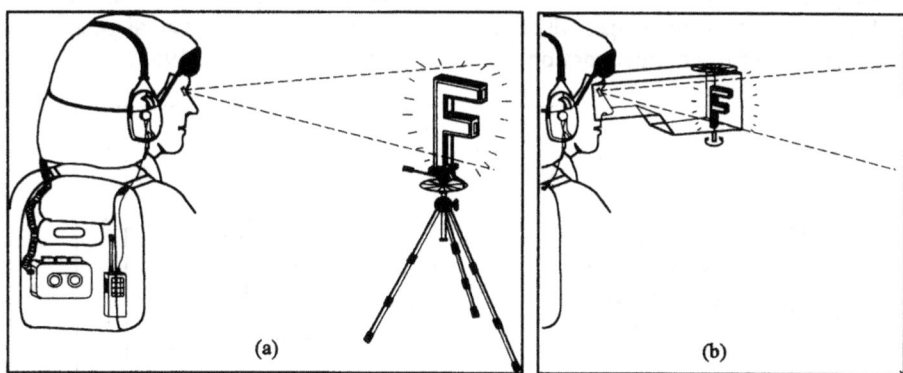

Fig. 2. (a) Observer with headset (blindfold, earphones, white noise tape player, HF receiver) memorizing the target orientation. The observer, with eyes closed and blindfold on, keeps the target orientation in memory, and after a signal opens the eyes with the blindfold removed. Then (b) the observer adjusts the 1:5 scale small 'F' manually.

Subjects. Twelve did participate in the experiment (5 men and 7 women). All but 2 of them were undergraduate students in their early twenties and were paid to participate. All subjects had normal or corrected-to-normal vision and were naive to the purpose of the experiment.

2.1 Procedure

Preliminary task. The purpose of this preliminary experiment was to determine how accurately subjects could discriminate object orientation under reduced-cue conditions (the objects borders were glowing in darkness). This calibration task preceded each of the two conditions (OC and TC task sessions) of the experiment. After listening to the instructions, the subjects had one practice run. To exclude auditory cues, subjects wore earphones carrying white noise from a portable tape player. However, instructions from the experimenter were also delivered to their right ear through a HF transmission channel. Each subject was asked to look at and memorize the orientation of the large 'F' presented at a distance of 1.84 m, 3.5 cm below the eye-level (see Figure 2a). Then, they were instructed to close their eyes and put on a blindfold, while the object was hidden (covered with a black paper). After 15 sec., they were asked to open their eyes and remove the blindfold. They could then look at the 1:5 scale miniature of the object (target) and rotate it until it matched the memorized orientation of the target (see Figure 2b). Only 8 object orientations were tested with 2 repetitions of each. The instructions further requested that the direction pointed by the letter 'F' be determined by its two horizontal segments. Accuracy was encouraged. After having oriented the small F, the subjects closed their eyes, put on the blindfold, and waited for the next trial. The orientation of the small 'F' was recorded by the experimenter.

Main task. In addition to the previously described headset, subjects wore a helmet bearing two antenna-like bars oriented in the head sagittal plane, with an infrared-reflective marker at their tips, which allowed recording of the 3D spatial position of the head by an ELITE video motion analyzer (Ferrigno and Pedotti 1985).

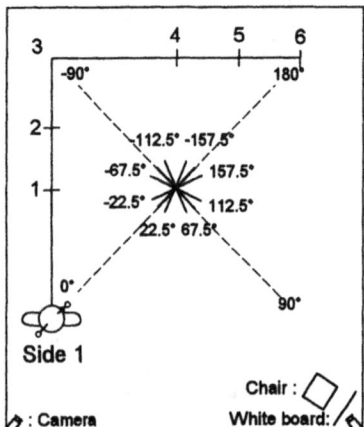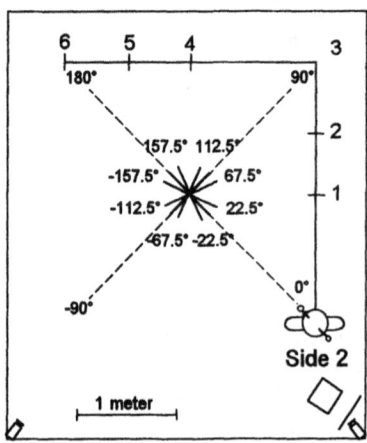

Fig. 3. Main task : Arrival Points (1 to 6) for both departure Sides (1 and 2), and observer memorizing one of the 12 target possible orientations (in viewer-centered coordinates).

The subject entered the darkened experimental room blindfolded and eyes closed, and was led to a rest chair. After the room lights were turned on, she/he was allowed to open the eyes and faced a uniform white very bright surface (luminance = 115 cd/m2) fixed on the wall that maintained the subject in a state of light adaptation. When the room lights were turned off, the subject closed her/his eyes, put the blindfold on, stood up, and was led to an initial position (see Figure 3) facing terminal position 3. Following instructions, the subject opened her/his eyes (blindfold removed), turned the head toward the object location, and memorized its orientation. The object pointed to one of 12 different viewer-centered directions illustrated in Figure 3 in viewer-centered coordinates. Once the object orientation was memorized, the subject closed her/his eyes, turned the head toward terminal position 3 and indicated when ready. E2 covered the object and, following the signal from E1, guided (by hands on shoulders) the subject along a path to one of 6 possible terminal positions (distance from initial position: 1=1.30m; 2=1.95m; 3=2.60m; 4=3.90m; 5=4.55m; and 6=5.20m) of each departure side (see Figure 3). As soon as the terminal position was reached, the ELITE recording began and the subject was instructed to turn toward the object location (as defined by its vertical bar) and visualize its orientation from this new vantage point. The subject indicated when she/he got a clear mental image of the object orientation, and the ELITE was stopped. The subject then opened her/his eyes (blindfold removed) and adjusted the 'small F' so that it matched her/his object mental representation. The subject was then instructed to close her/his eyes and was guided back to the rest chair until the next trial began. Figure 4 shows the main experimental events as well as the time measurements we performed. The duration of the walk from the initial to the terminal position was measured with a chronometer. Object localization time and object orientation latency were measured from the ELITE recordings.

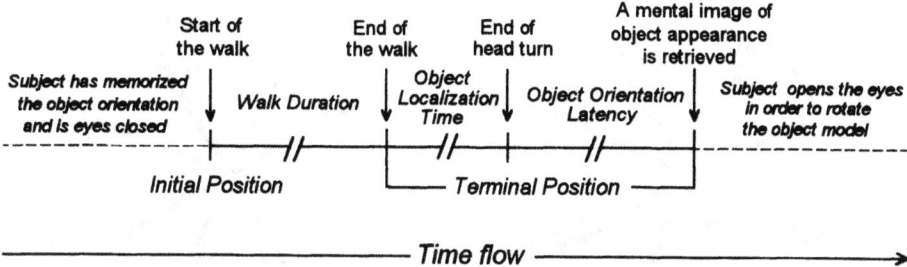

Fig. 4. Experimental events flow chart for the main task.

Depending on the experimental session, the cognitive task was different. In the *Object-Centered* (OC) *task*, the subject focused attention on the object during the walk. Following a given code (one number per object side), the subject indicated loudly, during the walk, which side of the object would be "seen" from her/his current position. In the *Trajectory-Centered* (TC) *task*, the subject was instructed to pay attention to the walking trajectory, count out loud the number of steps, and tell the direction (either left or right) of the turn if she/he made any. From the terminal position, she/he was asked to reconstruct the visual appearance of the hidden object by recalling its perspective from the initial position and taking into account the walked trajectory. Subjects were allowed only one practice trial before the experiment started, and were asked to walk at self-judged optimal speed while performing the tasks.

Half (n=6) of the subjects did OC task first, and then the TC task ("OC-TC" order), whereas for the 6 others the order was inverted ("TC-OC" order). Each session comprised 24 trials (not including the practice trial) : 2 repetitions of 6 terminal positions by 2 departure sides. The 12 viewer-centered object orientations for each combination of terminal position by departure side were pseudorandomly counterbalanced accross all the subjects, in an incomplete balanced design (Cochran and Cox 1957). Each subject kept the same set of 'object orientation by terminal position by departure side' combination for both tasks. For statistical analytic purposes, we will consider a 2 levels object orientation factor: sagittal/frontal (0, 90, 180, and 270 deg in viewer-centered coordinates) and other orientations.

2.2 Results and Discussion

All the results were analyzed using analysis of variance on the different dependent variables. In order to simplify the presentation of the experimental results, the p values of the significant main effects will only be mentioned between parentheses.

Preliminary task. Figure 5 shows the means and one standard error of the mean for each object orientation in a polar (map-view-like) representation. Averages and dispersions were calculated over 12 values (one mean per subject per object orientation). Results show evidence for a significant (p<0.0001) *perceptual bias* toward diagonal orientations (i.e., -45/135 and 45/-135 deg).

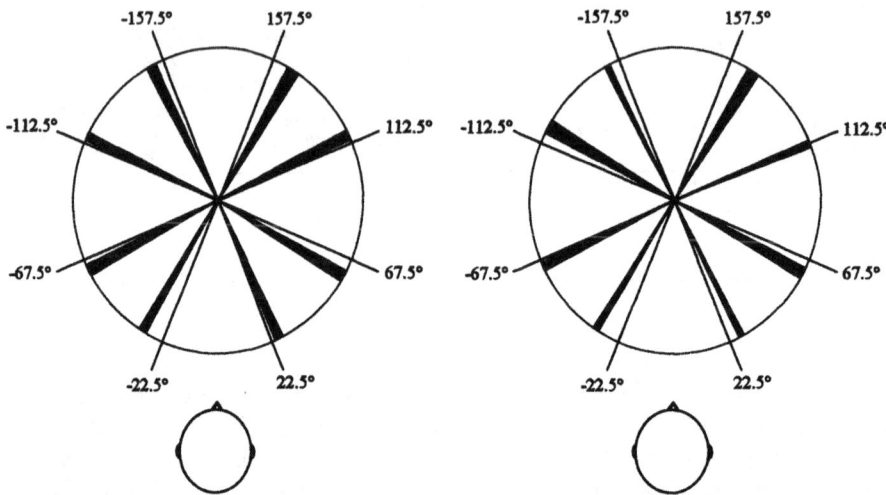

Fig. 5. Polar representation of the results of the preliminary task as a function of each target orientation and walking task. Each circular sector is centered on the mean adjustment response. Shaded areas represent ± one standard error of the mean. The respective correct object orientation is indicated by the prolonged lines next to each sector. Left circle shows results for the OC session, and right circle for the TC session.

The results for the walking experiment follow.

Walk Duration. Figure 6a shows the results for the OC (object-centered) and TC (trajectory-centered) tasks as a function of each level of the terminal position factor. Subjects walked significantly (p=0.0009) faster in the TC task than in the OC task. There is also a significant (p<0.0001) effect of terminal position (walked distance) on walk duration and a significant (p<0.0001) task by terminal position interaction, reflecting the fact that subjects did turn slower at the corner (terminal position 3) resulting in even larger duration differences between OC and TC tasks for terminal positions 4,5,6.

Object Localization Time. There was no significant difference between both tasks on object localization time. This result suggests that object localization process relies on the same information for both tasks : self-to-object distance and direction, as sketched in Figure 1 model.

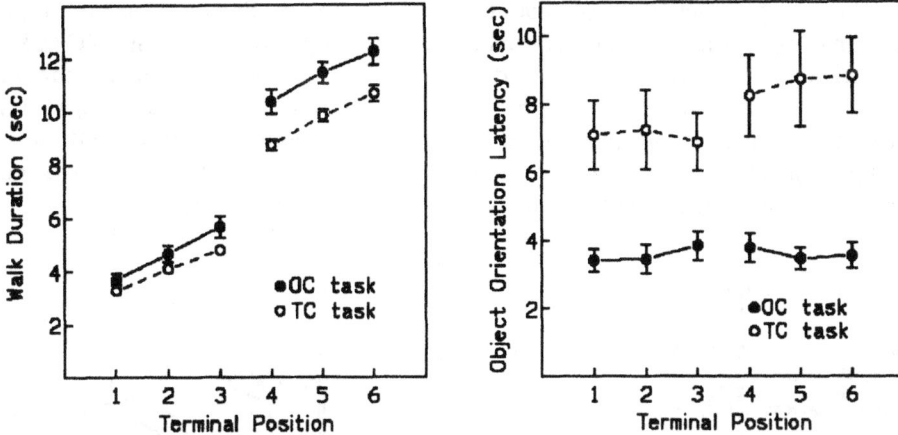

Fig. 6. Mean (a) Walk Duration and (b) Object Orientation Latency results and their standard
error bars plotted as a function of each level of terminal position for OC and TC tasks.

Object Orientation Latency (Figure 6b). An analysis of variance on reaction time
showed a effect of processing mode, whereby There was significantly (p=0.002)
smaller latencies in the OC task ($M = 3.59$ sec ; $SD = 1.97$ sec) than in the TC task
($M = 7.84$ sec ; $SD = 5.57$ sec). Close inspection of Figure 6b suggests that, for TC
task, terminal positions 4, 5, and 6 led to higher RTs as compared to the other
terminal positions, whereas there is no such jump in RTs for the OC task. This is
translated by a significant (p=0.006) interaction between terminal position and
processing mode on RT when comparing terminal positions 1,2,3 together against
4,5,6. This result is probably a consequence of the corner in the path which causes,
in the final inference process of TC processing mode (see Figure 1 model), one
rotation more to be taken into account.

Object Localization and Orientation Errors. The following analyses of object
localization and orientation errors were conducted under the two general
assumptions that observers correctly perceived (a) their course -- 'course' refers to the
direction of one's travel, whereas 'heading' refers to the direction one is facing (see
Fukusima et al. 1995) -- and, (b) the right angle turn between the two path segments
(the one containing terminal positions 1,2,3 and the other containing terminal
positions 4,5,6) while walking blindfolded. The last assumption is supported by the
'rectilinear normalization' effect whereby navigation and orientation judgements tend
to be carried out in (and distorted toward) a normalized, right-angle-grid world
(Stevens and Coupe 1978).
Three hypotheses were formulated and tested, regarding the sources of error.

Hypothesis 1, Correct Initial Perception of Object Location and Walked Distance:
Here, we analyze the heading and object orientation errors assuming correct walked
distance and initial object location perceptions. Heading error (θ error) was
computed (see Figure 7a) by subtracting the correct heading θ toward the object

location in the room from the subject's actual heading θ'. Figure 8a reports mean heading (θ) error as a function of each terminal position level for OC and TC tasks. Negative errors indicate an underestimation (subject has not turned enough toward the object location to face it) and positive errors an overestimation (the subject has turned too much). θ underestimation error increased significantly (p<0.0001) with terminal position similarly for terminal positions 1,2,3 and 4,5,6. There was no effect of the other main factors on θ error.

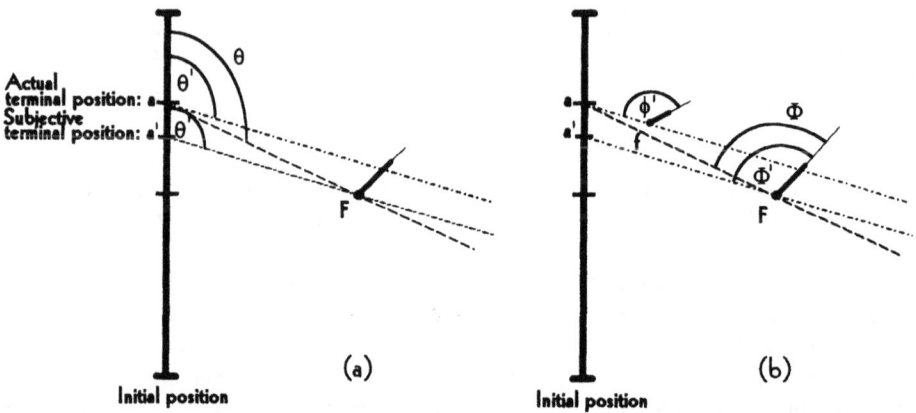

(a)   (b)

Fig. 7. (a) Map-view of the first path segment of departure side 1, illustrating heading error : θ error = θ' - θ ,i.e., subject's heading angle - correct heading angle; Walked Distance Error: WDE = a' - a . (b) Object Orientation Error: OOE = ϕ' - Φ ,i.e., subject's response - correct response from a; Subjective Orientation Error: SOE = ϕ' - Φ' ,i.e., subject's response - correct response from a'.

Considering that θ error reflects an error in *heading execution*, which implies that the subject supposes to face the object correctly, it is possible to analyze the errors on object orientation. Figure 8b reports the mean results and one standard error of the mean for the *object orientation error* (OOE) as a function of each level of terminal position for OC and TC tasks. OOE is the difference between observer's response and the one expected from the current terminal position, irrespective of θ error. Accordingly, positive values refer to overestimations, i.e., small 'F' is rotated more than expected (ClockWise for Side2, counterCW for Side1); whereas negative values refer to underestimations, i.e., small 'F' is rotated less than expected (CW for Side1, CCW for Side2). Accordingly, OOE are comprised between -180 and 180 degrees. OOE is significantly (p<0.0001) related to the terminal position. Descriptively, the incremental underestimation per 'terminal position mean unit' (0.87 meters) is -5.35 degrees, i.e., -6.15 deg/m. A trend analysis showed that the linear relation between mean OOE and terminal position is significant (p=0.0013); and that the deviation from this linear trend is non significant.

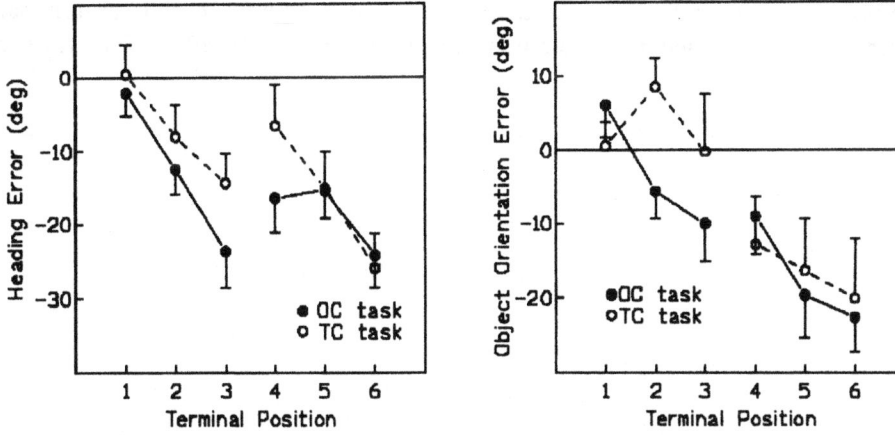

Fig. 8. Mean results for (a) Heading Error (θ Error) and (b) Object Orientation Error, and their standard error bars plotted as a function of terminal position for OC and TC tasks.

In order to *correct* OOE data for possible initial object orientation misperception, we subtracted the errors in the preliminary experiment adjustment task from the OOE data for each subject and her/his corresponding object orientation response. As a consequence, data corresponding to sagittal/frontal orientations of the object are not modified since there was no adjustment value for these orientations. An analysis of variance on those corrected OOE data showed *no* main difference with respect to the initial OOE data. Therefore, errors on initial object orientation misperception do *not* account for the distribution of OOE errors observed at the different terminal positions.

However, the observed *incremental underestimation* error on object orientation estimation with walked distance is rather surprising because in contradiction with Pinker and Finke (1980) demonstration that mental images rotating in depth (around the vertical gravity axis) 'move' a *constant* fraction farther than the amount defined by the physical rotation of the counterpart object. So, either there was an error on initial perception of object location (hypothesis 2) or an error on walked distance estimation (hypothesis 3).

Hypothesis 2, Correct Heading Execution Toward the Object After Initial Misperception of its Location: Assuming no error in the heading toward the object from the terminal position, initial perceived object location may be computed by triangulating mean θ' at corresponding terminal positions for each path segment, e.g., triangulating mean θ obtained at terminal position 2 with that at terminal position 5 (see Figure 9a). Similar *crossing points* were also computed by triangulating mean θ' at terminal positions 1-4 and 3-6 for each task condition. We examined both lateral and depth deviations of object subjective (initially misperceived) location as compared to object actual location within an equivalent *self-to-object* coordinate system for both departure sides. In Figure 9b, the error bars denote ± one standard error of the mean lateral and depth errors represented as a function of task condition for the crossing points derived from mean θ' at terminal positions 1-4, 2-5, and 3-6. There was a significant (p<0.0001) effect of θ' crossing points on lateral error.

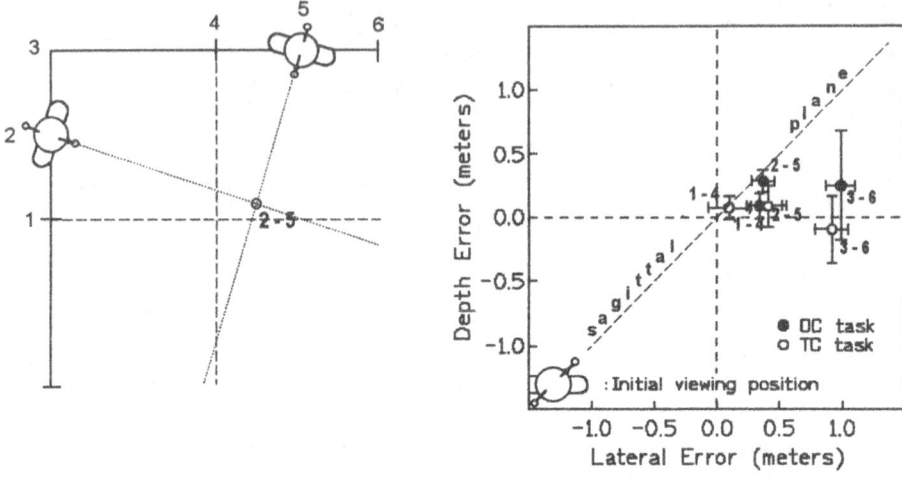

Fig. 9. (a) Example of a crossing point obtained by triangulating the headings (θ') from terminal positions 2 and 5. (b) Mean lateral and depth errors as a function of task condition for the crossing points derived from mean θ' at terminal positions 1-4, 2-5, and 3-6. Error bars denote ± one standard error of the mean error with respect to the correct object location (0 , 0 coordinates) and initial viewing position (-1.3 , -1.3 coordinates) whatever the departure side.

Heading triangulations led to inconsistent results with respect to object location initial perception. If there was an initial misperception of self-to-object distance, the triangulated object locations should have stood along the viewer's head *sagittal* plane (Philbeck and Loomis 1995) with respect to the initial viewing position. However, triangulating the heading of the subjects from the endpoints of each path segment (i.e., terminal positions 3 and 6) led to subjective locations of the object which showed a deviation in the initial *frontal* plane of the viewer; a result clearly inconsistent with hypothesis 2. Since the various θ' crossing points are related to the walked distance, it is necessary to re-evaluate the data under the assumption that subjects did errors on walked distance perception, rather than an error on inital misperception of object location.

Hypothesis 3, Correct Initial Perception of Object Location, but Misperception of Walked Distance: Under the assumption that the observer has correctly perceived the object location from the initial viewing position and made no heading execution errors, θ error would reflect an error on the walked distance estimation (Walked Distance Error : WDE). Consequently, the observer subjective position along the path can be directly derived from θ' (see Figure 7a) as reflected by a similar distribution (pattern) of the results displayed in Figure 7a & 9a. However, since the correct heading angle θ varies with the terminal position, two identical θ errors will not give identical WDE values. This is the reason that the pattern of θ errors does not match completely the pattern of WDE. We found a significant (p=0.03) effect of terminal position on walked distance error WDE but no effect of the other main factors.

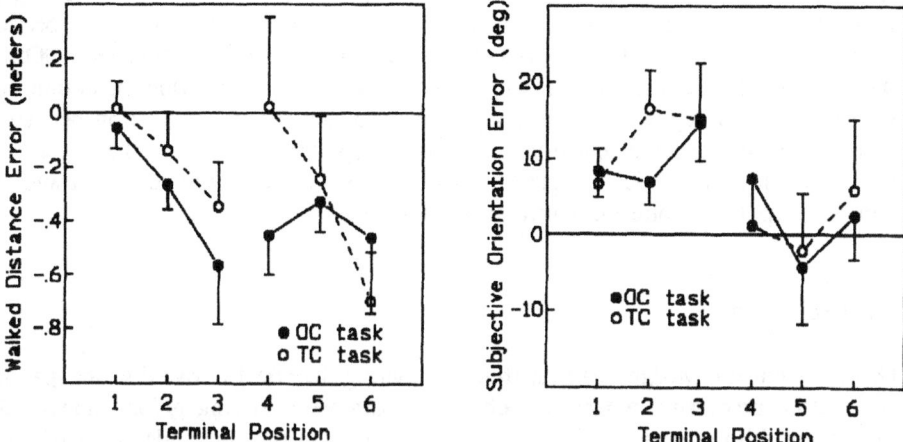

Fig. 10. Mean results for (a) Walked Distance Error, and (b) Subjective Orientation Error, and their standard error bars plotted as a function of each level of terminal position for both OC and TC tasks.

However, under the assumption that the observer had misperceived the walked distance but correctly perceived object location from the initial viewing position, and correctly performed her/his intended heading toward the object, an analysis was conducted on the difference (Subjective Orientation Error , SOE) between observer and expected object orientation response at observer's subjective position a' (see Figure 7b). There was a significant (p=0.025) effect of terminal position, but no effect of the other main factors. More precisely (see Figure 7b), there is a significant (p=0.0031) object rotation overestimation (i.e., farther rotation) for terminal positions 1,2,3 as compared to more or less correct responses at terminal positions 4,5,6 possibly due to the turn between the two path segments. Also, initial sagittal/frontal (0°, 90°, 180°, 270°) object orientations led to significantly (p=0.004) smaller subsequent SOE ($M = 1.9$ deg ; $SD = 31$ deg) as compared to the other initial object orientations ($M = 8.9$ deg ; $SD = 35.3$ deg). This suggests that when the perception of the object orientation from the initial position is free from ambiguity, the error at the new *subjective* viewing *position* is smaller.

In summary, under the current hypothesis, errors on walked distance and object estimated orientation from the current terminal position are compatible with what is known from literature on mental images rotated in depth (Pinker and Finke 1980). The turn between the two paths segments is an additional source of error. However, it remains to investigate what is the contribution of the initial misperception of object orientation, as demonstrated in the preliminary adjustment task, on the error from the terminal position. An analysis of variance on corrected SOE data for initial misperception (as for the above corrected OOE) show *no* main difference in the results as compared to the initial SOE data. Therefore errors on initial object *orientation* misperception do not account for the distribution of errors observed here.

Absolute Errors on Object Orientation. Mean |OOE| was significantly (p=0.03) higher for the TC (M = 28.4 deg ; SD = 27.6 deg) than for the OC (M = 22.5 deg ; SD = 18.5 deg) task due to 90°, 180° and 270° errors in the TC inference process. There was also a significant (p=0.002) effect of terminal position on |OOE|. Descriptively, the additional mean angular error per 'terminal position mean unit' is 3.1 deg, i.e., 3.6 deg/m. A trend analysis showed that the linear relation between mean |OOE| and terminal position is significant (p=0.003); and that the deviation from this linear trend is non significant. An ANOVA on |SOE| only showed a significant (p=0.002) effect of terminal position on |SOE|.

3 Conclusions

This is, to our knowledge, the first study comparing the effect of two distinct spatial information processing modes on behavioral parameters during guided nonvisual navigation, following an object location and orientation preview under reduced-cue condition. In the object-centered (OC) task, observers continuously kept track of the object appearance during the walk, whereas in the trajectory-centered (TC) task, observers deduced object attributes at a final viewing position from continuous trajectory-mapping and knowledge of the initial object perspective.

   The results of the present study are consistent with the Figure 1 cognitive model, i.e., in the OC task, since observers have continuously updated the object's visual appearance during their walk, the response is permanently available in the visuo-spatial buffer where image transformation is supposed to occur (Kosslyn 1981). The high cognitive load of this continuous object orientation updating likely explains the lower spontaneous walk velocity. In contrast in the TC task, observers only need to keep in memory a static mental image of the initial object perspective view and mentally map the walked trajectory, which constitutes a lower cognitive load with respect to the other condition. However, further processing will occur during the subsequent phase in which the visual appearance of the object has to be deduced and reconstructed from the new vantage point by retrieving the memorized object initial view and the mapped trajectory. This last "inferential process" would generate typical misalignment errors when using directional information from a vantage point different from the one in which it was learned (Levine et al. 1982).

   The effects of the processing modes described above on updating of object localization and orientation were analyzed under the general assumption that observers correctly perceived their course while walking blindfolded. Then, we tested *three* different *hypotheses* on the origin of the updating errors to find out which one would account better for the observed results. Hypothesis 1 assumed that subjects perceived correctly the distance walked, but made errors when turning toward the object and supposed to face the object correctly. Accordingly, we examined the difference between the subject's chosen object orientation and the correct one expected from the current viewing position. Under this hypothesis 1, we observed an incremental underestimation error of object orientation with walked distance. This result is in contradiction with Pinker and Finke (1980) demonstration that mental images rotating in depth (around the vertical gravity axis) 'move' a constant fraction farther than the amount defined by the physical rotation of the counterpart object.

Consequently, we tested hypothesis 2 which assumes that walked distance and heading estimation were correct, but subjects misperceived object location from the initial viewing position. The object location perceived initially was computed by triangulating mean headings at equivalent terminal positions for each path segment. Since the subjective object locations, under hypothesis 2, were related to the walked distance, we re-evaluated the data under the assumption that subjects did errors on walked distance perception, rather than on initial perception of object location. Accordingly, under hypothesis 3, the subject's heading toward the object location would reflect her/his subjective viewing position along the path, after accurate initial object location perception and correct execution of the intended head turn toward object location. Results, under hypothesis 3, showed that object was rotated a constant amount farther than expected from viewer's subjective position, which is consistent with previous research (Pinker and Finke 1980). Under this hypothesis, we also found that when object orientation from the initial position is free from perceptual ambiguity, the error at object orienting from the new subjective viewing position is negligible. In summary, estimation of walked distance as well as spatial inference processes are two main sources of errors when updating object location and orientation while walking blind under external guidance.

Congruent pieces of evidence are supplied by research on multisensor integration modeling (Durrant-White 1990) and autonomous mobile robots navigation (Brown et al. 1989). They suggest that when a sensor information, e.g., vision, is missing, cooperative information (Durrant-White 1990) will be supplied by other sensors in order to update an internal model of the missing sensor input. However, due to errors that increase over time in the sensor model, new visual input will be periodically necessary (Brown et al. 1989). Similarly, we suggest that during nonvisual navigation, available sensors cooperate to allow *partial* re-creation of the missing visual information through mental imagery. However, at certain steps, spatial inference (computational) processes would be necessary to allow recontruction of the *global* perspective structure representation. Accordingly, during everyday life locomotion we may only keep track of part of the spatial layout during our displacement, and then regenerate the remainder from the new vantage points. Such 'updating-reconstruction-updating' cycle would constitute the basis of perspective structure updating during nonvisual navigation.

Finally, further work has to be done in order to evaluate which processes (cognitive, motoric, proprioceptive etc.) are open to conscious manipulation and which rather update the changing information automatically, are not introspectable, very robust, etc.

References

Amorim M-A, Loomis, JM, & Fukusima, SS (1995) Spatial updating of objet shape during real and imagined perspective change following visual preview. Proceedings of the ICPA-8 Conference Poster Session, Marseilles, France

Brown C, Durrant-White H, Leonard J, Rao B, Steer B (1989) Centralized and decentralized Kalman filter techniques for tracking, navigation, and control. Revised Technical Report 277, Computer Science Department, University of Rochester NY

Cochran WG, Cox GM (1957) Experimental designs. Toronto: John Wiley & Sons

Durrant-White HF (1990) Sensor models and multisensor integration. In Cox IJ, & Wilfong GT (eds) Autonomous robot vehicles. New York: Springer-Verlag, pp 73-89

Ferrigno G, Pedotti A (1985) ELITE: A digital dedicated hardware system for movement analysis via real-time TV signal processing. IEEE Trans on Biomed Eng 32:943-950

Fukusima SS, Loomis JM, Da Silva JA (1995) Visual perception of egocentric distance as assessed by triangulation. Manuscript submitted for publication.

Gibson JJ (1979) The Ecological Approach to Visual Perception. Boston MA: Houghton-Mifflin

Glasauer S, Amorim M-A, Vitte E, Berthoz A (1994) Goal-directed linear locomotion in normal and labyrinthine-defective subjects. Exp Brain Res 98:323-335

Golledge RG, Klatzky, RL, Loomis, JM (1994) Cognitive mapping and wayfinding by adults without vision. In J. Portugali J (ed) The construction of cognitive maps

Huttenlocher J, Presson CC (1973) Mental rotation and the perspective problem. Cogn Psychol 4:277-299

Kosslyn S (1981) The medium and the message in mental imagery: A theory. Psychol Rev 88:46-66

Lee DN (1980) Visuo-Motor Coordination in Space-Time. In Stelmach GE, Requin J (eds) Tutorials in motor behavior. Amsterdam: North-Holland, pp 281-293

Levine M, Jankovic IN, Palij M (1982) Principles of spatial problem solving. J Exp Psychol: Gen 111:157-175

Loarer E, Savoyant A (1991) Visual imagery in locomotor movement without vision. In Logie RH, Denis M (eds) Mental Images in Human Cognition. Elsevier Science, pp 35-46

Loomis JM, Da Silva JA, Fujita N, Fukusima SS (1992) Visual space perception and visually guided action. J Exp Psychol: HPP 18:906-921

Loomis JM, Klatzky RL, Golledge RG, Cicinelli JG, Pellegrino JW, Fry PA (1993) Nonvisual navigation by blind and sighted: Assessment of path integration ability. J Exp Psychol: Gen 122:73-91

Mittelstaedt ML, Glasauer S (1991) Idiothetic navigation in gerbils and humans. Zool Jahrb Abteil Zool Physiol der Tiere 95:437-435

Mittelstaedt H, Mittelstaedt M (1980) Homing by path integration in a mammal. Naturwiss 67:566

Philbeck JW, Loomis JM (1995) A comparison of two indicators of perceived egocentric distance under full-cue and reduced-cue conditions. Manuscript submitted for publication.

Pinker S, Finke RA (1980) Emergent two-dimensional patterns in images rotated in depth. J Exp Psychol: HPP 6:244-264

Rieser JJ (1989) Access to knowledge of spatial structure at novel points of observation. J Exp Psychol: Learn, Mem, & Cogn 15:1157-1165

Rieser JJ, Rider, EA (1991) Young children's spatial orientation with respect to multiple targets when walking without vision. Dev Psychol 27:97-107

Stevens A, Coupe P (1978) Distortions in judged spatial relations. Cogn Psychol 10:422-437

Path Selection and Route Preference in Human Navigation:
A Progress Report[1]

Reginald G. Golledge

Department of Geography

and

Research Unit in Spatial Cognition and Choice

University of California Santa Barbara

Santa Barbara, California 93106-4060

Phone: (805) 893-2731

Fax: (805) 893-3146

e-mail: golledge@geog.ucsb.edu

Abstract. Two critical characteristics of human wayfinding are destination choice and path selection. Traditionally, the path selection problem has been ignored or assumed to be the result of minimizing procedures such as selecting the shortest path, the quickest path or the least costly path. In this paper I draw on existing literature from cognitive mapping and cognitive distance, to define possible route selection criteria other than these traditional ones. Experiments with route selection on maps and in the field are then described and analyzed to determine which criteria appear to be used as the environment changes and as one increases the number of nodes along a path (i.e., as trip chaining replaces a simple Origin-Destination (O-D) pairing.

1 Introduction

Not only do we select and follow a limited set of paths through the complex environments in which we live, but we have developed many models capable of finding solutions to these path selection problems (e.g., linear programming; traveling salesmen; shortest path). The question is, however, are these the criteria used by humans to solve their own movement problems - or are they methods best suited to mathematical or computer determination of optimal paths through complex multi-node networks to ensure economic efficiency of commercial or fleet traffic, but yet using criteria of which people in general are unaware, or are incapable of using? To explore this question, we examine the process of human navigation and report on pilot experiments that provide insights into the variety of Path Selection criteria used in different contexts.

---

[1]This project was supported by NSF Grant #SES-9207836 and UCTC Grant DTRS 92-G-0009

2   Background

Navigation seems to be one of the primary functions of vision in virtually all biological systems. The processes involved includes cue or landmark recognition, turn angle estimation and reproduction, route link sequencing, network comprehension, frame of reference identification, route plotting strategies (e.g., dead reckoning, path integration, environmental simplification and en-route choice, shortcutting). These processes are used in encoding environmental information for internal processing and use in wayfinding situations. Because of human inaccuracies and errors in recognizing places and coding geometrical components of landscapes, history has seen the development of a variety of technical aids designed to substitute for these human frailties. For example, the prismatic compass was developed to provide greater accuracy than was possible by visually estimating direction. Distances were not measured accurately until the development of distance units and devices such as surveyors' chains, theodolites, range finders, and now ultrasonic laser beams. To find one's way efficiently through complex network structures, computer programs focusing on criteria such as shortest path, minimizing total distance or time traveled, or maximal covering (Church & ReVelle, 1976) now replace the human interrogation of the network for destination choice and for optimal or feasible path selection in most transportation planning interactions where aggregate flows are allocated to routes. But what of the navigation and wayfinding activities of individuals? Do they conform to such principles?

Human navigation usually involves vision which in turn implies the use of inexact measurements and error prone or distorted cognitive maps. This is in contrast to the computerized algorithms for solving navigational problems that rely on explicit quantitative models and exact solution procedures. Some critical features of human navigation and wayfinding that have recently been highlighted are:

(1)   The human navigation system interacts with and adapts to the environment in which it is navigating (Golledge, 1995).

(2)   Navigation proceeds by initiating body motion and receiving and translating sensory feedback received from self perception of motion over time (Loomis, et al. 1992).

(3)   The imagery developed by sensing the environment constrains the nature, type, speed and direction of motion (Golledge, 1992; Kitchin, 1994; Gärling, et al. 1984).

(4)   Potential routes are imaged as larger or shorter depending on whether they proceed towards or away from a primary node or reference point (Sadalla, Burroughs, & Staplin, 1980).

(5)   Many route-distances are imaged as being non-symmetric (Montello, 1992).

Thus, human navigation is often conceived of as a suboptimal system, as compared to vehicle navigation which is often considered as optimized movement in a precisely specified networked environment.

3   Research Questions

We wished to examine questions about: (a) how characteristics of the global stimulus environment affected route choices overall; (b) how the differences between pairs of points affected route choice within a given environment; and (c) how varying network properties influenced path selection criteria.

Questions investigated included the following:

- Do people try to retrace routes when the task involves using more than a single origin or destination?
- How consistent are people in terms of their criteria for route selection as the environment changes (e.g., from simple grid to grid with curves or grid with diagonals)?
- How often do people retrace the same route when traversing between origins and destinations?
- How often is the same criteria chosen when traveling routes of different complexity?
- What criteria do people usually think they use when they are performing route selection tasks in the laboratory and in the field?
- What criteria do people feel they use most frequently when choosing routes in their normal everyday movements through real world environments?

4   Hypotheses

Specific hypotheses to be examined were:

- (a) The dominant route selection criteria will change as the environment changes.
- (b) The dominant route selection criteria will change as trip complexity changes from a single origin-destination pairing to a multiple stop trip.
- (c) As the number of potential "stops" increase in a trip chain, the probability of retracing a route will decrease.
- (d) Traditionally accepted criteria such as shortest path or least time will dominate as route selection criteria.
- (e) Route selection criteria will not change as orientational perspectives change.
- (f) Route selection criteria will not differ between map base or laboratory conditions and real world route following conditions.

5   Experiment #1

5.1   The Laboratory Tasks: Route Selection from Maps

In this project we studied the kinds of routes that people select when navigating through a given environment. Experiments were undertaken in the laboratory to observe routes taken and then inferences were made about the criteria that was used. Initially, subjects were given a series of maps on which two locations were marked. These maps consisted of simple rectangular grids. Three different routes were laid out from a common origin a to common destination. Subjects were asked to imagine that they lived in a town built around the grid network shown on each map, and to imagine that moving from the origin to the destination represented a daily home-work or work-home activity. They were asked to decide which of three routes they would take. The routes allowed them the choice of taking the longest leg first, the shortest leg first, or a stepwise route that approximated a diagonal join between origin and destination (simulating most direct, least effort or least time). Given the regularity of the grid, however, each route was exactly the same distance and varied only in its configurational properties. Maps and routes were configured so that trips were undertaken either as one travels from South to North in conventional coordinate terms or from North to South. Different configurations of O-D paths were provided while actual distances were kept constant. When choosing a route, subjects were required to place or hold the maps horizontally with the northern edge being furthest from the body. No rotation or translation of a given map (or subject) was permitted. However, by rotating a map 90° in either a clockwise or counter-clockwise direction and re-labeling the furthest edge as north, the same geometric configuration can be maintained while orientation and perspective changes. This procedure was followed for all map types.

A second task involved route selection after the number of nodes to be visited enroute was increased (i.e. trip chaining). Again, routes were configured so that travel took place either from South to North or North to South. In this task the environment was changed from a regular grid to one with some diagonal linkages.

A third task involved changing the regular grid to include curved roads and nonorthogonal and intermittent intersection blockages. Polygons representing either negative or positive externalities (e.g., waste dumps or parks) were interspersed throughout the maps. Blockages were described on different trials as parks (a positive attractor) or waste dumps (a negative attractor). The same route choice task was repeated controlling directional components and total length of trip. In this task the number of places to be visited was again increased to see if criteria were used that differed from simple barrier-free origin-destination selection. After each map trial was completed, individual suggestions were solicited regarding what route choice criteria were perceived as being used on these tasks, and what criteria the subject "usually" used in daily real world interactions. Such variables were examined to isolate the type of reasoning or inference that underlies path selection.

5.2 Subjects

Subjects consisted of 32 adults, 16 women and 16 men. Most were students. Ages ranged from 20-35 years of age. Approximately 50% were geographically trained.

5.3 Data Collection

The type of route chosen in the map computed by subjects was entered into a spreadsheet. Maps were examined to disclose what type of criteria were used to select routes. Results of matching these route types with routes actually chosen by subjects (i.e. percentage time each route was chosen) were tabulated (Table 1) which lists examples of path selection criteria.

Criteria	Rank
Shortest Distance	1
Least Time	2
Fewest Turns	3
Most Scenic/Aesthetic	4
First Noticed	5
Longest Leg First	6
Many Curves	7
Many Turns	8
Different from Previous	9
Shortest Leg First	10

Table 1: Ranking of Criteria Most Often Used in Route Selection

(i)   **Fewest Turns:** For each environment, the total number of people who chose a route with the fewest possible turns between each pair of points was recorded. If there was more than one unique route on the compiled map that had the fewest turns possible, then all such numbers were aggregated and the number of people using all such routes was recorded. The actual number of turns that defines "the fewest" for each pair of points was also recorded. The proportion of people in the particular stimulus group who chose a route with the fewest turns was calculated.

(ii)   **Longest Leg First:** This spreadsheet was prepared in a manner similar to Fewest Turns. Here the total number of people who chose a route in which the longest leg of their chosen route was the first segment of the route was first recorded. "Longest" was defined in terms of total distance (not number of blocks). If no one chose a route in which the longest leg was first, then the number of people entered was zero. The number of legs of each route was also recorded.

(iii) **Preference for Curves:** The question here was whether people had a preference for routes involving curves. For each pair of points, the number of people who indicated routes including at least one curved portion were averaged. Each unique route was recorded. The overall preference for curves was quite high. There was quite a bit of variation between routes. However, this measure does not take into account how many curved routes were possible between each pair of points.

(iv) **Preference for Diagonals:** This was similar to the Preference for curves spreadsheet. Again, the overall preference for taking a diagonal was quite high.

(v)   **Shortest Route:** For the diagonal and curve maps, actual distance was measured to determine the true shortest routes. For the regular Grid maps, since all routes that remain within the boundaries of the two points are necessarily of equal length, the question was whether subjects chose a route that would seem to minimize Euclidean distance by traveling "through the middle".

(vi) **Most Aesthetic:** This criteria could only be used with the final set of maps in which polygons representing parks and waste dumps were included. Routes heading away from waste dumps and/or following an edge of a park were labeled most aesthetic.

(vii) **Other Criteria:** Other criteria were defined in similar ways by observing characteristics of the chosen route and inferring what might have prompted its selection.

Detailed results of this study are published elsewhere (Golledge, 1995) but some of the more interesting results are reviewed here as being pertinent to several of the hypotheses offered earlier.

5.4 Route Selection Criteria

(i) **Fewest Turns:** It was apparent that as the environment changes, so does the popularity of this criteria, dropping from a high of 67% in a simple regular grid environment to 25% in a curvilinear environment. Data were reported for each of three environments (Grid, Diagonal, Curves). Path selection criteria changed when perspective changed , i.e. when travel was from a distant origin or to a distant destination. In the case where perspectives differ, there is a remarkable difference in choice of this strategy when the path to be traveled heads from Sth to Nth (65%) as opposed to heading from Nth to Sth (7%). A significant difference occurred in the diagonal environment also, but not in the curvilinear one.

With regard to the more complicated situation in which an intervening point was included on the trip (e.g., from home base A to intermediate point E to destination point C) substantial differences were found in path selection criteria in each type of environments. Focusing still on the fewest turns criteria, for the simple orthogonal grid map where the origin was in the Nth, 46% used the fewest turns as a strategy but only 38% used it when the origin was in the South. For the map with diagonals, 9% and 4.5% used fewest turns when A was in the Nth and Sth respectively; for the map with curves, 12% used it when A was in the Nth, while 21% did so when A was in the Sth. Similarly variable results were obtained for all the different criteria selected.

(ii) **Shortest Path:** Because of the way the simple regular grid was configured, all routes were of equal distance. Shortest path criteria thus could only be examined in the grid with diagonals, and·grid with curves cases. This criterion is the one generally accepted as dominant in most network flow or routing models. It makes sense that it should be so if one is trying to maximize economic utility or minimize costs or time expended in travel. In these experiments however, we again found inconsistencies in criterion use. For example, in the diagonals case, with a single O-D path, 58% used the strategy, while 84.5% used it in the trip chaining cases. Sixty-eight percent used the strategy when the origin was in the Nth, while 80% used it when the origin was in the Sth. For the environment with curves, 74% used it when A was in the Nth, while 90% used it when the origin was in the Sth. Eighty percent adopted it in the trip chaining case, but 54% used it for single O-D pairings.

We next considered situations where individuals were required to travel between A and B in each direction. Here we were concerned with the question of whether the same route was retraced, and if so, what this did to the route selection criterion. As an example, results are presented for the "longest leg first" criterion.

First in the simple grid environment, route retrace was not usually followed. For example, 44% subjects chose longest leg first when traveling from A to B when A was located in the Nth. However, 61% chose this strategy on the return route. This means the return route could not have been a retrace of the original! More confusion occurs when we change perspectives and pursue a path when A is in the Sth to a northerly located B. Here, only 29% used this criterion. In the reverse task, however, 64% chose the strategy!

On the map with curves, 35% chose this strategy when traveling from a distant origin to a close destination, but only 12% chose the strategy on the retrace task. When the origin was close and the destination distant, 13% chose it on the outbound journey and zero chose it on the retrace. When diagonals were included, a similar outbound and retrace pattern occurred, but with a close origin, differences again fluctuated widely from 7% to 20%.

When considering trip chaining, differences in criteria selection become marked depending on orientation. In a simple grid, 33% chose longest leg first when traveling from a distant origin towards a close destination, but zero percent did this on the return trip. When traveling from a close origin to a distant destination, 14% chose the strategy, but zero percent chose it when traveling the reverse route.

On the map which included some diagonals and again required traveling through an intermediate point, when the origin was distant, 35% used longest leg first, but on the return trip zero percent used that strategy. When the origin was in the Sth, 33% used longest leg first and again on the return trip zero percent used it. In the curvilinear condition 15% chose the strategy when A was distant while zero selected it on the return. It might be suggested that in these cases, a pure retrace strategy may have been used, thus precluding any "longest leg first" strategies from being implemented. Visual examination of subjects' maps tends to confirm this explanation. The occurrence of zero percent choice on the return trip does indicate that exact route retracing was a possible option as a route selection strategy.

Although there have been questions raised regarding the suitability of using maps in wayfinding tasks (Lloyd & Cammack, 1995), this set of exercises provides evidence that human path selection may not be the simple process that is usually assumed in network flow solution algorithms. While shortest path and least time were most highly ranked, it was also obvious that as one changed the complexity of the environment, and as trip making became more complex because of chaining several nodes together, path selection criteria changed. Also, there was no clear evidence that trip retracing was carried out except in some complex environments where chaining was required. Thus, it seems that some accounting for well known behaviors such as taking different routes to and from a given destination, or perceiving that routes heading in some direction are more acceptable than those heading in different directions (i.e., that there is an orientation bias in selecting routes) that can be partly accounted for by changing route selection criteria.

Given these laboratory based results, we now turn to a field experiment to see if they are duplicated or whether the experimental situation produced "artificial" behaviors.

6   Experiment #2

6.1   Path Selection in a Real Environment:

A second study was consequently undertaken to examine path selection criteria in a real world rather than laboratory setting.

Using information derived from the laboratory experiment, possible routes between two pairs of origins and destinations on a Western United States campus were used. Subjects were all familiar with the study area and were asked to select routes in both forward and reverse directions between the chosen points. Paths conforming to the criteria types identified in the laboratory experiment were defined and matched against the routes actually selected by subjects. Research questions again focused on inferring which criteria were used in path selection, whether route retraces were used, and what criteria were used most frequently. Only single O-D pairs were used; no trip chaining was investigated.

The principal hypotheses were similar to those examined in the map experiment. It was hypothesized that: (i) shortest distance and shortest time would be the two primary criteria; (ii) route retraces would occur frequently on both routes; and (iii) people will use the same criteria in this real world experiment that they use in everyday activities.

6.2   Subjects/Environment

The study was conducted on the campus of a Western United States university in the area between Ellison and Cheadle Hall (see Figure 1). The environment consists of a central open courtyard containing large regularly spaced planters. The courtyard adjoins Ellison Hall in an area divided by pathways and grassy areas. Two routes were selected for the study. The Stairs Route (A-B) consisted of the origin/destination pair of the flagpole at the north east corner of Cheadle Hall and the stairway door at the west end of the north wing of Ellison Hall. The Elevator Route (X-Y) consisted of the origin/destination pair of the flagpole at the north east corner of Cheadle and the elevator entrance at the east end of the north wing of Ellison .Each of these round trip routes was subdivided into forward and reverse components resulting in four route conditions:

- *forward stairs:* here the subjects' first task began at the flagpole, traveled to the stairs and returned; his or her second task began at the flagpole and traveled to the elevator and returned.
- *reverse stairs:* here the subjects' first task began at the stairs, traveled to the flagpole and returned; his or her second task began at the elevator and traveled to the flagpole and returned.
- *forward elevator:* here the subjects' first task began at the flagpole and traveled to the elevator and returned; his or her second task began at the flagpole and traveled to the stairs and returned.
- *reverse elevator:* here the subjects' first task began at the elevator and traveled to the flagpole and returned; his or her second task began at the stairs and traveled to the flagpole and returned.

All subjects were university staff or students (both graduate and undergraduate). An equal number of men and women, and geography, non-geography students were selected.

Subjects were chosen by convenience from responses to fliers advertising the study.

215

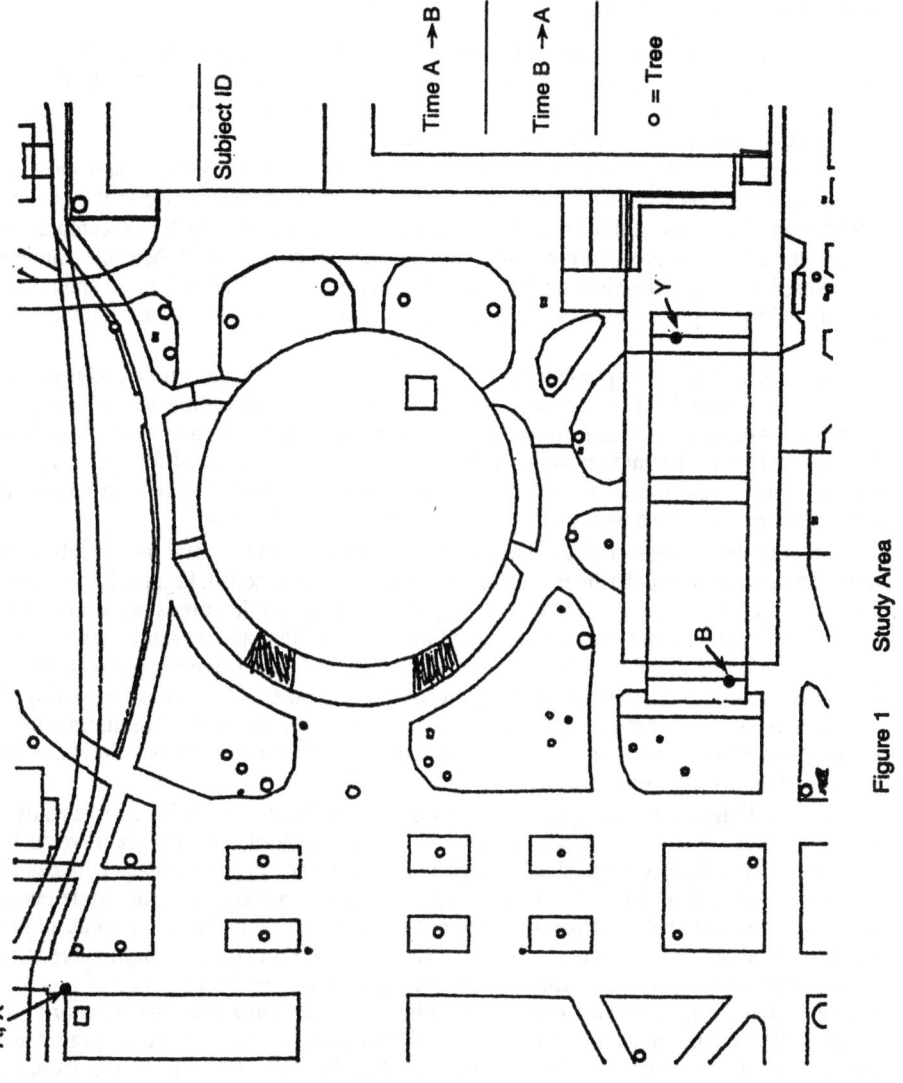

Subject ID

Time A →B

Time B →A

o = Tree

Figure 1   Study Area

6.3   Methods/Procedure

Thirty-two subjects were scheduled for the experiment during daytime hours. All subjects were very familiar with the study area. Subjects were randomly assigned to the four different conditions while ensuring that equal numbers of male and female, and geography non-geography students were placed in each condition.

Subjects were taken to the origin for their assigned route condition and then were read the appropriate directions. They then began to walk a route of their choice to the assigned destination. This route and the time taken to travel it was recorded by the researcher on a map of the area. This procedure was repeated for the reverse section of the route. Subjects then completed a questionnaire on the criteria they used in selecting their route and normal activity behavior, plus evaluations of self confidence in spatial tasks and normal modes of travel.

The average group response for rating route choice criteria usually used and perceived to be used in this field experiment were also examined. According to questionnaire responses, subjects rated shortest route, route taking the least time, and route proceeding in the direction of destination as being the most important. Criteria of fewest turns, first noticed, and "usual route" were next in importance. In general the criteria values are consistent between those used on the task and those commonly used.

To analyze the route choice behavior based on traveling in the environment, all routes used between origin and destination pairs were determined and coded. Figures 2 and 3 show routes chosen between A and B, and X and Y. All possible routes were coded by identifying segments and choice points, and a separate route code was provided for each possible route that could be taken on each task. The number of times a given route was taken was recorded. The maps produced by recording the routes subjects traveled during the experiment were then used to produce Table 2 which shows the route chosen, time taken to complete, whether the same route was taken in the forward and reverse directions and which direction was traveled more quickly for each subject.

For the Flagpole to Stairway route, 62.5% of the subjects traveled the same route in both directions. For the Flagpole to Elevator route, 15.6% of the subjects traveled the same route in both directions. This is a significant difference in route retrace between the two origin/destination pairs. This is apparently due to the existence of some route choice criteria present in this environment that produces a distinctly different route choice decision to be made depending on the direction of travel. Of particular importance is the layout of features near the elevator at Ellison Hall, including the presence of a central grassy area dividing travel into one of two paths. While traveling from the elevator to the flagpole 75% of the subjects chose a route that took them to the north of the grassy area that is encountered when leaving Ellison for Cheadle Hall. While traveling from the flagpole to Ellison Hall 75% of the subjects chose a route that took them south of this same grassy area. (i.e., route choice was dependent on direction of travel). One interpretation of this route choice behavior is that subjects chose a route that took them away from Ellison Hall as soon as possible when leaving the elevator and took them close to Ellison Hall as quickly as possible when approaching their destination. In this

217

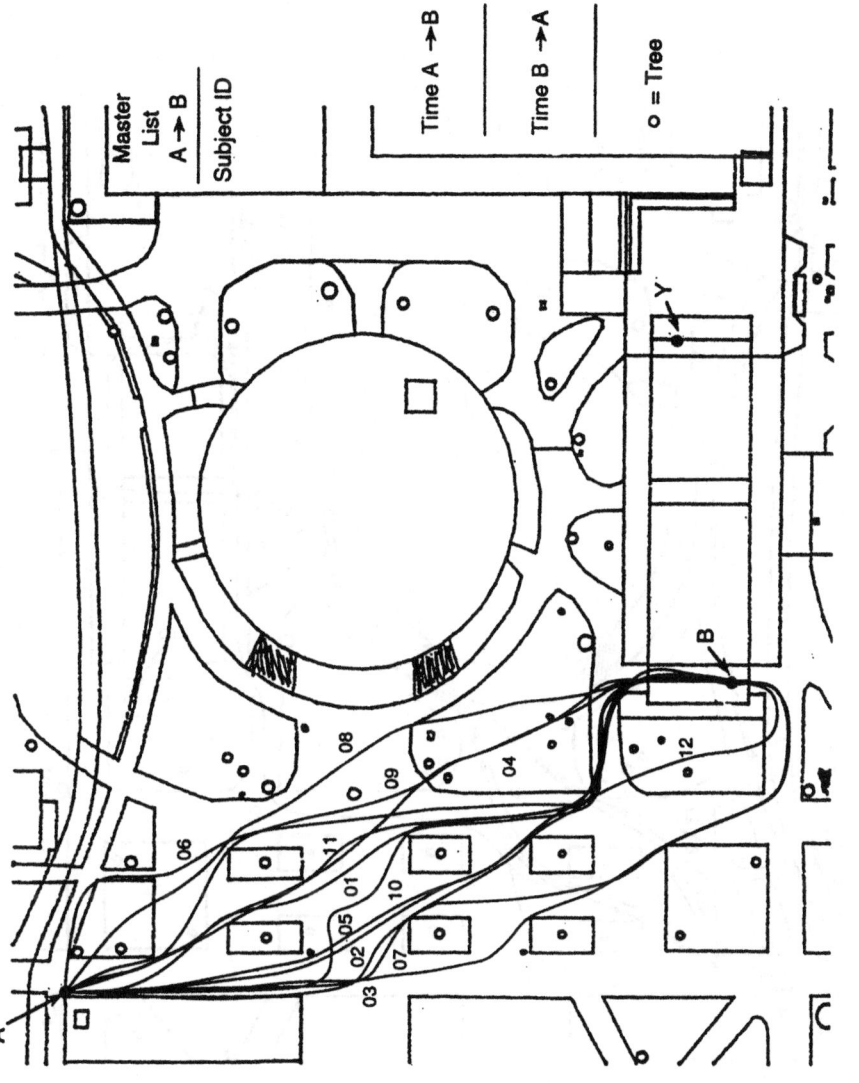

Master
List
A → B
Subject ID

Time A → B

Time B → A

o = Tree

Figure 2   Routes Taken A-B

218

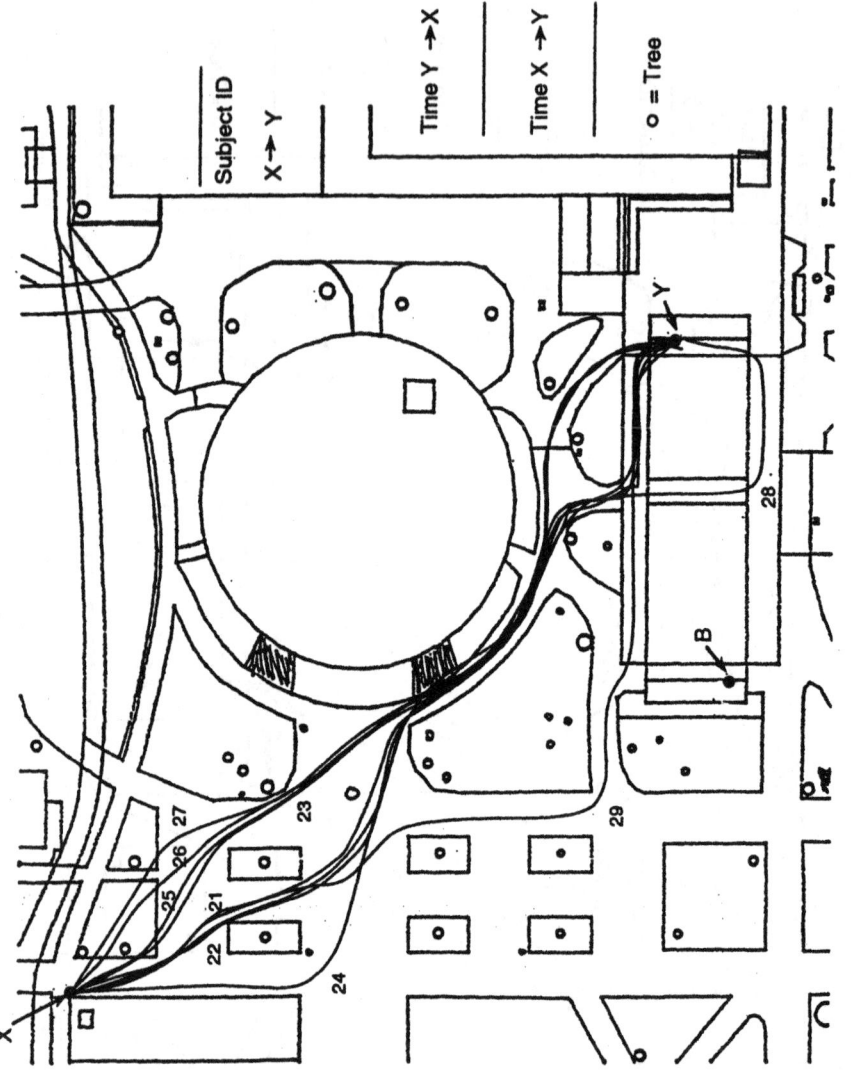

Subject ID
X → Y

Time Y → X

Time X → Y

o = Tree

Figure 3   Routes Taken X-Y

Table 2
Route Choice and Travel Times

Code	Geog/Non	M/F	A-B	B-A	X-Y	Y-X	A to B Route Chosen	Time (mins)	B to A Route Chosen	Time (mins)	X to Y Route Chosen	Time (mins)	Y to X Route Chosen	Time (mins)	Route Same?*	Fastest Route**
✓ 1	Geog	M	X				4	1:19	4	1:18	25	1:33	22	1:34	0	2
✓ 2	Geog	M	X				1	1:10	4	1:04	22	1:18	21	1:20	0	2
✓ 3	Geog	F	X				4	1:15	4	1:15	23	1:26	22	1:28	0	2
✓ 4	Geog	F	X				4	1:15	4	1:13	26	1:27	27	1:28	0	2
✓ 5	NonGeog	M	X				4	1:23	4	1:22	23	1:38	23	1:39	1	2
✓ 6	NonGeog	M	X				4	1:16	5	1:21	22	1:30	22	1:30	0	0
✓ 7	NonGeog	F	X				1	1:30	3	1:32	23	1:49	22	1:50	0	2
✓ 8	NonGeog	F	X				1	1:24	4	1:30	23	1:34	22	1:37	0	2
✓ 9	Geog	M		X			4	1:11	1	1:11	26	1:24	22	1:37	0	1
✓ 10	Geog	M		X			3	1:36	4	1:27	23	1:39	24	1:49	0	1
✓ 11	Geog	F		X			4	1:15	4	1:17	23	1:29	25	1:30	0	1
✓ 12	Geog	F		X			8	1:33	4	1:30	25	1:42	27	1:43	0	1
✓ 13	NonGeog	M		X			4	1:12	7	1:14	25	1:27	22	1:27	0	0
✓ 14	NonGeog	M		X			6	1:40	1	1:45	23	1:50	21	1:51	0	0
✓ 15	NonGeog	F		X			1	1:19	9	1:15	23	1:25	21	1:27	0	1
✓ 16	NonGeog	F		X			10	1:31	4	1:11	23	1:30	28	1:50	0	1
✓ 17	Geog	M			X		4	1:20	4	1:19	23	1:44	22	1:48	0	2
✓ 18	Geog	M			X		1	1:17	2	1:16	23	1:31	22	1:31	0	0
✓ 19	Geog	F			X		2	1:30	4	1:33	25	1:39	22	1:41	0	2
✓ 20	Geog	F			X		4	1:25	1	1:23	23	1:43	27	1:41	0	1
✓ 21	NonGeog	M			X		1	1:28	11	1:35	29	1:52	21	1:34	0	0
✓ 22	NonGeog	M			X		2	1:22	4	1:12	23	1:33	22	1:45	0	1
✓ 23	NonGeog	F			X		8	1:03	12	1:04	25	1:19	22	1:27	0	2
✓ 24	NonGeog	F			X		4	1:15	4	1:21	25	1:32	25	1:22	0	1
✓ 25	Geog	M				X	4	1:18	2	1:19	23	1:45	27	1:33	1	1
✓ 26	Geog	M				X	4	1:29	1	1:28	21	1:48	21	1:39	0	2
✓ 27	Geog	F				X	1	1:28	4	1:33	21	1:55	22	1:42	1	2
✓ 28	Geog	F				X	1	1:25	1	1:22	23	1:41	23	1:49	0	2
✓ 29	NonGeog	M				X	4	1:11	4	1:11	23	1:24	22	1:38	1	2
✓ 30	NonGeog	M				X	4	1:29	4	1:26	23	1:42	22	1:27	0	1
✓ 31	NonGeog	F				X	4	1:26	4	1:24	21	1:42	25	1:45	0	1
✓ 32	NonGeog	F				X	1	1:31	1	1:33	23	1:34	25	1:45	0	1

* 0 = different route
1 = same route

** 0 = neither faster
1 = first traveled faster
2 = second traveled faster

interpretation one could hypothesize that the building represented the destination on a larger scale of route planning and that leaving Ellison Hall represents leaving the elevator and conversely reaching Ellison Hall represents reaching the elevator.

The two routes further produced interesting differences when retracing is considered. On route A-B (flagpole to stairs), 62.9% took the same route in both a forward and reverse direction. On route X-Y (flagpole to elevator), only 15.6% took the same route.

For both routes 43.7% of the subjects traveled the first direction faster than the return direction. For the stairway route 46.9% of the subjects completed the return portion of the route faster than the first. For the elevator route, 43.7% of the subject completed the reverse segment faster than the first traveled. This doesn't support the intuitive position that subjects would travel the return route faster after having learned the route and environment on the first leg.

Between the flagpole and stairs, two routes (#16 and 9) accounted for 75% of subject's route choices regardless of direction traveled. Between the flagpole and the elevator five routes (#23, 25, 22, 24, and 26) accounted for 75% of subject's route choices regardless of direction traveled. However, a total of twelve routes were needed to account for all travel between the flagpole and stairs while only nine routes were needed to account for all travel between the flagpole and the elevator. It is interesting to consider this data in light of the differences in the spatial layout of the two route areas. The stairway route is primarily across a plaza that has regularly spaced planters which are obstacles to travel. These planters allow a generally straight line route between origin and destination but to some extent force the traveler to choose 'channels' to a destination. The elevator route differs in that only a portion consists of the plaza with planters. The rest of the route area consists of pathways restricting travel between buildings and around grass areas. Furthermore, these pathways radiate out from the elevator, causing diverging paths away from the elevator and converging paths toward the elevator.

7   Discussion

Other researchers have pointed to the facts of asymmetric distance cognition (e.g., Sadalla, Burroughs & Staplin 1980). These experiments add to their findings by focusing on the paths actually chosen, the criteria apparently most relevant to that route choice, and noting if there are differences between what criteria were used in a field experiment versus those used in daily travel. Some interesting results developed.

First, when comparing the two experiments, laboratory and field, one notices the similarity between the rating of criteria used in the experiments. In the field, minimizing time was given more support belying the result from the lab experiment in which subjects claimed they did not minimize time in everyday activities.

When considering route retracing, two things stand out. Even in this restricted environment, choice of routes varied depending on direction traveled and with respect to the nature of the environment. The fact that on one route (A-B) 62.9% took the same route both ways was significantly different from the result obtained from the other route (X-Y) when only 15.6% took the same route both ways. In the former, minimizing distance or time or turns could provide reasonable explanations for the observed behaviors. For the other route it appeared that route selection criteria changed indicating that a single selection criteria would seriously under predict the paths chosen. No significant differences were observed among males/females and geography/non-geography groups. Also, it did not appear that any one of the end points (flagpole, stairs, elevator) was considered to be a primary reference point and the others secondary. This implies that, in addition to the previously discovered asymmetry of distance perception among anchoring and other nodes, perceptions of the configuration of the environment itself (particularly different perspectives as one changes direction) may influence route choice. Thus, a route that seems shorter or quicker or straighter from one end may not be so perceived from the other end, thus inducing a change of route. The real question is whether the route selection criteria also change; from examining the actual paths taken and recording response times and other variables, it seems that they often do.

Although the field experiment did not directly test the influence of orientation direction as did the map test, there is room to infer that once again orientation direction played a part in route choice and the criteria used to select that route. Certainly the commonly used assumption that trips will be retraced and that the same criteria will be used for different trips, must be brought into question.

8 Directions for Further Analysis

Further study in this research project is designed to develop route classification procedures for the various routes actually taken by people in their everyday travel activities. This will determine if the route choice criteria listed in our questionnaire (shortest distance to travel, has fewest turns, longest leg first, most aesthetically pleasing, shortest leg first, has many curves, takes least amount of time, first noticed, has most turns, usual route, alternative to usual route, and always proceeds in direction of travel) are comprehensive or partial. However, it may not be possible objectively to classify routes based on some of the criteria such as: usual route, alternative to usual route, most aesthetic, and first noticed without extensive survey research. However, classification using the other criteria would allow comparisons to be made between the stated criteria used and the actual criteria used. This could be used to answer various questions including: what was the varying importance of the choice criteria when actually traveling in the environment? How does this rating vary for the different conditions? How does this compare to the varying importance of perceived criteria? For non-route retrace what was the criteria that caused a different route choice for the return trip? What difference does it make to predicting travel when one uses different route selection criteria for outbound and inbound trips? Does route selection criteria change with every change of trip purpose? Travel mode? And whether simple or chained trips are anticipated?

I think it would also be interesting to pursue what characteristics of the route areas have caused the differences between route retrace and differentiate route selection between origins and destinations. With this information as a knowledge base, it would then be appropriate to extend this work to a driving situation (i.e. using motorists as subjects). One could also determine if one or more trip purposes tend to produce route retraces more than others, or if increasing complexity of trip chains produced simple or multiple criteria for each route segment or for the entire trip sequence. A final problem would be to evaluate the degree of realism that can be attributed to conventionally used path selection criteria built into transportation models or the network models built into today's GIS.

9   Acknowledgments

Erika Ferguson, Graduate Student in Psychology; Amy Ruggles, Joanna Schulman, and John Dutton, Graduate Students in Geography, University of California Santa Barbara, for help in running experiments and preparing data for analysis.

10   References

Church, R.L., and ReVelle, C.S. (1976) Theoretical and computational links between the p-median, location set-covering, and the maximal covering location problem. Geographical Analysis, 8, 406-415.

Gärling, T., Böök, A., and Lindberg, E. (1984) Cognitive mapping of large-scale environments: The interrelationship of action plans, acquisition, and orientation. Environment and Behavior, 16, 3-34.

Golledge, R.G. (1992) Place recognition and wayfinding: Making sense of space. Geoforum, 23, 2: 199-214.

Golledge, R.G. (1995) Defining the Criteria Used in Path Selection. Paper presented to the International Conference on Activity Scheduling, Eindhoven, The Netherlands, May.

Golledge, R.G., and Zannaras, G. (1973) Cognitive approaches to the analysis of human spatial behavior. In W.H. Ittelson (Ed.), Environment and cognition. New York: Seminar Press, pp. 59-94.

Kitchin, R.M. (1994) Cognitive maps: What are they and why study them? Journal of Environmental Psychology, 14, 1: 1-19.

Lloyd, R., and Cammack, R. (1995) Constructing cognitive maps with orientation biases. Unpublished manuscript, Department of Geography, University of South Carolina.

Loomis, J.M., Da Silva, J.A., Fujita, N., and Fukusima, S.S. (1992) Visual space perception and visually directed action. Journal of Experimental Psychology: Human Perception and Performance, 18, 4: 906-921.

Montello, D.R. (1992) The perception and cognition of environmental distance: Mechanisms and information sources. Department of Geography, University of California, Santa Barbara.

Sadalla, E.K., Burroughs, W.J., and Staplin, L.J. (1980) Reference points in spatial cognition. Journal of Experimental Psychology: Human Learning and Memory, 5, 516-528.

How Spatial Information Connects Visual Perception and Natural Language Generation in Dynamic Environments: Towards a Computational Model

Wolfgang Maaß

Department for Computer Science
Universität des Saarlandes
Im Stadtwald 15
D-66041 Saarbrücken 11, Germany

E-Mail : maass@cs.uni-sb.de
Phone: (+49 681) 302-3393
Fax: (+49 681) 302-4421

Abstract. Suppose that you are required to describe a route step-by-step to somebody who does not know the environment. A major question in this context is what kind of spatial information must be integrated in a route description. This task generally refers to two cognitive abilities: Visual perception and natural language. In this domain, a computational model for the generation of incremental route descriptions is presented. Central to this model is a distinction into a visual, a linguistic, and a conceptual-spatial level. Basing on these different levels a software agent, called MOSES, is introduced who moves through a simulated 3D environment from a starting-point to a destination. He selects visuo-spatial information and generates appropriate route descriptions. It is shown how MOSES adopts his linguistic behavior to spatial and temporal constraints. The generation process is based on a corpus of incremental route descriptions which were collected by field experiments. The agent and the 3D environment are entirely implemented.

1 Introduction

What kind of spatial information is necessary for the provision of incremental route descriptions? This question combines two important cognitive modules: Visual perception and natural language generation. We present a computational model of a situated agent[1], called MOSES. He[2] moves through an unknown simulated urban-like 3D environment (see figure 2). His task is to select a path from a map and to describe appropriate actions step-by-step to a virtual listener moving along this path. In contrast to comparable models, MOSES does not simply access information about the environment from a database. MOSES has rather a visual perception module which allows him to perceive and select information from the simulated environment. This approach is grounded on research results gained during a cooperation with the visual perception group of the IIFB at the Fraunhofer Institute, University of Karlsruhe. In joint research with this group we investigated how a model-based approach for visual object selection can be used to automatically recognize 3D object representations. In several domains, we examined how real world data can be used in natural language description systems, e.g. in a soccer domain (cf. [André et al. 89]) and in a traffic domain (cf. [Schirra et al. 87]). The model presented here is based on these investigations.

Studies in visual perception, such as Marr's influential work (cf. [Marr 82]), investigate how visual information is used to construct an internal spatial representation of visible objects. Marr did not, however, show any links between 3D model representations and other processes using spatial information such as natural language processing. Researchers working in this area have mainly been concerned with problems related to the recognition of single objects and object parts (e.g., [Marr & Nishihara 78; Binford 71]). The main purpose of visual perception is to select and group information units in order to make sense out of basic sensor stimulations. Although selection of information is a major issue, most systems only investigate this at early levels of visual processing. Whenever we perceive our environment we select information. The process of visual selection presumes that information provided by the environment is much richer and more complex than what a perceptual system is able to process. Hence, it is assumed that our cognitive system constructs a spatial mental model of the current environment. From this perspective, visual perception is important as input for independent conceptual-spatial representations (e.g., [Johnson-Laird 83]). On the other hand, approaches from linguistics consider the linguistic structure used for describing configurations as being fundamental for spatial cognition (e.g., [Talmy 83; Lakoff 87; Herskovits 86]). Herskovits, for instance, recognizes the distinction

---

[1] We define a situated agent as a computational module which acts in virtual or real environments. It consists of one or more decision making modules and knowledge bases. Its reasoning and planning abilities mainly depend on perception and on self-obtained knowledge. Therefore its knowledge is generally incomplete and inconsistent. But a situated agent is able to adapt its behavior to changes in given situations of the environment.

[2] For readability reasons we use male forms while referring to MOSES throughout this article

between a spatial level and a linguistic level of spatial terms but she does not investigate the relation between both levels in detail (cf. [Herskovits 86, p.102]).

What is known though about a mode-independent spatial level? Baddeley and Hitch proposed in their working memory theory an independent module, called the *visuo-spatial sketch pad* (cf. [Baddeley & Hitch 74; Baddeley 86]). It is concerned with the temporary storage of visuo-spatial information. Another advocator who adopts a linguistic perspective is Jackendoff. Although his *conceptual structure* is strongly influenced by linguistic considerations he writes: "There is a *single* level of mental representation, *conceptual structure*, at which linguistic, sensory, and motor information are compatible"[3] (cf. [Jackendoff 83, p.17]). Another approach is proposed by Johnson-Laird, who states that Marr's general assumption that "all our knowledge of the world depends on our ability to construct models of it" is the basics for all computational models of cognitive processes (cf. [Johnson-Laird 83, p.402]). By his *mental model* approach, Johnson-Laird also suggests an independent knowledge structure between cognitive modules. As Johnson-Laird points out, "we have no way of knowing what the structure is (or even whether the notion makes sense) that is independent from the way in which we conceive the world" (cf. [Johnson-Laird 83, p. 402]). Based on ideas of mental models, Bryant outlined a spatial representation system (SRS) in which he stressed the importance of differernt kinds of frames of reference (cf. [Bryant 92]). Couclelis presents in her proposal how pre-conceptual schema representations, mental models, and cognitive maps can be seen as based on one another (cf. [Couclelis 95]).

A fundamental question for a complete computational theory dealing with the integration of natural language and visual perception is what kind of processes and representations lie in-between[4]. It is fairly well established that visual perception and natural language are independent, cognitive modules and that they have their own representations and processes. An implicit assumption for combining both systems is to look for well-suited interfaces. Similar to retinotopic projections, visual information obtained from a given situation first of all provides two-dimensional information projected onto a plane orthogonal to the direction of movement (see figure 1 which illustrates a crossing scenario). It can be directly distinguished between those objects on the left, those on the right, and those in front. The same holds for top and bottom. More complex to obtain is information about how items are ordered relative to one another. For instance, the relation that item A is behind item B generally requires commonsense knowledge about these items as well as stereo-vision.

Central to the model proposed here is a distinction between mode-specific and mode-independent representations of spatial knowledge. Representations associated to visual perception and natural language are mode-dependent. Conceptual and in particular spatial information is assumed to be processed and represented at a mode-independent level in-between. Representations and processes at this

---

[3] Italics are from the original text.
[4] The integration of visual processing and natural language processing is currently a new and hotly discussed topic in AI (cf. [McKevitt 94b; McKevitt 94a]).

level are not understood in detail. In experiments data is almost exclusively obtained by verbal descriptions (e.g., [Linde & Labov 75; Ehrlich & Johnson-Laird 82]). We discuss here how visual, spatial, and linguistic knowledge structures can be combined with one another to accomplish the task of incremental route descriptions. Therefore, a flow of information is followed from visual perception towards natural language. The advantage of three representation levels is that there is still a clear distinction between perceptual and linguistic processes and representations. This is in particular important in computational models for distinguishing between spatial relations on the conceptual-spatial level and spatial prepositions on the linguistic level. As a domain for investigating the relation between visual perception, natural language, and intermediate processes and representations, we use incremental route descriptions. Route descriptions can be distinguished into *complete* and *incremental route descriptions* (cf. [Maaß 93]). Incremental route descriptions are given step-by-step while moving along the path towards the destination, as from a co-driver. Hence, incremental route descriptions in combination with processes of visual perception are ideal for investigating different representation levels of spatial information. Complete route descriptions are given in advance by using spatial knowledge stored in long-term memory, which generally relates to research about 'cognitive maps' (cf. [Lynch 60; Downs & Stea 73; Allen & Kirasic 85; Hirtle & Jonides 85; McNamara et al. 92; Tversky 92]). Research in this domain is primarily interested in how people represent and retrieve spatial information. The uniform linguistic structure of German route descriptions is the reason why syntactic and semantic structures of route descriptions have been investigated by several linguistic studies (e.g., [Klein 82; Wunderlich & Reinelt 82; Habel 87; Meier et al. 88; Hoeppner et al. 90]). A comparison of complete and incremental route descriptions shows that in the incremental case linguistic structures depend more on descriptions of actions. But a viewer/speaker[5] has the additional tasks of moving through and anticipating changes in the environment. What has not been generally considered in this context are temporal dependencies. We outline how temporal dependencies are integrated in the proposed computational model to achieve adaptive and appropriate behavior.

In the proposed model we distinguish between three different types of objects: a person (unity of viewer and speaker), street items (street segments, decision points), and landmarks (buildings, cars, trees, signs, etc.). Each street item and each landmark are related to MOSES by one spatial relation. MOSES can always describe the position of a visible item in relation to his egocentric frame of reference. The motivation for the distinction between objects and spatial relations is that objects do not appear to "fly" around in space. If we perceive a situation, objects are spatially related to one another. In most models spatial relations between objects are defined on the basis of coordinates in an Euclidean system (e.g., [Müller 88; Gopal et al. 89; Hoeppner et al. 90]). In these models

---

[5] Before you can give a incremental route description you must visually obtain information from the environment. Therefore, MOSES is a combination of a viewer and a speaker.

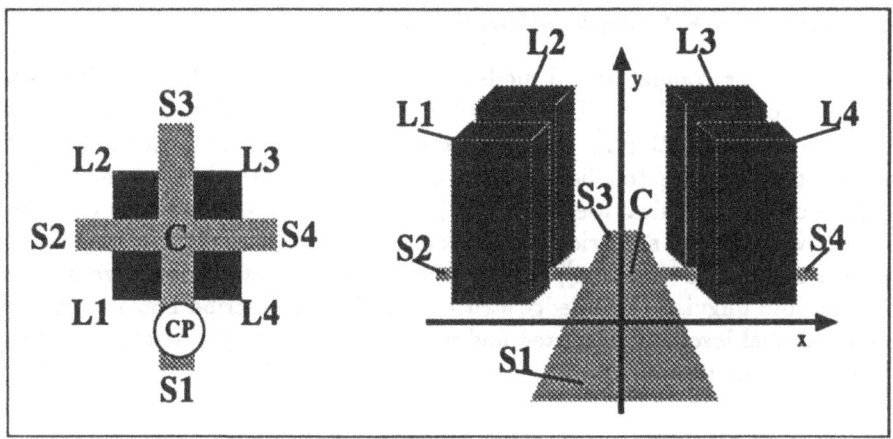

Fig. 1. 2D- and 3D view of a crossing

it is taken for granted that exact positions of objects are provided. Therefore, they are closely related to Geographical Information Systems (cf. [Frank 87; Goodchild 88]). This does not, however, seem to reflect how object locations are represented by the human mind. Research about 'cognitive maps' indicates that mental representations of space are quite inaccurate (e.g., [Tversky 92]), either because the representations themselves are fuzzy or because inference processes on these representations are not as exact as coordinates. From an efficiency perspective it is unreasonable to assume that we first obtain highly accurate geometric information and then transform this during subsequent steps into fuzzy long-term representations. A complementary approach is to use representations based on qualitative spatial relations. Kuipers and Freksa, for instance, propose mechanisms for interrelating places, streets, and the viewer to one another by qualitative spatial structures (cf. [Kuipers 78; Freksa 91]).

We asked people in computer-simulated and real-world environments to give incremental route descriptions. Similar to the results for complete route descriptions, we found that the structure of incremental route descriptions are quite schematic. The schematic structure is important for the process model here. These findings strongly relate to Neisser's visual perception cycle (cf. [Neisser 76]) and his use of schemata, Johnson-Lairds mental models (cf. [Johnson-Laird 83]), and Herrmann's *HOW schemata* (cf. [Herrmann & Grabowski 94]). In AI there are several approaches for formalizing the idea of schemata, such as Minsky's FRAMES or Schank and Abelson's SCRIPTS (cf. [Minsky 75; Schank & Abelson 77]). Schemata are *compiled knowledge* about generally limited domains. FRAMES and SCRIPTS provide a framework for expectations which represent situations compatible with the structure of the domain. The use of schemata is only appropriate in domains with clear-cut structures.

2 Towards a computational model

Central to our computational model is a situated agent, called MOSES, who
moves through simulated 3D environments (for details see [Maaß 93; Maaß 94]).
MOSES selects a path from a map. His task is to describe this path and the environment step-by-step to a listener, who is assumed to follow him (see figure 2 for
a view on the graphical user interface[6]). This can be metaphorically described
as a driver co-driver scenario (for a review of different computational models for
navigation refer to [Maaß 94]). At the linguistic level, spatial knowledge is transformed into linguistic knowledge structures. We first describe how information
at the spatial level is constructed and modified, followed by a description of the
transformation process.

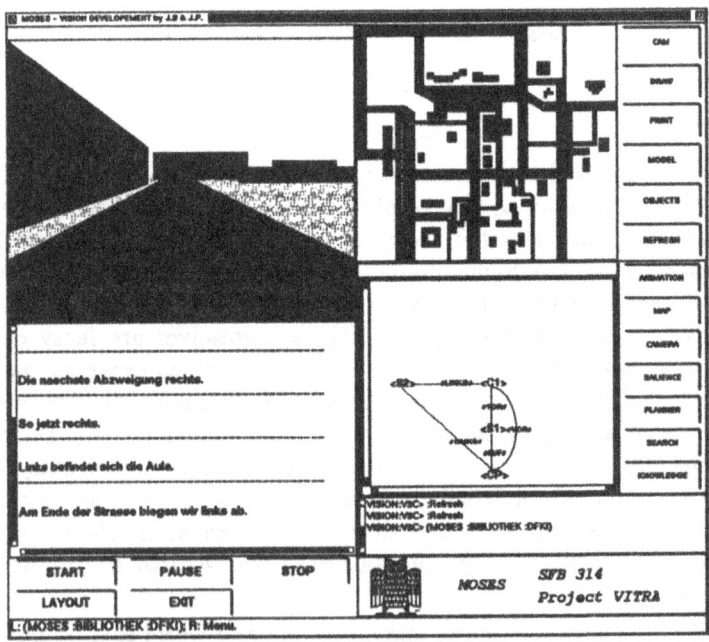

Fig. 2. A View on MOSES

Following Marr, we assume that the visual system generates 3D representa-

---

[6] The current version of MOSES is implemented in CommonLisp and CLOS with the
graphical user interface written in CLIM. The system has been completely developed
on Hewlett Packard Series 700 and SPARC workstations

tions of items obtained from the environment. How we construct 3D representations is, however, beyond the scope of this article (for details see [Herzog et al. 89; Koller et al. 92; Rohr 94]). MOSES has 3D-representations for different types of objects, such as buildings, streets, and cars. The interesting point is which objects and relations are selected from a input stream of visual information. We have determined a computational model for selecting objects by *visual salience* which is based on Treisman's *feature integration theory* (cf. [Maaß 95b]). Visual features, such as color, size, direction of movement, and orientation, are grouped into *feature clusters*. Only those entities which are 'indexed' by features and feature clusters are considered for the identification and categorization of objects. Path-related intentions which determine whether to turn right, left, or to go straight on at the next decision point guide MOSES' focus of *spatial attention area* (see figure 3). Items which lie in the spatial attention area are preferred. If an entity is considered to be salient within a given context it is identified by matching it with object schemata (for more details see [Maaß 95b]).

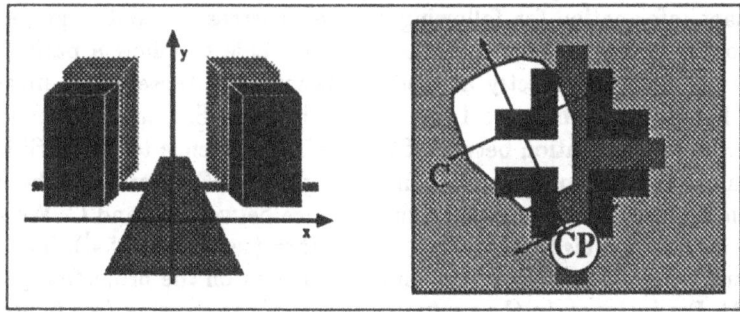

Fig. 3. Focus of spatial attention – top-down and perspective view

Once the objects have been selected, a set of spatial relations between them is determined. Therefore, MOSES transforms the perspective view of a situation into a two-dimensional representation adopting a top-down view. Objects and corresponding spatial relations are integrated in a coherent structure, called a *configuration description*. Here objects are related to one another and to MOSES by geometric spatial relations. Configurational descriptions, which are networks of spatial relations between objects (MOSES' egocentric frame of reference, landmarks, and street items), are divided into two categories: *minimal* and *extended configuration descriptions*. As will be described later, this distinction is mainly motivated by the consideration of temporal and situative constraints. Minimal configurational descriptions only include MOSES' location, street items, and the spatial relations between them. Hence, a minimal configurational description is the minimal amount of information required about the environment which enables MOSES to follow the path. If there is enough time, MOSES also determines landmark information, i.e., he selects landmarks which can be used for describing the next action. If a landmark is integrated in the configurational description we call it *extended configuration description*. For representations on the conceptual-spatial level, we use a restricted set of binary spatial relations, i.e. #left-of#, #right-of#, #in-front-of# and #behind#[7], but obviously not all possible spatial relations between objects, street items, and the viewer are actually determined. As indicated by figure 4, a complex configuration emerges if MOSES only determines one spatial relation between MOSES (CP), street items and landmarks, between landmarks and nearest street items, and between connected street segments and decision points[8].

We asked people to describe turn actions in computer-animated crossing scenes. We found that in time-restricted situations people tended to limit the length of their descriptions. If they had enough time they also referred to salient landmarks. Two classes of spatial relations can be distinguished. First, all street items are related to MOSES' egocentric frame of reference. Second, street segments and decision points are related to one another. Street items, such as street segments and decision points are of primary interest. Landmarks do not provide important information for following a path, whereas without a proper representation of street information, MOSES is not able to follow a path. There is a difference between directly accessible relations and those which must be inferred. For instance, in figure 1 the relation between S2[9] and L4 is not as easy to describe as the relation between S2 and C. In reference to MOSES' location, S2 is in the left half plane and L4 in the right one. Furthermore, the distance between S2 and L4 is greater than the distance between S2 and C. We say that S2 and L4 are not *visually near* to one another (see [Maaß 95a]). Two objects are visually near if they share the same visual area on the projection plane (see figure 1). For instance, in the perspective view of the crossing (see figure 1), L1, S2, and L2 share a similar area on the projection plane, i.e. these objects are visually near.[10]

For efficiency reasons, MOSES only evaluates a minimal set of spatial relations. It is inefficient to evaluate all spatial relations in every situation, especially in the case of moving objects. Therefore, a procedure incrementally adjusts

---

[7] The # indicates that these spatial relations are distinct from spatial prepositions, such as "left of". At the moment it is unclear whether the type of listed set of conceptual spatial relations is appropriate, but it is quite obvious that the four relations are not sufficient to represent all configurations.

[8] A decision point is a location on a street where the viewer has to decide how to continue a path. At a decision point MOSES might turn left or right or go straight on.

[9] CP is the current position of MOSES, L1 to L4 are landmarks, S1 to S4 are street segments, and C is the crossing section as indicated in figure 1.

[10] At the moment we do not consider depth information and experience of the viewer. Currently the distinction into a left and a right visual plane is important for the determination of visual nearness. The next step is to evaluate whether visual nearness must also refer to depth information.

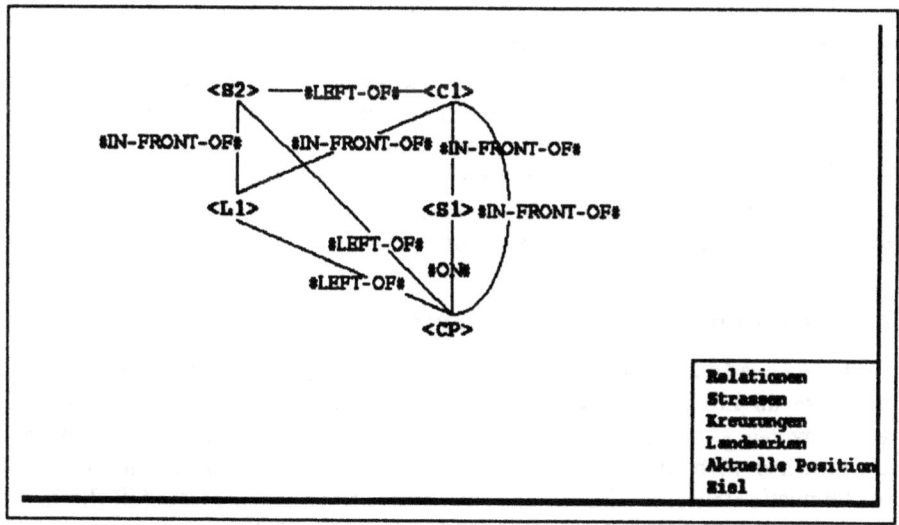

Fig. 4. Example for an extended configurational description

the configuration description to the environment. When a new configuration is formed, first of all the spatial relations between all items and the viewer need to be determined. Then spatial relations between street items can be established. The resulting structure is called a *minimal configurational description*. When MOSES selects a new landmark it is first related to MOSES' egocentric frame of reference by computing the best applicable spatial relation between MOSES and the landmark. The next step is to establish all spatial relations between the landmark and visually near items. Finally, a landmark or a street item is deleted if it is no longer visible or MOSES turns into a new street segment (cf. [Gopal et al. 89] for an initial approach to modeling the decay of spatial knowledge). If landmark information is selected from the environment, the representation structure is called an *extended configurational description* (see landmark L1 in figure 4). In time-restricted situations, MOSES is forced to depend on a minimal amount of information obtained from the environment. Therefore the information represented by the minimal configuration is the basics for MOSES to be able to orientate himself in complex environments. A spatial relation between MOSES's current position and the current street segment S1 is defined by the spatial relation #ON#(CP, S1). In the same way, the crossing C and street segment S2 are related to MOSES (see figure 4) by: (#IN-FRONT-OF#(C, CP) $\wedge$ (#LEFT-OF#(S2, CP)). Besides these relations, MOSES determines a minimal set of relations between street items. In order to avoid a combinatorial explosion not all possible spatial relations between street items are evaluated, only those between physically connected street items, e.g., (#LEFT-OF#(S2, C) $\wedge$ (#IN-FRONT-OF#(S1, C)). In MOSES we have a set of configuration schemata for a sample set of decision point situations with one additional landmark and procedures for combining schemata.

In summary, if an object is salient in a given situation it will be identified by a visual selection process. This triggers a process which integrates this landmark into a configurational description by determining geometric spatial relations between MOSES' current position, street items, and selected landmarks. As we will show next, configurational descriptions are important for the determination of approriate incremental route descriptions.

2.1 Selection of description schemata

We have already mentioned that spatial relations used in configuration descriptions are an initial approach, and mainly coined by verbal descriptions collected by our experiments. Now we describe how configurational descriptions are matched with linguistic structures. This mainly depends on findings that the linguistic structure of German route descriptions is schematic (cf. [Klein 82; Habel 87; Wunderlich & Reinelt 82; Meier et al. 88; Müller 88; Hoeppner et al. 90]). In familiar urban environments, we depend on experience and schemata about how particular objects are expected to be distributed in space. For instance, if we reach a crossing we expect to see buildings on the left and on right hand side and a street going in-between (New York is a master example for that). A ship in the middle of the crossing would cause us to hesitate because it does not fit into our general expectations about traffic situations. By experiments in computer-simulated and real world environments we collected a corpus of incremental route descriptions. In the first experiment, we asked test persons to describe turn left, turn right or go-straight actions in a computer simulated crossing scenario (the scenes presented in these experiments have been similar to the one presented in figure 1). In the first setting a simulated car was driven through an environment with medium speed. Most test persons only described the turning action itself. In settings with lower speed the test persons also included salient landmarks in their descriptions. In settings where they were asked to include a particular landmark they had difficulties in giving a correct description when the landmark was on the opposite side of the y-axis at the next street segment (see figure 1). In this setting most test persons described that the action ("An der nächsten Kreuzung biegst du links ab." ["At the next crossing turn left."]) followed by an extension of the description ("... dort, am ersten Gebäude auf der rechten Seite." ["... there, by the first building on the right-hand side."]). Some persons were not even able to integrate the indicated landmark. One possible conclusion is that in the second setting the landmark on the opposite side does not fit into the preferred schema of describing a situation. By examining the corpus of descriptions, we found that most descriptions can be categorized by a small set of syntactic schemata (for details see [Maaß 95a]). In particular, the categorization into 'WHAT', 'WHEN', 'WHERE', and 'WHERE TO'-phrases is helpful in understanding the structure of route descriptions. A 'WHAT'-phrase describes an action and is usually a verb phrase, e.g., "... mußt du abbiegen ..." (... you must turn ...). Temporal descriptions are introduced by 'WHEN'-phrases, e.g., "... jetzt ..." (... now ...). 'WHERE'-phrases describe the location of a landmark or a location where an action must be performed, e.g., "Da vorne ..." ["There in front ..."] or "Zwischen den beiden Häusern ..." ["Between those two buildings ..."]. An extension of a 'WHERE' phrase is a 'WHERE TO' phrase. The direction of an action is indicated by referring to locative information, e.g., "... nach links ..." ["... left ..."]. 'WHERE TO' phrases are commonly connected to 'WHAT' phrases, e.g., "... nach rechts abbiegen ..." ["... turn right ..."]. For instance, a typical description is: "Bitte gleich rechts abbiegen... hinter dem braunen Gebäude... Jetzt bitte" ["Please turn right ... after the brown building ... Now please."]. The structure of this corpus of German descriptions[11] can be described as a sequence of WHAT-WHERE-WHEN-WHERE TO phrases. We found that the test persons used in 70 percent of cases, one of the following phrase structures: WHERE+WHERE TO+WHAT, WHERE+WHERE TO, or simply WHERE. Based on these sequences we extracted a set of linguistic schemata, called *description schemata*. On the one hand, they reflect the linguistic structure of route descriptions and on the other hand, they correspond to spatial information represented by configuration descriptions. A configurational description provides explicit information about the spatial structure of a situation. Route descriptions mainly depend on the spatial structure represented by configurational descriptions. MOSES considers the given configurational description, intentions, the temporal structure of the situation, his linguistic abilities and knowledge about the listener to select an appropriate description schema (see figure 5). The temporal structure of a situation is constrained by the speed of MOSES and the distance to the next decision point. MOSES makes assumptions about how long it will probably take to reach the next decision point. According to this time interval, only those schemata which can be used to generate a description in time are selected. The next filter selects from these schemata those which correspond to the intended action at the next decision point. For this, only simple path-related intentions are considered, i.e. intentions to turn right, to turn left, or to go straight on. During the next selection step, those schemata are extracted which assume a similar spatial structure to that given by the configurational description. If there are objects selected by the object selection process, then those schemata are prefered which include a reference to salient objects at appropriate places. Most of all, MOSES descriptions depend on his type of movement. When he moves for instance with average car speed, intervals between decision points are sometimes quite short. In these situations, he only refers to route knowledge. If he moves at walking speed, he has more time and can refer to objects. For instance, if a salient object is on the right and his intention is to turn right he gives the description in two parts: "Please, turn right after the red building on the right." .... "Now, turn right please." First, he gives a complete description of the intended action by

---

[11] It is interesting to note that the phrases in the corresponding English descriptions are very similar in their structure. However, our corpus exceptionally consists of German descriptions. Hence we cannot draw any conclusions for other languages, although it seems that there are strong correlations.

referring to objects. Then, just before the action needs to be accomplished, he gives an additional hint. During the last two selection steps those schemata are selected which correspond to the properties of the speaker and the listener (for more details see [Maaß 95a]). This selection process extracts and instantiates one or more description schemata. If there are more than one schemata, MOSES uses the first one. It is clear that a more sophisticated conflict resolution procedure would be helpful, but in our domain we found that this simple strategy serves quite well.

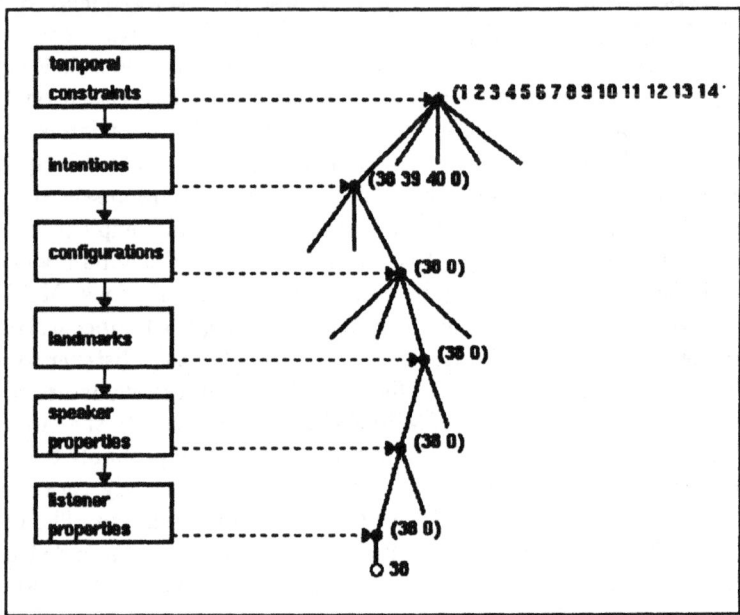

Fig. 5. Selection of a description schema

A description schema represents the semantics of a particular incremental route description. The structure of a schema is based on Jackendoff's *conceptual semantics* (cf. [Jackendoff 83]) and because these are based on simple utterances we carefully extended his formalism (for an example see figure 6). MOSES has a repertoire of almost 60 description schemata. Basic constituents of a description schema are *things* (persons), *locations* (places), and *paths*. They are used in higher-order structures, such as *actions* and *states*. The general structure of an action consists of a reference to the listener's reference frame followed by a description of a path and a place. Hence we can represent utterances such as: "Please, turn right after the building on the right." Figure 6 shows the conceptual structure of this description.

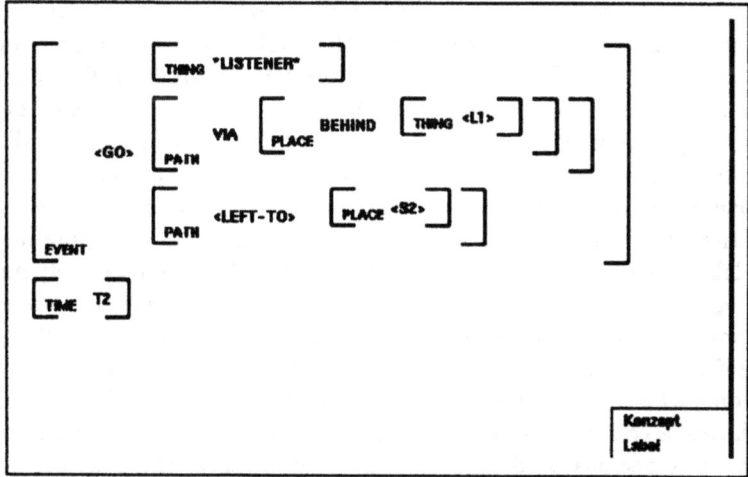

Fig. 6. Example for an *action* description schema

The marker <GO> indicates that schema represents a description of an action. The description is given by adopting the listener's perspective, i.e., the listener's egocentric frame of reference. Then two substructures of type *path* follow. L1 is the pointer to the structure of this landmark. If this location behind L1 is reached, the listener must turn left into a place referred to by S2. Finally, the temporal marker indicates the temporal interval (t2) when the description must be given. These time intervals (t1 to t5) are extracted from experiments in real environments, where we found that five intervals can be distinguished. Most descriptions are given during the last 10 seconds before a action need to be performed. Two time intervals correspond to this time interval (t4 and t5). The environment are presently restricted to crossings with rectangular street configurations. In the future, however, this will be extended so that more complex decision point configurations can also be described. Then, a description schema is transformed into surface structures which serves as input for the natural language process. (For a description of input structure for the natural language generator see [Maaß 95b]). The visual object selection process provides information about landmarks, such as color, height, and width. This kind of information is used for referring to physical attributes of landmarks (cf. [Maaß et al. 95]). Finally, MOSES generates the following description: "Biegen Sie hinter dem günen Haus links ab." ["Turn left after the green building on the left-hand side."].

2.2 Adaptation to the environment

Most AI systems are only built to do *something*. A recently emerging constraint is that they are also required to do something *at anytime*. For instance, robots should not stop on railway tracks to reason about what to do next. An important constraint of anytime algorithms is that the quality of behavior increases with the quantity of the limited resource (cf. [Russel & Wefald 91]). Relating this to the domain of route descriptions means that a description should asymptotically increase in quality with the available amount of time. Most models dealing with spatial knowledge presuppose that the processing time is small compared to the time span of an event. On one hand, temporal constraints for the agent are given by external events in the environment. On the other hand, and more important for models of cognitive processes, temporal constraints are subjectively measured by the agent. When MOSES approaches a decision point he makes assumptions about how long it will probably take to reach this point. Other internal processes refer to this temporal constraint. By measuring time intervals, we found that people who where asked to incrementally describe a route in a real environment showed a common pattern. In situations where the next decision point was far away, test persons tended to give the description of the next action about 10 seconds before arriving at the decision point. In some cases the test person explicitly mentioned that he/she had waited to give the description at the 'right' point of time. In situations where the next decision point was quite near, he/she reduced the complexity of the description, i.e. by only referring to street items. This motivates the distinction between minimal and extended configuration descriptions. First the minimal configuration is generated and used as input for the language generation system. If landmark information and additional path information is obtained, MOSES extends the minimal configuration. This allows MOSES to describe a situation after a short initialization phase. Because MOSES moves through a simulated environment, he adjusts his descriptions to his own movements and to changes in the environment. In situations with little time he only selects and describes a restricted set of visual items. For instance, if MOSES turns left at a crossing and the time interval to the next decision point is only about 10 seconds, then he does not have enough time to analyze the whole scene in detail. Therefore, the description is adapted to this temporal limitation. The main task is to give the appropriate descriptions at the 'right' point of time so that the listener knows where to perform which kind of actions.

3   Summary and conclusion

Incremental route descriptions are ideal for investigating representation levels of spatial knowledge. We outlined a three-level approach for representing spatial information consisting of a visual level, a conceptual-spatial level, and a linguistic level. We focused on the interrelation of representations on the conceptual-spatial level and the linguistic level. Incremental route descriptions provide a well-structured domain for the investigation of the distinction between these three levels. Fundamental to the model is the dissection into mode-specific and mode-independent representations of space. Spatial information obtained by visual perception is represented by 3D models. There are evidences to assume a mode-independent representation structure on a conceptual-spatial level between visual perception and natural language processes. Spatial information about the environment, which is stored in configurational descriptions, is used as input for a description schema selection procedure. Central to MOSES is the inherent schematic structure of incremental route descriptions. What has generally not been considered up to now in the context of route descriptions is the influence of temporal constraints. Therefore, we indicated the importance of temporal constraints, as well as their integration into MOSES to achieve 'anytime' behavior. We are currently examining temporal constraints in more detail and how they affect the verbal behavior of MOSES. We are also investigating the hierarchical organization of spatial configuration descriptions and description schemata. By our corpus of route descriptions and further experiments, we hope to gain more insights into the underlying cognitive and spatial structures.

Since visual perception and natural language are two complex research areas on their own, we are far from having anything like a complete theory which integrates both fields. Nevertheless further efforts focusing on the integration of both areas are required for a better understanding of cognitive processes and representations and also for their use in computational systems.

4 Acknowledgements

I would like to thank Jörg Baus and Joachim Paul for taking care of the implementation issues and for fruitful discussions. In addition, for comments from reviewers which helped to improve the quality of this article in many ways. This work is supported by a grant of the Graduiertenkolleg *Kognitionswissenschaft* at the University of the Saarlandes, Saarbrücken.

References

[Allen & Kirasic 85] G. L. Allen and K. C. Kirasic. *Effects of the cognitive organization of route knowledge on judgments of macrospatial distance.* Memory and Cognition, 13(3):218–227, 1985.

[André et al. 89] E. André, G. Herzog, and T. Rist. *Natural Language Access to Visual Data: Dealing with Space and Movement.* In: F. Nef and M. Borillo (eds.), Logical Semantics of Time, Space and Movement in Natural Language. Proc. of 1st Workshop. Hermès, 1989.

[Baddeley & Hitch 74] A. D. Baddeley and G. J. Hitch. *Working Memory.* In: G. Bower (ed.), Recent advances in learning and motivation. New York: Academic Press, 1974. Vol. VIII.

[Baddeley 86] A. D. Baddeley. *Working Memory.* Oxford: Oxford University Press, 1986.

[Binford 71] T. O. Binford. *Visual Perception by Computer.* In: Proc. IEEE Conf. on Systems and Control, 1971.

[Bryant 92] D. J. Bryant. *A Spatial Representation System in Humans.* Journal of Memory and Language, 31:74–98, 1992.

[Couclelis 95] H. Couclelis. *Verbal directions for way-finding: space, cognition, and language.* In: J. Portugali (ed.), The Construction of Cognitive Maps. Kluwer Publishers, 1995. in print.

238

[Downs & Stea 73] R. M. **Downs** and D. **Stea**. *Cognitive Maps and Spatial Behaviour: Process and Products*. In: R. M. Downs and D. Stea (eds.), Image and Environment. Cognitive Mapping and Spatial Behaviour, pp. 8–26. Chicago: Aldine, 1973.

[Ehrlich & Johnson-Laird 82] K. **Ehrlich** and J. N. **Johnson-Laird**. *Spatial descriptions and referential continuity*. Journal of Verbal Learning and Verbal Behavior, 21:296–306, 1982.

[Frank 87] A. **Frank**. *Towards a Spatial Theory*. In: Proc. of the International Symposium on Geographic Information Systems: The Research Agenda, pp. 2:215–227, Crystal City, Virginia, 1987.

[Freksa 91] Ch. **Freksa**. *Conceptual neighborhood and its role in temporal and spatial reasoning*. In: M. Singh and L. Trave-Massuyes (eds.), Decision support systems and qualitative reasoning, pp. 181–187. Amsterdam: North-Holland, 1991.

[Goodchild 88] M. F. **Goodchild**. *Towards an Enumeration and Classification of GIS Functions*. In: Proc. of the International Geographic Information Systems Conference: The Research Agenda, pp. II:67–77, Washington, 1988. NASA.

[Gopal et al. 89] S. **Gopal**, R. **Klatzky**, and T. **Smith**. *NAVIGATOR: A Psychologically Based Model of Environmental Learning Through Navigation*. Journal of Environmental Psychology, 9:309–331, 1989.

[Habel 87] Ch. **Habel**. *Prozedurale Aspekte der Wegplanung und Wegbeschreibung*. LILOG-Report 17, IBM, Stuttgart, 1987.

[Herrmann & Grabowski 94] T. **Herrmann** and J. **Grabowski**. *Sprechen: Psychologie der Sprachproduktion*. Spektrum, Akademischer Verlag, 1994.

[Herskovits 86] A. **Herskovits**. *Language and Spatial Cognition. An Interdisciplinary Study of the Prepositions in English*. Cambridge, London: Cambridge University Press, 1986.

[Herzog et al. 89] G. **Herzog**, C.-K. **Sung**, E. **André**, W. **Enkelmann**, H.-H. **Nagel**, T. **Rist**, W. **Wahlster**, and G. **Zimmermann**. *Incremental Natural Language Description of Dynamic Imagery*. In: Ch. Freksa and W. Brauer (eds.), Wissensbasierte Systeme. 3. Internationaler GI-Kongreß, pp. 153–162. Berlin, Heidelberg: Springer, 1989.

[Hirtle & Jonides 85] S. **Hirtle** and J. **Jonides**. *Evidence of hierarchies in cognitive maps*. Memory and Cognition, 13(3):208–217, 1985.

[Hoeppner et al. 90] W. **Hoeppner**, M. **Carstensen**, and U. **Rhein**. *Wegauskünfte: Die Interdependenz von Such- und Beschreibungsprozessen*. In: C. Freksa and C. Habel (eds.), Informatik Fachberichte 245, pp. 221–234. Springer, 1990.

[Jackendoff 83] R. **Jackendoff**. *Semantics and Cognition*. Cambridge, MA: MIT Press, 1983.

[Johnson-Laird 83] P. N. **Johnson-Laird**. *Mental Models: Towards a Cognitive Science of Language, Inference, and Consciousness*. Cambridge University Press, 1983.

[Klein 82] W. **Klein**. *Local Deixis in Route Directions*. In: R. J. Jarvella and W. Klein (eds.), Speech, Place, and Action, pp. 161–182. Chichester: Wiley, 1982.

[Koller et al. 92] D. **Koller**, K. **Daniilidis**, K. **Thórhallson**, and H. H. **Nagel**. *Model-based Object Tracking in Traffic Scenes*. In: G. Sandini (ed.), The Second European Conference on Computer Vision, pp. 437–452, Berlin, Heidelberg, 1992. Springer.

[Kuipers 78] B. **Kuipers**. *Modeling Spatial Knowledge*. Cognitive Science, 2:129–153, 1978.

[Lakoff 87] G. Lakoff. *Women, Fire, and Dangerous Things. What Categories Reveal about the Mind.* Chicago: Chicago University Press, 1987.

[Linde & Labov 75] C. Linde and W. Labov. *Spatial Network as a Site for the Study of Language and Thought.* Language, 51:924–939, 1975.

[Lynch 60] K. Lynch. *The Image of the City.* MIT Press, 1960.

[Maaß et al. 95] W. Maaß, Jörg Baus, and Joachim Paul. *Visual Grounding of Route Descriptions in Dynamic Environments.* In: AAAI Fall Symposium on "Computational Models for Integrating Language and Vision", MIT, Cambridge, MA, 1995. AAAI. in print.

[Maaß 93] W. Maaß. *A Cognitive Model for the Process of Multimodal, Incremental Route Description.* In: Proc. of the European Conference on Spatial Information Theory. Springer, 1993.

[Maaß 94] W. Maaß. *From Visual Perception to Multimodal Communication: Incremental Route Descriptions.* Artificial Intelligence Review Journal, 8(5/6), December 1994. Special Volume on Integration of Natural Language and Vision Processing.

[Maaß 95a] W. Maaß. *Ein situierter inkrementeller Wegbeschreibungsagent in 3D-Umgebungen.* PhD thesis, Universität des Saarlandes, 1995. in preparation.

[Maaß 95b] W. Maaß. *Selection of objects by evaluation of visual features.* in preparation, 1995.

[Marr & Nishihara 78] D. Marr and H. K. Nishihara. *Representation and Recognition of the Spatial Organization of three-dimensional shapes.* In: Proc. Royal Society of London B, pp. 269–294, 1978.

[Marr 82] D. Marr. *Vision: a computational investigation into the human representation and processing of visual information.* San Francisco: Freemann, 1982.

[McKevitt 94a] P. McKevitt (ed.). *Integration of Natural Language and Vision Processing.* AAAI-94 Workshop. Seattle, WA, 1994.

[McKevitt 94b] P. McKevitt (ed.). *Special Volume on the Integration of Natural Language and Vision Processing*, volume 8: Artificial Intelligence Review Journal. Dordrecht: Kluwer, 1994.

[McNamara et al. 92] T. McNamara, J. Halpin, and J. Hardy. *The representation and integration in memory of spatial and nonspatial information.* Memory and Cognition, 20(5):519–532, 1992.

[Meier et al. 88] J. Meier, D. Metzing, T. Polzin, P. Ruhrberg, H. Rutz, und M. Vollmer. *Generierung von Wegbeschreibungen.* KoLiBri Arbeitsbericht 9, Fakultät für Linguistik und Literaturwissenschaft, Universität Bielefeld, 1988.

[Minsky 75] M. Minsky. *A Framework for Representing Knowledge.* In: P. H. Winston (ed.), The Psychology of Computer Vision. New York: McGraw-Hill, 1975.

[Müller 88] S. Müller. *CITYGUIDE: Ein System zur Wegplanung und Wegbeschreibung.* Diplomarbeit, Fachbereich Informatik der Universität des Saarlandes, 1988.

[Neisser 76] U. Neisser. *Cognition and Reality.* San Francisco: Freeman, 1976.

[Rohr 94] K. Rohr. *Towards Model-based Recognition of Human Movements in Image Sequences.* Computer Vision, Graphics, and Image Processing (CVGIP): Image Understanding, 59(1):94–115, 1994.

[Russel & Wefald 91] S. Russel and E. Wefald. *Do the Right Thing: Studies in Limited Rationality.* Cambridge, MA: MIT Press, 1991.

[Schank & Abelson 77] R. C. Schank and R. P. Abelson. *Scripts, Plans, Goals and Understanding.* Hillsdale, NJ: Erlbaum, 1977.

[Schirra et al. 87] J. R. J. **Schirra**, G. **Bosch**, C.-K. **Sung**, and G. **Zimmermann**. *From Image Sequences to Natural Language: A First Step Towards Automatic Perception and Description of Motions*. Applied Artificial Intelligence, 1:287–305, 1987.

[Talmy 83] L. **Talmy**. *How Language Structures Space*. In: H. Pick and L. Acredolo (eds.), Spatial Orientation: Theory, Research and Application, pp. 225–282. New York, London: Plenum, 1983.

[Tversky 92] B. **Tversky**. *Distortions in cognitive maps*. Geoforum, 23:131–138, 1992.

[Wunderlich & Reinelt 82] D. **Wunderlich** and R. **Reinelt**. *How to Get There From Here*. In: R. J. Jarvella and W. Klein (eds.), Speech, Place, and Action, pp. 183–201. Chichester: Wiley, 1982.

On the Determination of the Optimum Path in Space

Emmanuel Stefanakis

Dept. of Electrical and Computer Engineering
National Technical University of Athens
Zographou, Athens, Greece 15773
E-mail: stefanak@theseas.ntua.gr

Marinos Kavouras

Dept. of Rural and Surveying Engineering
National Technical University of Athens
Zographou, Athens, Greece 15773
E-mail: mkav@theseas.ntua.gr

Abstract: Various algorithms have been proposed for the determination of the optimum paths in line networks. Moving in space is a far more complex problem, where research has been scarce. An example would be the determination of the shortest sea course between two given ports. This paper presents an examination of the problem, states the weaknesses of the existing solutions, and introduces a new approach, which can be easily applied to a variety of spaces, while considering different travel cost models. The implementation of the algorithm for movements on the plane surface, in the three-dimensional space, and on the spherical surface as an approximation of the earth, has been examined. The results are illustrated through several examples.

1. Introduction

The determination of the *optimum path* between two physical locations is a very common problem in applications such as Cartography, Robotics and Geographic Information Systems (GISs). Optimum in this context refers to a minimal accumulation of what amount to incremental travel costs associated with different media. It may be the shortest, fastest, least-expensive, or least-risky path.

When movement is restricted to the chains of a linear network, like road or aircraft corridor networks, a weighted graph can be used as a model and associated algorithms [Gibb85, Sedg90] may be applied for the determination of the optimum path.

Moving in space is a far more complex problem, where research has been scarce. Examples would be the determination of the fastest path between two villages (Figure 1a); the shortest sea course between two ports (Figure 1b); the most regular gradient on ground path over a mountainous terrain (Figure 1c); the least-risky path in a hostile environment, for instance, the path with the maximum concealment time vis-à-vis an enemy or an observer (Figure 1d).

Clearly, the space under study has its own peculiarities. For instance, it may be composed of a number of regions or volumes with different travel cost values assigned to them (e.g., walking on grass or sand); it may involve various means of travel (e.g., walking or driving); the direction of movement may introduce a variable cost value (e.g., moving with or against the wind).

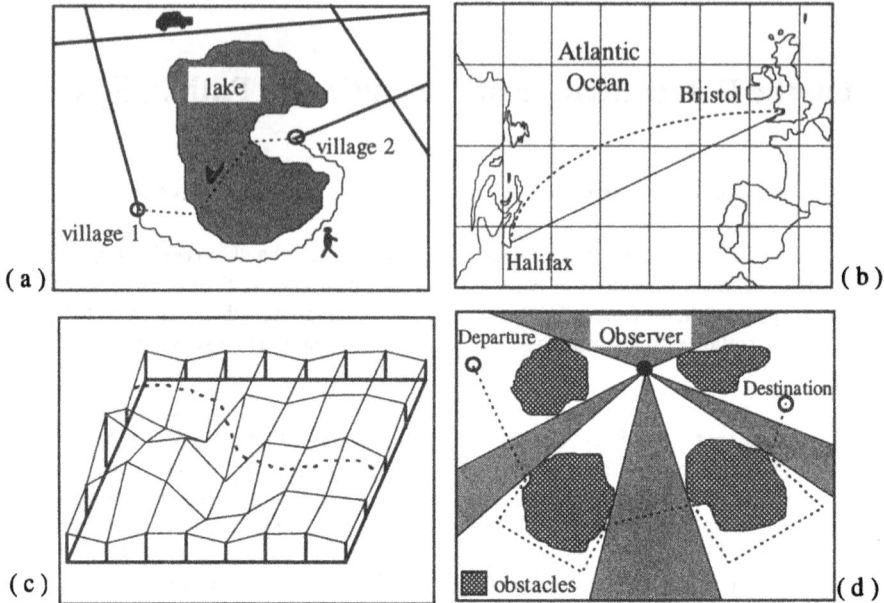

Fig. 1. Examples of optimum paths in space.

Previous work in the area of optimum path finding has been mostly confined to the examination of the problem on a plane surface consisting of zones characterized by variable costs of movement [Warn57,61, Lind67,69ab, Doug94ab]. The proposed solutions suffer from several weaknesses, which have a negative impact on the determination of the optimum path. The objectives of this paper are twofold: first, to introduce a new approach to the optimum path finding problem; and second, to examine its implementation to a variety of spaces.

Section 2 briefly reviews existing solutions to the optimum path finding problem. Section 3 introduces the new approach for the determination of the optimum path connecting two physical locations. Section 4 examines its applicability to a variety of spaces. Finally, Section 5 concludes the discussion by summarizing the contribution of the paper and giving hints for future research in the area of optimum path finding.

2. Existing Solutions to the Optimum Path Finding Problem

The problem of the optimum path finding has been discussed in the past and several solutions have been proposed. The research has been mostly confined to movement on a plane surface consisting of zones characterized by different travel cost values.

An early proposed solution [Warn57] adopts the analogy of *light refraction*. It can be shown that the optimum path connecting two physical locations lying on differently weighted zones is similar to the path of light as expressed in the law of refraction, otherwise known as the Fermat principle of the Snell's (or Descartes') law (e.g., dashed line in Figure 1a). This solution becomes difficult to apply when the boundary between the zones is wiggly (i.e., other than a straight line) and where the cost of movement varies gradually over the surface (i.e., fuzzy boundaries). This inconvenience may be overcome by approximating the surface under study with a sample surface, specifically a regular grid (i.e., generalization process).

More recent research [Warn61, Lind67,69ab] led to a three step approach for the determination of the optimum path(s) on a surface [Aron89]:

Step 1: Generation of the friction surface.
The purpose of this step is to generate a data layer in raster format (i.e., a dasymetric map) whose cells are assigned values that represent the cost of movement across them (e.g., Figure 2b). This layer is called *friction surface* and can be derived from a vector map of land uses (Figure 2a) by applying a sampling technique (vector-to-raster conversion). Based on testing, previous experience, or other information, the cost to traverse one cell is determined for each land use type.

Step 2: Generation of the accumulation cost surface.
An *accumulation cost surface* [Warn61] is defined as one that represents by the Z coordinate value, the cost of movement to (from) any point X_i, Y_i from (to) a *point of reference X_o, Y_o* (clearly, $Z_o=0$). The point of reference may be either the departure or the destination point of the trip. On a homogeneous plane the accumulated cost surface is an inverted cone, centered on the point of reference and fully described by a mathematical model.

When the surface under study consists of barriers or zones where cost of movement differs from zone to zone, the accumulation cost surface is derived from the friction surface by applying a *spread function* (i.e., the principle of the advancing wave front). A spread function evaluates phenomena that accumulate with distance [Jens85, Toml90]. Its operation can be thought as progressing step-by-step outward in all directions from a starting point and adding the travel cost of each successive step to the accumulated cost back to that point. After determining the cell on the friction surface which represents the point of reference, the spread function searches the neighbouring eight cells, stopping on the one not previously assigned an accumulated travel cost value; and assigns a cost value to it. From that cell another search begins in a similar way. This operation results in the accumulation cost surface in raster format (Figure 2c).

Step 3: Determination of the optimum path(s).
The accumulated cost surface generated in the previous step forms the basis for the determination of the optimum paths from (to) the point of reference to (from) any point of the surface under study. Two techniques have been proposed so far.

The *first technique* is based on the property of *slope lines* (i.e., lines that cut perpendicular the contour lines of a surface) to be the shortest lines connecting the summit to any point downhill. What is suggested [Warn61, Lind69ab] is to represent the accumulated cost surface by isodynamic (contour) lines and trace the slope lines from the point of reference to any point of interest (Figure 2d). The problem faced by this technique is that the spread function applied to generate the accumulated cost surface results in eight-sided pyramids instead of inverted cones around the points of reference and as a consequence the surface is represented by octagonal contours, while circles would be the desired result. This problem may be overcome by applying a more complex search in spread function [Doug94ab].

The *second technique* makes use of a *seek function* (also termed a stream function). The seek function performs a directed search outward in a step-by-step manner from a start location using a specified rule. This procedure is performed recursively until further movement would violate the decision rule. For the determination of the optimum path the seek function operates as follows. Starting from the cell that represents the location of interest on the accumulated cost surface, it progresses to the adjacent cell(s) with the smallest accumulated travel cost values. This operation is repeated until the point of reference is reached (i.e., this is the violation of the rule) (Figure 2e; dashed line).

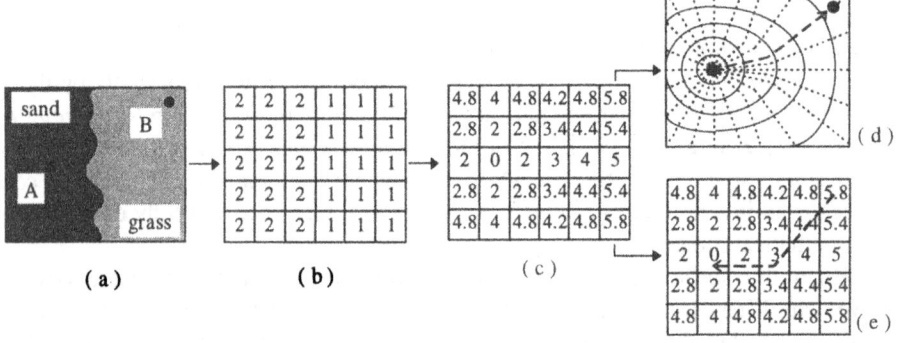

Fig. 2. The three step approach to the determination of the optimum path on a plane surface (from A to B)

The three step approach to the determination of the optimum path suffers from several weaknesses, such as:

- The generalization process as a result of the sampling applied to the formation of the friction surface is accompanied by a loss of information regarding the structure (i.e., components) of the surface under study.
- The generation of the accumulated cost surface is based on movements confined to eight directions (i.e., the number of direct and indirect neighbours of each cell on the regular grid).
- The friction surface is unable to represent a surface where the direction of movement introduces a variable travel cost value (e.g., walking on a mountainous surface), unless eight friction surfaces are maintained (i.e., one for each direction of movement).

These weaknesses, which may have a negative impact on the determination of the optimum path, led to the development of a more general solution, introduced in the following section.

3. A New Approach to the Optimum Path Finding Problem

This section introduces a new approach to the determination of the optimum path in space. The concept behind this approach is to establish a network connecting a finite number of locations (including departure and destination points) in space (Figure 3), so that effective algorithms coming from the weighted graph theory [Gibb85, Sedg90] can be adopted to indicate the optimum path for the desired trip.

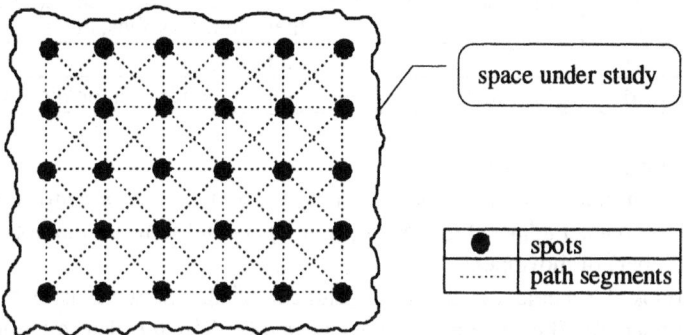

Fig. 3. The concept of the new approach on a plane surface.

In summary, the new approach consists of five steps:
1. Determination of a finite number of spots in space.
2. Establishment of a network connecting these spots.
3. Formation of the travel cost model.
4. Assignment of accumulated travel cost values to these spots from the point of reference (i.e., the departure or destination spot).
5. Determination of the optimum path(s).

These steps are discussed successively in the following subsections.

3.1 Determination of a Finite Number of Spots in Space

The inconvenience that characterizes the movement in space is the infinite number of *spots* (i.e., locations) involved in the determination of the optimum path. The proposed solution to overcome this problem is based on the technique of *discretization of space* under study. Discretization is the process of partitioning the continuous space into a finite number of disjoint areas or volumes (cells), whose union results in the space. By representing each of these cells with one spot (e.g., its center point), a finite set of spots is generated to be involved in the process of the determination of the optimum path(s).

A wide variety of *tessellations* (also termed meshes) are available for obtaining the desired partitioning of the space under study [Laur92]. Tessellations may consist of regular or irregular cells. In the former case, space is partitioned by a repeatable pattern of regular polyhedra (regular polygons on plane surface); while in the latter case, space is partitioned by an extending configuration of polyhedra with variable shape and size.

In order to achieve a uniform distribution of spots over the space under study a regular tessellation should be adopted. Table 1 lists some possible tessellations for several spaces.

Table 1. Available Tessellations

Space	Possible Tessellations
Plane Surface	regular grid, hexagonal grid, triangular grid, etc.
3-d Space	cubic blocks, other platonic solids, etc.
n-d Space	n-d blocks
Spheroid	geographic grid, polyhedral tessellations, cubic blocks, etc.

3.2 Establishment of a Network

After the determination of the spots over the space under study, a *network* should be established to connect adjacent or non-adjacent spots and indicate the possible paths (finite in number) of movement.

The proposed scheme for the establishment of the network is based on the tessellation used for the generation of spots involved in the determination of the optimum path. Specifically, each spot (assigned to a cell) is connected through network edges to the spots of the neighbouring cells. Independently of the tessellation adopted to partition space, each cell has three types of neighbour cells: a) *direct,* i.e., neighbours with shared edges; b) *indirect,* i.e., neighbours with common vertices; and c) *remote neighbours*. Remote neighbours are characterized by the level of proximity to the cell of reference. For instance, level-one (level-two) remote neighbours are the cells which are direct or indirect neighbours of the direct or indirect neighbours of the cell of reference (of the level-one remote neighbours of the cell of reference). Figure 4 illustrates an example for the regular and triangular tessellations on the plane surface. Note that, by increasing the number of neighbours considered, the directions of movements are augmented. For instance, in a regular tessellation on the plane surface, the direct neighbours introduce a set of four directions, the indirect neighbours another set of four directions, and the level-one remote neighbours another set of eight directions of movement. An exhaustive network would consider all direct, indirect and remote (of any level) neighbours.

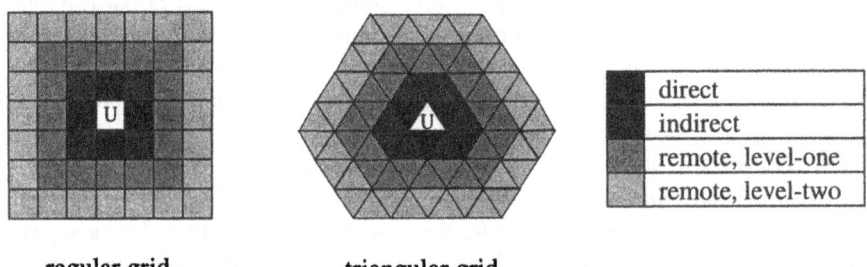

regular grid   triangular grid

	direct
	indirect
	remote, level-one
	remote, level-two

Fig. 4. Types of neighbour cells (plane surface).

3.3 Formation of the Travel Cost Model

The *travel cost model* assigns weights to the edges of the network established in the previous step. Its form depends on both the space under study and the application needs. Examples of travel cost models are:
- the model of the shortest path (minimum distance)
- the model of the fastest path (minimum time)
- the model of the least expensive path (minimum expenses)
- the model of the least risky path (minimum risk)

The *model of the shortest path* is examined next in more detail. This is the simplest model to understand and it is mostly used in the examples of Section 4. The space under study is represented by a set of spots which may be *accessible* or *non-accessible* (i.e., lying on obstacles; usually not considered). For instance, spots lying on the sea are accessible, while those lying on the continents are non-accessible for a ship (Figure 1b). The travel cost between two accessible spots is equal to the length of the shortest line connecting them, if they are *intervisible* [Aron89]. Two spots are intervisible, if the shortest line connecting them passes through no obstacle. If this is not the case, the travel cost between the two spots is computed implicitly passing through intermediate spots.

Figure 5 provides the pseudo-code of the travel cost function for the model of shortest path. Note that when two spots (U_from,U_to) are not intervisible (i.e., function intervisible() returns False) or the destination spot (U_to) is not accessible (i.e., function accessible() returns False), the edge connecting them is assigned a cost of infinity. This indicates that there is no direct path connecting the two spots. The function distance() returns the length of the shortest path between two spots that satisfies the rules of movement. For instance, when movement is restricted on a plane surface, it returns the length of the straight line connecting two spots; when movement is restricted on the surface of a sphere (e.g., as an approximation of the earth's surface), it returns the length of the great circle connecting the two spots.

```
travel_cost(U_from,U_to)
begin
   if accessible(U_to) and intervisible(U_from,U_to)
      then
      return distance(U_from,U_to);
   else
      return ∞;
   endif;
end;
```

Fig. 5. Pseudo-code of the travel cost function for the model of shortest path.

3.4 Assignment of Accumulated Travel Cost Values

After the establishment of the network and the formation of the travel cost model, *accumulated travel cost values* from a spot of reference (i.e., departure or destination spot) to all spots of the space under study (as chosen in step 1; Section 3.1) can be assigned. This task is performed by adopting techniques from graph theory [Gibb85,

Sedg90], concerning the determination of the travel cost values to the nodes of the graph from the node of reference; which are analogous to that applied by the spread function (Section 2).

Figure 6 lists a modified pseudo-code of the *Dijkstra's algorithm* [Dijk59] for the assignment of accumulated travel cost values to the spots of the space under study. After indicating the spot of reference (Ur) the algorithm works as follows. Initial travel cost values C are assigned to all spots of the space including the spot of reference. The latter is assigned the initial cost of movement (usually a nil cost), whereas the other spots are assigned a "huge" cost (i.e., a value larger than the most costly path expected), denoted by the symbol of infinity in Figure 6. A balanced tree (denoted by UV) should be maintained to keep track of all candidate spots to be visited next at any time during the execution of the code. Initially the tree is empty and the spot of reference is inserted to it. While the tree is not empty the following procedure is executed repetitively. The spot U, with the lowest accumulated cost value assigned to it, is retrieved and removed from the tree. All its neighbour spots (Un ∈ UN) are determined, based on the network established (Step 2; Section 3.2); they are inserted in the tree, if they are never visited before (i.e., their cost value is equal to infinity); and they are assigned a new cost value which is equal to the cost of U plus the travel cost from U to Un, if the sum is lower than the initial cost of Un.

```
accumulated_travel_cost(Ur)
begin
   C[Ur] ← 0;
   for each U ∈ UM - {Ur} do C[U] ← ∞;
   UV ← {Ur};
   while UV ≠ 0 do
   begin
      for U ∈ UV : C[U] = min do
      begin
         UV ← UV - {U};
         UN ← neighbours(U);
         for each Un ∈ UN do
         begin
            if accessible(Un) and C[Un] = ∞  then
            UV ← UV ∪ {Un};
            C[Un] ← min{C[Un], C[U]+travel_cost(U,Un)};
         end;
      end;
   end;
end;
```

Fig. 6. Pseudo-code of the accumulated travel cost function (based on Dijkstra).

When the loop stops (i.e., empty tree) the algorithm terminates. All spots over the space under study are assigned the accumulated travel cost values back to the spot of reference, as desired. Proofs regarding the rightness of the algorithm and its complexities are similar to those provided in graph theory concerning the Dijkstra's algorithm [Dijk59, Gibb85, Sedg90].

The storage and processing complexity of the algorithm are $O(N_S)$ and $O(N_{NEIGH}*N_S)$ ($O(N_S^2)$ for exhaustive networks) respectively, where N_S is the number of spots of the network and N_{NEIGH} is the average number of neighbours considered for each spot. For special cases of graphs, improved versions of the algorithm [John77, Fred87] provide a lower upper bound regarding the processing complexity.

3.5 Determination of the Optimum Path(s)

After the assignment of the accumulated travel cost values to the spots (nodes) of the network, the procedure for the *determination of the optimum path(s)* from the spot of reference to a spot of interest (or the reverse) is analogous to that adopted by the seek function (Section 2), and consists of the following operation. Starting from the spot of interest, all its neighbours are examined and the spot(s) with the smallest cost value assigned to it (them) are considered as the previous spot(s) of the optimum path(s). Starting from these spots, the operation is repeated, until the spot of reference is reached.

Figure 7 presents the pseudo-code of an algorithm for the determination of a unique optimum path between the spot of interest Ud to the spot of reference Ur. The algorithm can be easily extended for determining multiple optimum paths, if desired.

```
optimum_path(Ur,Ud)
begin
   accumulated_travel_cost(Ur)
   U = Ud;
   while U ≠ Ur do
   begin
        UN ← neighbours(U);
      for Un ∈ UN : C[Un] = min do
      begin
         U = Un;
         report U;
      end;
   end;
end;
```

Fig. 7. Pseudo-code for the determination of a unique optimum path.

The five step approach to the determination of the optimum path in space presents several advantages over the existing three step approach (discussed in Section 2):

- The new approach does not make use of a friction surface (i.e., raster data layer derived by applying a sampling technique) and as a consequence no information regarding the structure of the space under study is lost. For instance, on a non-homogeneous surface, like that shown in Figure 2a, the cost of movement along the edge of the network connecting two spots is directly computed by considering the edge on the vector map representing the land uses. Note that, while existing solutions are based on a generalization of both the spots considered and structure of the space under study, the new approach is based on the generalization of the former only.
- The number of directions of movement is not confined to the number of direct and indirect neighbours of each cell on the friction surface. Contrary, the new approach allows the consideration of a variable number of directions, which depends on the set of spots determined in the space under study and the network established over them.
- The proposed solution is based on the degeneration of the space under study into a network which can be simulated by a weighted graph. Therefore, powerful algorithms of graph theory can be easily adopted for the determination of the optimum path(s) in space.
- The new approach can be applied to a wide variety of spaces with various travel cost models associated to them. The following section presents several examples.

4. Implementation

The scope of this section is to examine the applicability of the new approach to a variety of spaces, such as the plane surface, the 3-D space, and the spherical surface as an approximation of the earth surface.

4.1 The Plane Surface

The plane surface is the simplest space to study. The finite number of spots can be easily determined by using one of the available two dimensional tessellations (Table 1) and locating one spot to the center point of each cell. The network is then established by connecting the spots through straight line segments, which are assigned weights depending on the structure of the surface and the travel cost model considered.

Figure 8 shows an example on the determination of the shortest path on a plane surface with obstacles (shaded areas). The tessellation used is the regular grid with a resolution of 10 units in both X and Y dimensions, while the established network connects each spot to its direct and indirect neighbour spots (Figure 8b) or direct, indirect and level-one remote neighbour spots (Figure 8c) by straight line segments (i.e., eight or sixteen direction of movements are considered). From the accumulated travel cost values assigned to the spots, the shortest paths are derived (solid lines in Figures 8b,c). Note that, by increasing the number of neighbours considered, a more accurate optimum path is derived, for the same set of spots (i.e., resolution of the tessellation) on the plane. The effects of both the number of spots and number of neighbours considered is one of the directions of our future research (see Section 5).

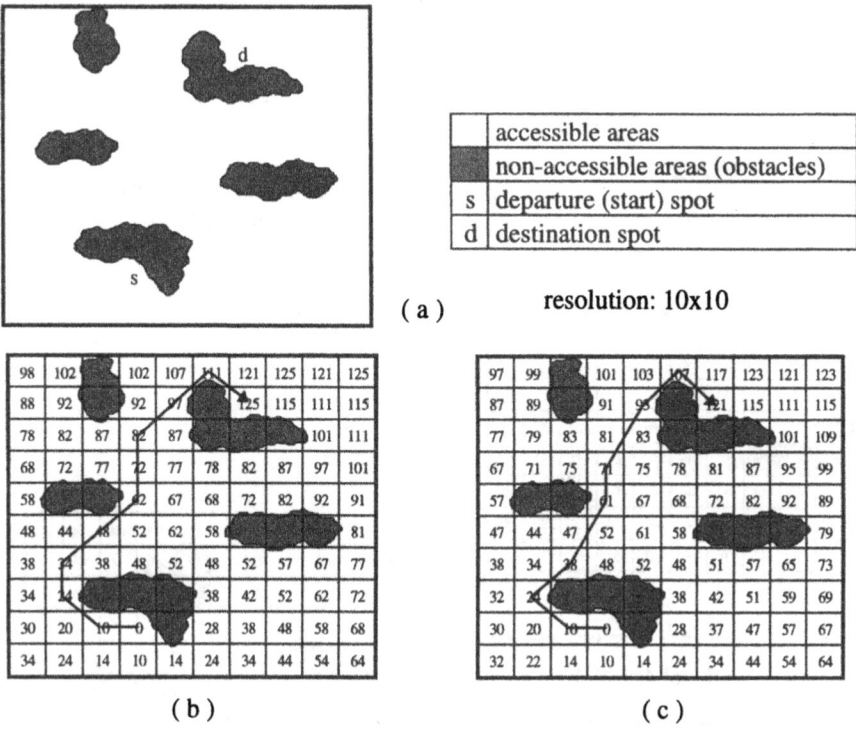

	accessible areas
▓	non-accessible areas (obstacles)
s	departure (start) spot
d	destination spot

(a)   resolution: 10x10

Fig. 8. The shortest path on a plane surface (values assigned to the grid cells refer to the accumulated travel costs of the corresponding spots).

4.2 The 3-D Space

Moving in the 3-D space is a generalization of movement on a plane surface. The spots are similarly determined using one of the available three dimensional tessellations (Table 1) and the network is established by connecting these spots through straight line segment, which are assigned weights depending on the structure of the space and the travel cost model considered.

Figure 9 presents an example on the determination of the shortest path in a 3-D space enclosing obstacles (shaded volumes). All obstacles have their basis on level 0, while their height is noted in Figure 9a over the shaded areas. The maximum height of the space is 40 units, while departure and destination spots have an altitude of 5 and 25 units respectively. The tessellation used is the regular grid (i.e., cubic blocks) with a resolution of 10 units in all X, Y and Z dimensions. As for the established network, it connects each spot with its direct and indirect neighbours. From the accumulated travel cost values assigned to the spots, showed in layers (Figure 9b,c.d,e) the shortest paths are derived (one of them is represented by the solid line, which jumps across the levels). The extension to spaces with higher dimensionality (n-D spaces) is considered as a simple generalization of the concept.

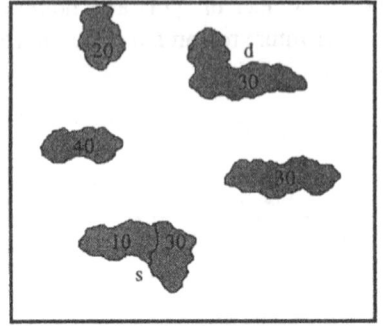

	accessible volumes
▓	non-accessible volumes (obstacles)
s	departure (start) spot (level 1)
d	destination spot (level 2)

maximum height = 40

(a)   resolution: 10x10x10

102	99	96	92	96	99	103	107	111	115
92	89	86	82	86	89	93	97	101	105
82	79	76	72	76	79	83	87	91	101
72	76	66	62	66	69	73	77	87	97
62			52	56	59	63	73	83	93
52	49	46	42	46	49	59	69	79	89
49	45	41	38	41	46	56	66	76	86
46	41	37	34	37	47	52	62	72	82
42	38	34	30	34	44	49	59	69	79
46	41	37	34	37	41	46	56	66	76

(b) : level 4: [30-40]

96	92		84	88	92	102	109	105	109
86	82		74	78		106	99	95	99
76	72	68	64	68				85	95
66	68	58	54	58	62	67	71	81	91
56			44	48	52	57	67	77	87
46	41	37	34	38	42				82
41	31	27	24	28	38	48	58	68	78
37	27	17	14		41	46	56	66	76
34	24	14	10		31	41	51	61	71
37	27	17	14	17	27	37	47	57	67

(d) : level 2: [10-20]

99	96	91	88	91	96	106	110	108	112
89	86	81	78	81		96	100	98	102
79	76	71	68	71				88	98
69	71	61	58	61	66	70	74	84	94
59			48	51	56	60	70	80	90
49	45	41	38	41	46				86
45	35	31	28	31	41	51	61	71	81
41	31	27	24		45	49	59	69	79
38	28	24	20		35	45	55	65	75
41	31	27	24	27	31	41	51	61	71

(c) : level 3: [20-30]

98	96		88	91	96	106	112	108	112
88	86		78	81		110	102	98	102
78	76	71	68	71				88	98
68	71	61	58	61	66	70	74	84	94
58			48	51	56	60	70	80	90
48	44	41	38	41	46				81
38	34	31	28	31	41	51	57	67	77
34	24					38	42	52	62	72
30	20	10	0		28	38	48	58	68
34	24	14	10	14	24	34	44	54	64

(e) : level 1: [0-10]

Fig. 9. The shortest path in 3-D space (values assigned to the grid cells refer to the accumulated travel costs of the corresponding spots).

4.3 The Spherical Surface

The movement on the surface of a sphere is examined separately due to the peculiarities of the underlying spherical geometry.

The problem can be approached by examining the movement on a 3-D space (Section 4.2), where the interior and exterior of the sphere are non-accessible (i.e., obstacles), while movement is allowed inside a spherical ring (can be thought as a buffer zone of the spherical surface in space). For instance, if cubic blocks are used for the generation of the spots on the three dimensional space, the set of the accessible spots will be located within the cubic blocks which intersect the spherical surface.

A more adaptive scheme to the spherical surface considers the movement on the surface itself. The determination of the spots is based on the available tessellations for the spheroid (Table 1). The tessellation which is closer to cartographers and geo-scientists, in general, is the geographic grid (Figure 10a). Figure 10b illustrates an example for the model of the shortest path on a sphere for a geographic grid of 30 degrees resolution (i.e., 62 spots). The direct and indirect neighbours are considered for each spot on the established network. The accumulated travel cost values are expressed in radius and refer to a spot (on the shaded grid cell) lying on the equator of the sphere. Contrary to the plane and 3-D space, the path connecting two spots is not a straight line segment any more, but the shortest arc of the *great circle* passing trough them. Note that all cells lying on the top (bottom) row of the matrix are assigned the same value, since they refer to a unique spot lying on the north (south) pole of the sphere.

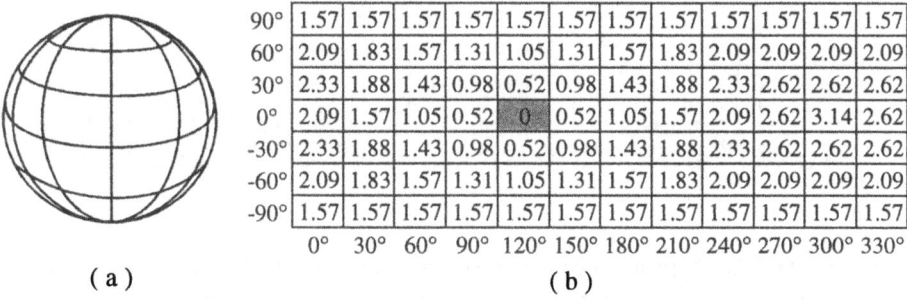

	0°	30°	60°	90°	120°	150°	180°	210°	240°	270°	300°	330°
90°	1.57	1.57	1.57	1.57	1.57	1.57	1.57	1.57	1.57	1.57	1.57	1.57
60°	2.09	1.83	1.57	1.31	1.05	1.31	1.57	1.83	2.09	2.09	2.09	2.09
30°	2.33	1.88	1.43	0.98	0.52	0.98	1.43	1.88	2.33	2.62	2.62	2.62
0°	2.09	1.57	1.05	0.52	0	0.52	1.05	1.57	2.09	2.62	3.14	2.62
-30°	2.33	1.88	1.43	0.98	0.52	0.98	1.43	1.88	2.33	2.62	2.62	2.62
-60°	2.09	1.83	1.57	1.31	1.05	1.31	1.57	1.83	2.09	2.09	2.09	2.09
-90°	1.57	1.57	1.57	1.57	1.57	1.57	1.57	1.57	1.57	1.57	1.57	1.57

(a)   (b)

Fig. 10. The accumulated cost values on a homogeneous sphere (expressed in rad).

Figure 11 illustrates an example on the determination of the shortest sea course on the earth surface approximated by a sphere. For visualization purposes a Plate Carrée projection has been used to represent the continents and oceans of the earth. The solid line shows the shortest path from Adelaide, Australia to New Orleans, USA. The line does not approximate sufficiently the curve of the great circle on the projection plane (as in Figure 1b), due to the low resolution of the grid (10 degrees) and the limited number of neighbours considered on the established network (direct and indirect).

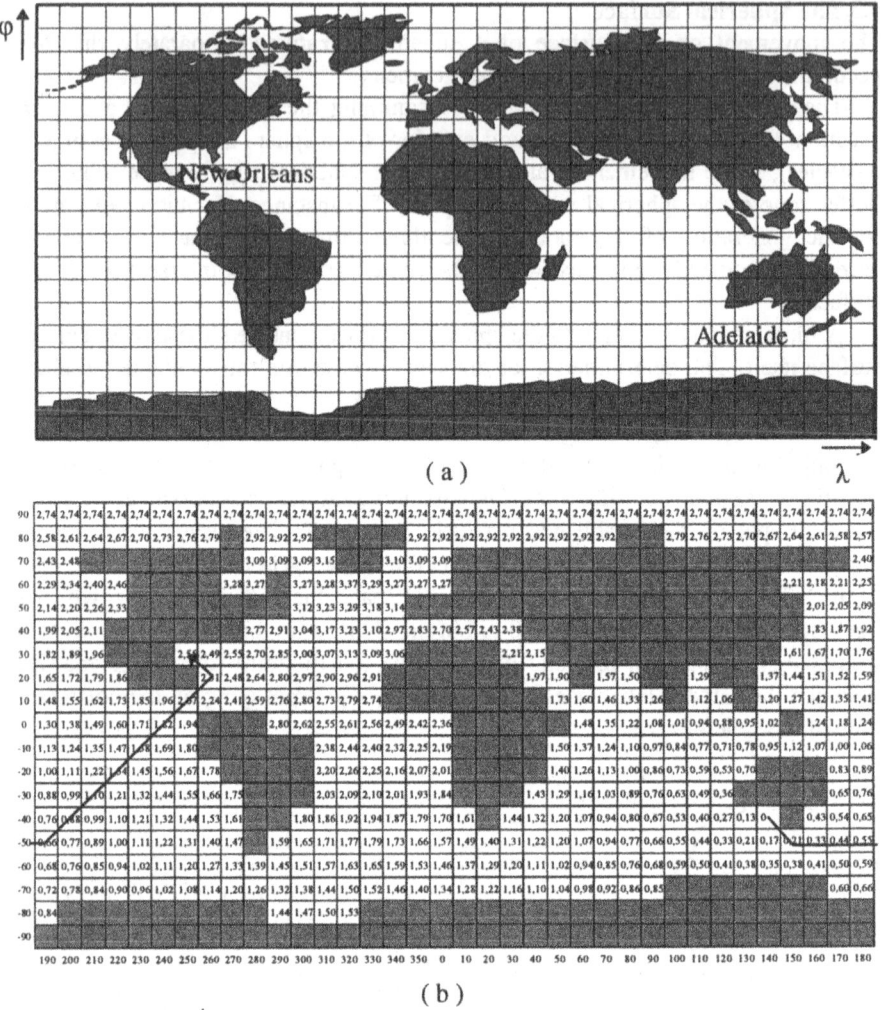

(a)

(b)

Fig. 11. The shortest sea course (distances expressed in rad; shaded cells in b correspond to non-accessible spots, i.e., land; continents are simplified).

The major problem faced by this tessellation is that the set of spots generated is non-uniformly distributed over the surface, because of the highly variable shape and size of the geographic grid cells; and as a consequence a variable accuracy on the determination of the optimum path is introduced, which depends on the region of the sphere studied. Clearly, there is a large concentration of spots around the poles which decreases by approaching the equator.

A more uniform distribution of the spots over the spherical surface can be obtained by adopting tessellations that are based on the recursive decomposition of regular polyhedra. A tessellation which is based on the recursive subdivision of the octahedron is considered here. This tessellation has been widely examined in the past as a hierarchical model for the representation of the earth surface on a Global Geographic Information System [Dutt84, Good90].

The surface of the earth is approximated by a octahedron whose vertices are located at the poles and at longitudes 0°, 180°, 90° E and 90° W around the equator. That is, the surface is firstly projected into eight triangular faces. Each triangle is then recursively subdivided into four by connecting the midpoints of its edges, and so on, until the predefined level of decomposition is reached. The advantage of this tessellation is that the triangular cells derived have a similar shape and size and as a consequence a uniform distribution of the spots over the spherical surface can be obtained.

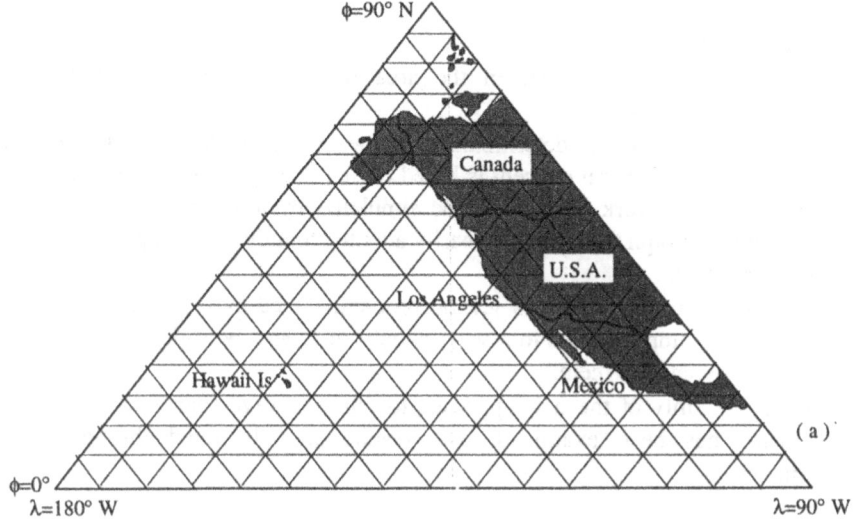

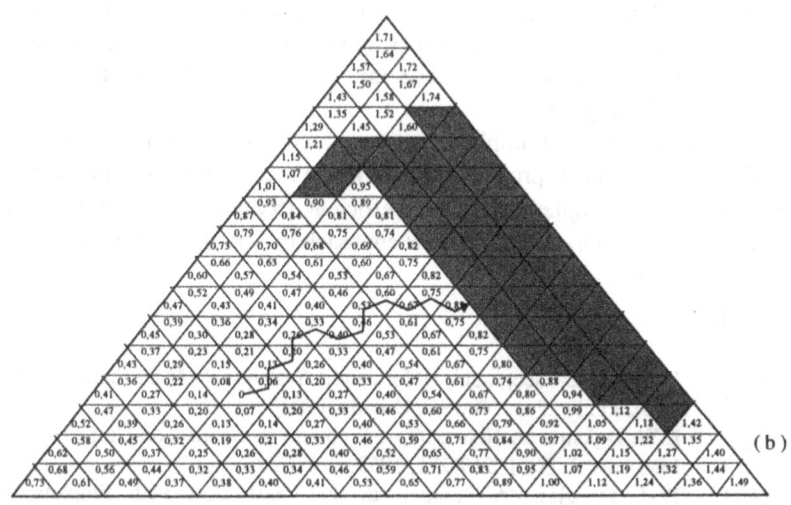

Fig. 12. The shortest sea course (distances expressed in rad; shaded cells in b correspond to non-accessible spots, i.e., land; continents are simplified).

Algorithms for the conversion of triangular cells to geographic coordinates as well as for the determination of their direct and indirect neighbours are available [Good91, Good92], and can be adopted for locating the spots over the surface and establishing the network over them.

Figure 12 shows a simplified example on the determination of the shortest sea course from Hawaiian Islands to Los Angeles, California. The quadrant 90° W to 180° W of the northern hemisphere has been only considered, while the established network is confined to the three direct neighbours of each spot. The tiling corresponds to the fourth level of subdivision and results in 256 triangular cells with an average resolution of 180 km [Dutt89].

5. Conclusion

The contribution of this paper in the area of optimum path finding can be summarized as follows:

- A new approach to the determination of the optimum path in space has been introduced. The general concept is based on the degeneration of the space under study into a network, which can be simulated by a weighted graph, so that algorithms of graph theory can be easily adopted to indicate the optimum path for the desired trip.
- The new approach overcomes several problems faced by existing solutions and can be easily implemented to a wide variety of spaces with different travel cost models associated to them.
- The applicability of the new approach to representative spaces of "real world" problems has been examined. The results are illustrated through several examples.

Our future research in the area includes:

- An extended analysis on the effects of the values assigned to the parameters of the new approach (i.e., the number of spots considered, how they are distributed in space, types of neighbours taken into account) on the determination of the optimum path in space.
- The development and implementation of complex travel cost functions for modeling "real world" problems related to the area of the optimum path finding.
- The design and implementation of efficient techniques and data structures to support the incorporation of the new approach in a production system.

References

[Aron89]   S. Aronoff: 'Geographic Information Systems: A Management Perspective', WDL Publications, Ottawa, Canada, 1989.

[Dijk59]   E.W. Dijkstra: 'A Note on Two Problems in Connection Graphs', Numerische Mathematik, No.1, 269-271, 1959.

[Doug94a]   D.H. Douglas: 'A Solution to the Least Cost Path Problem in GIS', Proceedings of the Canadian Conference on GIS, Ottawa, Canada, 1083-1089, 1994.

[Doug94b]   D.H. Douglas: 'The Parsimonious Path Based on the Implicit Geometry in Gridded Data and on a Proper Slope Line Generated from It', Proceedings of the International Symposium on Spatial Data Handling, Edinburgh, Scotland, 1133-1140, 1994.

[Dutt84]   G. Dutton: 'Geodesic Modeling of Planetary Relief', Cartographica, Vol.21, No.2,3, 188-207, 1984.

[Dutt89]   G. Dutton: 'Modeling Locational Uncertainty via Hierarchical Tessellation', in 'Accuracy of Spatial Databases', Edited by M. Goodchild and S. Gopal, Taylor-Francis Ltd, 125-140, 1989.

[Fred87]   G.N. Frederickson: 'Fast Algorithms for Shortest Paths in Planar Graphs, with Applications', SIAM Journal on Computing, Vol.16, No.6, 1004-1022, 1987.

[Gibb85]   A. Gibbons: 'Algorithmic Graph Theory', Cambridge University Press Ltd, 1985.

[Good90]   M.F. Goodchild, Y. Shiren: 'A Hierarchical Spatial Data Structure for Global Geographic Information Systems', Proceedings of the International Symposium on Spatial Data Handling, Zurich, Switzerland, 911-917, 1990.

[Good91]   M.F. Goodchild, Y. Shiren, G. Dutton: 'Spatial Data Representation and Basic Operations for a Triangular Data Structure', National Center for Geographic Information and Analysis, Report 91-8, 33 pp, 1991.

[Good92]   M.F. Goodchild, Y.Shiren: 'A Hierarchical Spatial Data Structure for Global Geographic Information Systems', Graphical Models and Image Processing, Vol.54, No.1, 31-44, 1992.

[Jens85]   S.K. Jensen: 'Automated Derivation of Hydrologic Basin Characteristics from Digital Elevation Model Data', Proceedings of the Auto-Carto 7, Digital Representation of Spatial Knowledge, 1985.

[John77]   D.B. Johnson: 'Efficient Algorithms for Shortest Paths in Sparse Networks', Journal of the Association of Computing Machinery, Vol.24, No.1, 1-13, 1977.

[Laur92]   R. Laurini, D. Thompson: 'Fundamentals of Spatial Information Systems', Academic Press Ltd, 1992.

[Lind67]   E.S. Lindgren: 'Proposed Solution for the Minimum Path Problem', Harvard Papers in Theoretical Geography, Geography and the Properties of Surfaces Series, Cambridge, No.4, 23 pp, 1967.

[Lind69a]   E.S. Lindgren: 'A Minimum Path Problem Reconsidered', Harvard Papers in Theoretical Geography, Geography and the Properties of Surfaces Series, Cambridge, No.28, 11 pp, 1969.

[Lind69b]   E.S. Lindgren: 'A Study of the Movement of a Point on a Plane and in Space', Harvard Papers in Theoretical Geography, Geography and the Properties of Surfaces Series, Cambridge, No.36, 16 pp, 1969.

[Sedg90]   R. Sedgewick: 'Algorithms', Addison-Wesley Publishing Company, Inc, 1990.

[Toml90]   C.D. Tomlin: 'Geographic Information Systems and Cartographic Modeling', Prentice Hall, Englewood Cliffs, N.J., 1990.

[Warn57]   W. Warntz: 'Transportation, Social Physics, and the Law of Refraction', The Professional Geographer, Vol.9, No.4, 2-7, 1957.

[Warn61]   W. Warntz: 'Transatlantic Flights and Pressure Patterns', The Geographical Review, Vol.51, 187-212, 1961.

A Unifying Framework
for
Multilevel Description of Spatial Data

Michela Bertolotto, Leila De Floriani, Paola Marzano

Department of Computer and Information Sciences, University of Genova
Viale Benedetto XV, 3, 16132 Genova, ITALY
Email: {bertmic,deflo,marzano}@disi.unige.it

Abstract. Defining a unifying model for describing spatial data at different levels of resolution is a relevant issue in several applications involving spatial data handling. Also, dimension-independence is becoming fundamental in many application contexts. We propose a unifying model for multiresolution description of spatial data which works in arbitrary dimension. Graph-based representations for encoding different instances of the abstract model are described.

1 Introduction

Current geographic information systems do not offer possibilities for multiresolution data handling. Apart from some hierarchical capabilities in raster modeling, which are based on structures derived from image processing, there is an almost total lack of features for manipulating and describing spatial data at different resolutions. Furthermore, a formal and general definition of the concept of resolution is lacking. In hypersurface modeling, since reduced subsets of large datasets are used for defining approximated representations, the term "resolution" is often intended as a measure of the distance between such a representation and the given dataset. In the context of spatial map representation, the term is used for expressing the levels of detail of topological entities described by the model [15, 22].

Moreover, not all tasks of a complex application necessarily require the same precision, and even a single task may need different precisions in different parts of the domain. Though, the development of a general-purpose model, which allows representation, analysis and manipulation of spatial data at variable resolutions, becomes a relevant issue. A multiresolution model should also support a representation on the basis of a reduced dataset, whenever the degree of precision and detail required by the application allows it.

Interestingly enough, most research efforts in the field of multiresolution modeling have been devoted to the representation of topographic surfaces, in the context of Geographical Information Systems (GISs) [16, 28, 8, 9, 23, 24, 25]. Some work has been done for the representation of multidimensional scalar fields in the three-dimensional case and for higher dimensions [30, 3]. Besides, most of the papers in the literature deal with strictly hierarchical structures, in which the link between two different levels implies a containment relation involving entities of such levels (see [10] for a survey). A hierarchical model is obtained by recursively refining portions of the underlying domain, each of which is considered independently of the others.

The application, however, of a local refinement criterion might not always be appropriate. In particular, when geometric properties are required to be globally satisfied at each level, hierarchical structures are not suitable. For example, if we consider models based on Delaunay complexes [2, 12, 18, 29], the empty sphere property characterizing such complexes can be only locally satisfied within the domain of those simplices that have been further refined in the hierarchy (see [4]).

Moreover, from a constructive point of view, hierarchical models only support a *top-down* construction paradigm, since their definition inherently provides a top-down criterion for progressively refining a given portion of the domain. In many applications, a *bottom-up* approach may be more convenient [26, 27, 17].

Multiresolution modeling offers important capabilities for spatial representation and reasoning in a more general framework. Little work has been done in this context: we can mention hierarchical modeling for a support to map generalization in automated cartography [15], as well as the development of hierarchical solutions in wayfinding and planning [7].

The aim of this work is to provide a formal and unifying definition of multiresolution models in arbitrary dimension. A basic problem in the context of multiresolution spatial data modeling is the lack of a systematic description, independent of both the dimension of spatial objects under consideration and of the specific application.

We face the problem from a general point of view, without focusing on a specific application so as to detect the essential properties such models should have as well as their potentials. For this reason, we will talk about *multilevel models* instead of multiresolution models, and about "levels of description" without associating any specific application-dependent meaning to them. In particular, we focus our attention on models based on simplicial complexes because of their extensive use in the context of spatial data modeling [14, 15, 20, 21]. The definitions and properties presented, however, can be easily generalized to the case of cell complexes.

The remainder of the paper is organized as follows. Section 2 reviews some of the basic definitions on Euclidean simplicial complexes. Section 3 introduces the multilevel simplicial model and describes some of its properties. In Section 4, we introduce spatial relations which seem to be interesting in the context of such model. Data structures for a multilevel model defined on graph-based representations are discussed in Section 5. Concluding remarks are briefly sketched in Section 6.

2 Euclidean Simplicial Complexes

In this Section, we introduce some basic definitions concerning simplicial complexes [1].

Let $V_\sigma = \{v_0, v_1, \ldots, v_d\}$ be a set of $d+1$ affinely independent points in the n-dimensional Euclidean space $\mathbb{E}^n$, with $d \leq n$. The subset σ of $\mathbb{E}^n$ formed by the points which can be expressed as a linear convex combination of the points of V_σ is called a d-simplex. The points of V_σ are called *vertices* of σ, while d is called the *order* of σ. Any s-simplex τ ($0 \leq s \leq d$) which is generated by a subset of $s+1$ vertices of σ is called an s-face of σ; if $s < d$, then τ is called a *proper face* of σ. The set $\{\sigma' \mid \sigma'$ s-face of σ, $s \leq d\}$ is called the s-skeleton of σ, and is denoted with $\sigma^{(s)}$. The 0-skeleton of σ is the set of vertices of σ. Two simplices σ and τ are said to be *coincident* if they have exactly the same 0-skeleton. The boundary of an s-simplex σ is the set of all k-faces of σ, for $0 \leq k \leq s-1$ (i.e., the union of all k-skeletons of σ, regarded as point-sets). In the following, the boundary of σ will be denoted by $b(\sigma)$; thus, the interior $int(\sigma)$ of σ is given by $\sigma \setminus b(\sigma)$. Whenever no ambiguity arises, a simplex will be interchangeably described as an algebraic entity or as the set of points belonging to it.

A finite collection Σ of simplices is called a *d-simplicial complex* (or, simply, a *d-complex*) when the following conditions hold:

(i) for each simplex $\sigma \in \Sigma$, all faces of σ belong to Σ;

(ii) for each pair of simplices $\sigma, \tau \in \Sigma$, either $\sigma \cap \tau = \emptyset$ or $\sigma \cap \tau$ is a face of both σ and τ;

(iii) d is the maximum among the orders of simplices belonging to Σ (d is called the *order* of Σ).

Figure 1 shows an example of a 3-complex.

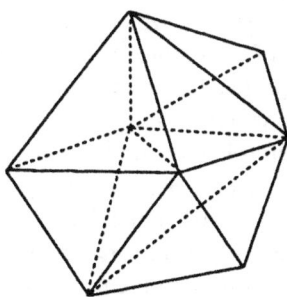

Fig. 1. An example of a 3-complex.

The set $\{\sigma' \mid \sigma'$ s-simplex of $\Sigma\}$, with $0 \leq s \leq d$, is called the s-skeleton of Σ and is denoted with $\Sigma^{(s)}$. In particular, the 0-skeleton of Σ is the set of vertices of Σ. The union of all s-simplices of Σ, $0 \leq s \leq d$, regarded as point sets, is the *domain* of Σ, denoted by $\Delta(\Sigma)$. The boundary and the interior of $\Delta(\Sigma)$ will be indicated with $b(\Delta(\Sigma))$ and $int(\Delta(\Sigma))$, respectively [1].

A d-complex Σ is *regular* if and only if, for each s-simplex τ, with $s < d$, there exists a d-simplex σ such that τ is a proper face of σ. Figure 2 shows an example

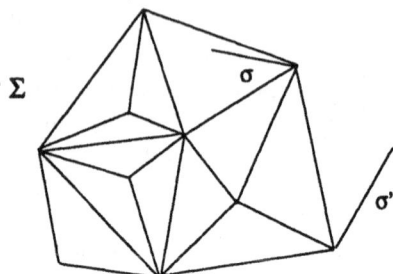

Fig. 2. An example of a non-regular 2-complex: simplices σ and σ' are not faces of any 2-simplex in the complex.

of a non-regular 2-complex. In the following, we will always consider regular complexes. A regular d-complex Σ is uniquely characterized by the collection of its d-simplices, since all other s-simplices, for $s < d$, can be obtained as a linear convex combination of vertices of d-simplices. In what follows, except when otherwise specified, a d-simplex, in a d-simplicial complex Σ, will be called a simplex. We will omit the dimension for complexes as well, i.e., we will use complex instead of d-complex, whenever no ambiguity arises.

Given a regular d-complex Σ, a collection of simplices $\Sigma' \subset \Sigma$ is called a *subcomplex* of Σ if and only if Σ' is a regular d-complex. Let $D' \subset \Delta(\Sigma)$; the set of all simplices of Σ contained in D' is called the *restriction* of Σ to D'. Note that the restriction of a regular complex Σ to a domain D' is not necessarily a subcomplex of Σ. Moreover, it may not completely cover D'. Figure 3 shows an example of restriction of a complex which is not a subcomplex.

Two simplices σ and τ of a complex Σ are said to be *incident* if either σ is a proper face of τ or τ is a proper face of σ; they are said to be *s-adjacent* if they share an s-face. In particular, two 0-simplices are *adjacent* if they are both faces of the same 1-simplex. A simplicial complex Σ is *s-connected* if, for each pair of simplices $\sigma, \tau \in \Sigma$, there exists a sequence $\{\sigma_0, \ldots, \sigma_k\}$ of $(s+1)$-simplices of Σ such that: (i) σ and σ_0 are either incident or coincident; (ii) $\forall i = 1, \ldots, k$, σ_{i-1} and σ_i are s-adjacent; (iii) τ and σ_k are either incident or coincident. In particular, we will say that a d-complex is connected if it is $(d-1)$-connected. Figure 4 illustrates examples of a 1- and a 0-connected 2-simplicial complexes.

There are several reasons to use simplicial complexes as discretization meshes instead of arbitrary cell complexes for several reasons. Any k-simplex σ has a constant number of s-faces ($s < k$) that can be retrieved in optimal time.

---

[1] Note that $b(\Delta(\Sigma))$ is composed of all $(d-1)$-simplices of Σ incident at only one d-simplex of Σ and of all their proper faces, while $int(\Delta(\Sigma)) = \Delta(\Sigma) \setminus b(\Delta(\Sigma))$.

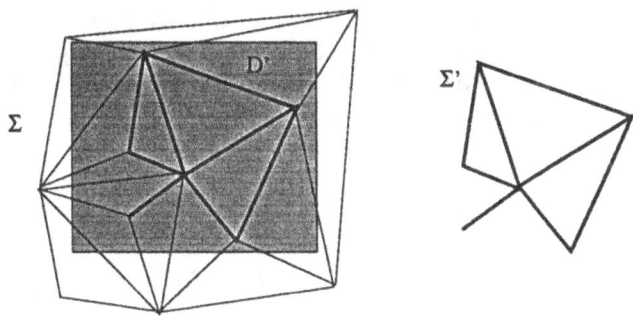

Fig. 3. A restriction of a complex which is not a subcomplex.

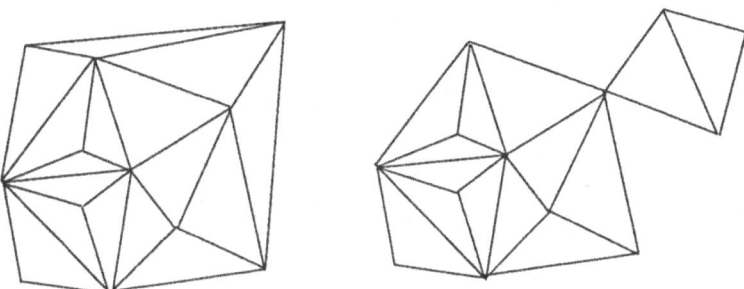

Fig. 4. A 1-connected (left) and a 0-connected triangulation (right).

Topological relations among adjacent d-simplices and among a d-simplex and its s-faces (with $s < d$) are constant: this makes data structures and algorithms for manipulating simplicial complexes more efficient.

3 Multilevel Simplicial Models

In this Section, we define multilevel models based on simplicial complexes. As pointed out in the introduction, such definitions can be easily extended to generic cell complexes. A multilevel simplicial complex (of order d) is a finite sequence of simplicial complexes having vertices at a set of points in a d-dimensional domain D; each complex of the sequence covers D and can be seen as a refinement of the previous complex or as a simplification of the next one.

Let $D \subset \mathbb{E}^d$ be a compact domain. Let V be a finite set of points in D. A *Multilevel Simplicial Complex (MSC)* having vertices at V is a collection $\mathcal{S} = [\Sigma_0, \dots, \Sigma_h]$ of regular connected d-simplicial complexes such that $\forall j = 0, \dots, h$, $\Delta(\Sigma_j) \equiv D$, $\Sigma_j^{(0)} \subseteq V$, and, $\forall k = 1, \dots, h$, $\Sigma_{k-1}^{(0)} \subsetneq \Sigma_k^{(0)}$[2].

Each $\Sigma_i \in \mathcal{S}$ is called the *complex at level i* in the model; thus, a simplex σ belonging to a complex Σ_i is said to be a *simplex belonging to level i*. In an

---

[2] Note that the Euclidean space in which each such d-simplicial complex results to be embedded is $\mathbb{E}^d$.

MSC, a simplex may belong to complexes at different levels. A simplex is said to be *created at level i* if i is the first level to which it belongs. Figure 5 shows a two-dimensional MSC with three levels: in this example, σ_1 belongs to both levels 0 and 1; thus, it is created at level 0.

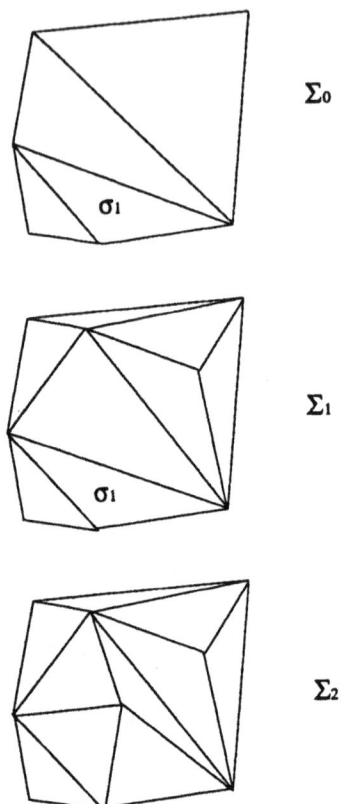

Fig. 5. An example of an MSC.

For each $i = 1, \ldots, h$, $\Sigma_i \not\equiv \Sigma_{i-1}$, since $\Sigma_{i-1}^{(0)} \subsetneq \Sigma_i^{(0)}$ and because different vertex sets generate different simplicial complexes. In particular, it can be shown that Σ_{i-1} and Σ_i differ for at least two d-simplices (see [5] for more details).

An important subclass of MSCs is represented by hierarchical simplicial complexes, in which, for each simplex σ_p at level i, there exists a set of simplices at level $i + 1$ partitioning σ_p. A hierarchical simplicial complex can be formally defined as follows.

An MSC $S = [\Sigma_0, \ldots, \Sigma_h]$ is said to be a *Hierarchical Simplicial Complex (HSC)* if, for every $i = 0, \ldots, h - 1$, and for each $\sigma_p \in \Sigma_i^{(d)}$, the restriction Σ' of Σ_{i+1} to σ_p is a subcomplex of Σ_{i+1} such that $\Delta(\Sigma') \equiv \sigma_p$. Note that Σ' can be formed by a single d-simplex. This is the case in which σ_p belongs to both

Σ_i and Σ_{i+1}. Figure 6 shows an example of a hierarchical simplicial complex: non-trivial restrictions (i.e., restrictions composed of more than one simplex) of Σ_1 and Σ_2 to simplices at levels 0 and 1, respectively, are shown.

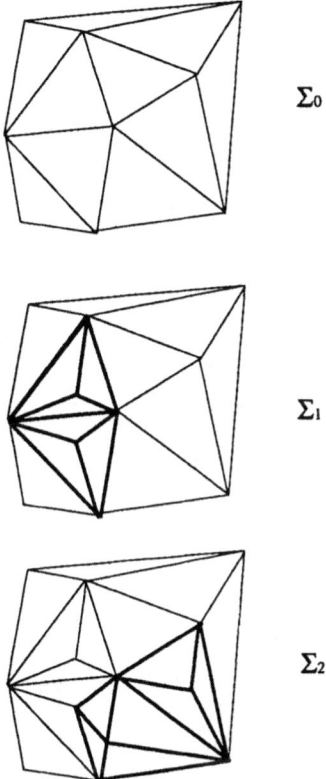

Fig. 6. An example of a hierarchical simplicial complex.

Since, as mentioned above, in an MSC a simplex may belong to different levels, we propose a more compact representation for MSCs that does not contain duplications of d-simplices among consecutive levels. Such representation is called a pruned multilevel simplicial complex since it is obtained by reducing the number of simplices at each level in an MSC.

A *Pruned Multilevel Simplicial Complex (PMSC)* associated with an MSC $\mathcal{S} = [\Sigma_0, \ldots, \Sigma_h]$ having vertices at V is a collection $\mathcal{S}' = [\Sigma'_0, \ldots, \Sigma'_h]$ such that

1. $\Sigma'_0 \equiv \Sigma_0$
2. for every $j = 1, \ldots, h$
   (a) $\Sigma'^{(d)}_j = \Sigma^{(d)}_j \setminus \Sigma^{(d)}_{j-1}$,
   (b) $\Sigma'^{(k)}_j = \bigcup^3_{\sigma \in \Sigma'^{(d)}_j} \sigma^{(k)}$ with $k = 0, \ldots, d-1$.

---

[3] Here the symbol $\bigcup$ is intended as union of sets of entities, not as a point-set union.

Each $\Sigma_i' \in S'$ is said to *correspond to level i* in the associated MSC. Note that Σ_i' is composed of all simplices created at level i.

Figure 7 shows the PMSC S' associated with the MSC of Figure 5. Since two

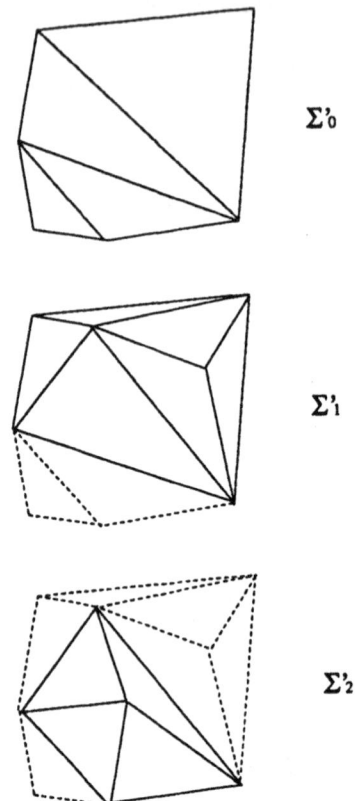

Σ'_0

Σ'_1

Σ'_2

Fig. 7. The PMSC S' associated with the MSC of Figure 5: simplices not belonging to S' are indicated with dotted lines.

consecutive complexes Σ_{i-1} and Σ_i in an MSC may differ for at least two d-simplices, $\Sigma_i'^{(d)}$ contains at least two elements. Moreover, $S' = [\Sigma_0', \ldots, \Sigma_h']$ is a collection of regular (not necessarily connected) simplicial complexes, since each Σ_i' is obtained by eliminating d-simplices from the corresponding Σ_i, but maintaining all its lower dimensional faces which are shared by simplices belonging to Σ_i' (as stated by condition 2.(b) in the definition of PMSC).

If S is an HSC, then we call S' the *Pruned Hierarchical Simplicial Complex (PHSC)* associated with S. From the definitions of HSC and PMSC it follows that, given a PHSC $S' = [\Sigma_0', \ldots, \Sigma_h']$, for each $i = 0, \ldots, h$ and for each $\sigma_p \in \Sigma_i'^{(d)}$, either the portion of domain covered by σ_p is not modified at any further level, or there exists a subcomplex at some deeper level "refining" σ_p. More formally, for each $\sigma_p \in \Sigma_i'^{(d)}$, two situations are possible: either for all $j > i$, $int(\Delta(\Sigma_j')) \cap int(\sigma_p) = \emptyset$ (in which case σ_p is called a *plain* simplex) or for some $j > i$ the restriction Σ' of Σ_j' to σ_p is a subcomplex of Σ_j' such that $\Delta(\Sigma') \equiv \sigma_p$ (in which case σ_p is called a *macrosimplex*). It is easy to show that if $k = \min \{z \mid i + 1 \le z \le h, \; int(\Delta(\Sigma_z')) \cap int(\sigma_p) \ne \emptyset\}$, then the restriction Σ_{σ_p} of Σ_k' to σ_p satisfies this condition (see [5] for a formal proof). Σ_{σ_p} is called the *direct refinement* of σ_p. Figure 8 shows the PHSC S' representing the HSC of Figure 6.

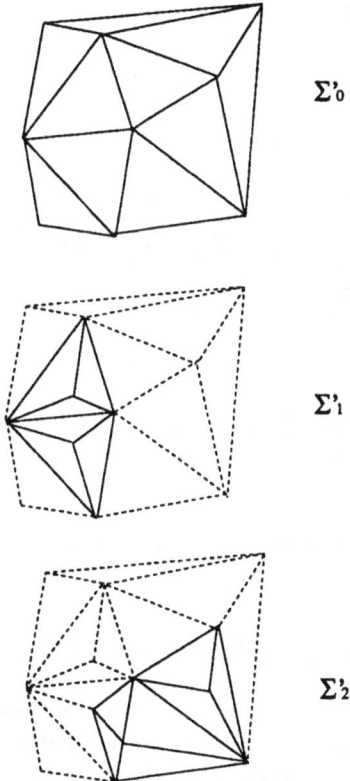

Fig. 8. The PHSC S' representing the HSC of Figure 6: simplices not belonging to S' are indicated with dotted lines.

Note that, in a PMSC, a simplex σ may appear at different, but non-consecutive, levels. Each simplex σ appearing at consecutive levels in S (MSC) is represented only once in S' (PMSC) (at its creation level). On the contrary, it might happen that a simplex σ belonging to level i does not belong to level $i+1$, but there exists some level $k > i + 1$, such that σ belongs to level k as well. In this case, we do not avoid duplication of σ at level i and k. Figure 9 illustrates this situation: simplex σ appears at levels 0 and 2. In the hierarchical case, a

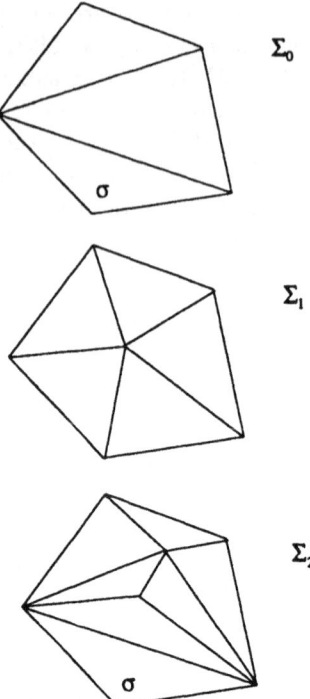

Fig. 9. An example showing that a simplex can be created at different levels: simplex σ is represented at levels 0 and 2.

simplex cannot appear at different levels; thus, duplications are avoided at both consecutive and non-consecutive levels.

4 Spatial Relations in a Multilevel Model

Sometimes, information concerning spatial relations among entities in a model need to be made explicit (although they are implicitly contained in the model itself). Here we consider such relations in the context of a multilevel model. In particular, we refer to the compact representation of an MSC (i.e., the PMSC), since the data structures we will present in Section 5 for an MSC are classified within such framework. The same relations, however, can be defined directly on the MSC model. We classify relations among entities in a PMSC into:

- *adjacency and incidence relations*, which involve entities belonging to the same level; adjacent and incident entities are characterized by non-proper intersection, i.e., they intersect at their boundaries;
- *interference relations*, which involve entities belonging to different levels and having a proper intersection.

We will refer to these two classes of relations as *horizontal* and *vertical* relations, respectively.

4.1 Horizontal Relations

Levels in a PMSC correspond to d-complexes. In a d-complex Σ we can define $(d+1)^2$ ordered topological relations between pairs of simplices. Such relations relate a k-simplex with its adjacent k-simplices or a k-simplex with the m-simplices incident at it, where $k, m = 0, 1, \ldots, d$, $m \neq k$. They can be subdivided into three classes:

- *adjacency relations*: they are relations R_{kk} $(k > 1)$ between a k-simplex σ and all k-simplices of Σ $(k-1)$-adjacent to σ. Relation R_{00} involves 0-simplices which are endpoints of the same 1-simplex.
- *incidence coboundary relations*: they are relations R_{km} between a k-simplex σ and all m-simplices of Σ incident at σ, with $m > k$.
- *incidence boundary relations*: they are relations R_{km} between a k-simplex and all m-simplices of Σ incident at σ, with $m < k$.

If $k = 0$, only adjacency and incidence coboundary relations exist, while for $k = d$ only adjacency and incidence boundary relations are defined. In a simplicial complex, boundary relations are constant: if the dimension of the Euclidean space in which the complex is embedded coincides with d (and here it is the case), the size of relations $R_{(d-1)d}$ and R_{dd} is 2 and $d + 1$, respectively.

Relation R_{d0} provides the minimum amount of information to be stored for representing regular simplicial complexes in order to be able to retrieve all other topological relations. A regular d-simplicial complex is fully characterized by the collection of its d-simplices, while all other s-simplices, for $s < d$, can be obtained as a linear convex combination of vertices of d-simplices. Thus, a minimal representation of a d-complex Σ must encode the set of d-simplices of Σ and, for each d-simplex in Σ, its $d + 1$ vertices.

4.2 Vertical Relations

Vertical relations in a PMSC $\mathcal{S}'$ involve entities belonging to different levels. Let Σ'_i and Σ'_j (with $0 \leq i \leq h - 1$ and $j > i$) be two complexes of $\mathcal{S}'$, and let $\sigma_p \in \Sigma_i'^{(m)}$ and $\sigma_q \in \Sigma_j'^{(n)}$ $(0 \leq m, n \leq d)$. We say that σ_p and σ_q *properly intersect* if and only if $int(\sigma_p) \cap int(\sigma_q) \neq \emptyset$.

Since we deal with regular d-simplicial complexes, we will consider representations explicitly encoding vertical relations between d-simplices, disregarding vertical relations between lower-dimensional simplices.

5 Data Structures

In this Section, we propose data structures for representing an MSC, which are based on its compact representation (the PMSC), by taking into account spatial relations (either vertical or horizontal) among its simplices. In order to provide an implementation-independent description for such structures, we will use a graph formalism.

5.1 Encoding Horizontal Relations

A data structure for encoding horizontal relations in a PMSC $\mathcal{S}'$ relies on the encoding of entities and topological relations within each of its complexes.

Each complex in a PMSC can be described by means of a graph-based representation, called the *relation graph*. A relation graph describing a complex Σ'_j $\mathcal{S}'$ is a graph $G_{Rj} = (\mathcal{N}_{Rj}, \mathcal{A}_{Rj})$ such that

- $\mathcal{N}_{Rj} \subseteq \cup_{i=0}^{d} \Sigma'^{(i)}_j$, and
- $\mathcal{A}_{Rj} \subseteq \cup_{0 \leq k, m \leq d} R_{km}$.

The resulting structure representing a PMSC $\mathcal{S}'$ is composed of $h + 1$ relation graphs, one for each complex in $\mathcal{S}'$. Note that, since complexes of $\mathcal{S}'$ may be disconnected, the corresponding relation graphs may be disconnected.

Different relation graphs may be used for a given complex, depending on the subset $\mathcal{N}_{Rj}$ of entities and on the subset $\mathcal{A}_{Rj}$ of relations encoded. However, as already pointed out in Section 4.1, a regular simplicial complex can be described by a minimal representation encoding its d-simplices, its 0-simplices as well as relation R_{d0}. A specific data structure for a regular simplicial complex Σ exploiting such property is the so-called *winged data structure* [13], for which $\mathcal{N}_{Rj} \equiv \Sigma^{(0)} \cup \Sigma^{(d)}$ and $\mathcal{A}_{Rj} \equiv R_{d0} \cup R_{dd}$. The space complexity of such structure is $O(dk^{\lfloor \frac{d+1}{2} \rfloor})$, where k is the total number of vertices in the complex: a complex with k vertices has $O(k^{\lfloor \frac{d+1}{2} \rfloor})$ d-simplices, in the worst case, each d-simplex has exactly $d+1$ vertices and at most $d+1$ adjacent d-simplices along a $(d-1)$-face, and each such face is shared by exactly two d-simplices.

Alternative data structures for encoding simplicial complexes can be obtained by simplifying existing structures for generic cell complexes and by taking into account properties of simplicial complexes. In the *incidence graph* proposed by Edelsbrunner [11], $\mathcal{N}_{Rj} \equiv \cup_{i=0}^{d} \Sigma^{(i)}$ and $\mathcal{A}_{Rj} \equiv (\cup_{k=1}^{d} R_{kk-1}) \cup (\cup_{j=0}^{d-1} R_{jj+1})$. Other data structures for cell complexes can be found also in [6] and [19].

5.2 Encoding Vertical Relations

In this Subsection, we propose a representation of vertical relations in a PMSC $\mathcal{S}'$ by means of a graph whose nodes are d-simplices in $\mathcal{S}'$ and whose arcs correspond to intersecting d-simplices belonging to different levels.

Nodes in such graph are organized into $h + 1$ levels: nodes at level i describe d-simplices of Σ'_i in $\mathcal{S}'$. Arcs encode vertical relations. The following definition formally specifies such representation.

Let $\mathcal{S} = [\Sigma_0, \ldots, \Sigma_h]$ be an MSC and let $\mathcal{S}' = [\Sigma'_0, \ldots, \Sigma'_h]$ be the associated PMSC. Let $\mathcal{N}_i = \Sigma'^{(d)}_i$ for $i = 0, \ldots, h$. The *Interference Graph* associated with $\mathcal{S}'$ is a directed graph $\mathcal{G} = (\mathcal{N}, \mathcal{A})$ such that

- $\mathcal{N} = \{\sigma_l \mid \sigma_l \in \mathcal{N}_i, \ i = 0, \ldots, h\}$
- $\mathcal{A} = \{(\sigma_k, \sigma_l) \mid \sigma_k \in \mathcal{N}_i, \ \sigma_l \in \mathcal{N}_j, \ 0 \leq i < h, \ i < j, \ int(\sigma_k) \cap int(\sigma_l) \neq \emptyset, \ j = \min\{z \mid i+1 \leq z \leq h, \ \exists \sigma \in \mathcal{N}_z, int(\sigma) \cap int(\sigma_k) \neq \emptyset\}$

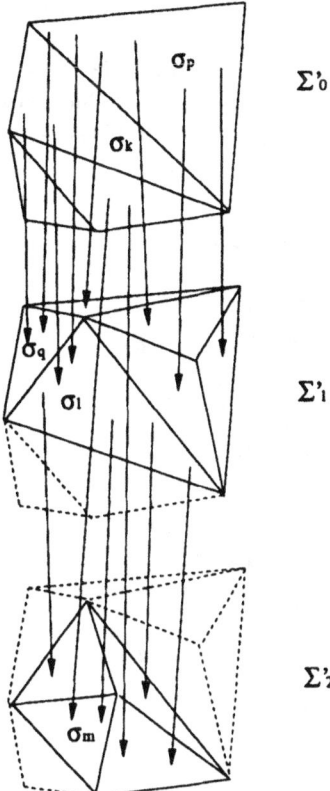

Fig. 10. The interference graph associated with the PMSC of Figure 7: simplex σ_p
appears as first element of six arcs joining Σ'_0 to Σ'_1; simplex σ_q appears as second
element of two arcs joining Σ'_0 to Σ'_1; simplex σ_k is linked to simplices intersecting it
in Σ'_1 (e.g., σ_l) and not to simplices intersecting it in Σ'_2 (e.g., σ_m), since the first level
at which σ_k disappears is level 1.

Figure 10 shows the interference graph describing interference relations in
the PMSC of Figure 7.

Note that, since a simplex σ_p belonging to a complex Σ'_i may have a non-
empty intersection with several simplices of a complex Σ'_j, σ_p may appear, as a
first element, in several arcs joining Σ'_i to Σ'_j (see σ_p in Figure 10). Similarly,
a simplex σ_q belonging to Σ'_j may have a non-empty intersection with several
simplices of Σ'_i; thus it may appear, as a second element, in several arcs joining Σ'_i
to Σ'_j (see σ_q in Figure 10). In an interference graph associated with a PMSC, we
have no trivial arc (i.e., no arc of the form (σ_k, σ_k)), since in PMSCs duplications
between consecutive levels are avoided. Condition $j = \min\{z \mid i + 1 \leq z \leq
h, \exists \sigma \in \mathcal{N}_z, int(\sigma) \cap int(\sigma_k) \neq \emptyset\}$ guarantees that simplex σ_k is connected to
simplices having non-empty intersection with σ_k and belonging to the first level
at which σ_k does not belong any more (see σ_k in Figure 10). Note that the
number of vertical links encoded in an interference graph is $O(hn_{max}^2)$, in the

worst case, where n_{max} is the maximum number of simplices generated at each level (see [4] for more details).

From the interference graph associated with a PHSC (see Figure 11 for an example), we can obtain a tree description of a hierarchical simplicial complex, which is more economical with respect to the interference graph in terms of space.

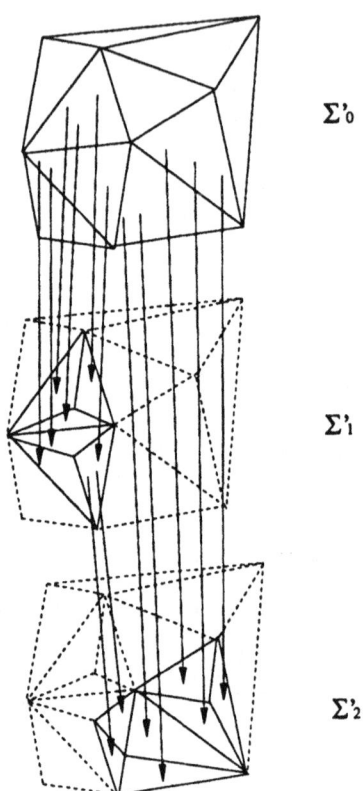

Σ'_0

Σ'_1

Σ'_2

Fig. 11. The interference graph representing the PHSC of Figure 8.

Given the interference graph associated with a PHSC, we can "group" into a single node simplices composing the direct refinement Σ_σ of a simplex σ. Only a single link between σ and Σ_σ is encoded, since each simplex has exact one "parent" in the interference graph (see Figure 12). By performing this process for each simplex in the model, we obtain the representation described below.

Let $\mathcal{S} = [\Sigma_0, \ldots, \Sigma_h]$ be an HSC and let $\mathcal{S}'$ be the corresponding PHSC. The *Simplified Tree (ST)* describing $\mathcal{S}$ is a tree $\mathcal{T} = (\mathcal{H}, \mathcal{E})$ such that

- $\mathcal{H} = \{\mathcal{H}_i^{(j)}, i = 0, \ldots, h, j \geq 1\}$ is the set of nodes, where
  - $\mathcal{H}_0^{(1)} \equiv \Sigma'_0$ is the root of $\mathcal{T}$;

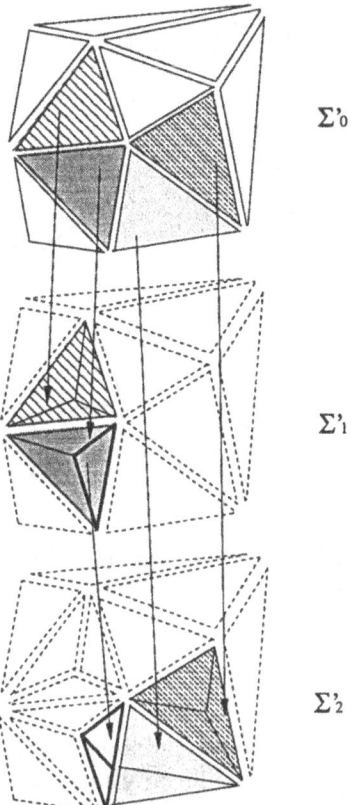

Fig. 12. Grouping simplices in the interference graph of Figure 9: each simplex has been coloured in the same way as its direct refinement.

- for $i = 1, \ldots, h$, $j \geq 1$, $\mathcal{H}_i^{(j)}$ is the subcomplex of Σ_i' such that $\exists \sigma_k \in \Sigma_m'^{(d)}$, $\Delta(\mathcal{H}_i^{(j)}) \equiv \sigma_k$ and $i = \min \{z \mid m+1 \leq z \leq h, \; int(\Delta(\Sigma_z')) \cap int(\sigma_k) \neq \emptyset\}$

- $\mathcal{E} = \{(\mathcal{H}_r^{(p)}, \mathcal{H}_i^{(q)}, \sigma_k) \mid \mathcal{H}_r^{(p)}, \mathcal{H}_i^{(q)} \in \mathcal{H}, \; \sigma_k \in \mathcal{H}_r^{(p)}, \; and \; \mathcal{H}_i^{(q)} \; is \; the \; direct refinement of σ_k\}$ is the set of arcs.

Note that each $\mathcal{H}_i^{(j)} \in \mathcal{H}$ is a subcomplex of Σ_i' that contains at least two d-simplices and for which there exists a unique d-simplex σ_k created at a previous level r such that σ_k coincides with the domain of $\mathcal{H}_i^{(j)}$ (see [5]). Besides, each node in $\mathcal{T}$, with the exception of the root, is a simplicial complex having a "simplicial" domain.

Figure 13 shows the ST representing the PHSC of Figure 8 obtained through the "grouping" process illustrated in the example of Figure 12.

Several hierarchical structures presented in the literature can be formally

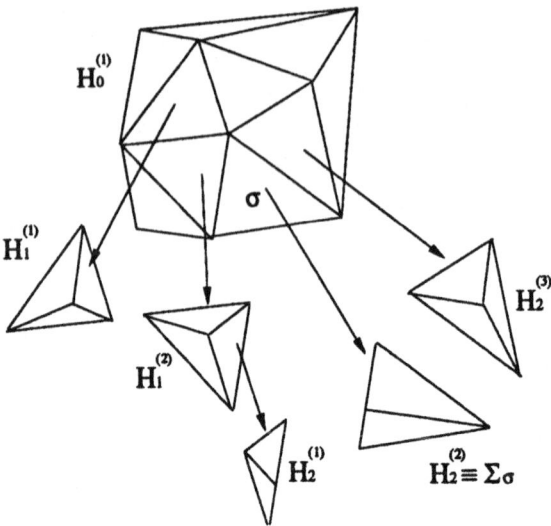

Fig. 13. The ST obtained from the interference graph of Figure 9. A simplex σ and its direct refinement Σ_σ are shown.

described by an ST[4]: the definition of ST corresponds to the definition of hierarchical model given in [10], which captures classical two-dimensional hierarchical structures. Also the d-dimensional model proposed in [3], which is, to our knowledge, the unique hierarchical proposal for arbitrary dimension, can be formally described by an ST. The number of vertical links stored in this structure is equal to the number of macrosimplices in the corresponding PHSC. As formally proven in [4], such number is $O(\frac{N_h}{m})$, where N_h is the number of plain simplices in the model and $m \geq 2$ is the minimum number of simplices in each node of the ST.

5.3 Encoding Horizontal and Vertical Relations

In order to obtain a complete geometric and topological description of a multilevel model, we can combine the two orthogonal graph-based representations, namely the *interference* graph (that plays the role of a "vertical" structure) and the *relation* graph (which, on the other hand, can be considered as a sort of "horizontal" structure). The idea is to have a sequence of graphs each corresponding to a single level and with interference links between nodes of different levels. The resulting representation is called *multilevel graph* and integrates each level of the interference graph with the relation graph associated with the corresponding complex. Formally, given a PMSC $\mathcal{S}'$, let $\mathcal{G} = (\mathcal{N}, \mathcal{A})$ be its associated interference graph and let $\mathcal{G}_{Ri} = (\mathcal{N}_{Ri}, \mathcal{A}_{Ri})$ (with $0 \leq i \leq h$) be the relation

---

[4] In particular, if we consider arbitrary cell complexes instead of simplicial complexes in the definition of HSC, all classical hierarchical models can be described within such framework.

graph corresponding to level i. The multilevel graph associated with $\mathcal{S}'$ is a pair $\mathcal{G}_M = (\mathcal{N}_M, \mathcal{A}_M)$ such that

- $\mathcal{N}_M = \cup_{i=0}^{h} \mathcal{N}_{Ri}$ and
- $\mathcal{A}_M = \mathcal{A} \cup (\cup_{i=0}^{h} \mathcal{A}_{Ri})$

Note that $\mathcal{N} \subseteq \cup_{i=0}^{h} \mathcal{N}_{Ri}$, since, as illustrated in Section 4.1, a minimal representation of horizontal relations in any complex Σ_i' of $\mathcal{S}'$ must encode the set of d-simplices of Σ_i' and, for each d-simplex, its $d+1$ vertices. Moreover, two kinds of arcs are presented in the multilevel graph:

- *horizontal* arcs (i.e., arcs in $\cup_{i=0}^{h} \mathcal{A}_{Ri}$) which correspond to horizontal relations;
- *vertical* arcs (i.e., arcs in $\mathcal{A}$) which correspond to vertical relations.

Clearly, the choice of the horizontal and the vertical data structure will depend on the specific application in which the integrated representation is going to be used and, thus, on the operations we need to perform on it. In [4], we have proposed a data structure specific for visualization purposes: such data structure can be used when a "non-topological" representation is required, but interference links are needed. The principles beyond such structure are to avoid duplication of simplices belonging to consecutive levels and to encode only minimal topological information. To this aim, given an MSC $\mathcal{S}$, vertical relations are represented through a data structure that directly implements the *interference* graph for the associated PMSC $\mathcal{S}'$; as far as horizontal relations are concerned, for each complex in $\mathcal{S}'$, we encode d- and 0-simplices plus relation R_{d0}. The resulting integrated data structure is called a *Sequence of Lists of Simplices (SLS)*.

The total number N_s of simplices stored in the data structure can be expressed in terms of the number N_h of simplices of the complex Σ_h at the maximum level of precision. Indeed, $N_s \leq N_h + h n_{max}$, where n_{max} is the maximum number of simplices generated at each level. The complexity of the SLS is $O(N_h + h n_{max} + h n_{max}^2)$, where the quadratic term is due to vertical links (see [4]): the cost of interference links dominates the total storage cost of the structure.

In the hierarchical case, if the model is described by an ST $\mathcal{T} = (\mathcal{H}, \mathcal{E})$, a representation encoding both vertical and horizontal relations can be obtained by integrating the relation graph associated with each node $\mathcal{H}_i^{(j)}$ (with $i = 0, \ldots, h, j \geq 1$) in $\mathcal{H}$ with the tree structure induced by $\mathcal{E}$. In this case, we do not have a complete description of the topological relations at each level, but only a "partial" description, which is local to each $\mathcal{H}_i^{(j)}$. Thus, while the tree-representation induced by an ST allows saving interference links, it implies a loss of topological information about entities at the same level in the model. A global topological description could be obtained by explicitly maintaining information about the relationship between different nodes in the tree [9].

6 Concluding Remarks

In the paper, we have formally defined a model for multilevel representation of spatial data. Such model provides both a dimension- and an application-independent framework for describing multiresolution models.

We have classified spatial relations among entities composing a multilevel model into *vertical* and *horizontal* relations and we have presented graph-based representations obtained by making information explicit which are implicitly contained in the abstract model. The tradeoff between storage cost for encoding explicit information and computation time for extracting implicit information characterizes each of such representations.

Moreover, the proposed model provides a unifying definition for multiresolution models, since both hierarchical and pyramidal models proposed in the literature for surface representation can be formally described within such framework. For example, the pyramidal structure proposed in [8] is a special case of interference graph for an MSC, in which all complexes are Delaunay triangulations. The other two-dimensional structures proposed in the context of surface approximation (see [16, 28, 9, 24, 25]), and classified as hierarchical models in [10], are special cases of the simplified tree for an HSC.

Future work involves investigating the problem of designing data structures that, unlike the SLS structure described in Section 5.3, encode topological relations among simplices belonging to the same level, besides minimal topological information (i.e., relation R_{d0}). This is fundamental when the extracted representation at a given level is required to encode also such relations, or whenever navigation across the model is needed (for instance, for performing spatial queries).

Another important issue is to define more economical encoding structures for the interference graph, which avoid storing the whole set of interference links.

Acknowledgements

This work has been supported by the Strategic Project "Knowledge through images: an application to cultural heritage" of the Italian National Research Council under contract N. 94.04221.ST74 and by the Project "Models and systems for handling environmental and land data" of the Italian National Research Council under contract N. 95.01057.CT12.

References

1. Agoston, M.K.,*Algebraic Topology: A First Course*, Pure and Applied Mathematics, Marcel Dekker (Ed.), New York, 1976.
2. Avis, D., Battacharya, B.K., "Algorithms for computing d-dimensional Voronoi diagrams and their duals", in *Advances in Computing Research*, 1, JAI Press, F.P. Preparata (Ed.), 1983, pp. 159-180.

3. Bertolotto, M., De Floriani, L., Puppo, E., "Hierarchical Hypersurface Modeling", *IGIS'94: Geographic Information Systems*, LNCS 884, J. Nievergelt, T. Roos, H. J. Schek, P. Widmayer (Eds.), Springer-Verlag, 1994, pp. 88-97.

4. Bertolotto, M., De Floriani, L., Marzano, P., "Pyramidal Simplicial Complexes", *Solid Modeling'95 - 3rd ACM Symposium on Solid Modeling and Applications*, Salt Lake City, Utah, 1995, pp. 153-162.

5. Bertolotto, M., De Floriani, L., Marzano, P., "A Unifying Framework for Multilevel Description of Spatial Data", *Technical Report* DISI, Department of Computer and Information Sciences, University of Genova, 1995 (in preparation).

6. Brisson, E., "Representing Geometric Structures in *d*-Dimensions: Topology and Order", *Proceedings 5th ACM Symposium on Computational Geometry*, Saarbruchen, 1989, pp.218-227.

7. Car, A., Frank, A., "Modelling a hierarchy of space applied to large road networks", *IGIS'94: Geographic Information Systems*, LNCS 884, J. Nievergelt, T. Roos, H. J. Schek, P. Widmayer (Eds.), Springer-Verlag, 1994, pp.15-24.

8. De Floriani, L., "A pyramidal data structure for triangle-based surface description", *IEEE Computer Graphics and Applications*, March 1989, pp.67-78.

9. De Floriani, L., Puppo, E., "A hierarchical triangle-based model for terrain description", *Theories and Methods of Spatio-Temporal Reasoning in Geographic Space*, LNCS N.639, A. U. Frank, I. Campari, U. Formentini (Eds.), Springer-Verlag, September 1992, pp. 236-251.

10. De Floriani, L., Marzano, P., Puppo, E., "Multiresolution Models for Topographic Surface Description", *The Visual Computer*, 1995 (accepted for publication).

11. Edelsbrunner, H., *Algorithms in Combinatorial Geometry*, Springer-Verlag, 1987.

12. Edelsbrunner, H., Preparata, F.P., West, D.B., "Tetrahedrizing point sets in three dimensions", *Journal of Symbolic Computing*, 10, 1990, pp. 335-347.

13. Ferrucci, V., Paoluzzi, A., "Extrusion and boundary evaluation for multidimensional polyhedra" *Computer Aided Design*, 3, 1, 1991, pp. 40-48.

14. Frank, A., Kuhn, W., "Cell graph: a provable correct method for the storage of geometry", *Proceedings SDH'86*, Seattle, WA, 1986.

15. Frank, A., U., Timpf, S., "Multiple representations for cartographic objects in a multi-scale tree - an intelligent graphical zoom", *Computer & Graphics*, 18, 6, 1994, pp. 823-829.

16. Gomez, D., Guzman, A., "Digital model for three-dimensional surface representation", *Geo-Processing*, 1, 1979, pp. 53-70.

17. Hoppe, H., DeRose, T., Duchamp, T., McDonald, J., Stuetzle, W., "Mesh Optimization", *Computer Graphics Proceedings*, Annual Conference Series, 1993, pp. 19-26.

18. Joe, B., "Construction of three-dimensional Delaunay triangulations using local transformations", *Computer Aided Geometric Design*, 8, 1991, pp. 123-142.

19. Lienhardt, P., "Topological Models for Boundary Representations: a Comparison with *n*-dimensional Generalized Maps", *Computer Aided Design*, 23, 1, 1991, pp. 59-82.

20. Pigot, S., "A Topological Model for a 3D Spatial Information System", *Proceedings SDH'92*, Charleston, SC, 1992.

21. Pigot, S., "Generalized Singular 3-Cell Complexes", *Proceedings SDH'94*, Edinburgh, UK, 1994, pp. 89-111.
22. Puppo, E., Dettori, G., "Towards a formal method for multiresolution spatial maps", *Proceedings SSD'95*, Portland, Maine, 1995, (to appear).
23. Samet, H., *The Design and Analysis of Spatial Data Structures*, Addison-Wesley, Reading, MA, 1990.
24. Samet, H., Sivan, R., "Algorithms for constructing quadtree surface maps", *Proceedings SDH'92*, Charleston, August 1992, pp. 361-370.
25. Scarlatos, L.L., Pavlidis, T., "Hierarchical Triangulation Using Cartographic Coherence", *CVGIP: Graphical Models and Image Processing*, 54, 2, March 1992, pp. 147-161.
26. Schroder, W., J., Zarge, J., A., Lorensen, W., E., "Decimation of Triangle Meshes", *Computer Graphics*, 26, 2, 1992, pp. 65-70.
27. Turk, G., "Re-tiling Polygonal Surfaces", *Computer Graphics*, 26, 2, 1992, pp. 55-64.
28. Von Herzen, B., Barr, A.H., "Accurate triangulations of deformed, intersecting surfaces", *Computer Graphics*, 21, 4, July 1987, pp. 103-110.
29. Watson, D.F., "Computing the n-dimensional Delaunay tesselation with applications to Voronoi polytopes", *The Computer Journal*, 24, 1981, pp. 167-171.
30. Wilhelms, J., Van Gelder, A.A., "Octrees for faster isosurface generation", *ACM Transaction on Graphics*, 11, 3, July 1992, pp. 201-227.

Updating Visibility Information on Multiresolution Terrain Models

Paola Magillo, Leila De Floriani and Elisabetta Bruzzone

Information and Computer Science Department (DISI), University of Genova, Viale Benedetto XV, 3, 16132 Genova, Italy

Abstract. We propose a new randomized incremental approach to visibility update on a terrain model when varying the level of resolution. In particular, we consider visibility update on multiresolution terrain models, which encode surface descriptions at different resolution degrees. Randomized dynamic algorithms for upper envelope of triangles and segments are used for updating the visible image and the horizon, respectively. In this paper, we propose a new randomized dynamic algorithm for upper envelope of triangles, which is an extension of the semi-dynamic algorithm by Boissonnat and Dobrindt [6]. A dynamic algorithm for computing the upper envelope of segments was proposed in a previous paper [14]. It is shown that, under suitable conditions, the update of visibility information by means of these algorithms is more convenient than a complete recomputation.

1 Introduction

Terrain models play a fundamental role in geographic information systems (GISs), providing a discrete representation of a terrain which can be manipulated by a computer. An elevation model of a terrain usually approximates a surface through a network of faces (described by analytical expressions from a predefined class) having their vertices at a set of sampled data points.

Since very large sets of data are often available, the representation of a terrain at different scale/resolution through a unified model is a topic of relevant interest in geographic data processing. To this aim, multiresolution terrain models have been developed, which provide a description of a surface at different levels of detail as well as a data compression mechanism. Multiresolution terrain models describe a surface through increasingly finer networks of faces. In the literature, multiresolution terrain models based on two different approaches have been proposed. *Hierarchical* terrain models are based on a recursive partition scheme, in which a face at a given level of resolution is expanded, independently from the other ones, into a collection of faces at the next level. In *pyramidal* terrain models, on the contrary, any subset of faces at a certain level can be replaced by another set of faces at the next level, in such a way that a face in the new set is not necessarily contained into a single face of the old set, but can interfere with several ones. Thus, hierarchical terrain models can be considered as a subclass of pyramidal ones. A multiresolution terrain model implicitly represents a family of terrain models, describing the same surface at different resolutions. A specific model of such family can be explicitly obtained once a target level of resolution is assigned.

Describing a terrain through visibility information has a variety of applications, such as geomorphology, navigation, terrain exploration. Problems which can be solved based on visibility are, for instance, the computation of the minimum number of observation points needed to view a given region, the computation of paths with specified visibility characteristics (e.g., hidden paths with respect to a predefined set of observers, scenic paths from which large portions of the terrain can be viewed), the computation of optimal locations for television transmitters [8, 11].

Visibility computation on multiresolution terrain models presents two aspects:

(1) *Compute* some visibility information, related to a terrain representation satisfying an application-specific degree of resolution.
(2) *Update* some visibility information when passing at a new level of resolution.

Here, we focus on problem (2). In particular, we consider the dynamic update of *visible images* and *horizons* on a multiresolution terrain model. Varying the accuracy of the representation means replacing a subset of terrain faces with new ones. We illustrate the application of randomized dynamic algorithms for maintaining the upper envelope of triangles and segments to update the visible image and the horizon, respectively, after insertions and deletions of terrain faces and edges.

The algorithm for dynamically updating the upper envelope of triangles is a new extension of a semidynamic algorithm (i.e., an algorithm in which only insertions of triangles are allowed), developed by Boissonnat and Dobrindt [6]. The dynamic algorithm for horizon maintenance was proposed by De Floriani and Magillo in [14]. Both algorithms use a data structure, called the *Influence Direct Acyclic Graph* (IDAG), developed for the design of geometric algorithms [10, 5, 7]. This technique provides bounds for the expected time and space complexity when averaging on all possible permutations of the input data.

Of course, dynamic algorithms can also be used for solving problem (1) (i.e., for computing visibility information from scratch), by incrementally inserting all faces and edges belonging to the terrain representation, encoded in the multiresolution model, which corresponds to the required level of resolution. Indeed, an on-line approach necessarily exhibits higher time complexity than a static one. Efficient algorithms for computing both horizons and visible images have been proposed in the literature, including methods which work with an optimal worst-case complexity [20, 19, 16] and output-sensitive ones [12, 21, 1, 18, 24]. Anyone of these algorithms can be used for visibility computation on a multiresolution terrain model, by explicitly building the model corresponding to the given resolution threshold in a preliminary step.

2 Multiresolution Terrain Models

A natural terrain can be mathematically modeled as a surface in $\mathbb{E}^3$, which is the graph of a bivariate real-valued function $z = \phi(x, y)$, defined over a subset D of the $x - y$ plane, called the *domain* of the surface.

In practice, *Digital Terrain Models* (DTMs) are used, which provide a discrete approximation of a terrain, built on the basis of a finite set S of sampled data points belonging to the surface. A *Digital Terrain Model* (DTM) built on S consists of a pair $\mathcal{D} \equiv (\Sigma, \Phi)$, where:

- Σ is a plane subdivision, having the set of points $\{(x, y) \mid (x, y, z) \in S\}$ as its vertices;
- Φ is a family of bivariate continuous functions, such that every function $\phi_i \in \Phi$ is defined over a closed region $f_i \in \Sigma$, and interpolates the elevations of the data points whose projections are the vertices of f_i.

A digital terrain model is indeed a special case of a mathematical model of a terrain: the domain D can be defined as the union $\cup\{f_i \mid f_i \in \Sigma\}$, while function ϕ consists of the union of all functions $\phi_i \in \Phi$ (up to conventions for defining ϕ on the common boundary of adjacent regions in Σ, when the corresponding functions do not set the same value). Surface patches, defined over the regions of Σ, are called *faces* of the DTM. The spatial complexity of a DTM linearly depends on the number n of sampled points in S.

Existing DTMs can be classified into *Regular Square Grids* (RSGs), and *Triangulated Irregular Networks* (TINs). In RSGs, the domain subdivision consists of a rectangular grid, and piecewise linear or quadratic functions are used. TINs are characterized by a triangulation of the domain and by linear interpolating functions. TINs are more flexible and better adapt to terrain features than RSGs, since they can deal with irregularily distributed data sets and include surface-specific points and lines. Often, a Delaunay triangulation [23] is used as a domain subdivision for a TIN, because of its good behavior in numerical interpolation.

In practical applications, a large quantity of sampled points is available for a terrain. Thus, a surface model capable of compressing spatial data according to an accuracy-based criterion is required. A DTM is called *approximate* when it is built based on a proper subset S' of the whole dataset S. Different norms can be used to evaluate the approximation error. The error at a point $P \equiv (x, y, z) \in S - S'$ is often evaluated as the infinite norm, i.e., the absolute value of the difference between the interpolated and the measured elevation values. The error on a face is defined as the maximum error of points in $S - S'$ whose vertical projections lie inside or on the boundary of the face. A DTM is said to approximate a terrain *at level ε of accuracy* if, for every face, the approximation error is less or equal to ε. While digital terrain models are widely used in commercial packages for terrain representation, organizing terrain representations at different levels of detail into a compact structure is a recent research issue.

Multiresolution, hierarchical and pyramidal terrain models are formally treated in [4]. Here, we recall the basic definitions in an informal style. At a high level of abstraction, a *multiresolution terrain model* is a sequence $\mathcal{N} \equiv \{\mathcal{D}_0, \ldots, \mathcal{D}_h\}$ of digital terrain models such that:

- each DTM $\mathcal{D}_k$, for $k = 0, \ldots, h$, covers the same domain D, and
- for every $k = 0, \ldots, h - 1$, the set of vertices of $\mathcal{D}_k$ is contained in that of $\mathcal{D}_{k+1}$.

Faces belonging to consecutive DTMs in sequence $\mathcal{N}$ can be characterized by an *interference relation*, which connects pairs of faces $\phi_i \in \mathcal{D}_k$, $\phi_j \in \mathcal{D}_{k+1}$ such that their horizontal projections have a non-empty intersection. In the following, we will say that a face *horizontally* intersects (contains, is contained into) another face when the corresponding projections intersect (or one is contained into the other one). An interference between two faces $\phi_i \in \mathcal{D}_k$ and $\phi_j \in \mathcal{D}_{k+1}$ may correspond to a proper horizontal intersection, to a horizontal containment of the ϕ_j into ϕ_i, or, finally, to the fact that $\phi_i \equiv \phi_j$.

We say that every face of $\mathcal{D}_k$ belongs to *level k*. Indeed, a face ϕ_i may belong to many consecutive levels: ϕ_i belongs to both levels k and $k+1$ if it is not affected by refinements occurred between the two levels. Based on such remark, we can provide a model in which faces belonging to different levels are not replicated: a face ϕ_i is represented only once, i.e., at the lowest level to which it belongs. Moreover, interference links between remaining faces of a generic level $k > 0$ (once replicated faces have been suppressed) can also be explicitly stored, thus leading to the definition of a pyramidal terrain model.

More formally, a *pyramidal terrain model* is a pair $\mathcal{P} \equiv (\mathcal{N}', \mathcal{A})$, where:

- $\mathcal{N}' = \{\mathcal{D}'_0, \ldots, \mathcal{D}'_m\}$ is a sequence of DTMs, such that $\mathcal{D}'_0 \equiv \mathcal{D}_0$, and, for every $k = 1, \ldots, h$, $\mathcal{D}'_k$ is composed of the faces of $\mathcal{D}_k$, except for those belonging to $\mathcal{D}_{k-1}$;
- $\mathcal{A}$ is a set of labelled arcs; there is an arc $(\mathcal{D}'_i, \mathcal{D}'_j)$, with $j > i$, for every pair of faces $\phi_i \in \mathcal{D}'_i$ and $\phi_j \in \mathcal{D}'_j$, such that ϕ_i and ϕ_j have a non-empty horizontal intersection; such arc is labelled with the pair (ϕ_i, ϕ_j).

A pyramidal terrain model is represented by a multigraph (i.e., a graph with parallel arcs), having $\mathcal{N}'$ as its set of nodes and $\mathcal{A}$ as its set of arcs, since several arcs $(\mathcal{D}'_i, \mathcal{D}'_j)$, labelled with different pairs of faces, may exist (see Figure 1 (a)). The size of a pyramidal terrain model is at most quadratic in the total number of sampled points used.

A special case of a multiresolution terrain model occurs when all interference relations reduce to a horizontal containment of a set of faces of $\mathcal{D}_{k+1}$ into a single face of $\mathcal{D}_k$. In this case, all faces of $\mathcal{D}_{k+1}$, which are contained into the same face ϕ_i of $\mathcal{D}_k$, can be grouped into a separate DTM, whose domain covers the $x - y$ projection of ϕ_i, and the horizontal containment of such DTM into ϕ_i can be explicitly represented. Moreover, only proper containment links are represented

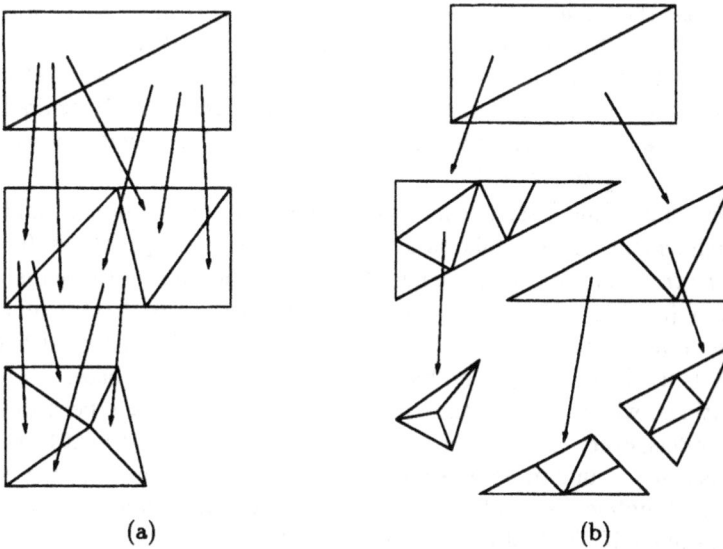

Fig. 1. An example of a pyramidal terrain model (a) and of a hierarchical one (b).

(i.e., as in pyramidal models, a face is represented only once). This process leads to the definition of a hierarchical terrain model.

More formally, a *hierarchical terrain model* (HTM) is a pair $\mathcal{H} = (\mathcal{N}', \mathcal{E})$, where:

- $\mathcal{N}' = \{\mathcal{D}'_0, .., \mathcal{D}'_m\}$ is a collection of digital terrain models, each containing more than one face (with the possible exception of $\mathcal{D}'_0$) such that, for every $j = 1, \ldots, m$, there is exactly one index $i < j$ and one face $\phi_j \in \mathcal{D}'_i$ such that the domain of $\mathcal{D}'_j$ covers the horizontal projection of ϕ_j; the vertices of ϕ_j must be vertices in $\mathcal{D}'_j$. $\mathcal{D}'_j$ is called the *direct expansion* of ϕ_j;
- $\mathcal{E}$ is a collection of labelled arcs. For every pair of DTMs $\mathcal{D}'_i$ and $\mathcal{D}'_j \in \mathcal{N}'$, such that the $\mathcal{D}'_j$ is the direct expansion of a face ϕ_j of $\mathcal{D}'_i$, $\mathcal{E}$ contains an oriented arc $(\mathcal{D}'_i, \mathcal{D}'_j)$, labelled with face ϕ_j.

A hierarchical terrain model can be represented by a tree with nodes in $\mathcal{N}'$ and labelled arcs in $\mathcal{E}$, where $\mathcal{D}'_0$ is the *root* and the children of a node $\mathcal{D}'_i$ correspond the direct expansions of its faces (see Figure 1 (b)). Faces refined in the HTM are called *macrofaces*, while the remaining faces are called *simple faces*. The total size of an HTM is linear in the number of simple faces, which, in turn, linearly depends on the total number of inserted data points [13].

Hierarchical terrain models have the advantage of a reduced space complexity with respect to pyramidal ones, but they do not allow global properties to be satisfied on the whole domain (as, for instance, the Delaunayhood of a triangulation).

Usually, a multiresolution terrain model is built based on a decreasing sequence $[\varepsilon_0, \ldots, \varepsilon_h]$ of *resolution thresholds*: the first DTM $\mathcal{D}_o$ of sequence $\mathcal{N}$

approximates the terrain at level ε_0 of accuracy, while each $\mathcal{D}_k \in \mathcal{N}$, with $k > 0$, is obtained from $\mathcal{D}_{k-1}$ by inserting new data points until the resulting approximation satisfies the resolution threshold ε_k. In this case, we say that the multiresolution model is *explicit*, otherwise, it is called *implicit*. Pyramidal terrain models provide an explicit multiresolution, while both implicit and explicit hierarchical terrain models exist.

In early hierarchical terrain models, multiresolution is implicit, and the recursive refinement process follows a predefined pattern. *Quadtrees* [9, 25] and *quaternary triangulations* [17] are based on the replication of the same geometric pattern (a rectangle and an equilateral triangle, respectively) at different scales; in *ternary triangulations* [22], a fixed topological pattern with variable geometry is used. More recently, hierarchical terrain models based on irregular triangulations, which provide an explicit multiresolution, have been developed. In such models, every triangle t is expanded into a DTM at the next level by iteratively inserting points in its interior or on its sides, until the required accuracy is reached. *Adaptive Hierarchical Triangulation* [26] and *Hierarchical Delaunay Triangulations* (HDTs) [13] are models of this type.

The only existing proposal for a pyramidal terrain model is the *Delaunay Pyramid* [3]. This model corresponds to an explicit multiresolution model composed of a sequence of Delaunay triangulations, each of which represents the terrain, over the whole domain, at a predefined level of detail.

Data structures for encoding hierarchical and pyramidal terrain models combine individual representations of the single DTMs involved in the model and information describing the interference (containment or intersection, respectively) links. In the following, we describe a data structure, which encodes the minimal information necessary for applying the randomized algorithms discussed in Section 4.

For our purposes, a single DTM $\mathcal{D}_i' \in \mathcal{N}'$ can be represented by the set of its faces: each face must store its vertices and the equation of the surface containing the face (for planar triangular faces, the three vertices define the supporting plane).

In a hierarchical model, a bidirectional link is maintained between each macroface and its direct expansion: for every arc $(\mathcal{D}_i', \mathcal{D}_j') \in \mathcal{E}$, labelled with a face ϕ_j, we encode a link between ϕ_j and $\mathcal{D}_j'$. In a pyramidal model, a bidirectional link is maintained between pairs of intersecting faces belonging to different levels: for every arc $(\mathcal{D}_i', \mathcal{D}_j') \in \mathcal{A}$, labelled with a pair (ϕ_i, ϕ_j) of faces, we encode a link between ϕ_i and ϕ_j.

If multiresolution is explicit, a threshold value in the sequence $[\varepsilon_0, \ldots, \varepsilon_h]$ is associated with each face. Since a face ϕ_i may belong to several levels, but it is stored only once in the structure, the associated threshold value of ϕ_i corresponds to the highest (i.e., most refined) level. Note that the actual approximation error on a face associated to level k can be less than ε_k: the approximation error of a face is also stored together with the face.

As mentioned above, multiresolution terrain models provide an *implicit* representation of a terrain at different levels of abstraction. Given a resolution

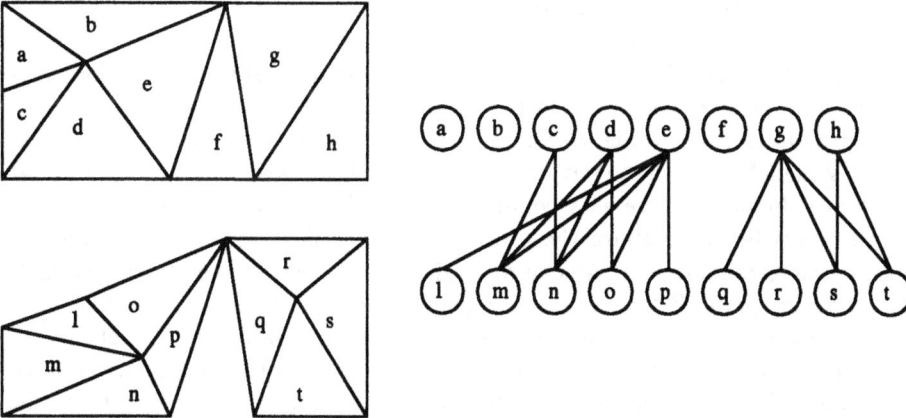

Fig. 2. Levels i and $i + 1$ in a pyramidal terrain model, and the corresponding graph $\mathcal{G}$. If the faces at level i corresponding to an approximation error greater than ε are d and e, then the extracted faces are $a, b, l, m, n, o, p, f, g, h$.

threshold ε, a digital terrain model corresponding to a level ε of accuracy, which we will call the *target* DTM at level ε, can be obtained through a traversal of the structure representing a hierarchical or a pyramidal terrain model. In the following, we describe methods for extracting just the faces composing the target DTM, without reconstructing topological relations between them, since topology is not relevant for the algorithms described in Section 4.

On a hierarchical terrain model, the target faces at level ε of accuracy can be extracted through a traversal of the tree describing the model. Starting from the root, every face ϕ_i is examined: if ϕ_i is simple or satisfies accuracy ε, then ϕ_i is sent to output; otherwise, the direct expansion of ϕ_i is recursively examined. The time complexity of the extraction algorithm is proportional to the number of extracted faces [13].

Let us consider a pyramidal terrain model based on a sequence $[\varepsilon_0, \ldots, \varepsilon_h]$ of tolerance values. Let ε be a target error tolerance, such that $\varepsilon_i \leq \varepsilon \leq \varepsilon_{i+1}$ for some $0 \leq i < h$. In order to extract all target faces at level ε of accuracy, we consider the graph $\mathcal{G}$ whose nodes are the faces belonging to levels i and $i + 1$, and whose arcs represent the interference links between such faces. We compute all the faces at level i whose approximation error is greater than ε, and the connected components of $\mathcal{G}$ containing such faces. Then, all the faces at level $i + 1$, which belong to some connected component, and all the faces at level i, which do not belong to any connected component, are reported (see Figure 2 for an example). The extraction algorithm has a quadratic worst-case time complexity in the number of output faces [3].

3 Visibility on Terrains

Two points P_1 and P_2 on a terrain are said to be mutually *visible* if, for every point $Q \equiv (x, y, z)$ belonging to the open segment $P_1 P_2$, either $(x, y) \notin D$, or $z > \phi(x, y)$. In other words, P_1 and P_2 are visible when the straight-line segment joining them lies above the terrain; such segment is allowed to touch it at most at its two endpoints. A necessary condition for visibility between two points is that both points lie on or above the terrain surface, if their vertical projection lies inside the terrain domain. We can formalize this concept by defining a *candidate point* as any point $P \equiv (x, y, z)$, such that either $(x, y) \notin D$ or $z \geq \phi(x, y)$.

In visibility problems on a terrain, the *viewpoint*, which we denote by V, is located either at infinity in some direction, or at a canditate point (x, y, z). We call a *viewsphere* any sphere centered at V and large enough to contain the whole terrain inside; if V lies at infinity, then the viewsphere is a plane normal to the view direction and lying beyond the terrain with respect to the viewpoint. Intuitively, the viewsphere is used as a "projection screen" when viewing the terrain.

Algorithms for visibility computation, developed in the computational geometry literature, operate on *Polyhedral Terrain Models* (PTMs), i.e., planar-faced digital models. Essentially, polyhedral models used in practice have triangular faces, being either TINs or RSGs in which every rectangular cell is subdivided into two square triangles. In the remainder of this paper, we assume to deal with polyhedral triangulated terrain models.

The *visible image* of a scene (possibly a terrain) is a partition of the viewsphere into maximally connected regions, such that only one face of the scene is visible in each region. Each region is labelled with the face visible inside it. The problem of computing visible images is generally referred as the *Hidden Surface Removal* (HSR) problem. The general quadratic upper bound to the space complexity of the visible image [15, 27] applies to a polyhedral terrain as well [11], thus giving a worst-case space requirement of $O(n^2)$ for a PTM with n vertices.

A characterization of the visible image can be given in terms of the upper envelope of a set of polygons in space. The *upper envelope* of a set of polygons in 3D defines a partition of the horizontal plane into maximally connected regions, each of which is labelled with a polygon, in such a way that, if a region R is labelled with a polygon p, then p is the polygon with maximum height over R. The visible image of a polyhedral terrain, with respect to a viewpoint V, is equal to the upper envelope of the set of its polygonal faces, when the viewsphere plays the role of the horizontal plane and the height of a point is the same as its distance from the viewsphere.

The HSR problem for a three-dimensional scene has been extensively studied in the literature. Existing HSR algorithms can be classified into worst-case optimal algorithms [19, 15], algorithms whose time complexity depends on the number of intersection points between projected edges on the viewsphere [27], and algorithms sensitive to the output size [24, 18, 21, 1, 12]. Some algorithms are specific for polyhedral terrains [18, 24]. The upper envelope of n triangles in 3D can be computed either with a divide-and-conquer algorithm, with optimal time complexity [16], or with an incremental one. A randomized semidynamic incremental algorithm, proposed in [6], presents an expected construction time equal to $O(n^2\alpha(n)\log n)$ in the general case, and $O(n^2\log n)$ for disjoint triangles, as those forming the surface of a terrain (α denotes the inverse of Ackermann's function, which can be considered as a constant in practice).

Intuitively, the *horizon* of a terrain with respect to a viewpoint V, provides, for every radial direction from V, the farthest point from V, lying on the terrain in specified direction and visible from V. On a polyhedral terrain, every point of the horizon belongs to an edge of the PTM. Thus, we can collect the maximal radial intervals in which the horizon corresponds to the same terrain edge, and represent the horizon as an ordered list of radial intervals, each labelled with the terrain edge corresponding to the horizon inside it.

Computing the horizon of a viewpoint on a polyhedral terrain can be transformed into the computation of the upper envelope of a set of possibly intersecting segments in the plane. Given a set $\mathcal{L}$ of p segments in the plane, the *upper envelope* of such segments consists of a partition of the horizontal axis into maximal intervals, such that in each interval I there is an only segment s having the maximum vertical height over I; interval I is thus labelled with segment s. The horizon on a polyhedral terrain is the same as the upper envelope of the set of segments obtained by projecting the terrain edges onto the viewsphere. It has been shown [11] that the complexity of the upper envelope of p segments in the plane is equal to $\Theta(p\alpha(p))$, and, thus, the size of the horizon on a polyhedral terrain with n vertices is equal to $\Theta(n\alpha(n))$.

The upper envelope of p segments in the plane can be computed either by a divide-and-conquer approach [2, 20], where a worst-case optimal $O(p\log p)$ time complexity can be achieved [20], or by an incremental one, with a complexity equal to $O(p^2\alpha(p))$. A dynamic incremental algorithm has been proposed in [14], which constructs the upper envelope of p segments in an expected time equal to $O(p\alpha(p)\log p)$, under the assumption of a random insertion sequence.

Here, we consider the problem of updating visibility information on a multiresolution terrain, when the target level of resolution changes from ε to ε'. A trivial way for producing an updated visible image or horizon consists of re-computing it from scratch, starting from the new terrain representation. Dynamic incremental algorithms, on the contrary, can maintain visibility information under deletions and insertions of faces and edges, and thus are more convenient if the change in the terrain involves a restricted number of entities.

4   A Fully Dynamic Approach to Visibility Computation

In this Section, we present a new dynamic algorithm for computing and updating the upper envelope of triangles, and briefly review the dynamic algorithm for maintaining the upper envelope of segments, which has been presented in [14]. Such algorithms can be used for updating visible images and horizons, respectively, on a multiresolution terrain model. We explicitly state under which conditions the dynamic update should be preferred to the complete reconstruction of such visibility structures.

Both algorithms follow the general scheme, proposed in [5, 7], for designing randomized dynamic incremental algorithms. The core of the method is in a special data structure, called the *Influence Directed Acyclic Graph* (IDAG). Such data structure increases the efficiency of locating the portion of the current upper envelope that is modified by the insertion of a new object (triangle or segment, respectively), when averaging on the insertion order. In addition, the deletion of an object affects the data structure only locally.

The triangle algorithm, which is described in details in the following, is essentially a dynamic extension of the on-line incremental algorithm proposed by Boissonnat and Dobrindt. While the algorithm in [6] only performs triangle insertions, our method allows randomized deletions of triangles, under the simplified hypothesis that triangles are pairwise disjoint. The algorithm for segments is a further simplification of the former one from three to two dimensions. This last algorithm is only briefly sketched here; we refer to [14] for a complete description.

Let $\mathcal{T}$ be a set of semi-disjoint triangles in space. The output of the algorithm is not exactly the upper envelope $Env(\mathcal{T})$ of $\mathcal{T}$, but another plane subdivision, which refines $Env(\mathcal{T})$. Such subdivision is called the *trapezoidal decomposition* induced on the $x - y$ plane by $Env(\mathcal{T})$, and denoted by $Trap(\mathcal{T})$. We classify the vertices of $Env(\mathcal{T})$ into *primary junctions*, corresponding to projected triangle vertices, and *T-junctions*, corresponding to points where an edge of a triangle disappears under another triangle. $Trap(\mathcal{T})$ is obtained from $Env(\mathcal{T})$ by drawing a vertical line through each primary junction, until another edge is encountered in both directions, and drawing in the same way a half-line through each T-junction, in opposite direction with respect to the orientation of the "T". An example of trapezoidal decomposition is shown in Figure 3. Each region of $Trap(\mathcal{T})$ is a trapezoid, possibly degenerating into a triangle, and it is fully contained in a region R of $Env(\mathcal{T})$; thus, it is labelled with the same triangle t_j labelling R. Since, for each vertex of $Env(\mathcal{T})$, at most two new vertices are added to $Trap(\mathcal{T})$, the spatial complexity of the two subdivisions is the same, i.e., $O(n^2)$ for n disjoint triangles. $Env(\mathcal{T})$ can be obtained from $Trap(\mathcal{T})$ by vertically merging adjacent trapezoids labelled with the same triangle.

Before introducing the IDAG and the algorithm, we give some preliminary definitions, which translate the general scheme of [5, 7] into the specific context. We say that a trapezoid tp of $Trap(\mathcal{T})$ is *defined by* at most five triangles $t_j, t_{\text{left}}, t_{\text{right}}, t_{\text{up}}, t_{\text{down}} \in \mathcal{T}$ (see Figure 4), where

- t_j is the triangle labelling tp;
- t_{up} and t_{down} are the triangles containing the edge whose projection contains the upper and lower edge, respectively, of trapezoid tp;
- if the left edge of tp is a vertical line drawn through a primary junction P, then t_{left} is the triangle containing the vertex whose projection is P; if the left edge of tp is a vertical line drawn through a T-junction Q, then t_{left} is the triangle containing the edge which disappears under another triangle at Q; symmetrically t_{right}.

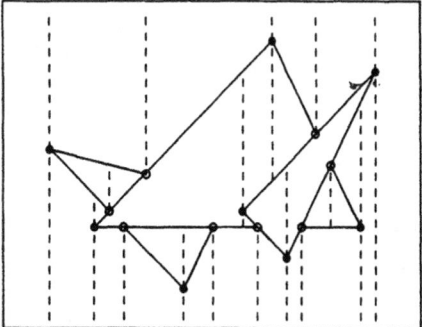

Fig. 3. The trapezoidal decomposition induced by the upper envelope of four triangles: primary junctions are marked in black; T-junctions in white.

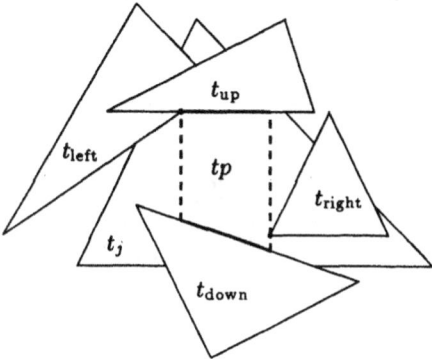

Fig. 4. The five triangles defining a trapezoid tp.

A triangle t is said to be *in conflict* with a labelled trapezoid tp if and only if t properly intersects the unbounded prism, which is the intersection between the locus of points (x, y, z), with $(x, y) \in tp$ and $z \in \mathbb{R}$, and the upper half-space defined by the plane containing triangle t_j. A labelled trapezoid belongs to $Trap(\mathcal{T})$ if and only if it is defined by triangles in $\mathcal{T}$ and is not in conflict with any triangle in $\mathcal{T}$.

Let $t_1, \ldots, t_n$ be the sequence of triangles inserted and not deleted, sorted by insertion time (i.e., t_i inserted after t_{i-1}); set $\mathcal{T} = \{t_1, \ldots, t_n\}$ is called the *current set*. The *Influence Direct Acyclic Graph* (IDAG) stores the current trapezoidal decomposition, plus the "history" of its construction. History does not contain trace of deletions: the current trapezoidal decomposition is thought as built by stepwise inserting triangles $t_1, \ldots, t_n$, even if the actual update sequence may contain extra triangles which have been inserted and then deleted.

For describing the IDAG, we use the same terminology used for trees, even if convergent paths may exist. Each node of the IDAG corresponds to a labelled trapezoid belonging to the trapezoidal decomposition of some subset $\mathcal{T}_k = \{t_1, \ldots, t_k\}$ of $\mathcal{T}$, with $k \leq n$. The root is a conventional unbounded trapezoid, covering the whole plane. At a generic step, the leaves correspond to

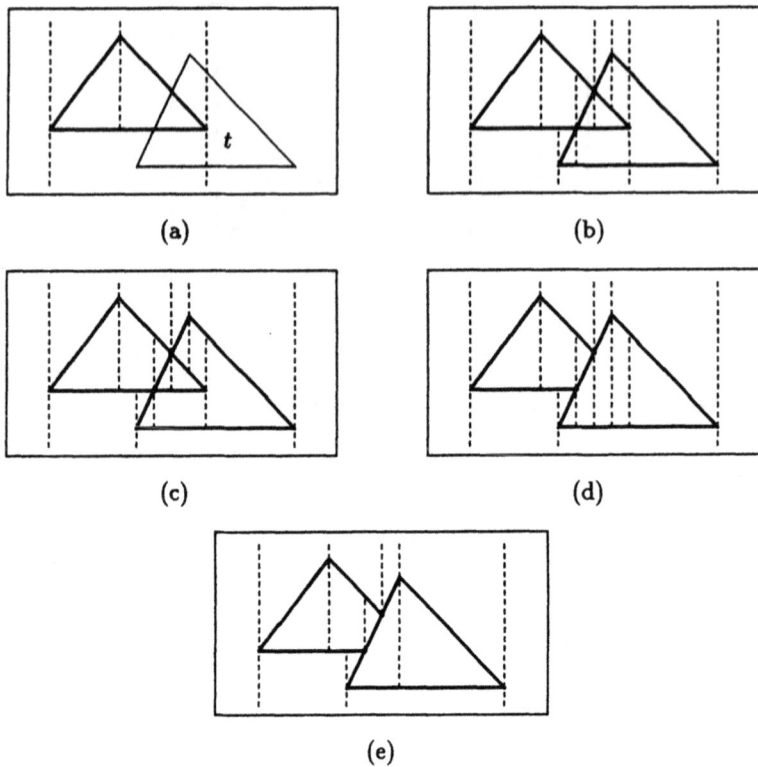

Fig. 5. Insertion of a new triangle t in the trapezoidal decomposition: (a) trapezoids of the current decomposition in conflict with t; (b) cutting step inside each trapezoid; (c) decomposition after external merge; (d) decomposition after internal merge; (e) final updated decomposition.

the trapezoids of the current decomposition. When a triangle t is added, leaves in conflict with t become internal nodes, and new leaves, whose labelled trapezoids are defined by t, are created as their children. A conflicting leaf l, whose associated trapezoid is tp, becomes parent of all new nodes whose trapezoid properly intersects tp. Every child is contained into the union of its parents. The number of children for each node is bounded by a constant value (10 for non-intersecting triangles).

In order to efficiently perform deletions, the IDAG is enriched by adding to each node the list of all triangles of the current set conflicting with it (*conflict list*). If a node is a leaf, then its conflict list is empty. Every node in conflict with a triangle t has at least one parent in conflict with the same triangle t.

The insertion method is basically the same algorithm proposed by Boissonnat and Dobrindt [6], with simplifications due to the fact that we are dealing with semi-disjoint triangles, and with some extra work represented by the update of conflict lists. The insertion of a new triangle t in $Trap(\mathcal{T})$ consists of a *location step* and of an *update step*:

- **Location step**
  All trapezoids of $Trap(\mathcal{T})$ that are in conflict with t are found through an IDAG traversal: starting at the root, all unvisited children of a node, which are in conflict with t, are recursively traversed; finally, visited leaves are collected (see Figure 5 (a)). While traversing the IDAG, we add t to the conflict list of each visited node. Leaves in conflict with t are said to be *killed* by t, since the corresponding trapezoids disappear from the current trapezoidal decomposition as soon as t is inserted. On the contrary, t is called the *creator* of nodes built during the subsequent update step.
- **Update step**
  The new trapezoidal decomposition is computed.
  - *Cutting substep*
    Each trapezoid tp, in conflict with t, is subdivided by drawing vertical lines from the projected vertices of t lying inside tp, and from the possible intersection points between projected edges of t and boundary edges of tp (see Figure 5 (b)). For each new trapezoid, a leaf node is created and linked to the killed node representing tp. Nodes created at this step (*temporary nodes*) are possibly not properly defined trapezoids, which must be merged with other temporary nodes. We can distinguish between external and internal merges:
  - *External merge substep*
    We examine pairs of vertically adjacent new trapezoids tp_1, tp_2, not covered by t: if the vertical edge between tp_1 and tp_2 does not contain any vertex of the upper envelope, we merge tp_1 and tp_2 (see Figure 5 (c)).
  - *Internal merge substep*
    We remove the edges of $Trap(\mathcal{T})$ covered by t. For each triangle edge l, supporting an edge of $Trap(\mathcal{T})$ (partially) covered by t, we collect in two lists all the killed trapezoids whose upper and lower transversal edge, respectively, is supported by l. While traversing each pair of lists, associated with an edge l, we extend the vertical lines corresponding to side edges of adjacent trapezoids above and below the line (see Figure 5 (d)). Finally, we perform vertical merges between vertically adjacent pairs of resulting trapezoids, as described in the external merging step (see Figure 5 (e)).

When a triangle t is deleted, the history of the incremental construction, stored in the IDAG, must be rebuilt as if t had never been inserted. History reconstruction is local to the part of the plane, called *critical region*, where triangle t appeared on the upper envelope. The deletion of a triangle t consists of a location step and of an update step:

- **Location step**
  As in the insertion step, this phase involves a top-down traversal of the

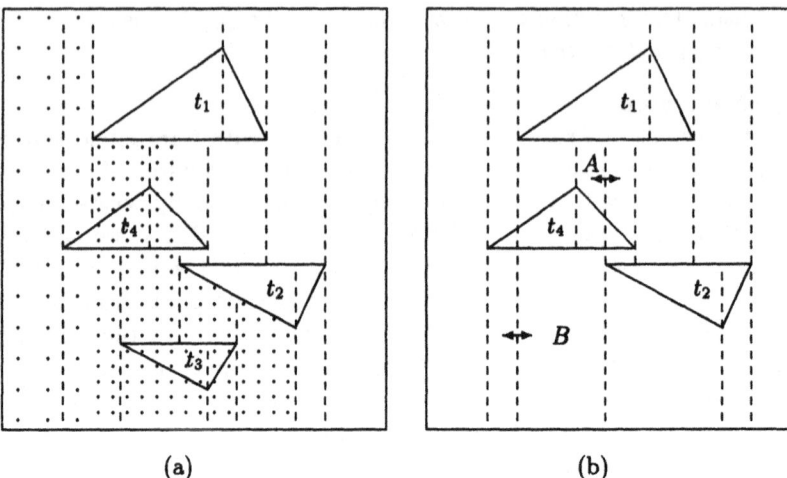

(a)   (b)

Fig. 6. (a) The critical (widely dotted) and co-critical (largely dotted) regions at the deletion of triangle t_3. (b) When triangle t_4 is reinserted, temporary node A must be merged at its right side (in the original construction, this operation created an unhooked node); temporary node B must be merged at its left side (in the original construction, this operation created a removed node).

IDAG. Three different types of nodes are collected:
- Nodes that were leaves when t was inserted, and have been killed by its insertion (*killed* nodes). These nodes form the initial critical region.
- Among the descendants of killed nodes, we collect all nodes defined by t. These nodes (*removed* nodes) must disappear in the updated envelope.
- Nodes not defined by t, but with at least one removed parent (*unhooked* nodes). These nodes remain in the updated envelope, but their history must be (partially) rebuilt.

While visiting the IDAG, we also delete t from the conflict list of traversed nodes. Nodes forming the critical region (called *critical nodes*) are stored in a priority queue, with the insertion time of the new killer (the older conflicting triangle after t) as its key. Unhooked nodes are stored in a dictionary. Moreover, unremoved parents of removed nodes not created by t are also collected and stored in a priority queue, with the insertion time of their killer as their key. Unlike unremoved parents of removed nodes created by t (i.e., killed nodes), these nodes, which we call *co-critical*, are outside the proper critical region, but they will gain new children during the update step.

- **Update step**

We consider all triangles, originally inserted after t, which are creators of removed nodes. In the update step, such triangles are inserted again, one by one, according to their original insertion time. The re-insertion of a triangle t' is local to critical and co-critical nodes.

Special attention must be paid to merge operations, which occur at the boundary of the critical region, i.e., between a temporary child tp_1 of a critical node and a temporary child tp_2 of a non-critical node. In the original IDAG construction, tp_2 was merged with a temporary child tp'_1 of a removed node, thus creating a removed or unhooked node tp_3 (see Figure 6). If tp_3 is a removed node, then the parent of tp_2 is a co-critical node, tp_2 is re-created, and the merge operation is performed in the usual way. If tp_3 is unhooked, the operation is more complicated:

- If the merge operation occurs at a vertical edge, then the corresponding unhooked node tp_3 is found as a child of the node adjacent to the parent of tp_1 along the disappearing edge.

- For merges occurring at transversal edges, there are no adjacency links to help us. During the location step, the unhooked children of a removed node tp, which are the result of a merge operation at a transversal edge of tp, are stored in an appropriate list (if a child tp' of tp results from a merge operation at the upper edge of tp, then tp' is stored in the upper list related to the triangle edge supporting the upper edge of tp). Separated lists are constructed for every creator triangle, and stored in a dictionary. During the reinsertion of a triangle t', the upper and lower lists of unhooked nodes created by t' are merged with those of temporary nodes newly created by t'. The unhooked node(s) tp_3 which must replace each temporary node tp_1 are thus found.

Note that, if a trapezoid tp, created during the reinsertion of t', is not in conflict with t, then no new node is created for tp, but the old one, already present among unhoocked nodes, is used. In this way, original arcs starting from and pointing to a node tp are preserved.

After reiserting t', the critical region is updated by removing nodes in conflict with t' (just processed) and adding new nodes created by t', whose trapezoid is in conflict with t'.

The algorithm for dynamically maintaining the upper envelope of segments works with the same high-level approach illustrated above, with some simplifications due to the reduced dimensionality of the problem (from three to two dimensions). We will not illustrate the algorithm in details here. For a full description, see [14].

The randomized complexity analysis of both algorithms provides an extimation of the expected space complexity of the IDAG, and of the expected time complexity for the insertion/deletion of a triangle or segment. Such analysis is randomized, and exploits general results for dynamic incremental algorithms (see [7]). The analysis of insertion refers to an average case on all possible insertion sequences of the given input objects, all considered equally probable (or, equivalently, the object inserted at each step is randomly chosen within the given data set). The analysis of deletion refers, in addition, to an average case on the object selected to be removed, which is, with the same probability, anyone of the present objects. The worst case is considered for the output size, i.e., $O(n^2)$ for the upper envelope of n semi-disjoint triangles and $O(n\alpha(n))$ for the upper envelope of n segments. Here, we just report the major results of the randomized analysis, which are summarized in Figure 7.

	Triangles	Segments
Worst-case output size	$\Theta(n^2)$	$\Theta(n\alpha(n))$
IDAG size	$O(n^2)$	$O(n\alpha(n))$
Conflixt lists size	$O(n^2)$	$O(n\alpha(n)\log n)$
Problem complexity	$\Theta(n^2)$	$\Theta(n\log n)$
Insertion time	$O(n\log n)$	$O(\alpha(n)\log n)$
Full construction time	$O(n^2\log n)$	$O(n\alpha(n)\log n)$
Deletion time	$O(n\log n)$	$O(\alpha(n)\log n)$

Fig. 7. Complexity results from the randomized analysis of the incremental algorithms for upper envelope computation. n denotes the input size.

For the triangle algorithm, the expected IDAG size is $O(n^2)$, that is worst-case optimal, and is not increased by the presence of conflict lists. The expected time complexity for the insertion of the i-th randomly-selected triangle is $O(i\log i)$, i.e., $O(n^2\log n)$ for inserting all triangles, and the expected time for a random deletion is $O(n\log n)$.

In the case of segments, the expected size of the IDAG, except for conflict lists, is $O(n\alpha(n))$, and it becomes $O(n\alpha(n)\log n)$ counting conflict lists as well. The expected time complexity for the insertion of the i-th segment is $O(\alpha(i)\log i)$, i.e., $O(n\alpha(n)\log n)$ for inserting all segments; the expected time for a random deletion is $O(\alpha(n)\log n)$.

Now, we consider the cost of updating an already computed horizon or visible image. We denote with $\mathcal{D}_1$ the old terrain model and with $\mathcal{D}_2$ the updated one. The sizes of $\mathcal{D}_1$ and $\mathcal{D}_2$ are denoted by n_1 and n_2, respectively. The new visible image or horizon can be either re-computed from scratch, or it can be derived from the old one through an update operation. In the following, we compare these two approaches.

Computing the new visible image by scratch takes a time equal, in the worst case, to $O((n_2)^2)$, if the optimal algorithm by Edelesbrunner, Guibas and Sharir [16] is used. A re-computation of the horizon costs $O(n_2\log n_2)$, if the optimal algorithm by Hershberger [20] is used.

In order to estimate update costs, we denote by k_{old} and k_{new}, respectively, the number of faces (edges) appearing in $\mathcal{D}_1$ and not in $\mathcal{D}_2$, and appearing in $\mathcal{D}_2$ and not in $\mathcal{D}_1$ (learly, $k_{old}\le n_1$ and $k_{new}\le n_2$). Updating the visible image costs $O(k_{old}n_1\log n_1)$ for deletions, and $O(k_{new}n_2\log n_2)$ for insertions. The update of the horizon costs $O(k_{old}\alpha(n_1)\log n_1)$ for deletions, and $O(k_{new}\alpha(n_2)\log n_2)$ for insertions. We want to quantify how "small" the "difference" between the two terrain models must be in order to make the dynamical update convenient. We assume that $\mathcal{D}_2$ is "so close" to $\mathcal{D}_1$ that n_2 and $n_2 = \Theta(n)$. The update of the visible image is more efficient than its direct recomputation if quantities $O(k_{old}n\log n)$ and $O(k_{new}n\log n)$ are both less than $O(n^2)$, i.e., if k_{old} and $k_{new} = O(\frac{n}{\log n})$. For the horizon, the update process is convenient when k_{old} and $k_{new} = O(\frac{n}{\alpha(n)})$ (note that α is very close to be a constant function).

5 Concluding Remarks

In this paper, we have considered the problem of updating the visible image and the horizon on multiresolution terrain models. These two problems reduce to the computation of the upper envelope of a set of disjoint triangles in space (in the hypothesis of dealing with a triangle-based terrain model) and of a set of segments in the plane, respectively.

We have proposed a dynamic approach to solve the above problems, based on randomized algorithms. In particular, we have described a new algorithm for determining the upper envelope of triangles, and briefly summarized an algorithm for computing the upper envelope of segments, fully described in [14]. Such randomized algorithms use an influence graph to store the history of the computation, according to the general framework of [7]; their advantage is in being computationally simple and having a good expected time complexity.

On multiresolution terrain models, varying the resolution level of the terrain representation means replacing a subset of terrain faces and edges. Thus, the dynamic algorithms proposed here allow updating the visible image or the horizon accordingly. This approach is useful when the modification in the terrain representation involves a rather small subset of faces, as shown in Section 4.

Acknowledgements

This work has been supported by the project "Models and Systems for Handling Environmental and Land Data" of the Italian National Research Council under contract N. 95.01057.CT12

References

1. Agarwal, P.K., Sharir, M.: Applications of a New Space Partitioning Technique. Discrete and Computational Geometry 9 (1986) 11–38.
2. Atallah, M.: Dynamic computational geometry. Proceedings 24th Symposium on Foundations of Computer Science (1983) 92–99.
3. Bertolotto, M., De Floriani, L., Marzano, P.: An efficient representation for pyramidal terrain models. Proceedings 2nd ACM Symposium on Advances in GIS (1994) 129–136.
4. Bertolotto, M., De Floriani, L., Marzano, P.: A unifying framework for multilevel description of spatial data. These Proceedings.
5. Boissonnat, J.D., Devillers, O., Schott, R., Taillaud, M., Yvinec, M.: Application of random sampling to on-line algorithms in computational geometry. Discrete and Computational Geometry 8 (1992) 51–71.
6. Boissonnat, J.D., Dobrindt, K.: On-Line construction of the upper envelope of triangles in $\mathbb{R}^3$. Proceedings 4th Canadian Conference on Computational Geometry (1992) 311–315.
7. Boissonnat, J.D., Yvinec, M.: Structures et algorithmes géométriques (1994). In preparation.

8. Cazzanti, M., De Floriani, L., Nagy, G., Puppo, E.: Visibility computation on a triangulated terrain. Progress in Image Analysis and Processing II (1991) 721–728.

9. Chen, Z.T., Tobler, W.R.: Quadtree representation of digital terrain. Proceedings Autocarto (1986) 475–484.

10. Clarkson, K.L., Shor, P.W.: Application of random sampling in computer geometry. Discrete and Computational Geometry 4 (1989) 387–421.

11. Cole, R., Sharir, M.: Visibility problems for polyhedral terrains. Journal of Symbolic Computation 17 (1989) 11–30.

12. De Berg, M., Halperin, D., Overmars, M., Snoeyink, J., Van Kreveld, M.: Efficient ray shooting and hidden surface removal. Algorithmica: An International Journal in Computer Science 12 (1994) 30–53.

13. De Floriani, L., Puppo, E.: A hierarchical triangle-based model for terrain description. Theories and Methods of Spatio-Temporal Reasoning in Geographic Space (1992) 236–251.

14. De Floriani, L., Magillo, P.: Horizon Computation on a Hierarchical Triangulated Terrain Model. The Visual Computer: An International Journal of Computer Graphics 11 (1995) 134–149.

15. Devai, F.: Quadratic bounds for hidden line elimination. Proceedings 2nd ACM Symposium on Computational Geometry (1986) 269–275.

16. Edelsbrunner, H., Guibas, L.J., Sharir, M.: The upper envelope of piecewise linear functions: algorithms and applications. Discrete and Computational Geometry 4 (1989) 311–336.

17. Gomez, D., Guzman, A.: Digital model for three-dimensional surface representation. Geo-processing 1 (1979) 53–70.

18. Katz, M.J., Overmars, M.H., Sharir, M.: Efficient hidden surface removal for objects with small union size. Proceedings 7th ACM Symposium on Computational Geometry (1991) 31–40.

19. McKenna, M.: Worst case optimal hidden surface removal. ACM Transactions on Graphics 6 (1987) 19–28.

20. Hershberger, J.: Finding the upper envelope of n line segments in $O(n \log n)$ time. Information Processing Letters 33 (1989) 169–174.

21. Overmars, M., Sharir, M.: Output-sensitive hidden surface removal. Proceedings 30th IEEE Symposium on Foundations of Computer Science (1989) 598–603.

22. Ponce, J., Faugeras, O.: An object-centered hierarchical representation for 3D objects: the prism tree. Computer Vision, Graphics and Image Processing 38 (1987) 1–28.

23. Preparata, F.P., Shamos, M.I.: Computational Geometry: An Introduction (1985), Springer Verlag.

24. Reif, J.H., Sen, S.: An efficient output-sensitive hidden-surface removal algorithm and its parallelization. Proceedings 4th ACM Symposium on Computational Geometry (1988) 193–200.

25. Samet, H., Sivan, R.: Algorithms for constructing quadtree surface maps. Proceedings 5th International Symposium on Spatial Data Handling (1992) 361–370.

26. Scarlatos, L.L., Pavlidis, T.: Hierarchical triangulation using cartographic coherence. Graphical Models and Image Processing 54 (1992) 147–161.

27. Schmitt, A.: Time and space bounds for hidden line and hidden surface algorithms. Proceedings Eurographics (1981) 43–56.

Theory for the Integration of Scale and Representation Formats: Major Concepts and Practical Implications

Bud P. Bruegger

InfoTerra, Via Michelangelo 39, 58100 Grosseto, Italy
bud@bruegger.telnetwork.it

Abstract. While GISs become ever more powerful and comfortable to use, spatial data integration has remained a difficult user responsibility with only sparse system support. A recent Ph.D. thesis has proposed a spatial theory that attempts to remove some major impediments to automation in this area. In particular, the work focuses on the integration of scale and format (raster/vector) differences. As a first of a planned series of publications, this paper describes the targeted problem, the major concepts of the solution, and discusses some practical implications that show potential benefits and difficulties of the described work.

1 Introduction

While GISs become ever more powerful and comfortable to use, the integration of spatial data has remained a difficult user responsibility with only sparse system support [Flowerdew, 1991]. Users are faced with difficult decisions about the choice of integration methods and parameters. The integration process is typically highly subjective and its consequences on the information content of data sets are hardly understood.

To remove some major impediments to automation in this area, a recent Ph.D. thesis [Bruegger, 1994] has developed a spatial theory that formally models the problems of integrating scale and representation formats. The theory encompasses issues of (i) representation of spatial information at limited resolution and thus level of detail, (ii) representation formats (raster/vector), (iii) relevant meta data, (iv) uncertainty introduced by the scaling process and finite approximation, and (v) conversions across representation formats and scales. The broad conception of the presented work follows Goodchild's second "hard challenge" [Goodchild, 1992a]: "To devise a system of theory, terminology, and meta data that will support improved sharing of spatial data".

As a first of a planned series of publications, this paper describes the *targeted problem*, the *major concepts* of the solution, and discusses some *practical implications* that show potential benefits and difficulties of the described work. Many other aspects are *excluded* from this paper but can be found in [Bruegger, 1994]. They include:

- *mathematical formalism* (in the form of an algebraic specification)
- *uncertainty model* (limited to uncertainty introduced by scaling and finite approximation)
- *visualization issues*
- *implementation issues*
- *application examples* (format-independent GIS user interfaces, automatic scale-conversions based on meta data, complete propagation of uncertainty during scale and format conversion, specification of minimal data quality requirements by user

and consequential automatic fitness for use assessment, sliver-free overlay algorithm based on proposed uncertainty model).

Data integration has found a fair share of attention in the GIS literature (e.g., [Rhind, 1984], [Abel, 1990], [Piwowar, 1990], [Flowerdew, 1991], [Shepherd, 1991], [Maguire, 1991], [Stephan, 1993]). The presented theory differentiates itself by (i) explicitly modeling the scaling process and its effects on spatial data, and (ii) its wide conception that includes representation, format differences, meta data, uncertainty model, and conversions.

The remainder of this paper is organized as follows: Section two describes the targeted problems of data integration, analyzes their origin, and specifies requirements for possible solutions. Section three through five show the approaches taken to satisfy these requirements: In particular, section three discusses format-independent modeling, section four the concept of "consistency" of a spatial theory, and section five the explicit model of the scaling process. Sections six and seven discuss practical implications at the example of the concept of scaling: namely, the potential impact of the proposed formalization on remote sensing and cartographic practice is explored. Finally, possibilities and problems of an implementation of the theory in GIS software are discussed in section six.

2 Problems of Data Integration

This section focuses on some fundamental data integration problems that have their origin in an incomplete conceptual understanding and can therefore be solved by a supporting theory. The discussion starts with a description of the integration process, gives examples for problems, and derives requirements for possible solutions.

Many GIS operations (such as overlay) require that their argument data sets are of the same representation format (raster or vector) and at comparable scale. Due to the diversity of data sources, this is usually not automatically the case. To satisfy the requirement, GISs therefore provide users with a toolbox of conversion routines that change the format or scale of data sets.

Such conversion routines come in a wide variety. Van der Knaap studied the diversity of vector-to-raster conversion routines and pointed out their semantic differences [1992]. The variety of scale-change or generalization methods is even greater: Scale-change can be performed in the raster [Monmonier, 1983], or vector domain (see [Buttenfield, 1989] for a survey); and in each, many different scale-reduction methods have been proposed. This variety is further emphasized by the parametrization of conversion methods, where a different choice of parameters changes the semantics of the method. For example, the Douglas-Poiker algorithm [Douglas, 1973] uses the width of a band to control the degree of generalization.

The variety of conversion methods causes two major problems: (i) a high degree of subjectivity of the integration process, and (ii) difficult decisions being left in the user's responsibility.

(i) The subjectivity of the integration process is evident in the fact that a single problem, starting from the same input data sets, will yield different results on different GIS platforms and with different human operators: Different GIS packages implement the same conceptual conversion with slightly different semantics and different human operators make different decisions about the most appropriate conversion routine and its parameters. In order for GIS to mature towards a science

[Goodchild, 1992b], such subjectivity has to be replaced by a formal theory in support of data integration.

(ii) The choice of appropriate conversion methods and parameters requires difficult decisions from users. The following examples illustrate this:

1. Two vector coverages shall be overlaid. One is digitized from a map at a scale of 1:25,000, the other at 1:100,000. The GIS provides a Douglas-Poiker algorithm for generalization. Decisions: What is the appropriate parameter of this algorithm? How much uncertainty is introduced by the generalization? Is there a danger of producing an inconsistent vector coverage with self-intersecting boundary lines [Beard, 1991]?

2. A raster and a vector data set of comparable scale shall be overlaid. Both, a raster and a vector implementation of the overlay operation are available. Decision: Shall the vector data set be converted to raster and the overlay then be performed in the raster domain, or shall the data sets be overlaid in the vector domain? What are the implications on data quality of the alternative decisions?

3. A large-scale vector data set shall be overlaid with a relatively small-scale raster data set. Decision: Shall the large-scale data set first be converted to raster and then generalized in the raster domain, or the other way around? Are raster and vector generalization methods conceptually equivalent? If so, which are the rules to find equivalent parameters of raster and vector methods?

4. A classified satellite image of 80m resolution shall be overlaid with a 1:10,000 map that was digitized and rasterized. Decision: How much does the digitized map have to be generalized to be of a comparable "scale"? What "scale" corresponds to 80m resolution?

The above examples demonstrate the following four basic problems: (i) incompatibilities between components of the "current spatial theory"[1], (ii) incompatible concepts between the raster and vector domain, (iii) incomplete understanding of scale and scale-change impedes integration, and (iv) the resolution concept of remote sensing and scale concept of cartography are separated by a conceptual gap.

(i) Example 1 illustrates the incompatibility between meta data and scale conversion methods, since the cartographic scale and the parameter of the generalization routine cannot be precisely related. Similarly, the incompatibility between uncertainty model and conversion routines is evident in the fact that the uncertainty increase caused by the scale-change remains unquantified. The example further hints at an incompatibility between representation and conversion routines, since large generalization steps can result in inconsistent representations with self-intersecting boundary lines.

(ii) Examples 2 and 3 document that scale conversions and operations such as "overlay" have different semantics in the raster and vector domains, although they should be equivalent from a conceptual point of view.

(iii) Examples 1, 3, and 4 illustrate the lack of a precise understanding of how data sets are affected (meta data, uncertainty) by a limitation of scale. Consequently, the

---

1 The entirety of concepts that support the current practise are called "current theory" in this paper--even if they have never been proposed as a "theory" or treated as a whole.

selection of an appropriate generalization algorithms is difficult since it cannot be supported by formal reasoning with a theoretical basis.

(iv) Example 4 illustrates the different schools of thinking about the scaling process in remote sensing and cartography, even if the same format (i.e., raster) is used.

The above considerations show that the problems originate from a lack of conceptual basis and can be solved with a spatial theory with the following properties:

- *format-independent definition of concepts* (see section 3)
- *consistency of theory* through mutual compatibility among representation, meta data, uncertainty model, and conversions (see section 4).
- an *explicit model off the scaling process* (see section 5).
- a concept of scale that can *approximate concepts of remote sensing and cartography* (see sections 6 and 7).

3   Format-Independent Modeling

Format-independence of concepts is one of the fundamental design considerations behind the proposed spatial theory. The approach to achieving this is described in the following.

Many of the concepts used in current practice are defined in the domain of a single format and fail to have exact equivalents in the other format. An example is the Douglas-Poiker algorithm that is defined in terms of the entities of vector representations. Similarly, the concept of "raster resolution" is obviously limited to the raster domain.

The approach taken in the presented work relies on the fact that geometry is inherently infinite [Smith, 1992]. This infinite character is for example evident in the infinite number of points in Euclidean space or in the arbitrarily complex shape that regions (i.e., point sets) in such a space can assume.

In contrast to geometry, both computer models and data acquisition methods necessarily have to be finite.   Goodchild has studied different possible "discretizations" of the infinite "geographic reality" that result in finite representations [Goodchild, 1990]. He demonstrates how different discretizations of the same infinite reality lead to different representation formats. It can thus be followed that every finite representation is necessarily formatted; and that a format-independent model is necessarily infinite.

Based on these considerations, all concepts of the proposed theory are defined in a format-independent infinite domain. Different discretizations are then used to derive finite, formatted approximations of the infinite concepts.

Figure 1 illustrates this approach: It shows two spatial regions (shown as a disk and a ring, respectively) and a "buffer" operation that maps one into the other. The two representations and the buffer operation are all defined in the infinite domain of Euclidean space (b) and can be finitely approximated in either raster (a) or vector format (c). It is obvious that this approximation introduces uncertainty. Note that also the formatted implementations of the buffer operation are only approximations of the infinite concept and thus also introduce uncertainty. Since raster and vector implementations differ only in their approximation methods, they can both be treated as special cases of uncertain concepts defined in the infinite domain [Bruegger, 1994].

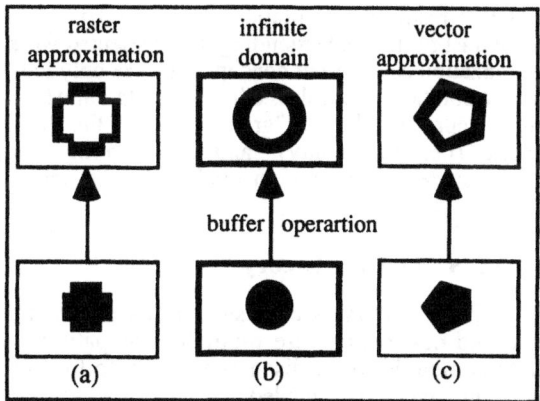

Fig. 1. Spatial region and derived buffer represented in an infinite geometric model (b), and in a raster (a) and vector approximation (c).

The example of Euclidean geometry shows that "modeling in the infinite" is natural and used on a daily basis. Note also, that explicit modeling of not directly implementable infinite concepts are common in other disciplines. For example, in image processing, linear filters are defined in the infinite domain as convolution integrals and then finitely approximated in the raster domain [Castleman, 1979].

The proposed theory uses Euclidean space as an infinite domain for "geographic reality" that represents the world at infinite resolution, free of scale limitations. For the modeling of resolution-limited representations, the infinite domain of "resolution-limited space" was introduced [Bruegger, 1994].

Explicit modeling of the infinite concepts directly supports (i) the design of format conversions, and (ii) the management of approximation uncertainty:

(i) The infinite model serves as a *framework for the design of format conversions*. For example, in figure 1, the infinite domain of Euclidean geometry contains both, the entities of the raster (i.e., cells), as well as those the vector domain (i.e., points, lines, polygons). While format conversion methods must obviously be finite in order to be implementable, the infinite model is indispensable for their definition.

(ii) Infinite models are also necessary to *capture uncertainty* caused by the approximation. Here, the infinite model represents the error-free standard against which finite approximations are measured. While the description of the uncertainty model is excluded from the scope of this paper, a detailed discussion can be found in [Bruegger, 1994].

The format-independent definition of representations, conversions, meta data, and uncertainty opens the way to a format-independent map algebra. It can be used for the implementation of GIS user interfaces that completely hide format issues from the user. The user specifies the desired analysis and manipulation operations at a format-independent conceptual level. The system is then responsible for selecting the appropriate formatted approximation of operations and execute format conversions where necessary [Bruegger, 1994].

Like all other concepts, also the scaling-process is modeled in the infinite domain. This is interesting from the point of view that the scaling process has often been understood to directly create finite representations. For example, the term "raster resolution" implies that resolution limitation also discretizes geometric space into discrete cells. In contrast to this understanding, the presented work treats resolution-limitation and discretization as different concepts that are only loosely related via the sampling theorem: "resolution" controls the level of detail at which knowledge about geographic reality is represented; the "granularity" of the discretization controls approximation quality (i.e., uncertainty); and the sampling theorem states that at lower resolution, the same approximation quality can be achieved using coarser granularity.

4 Consistency of a Spatial Theory

As stated above, a spatial theory is consistent if its representations, meta data, uncertainty model, and conversions are mutually compatible. An approach for achieving complete compatibility is discussed in the following. For reasons of simplicity, the uncertainty model is excluded.

Figure 2 illustrates the relationship between different components of the spatial theory. *Geographic reality* represents the world at infinite resolution with all its geometric detail. Due to a natural resolution-limitation of all sensors, only *resolution-limited representations*, such as R1 and R2, are actually perceivable. The scaling process that limits detail is modeled by an *abstraction process*. For example, A1 is the abstraction process that maps geographic reality to the representation R1. Representations can exist at different levels of detail, i.e., scale. In figure 2, R2 is less detailed than R1. The *conversion method* C12 maps R1 to R2 by further reducing detail of R1.

Geographic reality and the abstraction processes A1 and A2 are only of theoretical value. Namely, they serve for the definition of the representations (R1 and R2) and conversions (C12). R1, R2, and C12 have practical relevance since they are finitely implemented in a GIS as approximations of the infinite concepts.

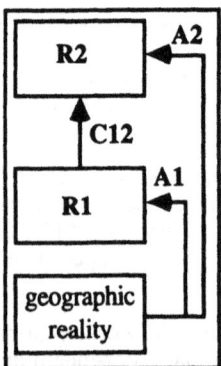

Fig. 2. Relationship between geographic reality, representations (R1 and R2), abstraction processes (A1 and A2), and conversions (C12).

The consistency of the proposed spatial theory is guaranteed by the following rules:

(1) $R_i = A_i$ (geographic reality)

(2) $metaData(R_i) = parameters(A_i)$

(3) $C_{ij}(A_i) = A_j$

(1) Every representation R_i is assumed to be derived from geographic reality by an abstraction process A_i. A_i can be considered a function of known type that is completely determined by a set of parameters. For example, "resolution" is such parameter in the proposed spatial theory.

(2) The parameters of A_i are directly used as meta data of the corresponding representation Ri. These meta data are a complete and precise specification of the information content of the representation. They describe exactly how the spatial knowledge was limited by the abstraction process (e.g., scaling).

(3) Similar to vector difference, conversions are defined as the "difference" of two abstraction processes (e.g., "$C12 = A2 - A1$"). For example, mapping geographic reality to R1 with the abstraction process A1 and then converting R1 to R2 with C12 must be equivalent to directly mapping geographic reality to R2 with abstraction process A2.

Rule (3) guarantees transitive closure of the theory: the result of every conversion is again a legal representation that follows rule (1). Consequently, the effect of conversions on the information content of target representations can be completely captured in terms of meta data. Further, the approach guarantees that *conversions are completely determined by their source and target meta data.*

A mathematically sound concept of "consistency" is the key to possible automation of data integration processes. Precise definitions allow machine interpretation of problems and easy implementation. For example, based on meta data, a GIS can automatically execute a necessary scale reduction of a data set before overlay. Such "intelligence" results in drastic improvement of user-friendliness. This benefit is possible thanks to the replacement of partly subjective processes with mathematically precise concepts.

5 Model of Scaling Process

A theory in support of spatial data integration must explicitly model the scaling process, i.e., how exactly detail is reduced between geographic reality and a scaled representation. According to the requirement of format-independence, the scaling process must be modeled in an infinite domain. To be part of a consistent spatial theory, it further takes the role of an abstraction process as described in the previous section.

As a point of reference, this section first describes "*geographic reality*" that features infinite resolution. The scaling process is then modeled in two steps: "*resolution limitation*" and "*feature abstraction*".

At the infinite resolution of "*geographic reality*", *points* of Euclidean space are the smallest resolvable spatial units. Assuming that geographic reality is inhabited by *discrete objects* such as buildings, lakes, or trees, every point in space belongs to a certain object; and the *geometry* of an object is the set of all points that belong to it. The geometry of objects can be arbitrarily complex. This is evident in the possibility of arbitrarily small disconnected components ("islands" and "holes"), and the potentially unlimited geometric complexity of boundary curves.

Geographic reality can neither be perceived nor represented: It is evident that no sensor exists that could inspect the properties of a single point--and that even if it existed, data acquisition effort would be infinite. Even if geographic variety was perceivable at its infinite level of detail, it would be impossible to represent it either with graphical (maps) or numeric (data set) means.

Resolution limitation has its origin in the *limited capabilities of sensors:* Instead of inspecting a single point, the smallest resolvable spatial units are *regions*. Within such a region, sensor sensitivity usually varies. Sensor resolution is therefore often specified by a *modulation transfer function* [Davis, 1991] that usually shows sensitivity in the form of a bell-curve.

A simplified model of sensor resolution is the *Instantaneous Field of View (IFOV)*. It assumes that the sensor sensitivity is homogeneous within a disk-shaped region and zero outside the disk. Hypothetical sensors that feature such a sensitivity distribution are usually called "ideal sensors".

The first step of the proposed scaling concept is called "*resolution limitation*" and is based on an perception with an ideal sensor (see figure 3): Several discrete objects of geographic reality fall inside the IFOV. The hypothetical sensor can directly observe a "*mixture*" of discrete objects (see box on the right of sensor). The size of the IFOV is called "*resolution*" and is the parameter of the according abstraction process. It is assumed to be constant over a data set.

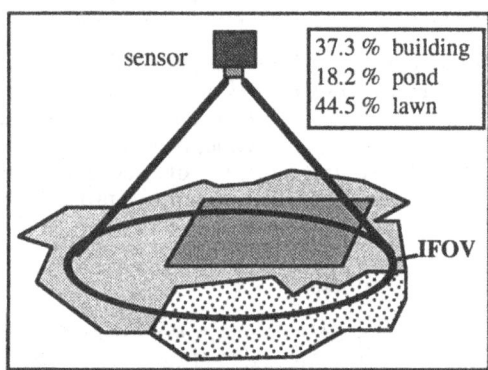

Fig. 3. Application of the IFOV resolution concept to a world that is inhabited by discrete objects.

The reduction of detail is modeled by using "mixtures" as attributes of IFOVs, rather than single objects as attribute of points. It is evident, that observed mixtures change continuously when the IFOV is moved. To express this continuous character, the representation composed of all possible IFOVs and their mixture attributes is called "*mixture field*".

Since in a continuous field lacks discrete entities with geometries, a second abstraction step, called "*feature abstraction*", is used to map "mixture fields" to discrete "*feature partitions*". Mixture fields can be seen as an intermediate step in the overall abstraction process from geographic reality to feature partitions.

Feature abstraction is based on the classification of mixtures based on a minimal percentage at which certain objects have to be present. For example, the *feature* "pond" could be defined as the set of all IFOVs that contain more than 85% of the object "pond". 85% is then called "level of homogeneity" or simply "*homogeneity*" and is the parameter of the feature abstraction process.

IFOVs that fail to reach the required homogeneity of any feature are united to form the *"transition zone"*. This transition zone is an expression of uncertainty that is introduced by the scaling process: A decrease of resolution typically increases the width of the transition zone between features. The transition zone is also suited for the modeling of gradual transitions in geographic reality, for example, that between two soil types or from forest to grassland.

Usually, relatively narrow transition zones separate large, highly homogeneous areas. The level of homogeneity is typically chosen to be high (i.e., 80% or above) to maintain an expressive feature attribute within the homogeneous areas.

Within features, homogeneities of less than 100% allow the absorption of "inhomogeneities" that are small compared to the resolution. For example, a small island in geographic reality can be absorbed in the feature "lake".

In analogy to the distinction of H- and L-resolution in remote sensing [Strahler, 1986], two kinds of features exist:
- A typical example for an H-resolution feature is a lake that is relatively large compared to resolution. The feature "lake" can then be seen as a generalized version of the object "lake". The feature's definition, and thus geometry, obviously depends on its resolution and homogeneity.
- A typical L-resolution feature is a forest that is composed of objects of type "tree" that are much smaller than the IFOV. Individual objects are thus not resolvable; and a feature corresponds to a collection of objects.

While objects exist in their own right, features exist only in the context of a given resolution and homogeneity. The distinction of objects and features in the proposed theory is closely related to similar concepts proposed by Couclelis [1992].

6 Comparison to Scaling in Remote Sensing

A formalization of a problem domain always results in stricter and narrower definitions of concepts and procedures. It therefore necessarily causes certain changes in practice. The present and following section illustrate such changes caused by the proposed stricter definition of scale.

This section compares the proposed model of scale to the scaling process in remote sensing that includes sensorial perception of images and their classification. The discussion shows that the proposed formalization requires some minor changes of practice, but is otherwise a good approximation of the scaling process used in remote sensing.

Classified remotely sensed images are an important data source of GISs. In order to use them within the framework of the proposed theory, they must satisfy the requirement of representations to be the result of an abstraction process as described previously in section four (rule 1). Since this is not directly the case, the following describes differences between remote sensing practice and proposed theory and how they can be resolved.

A first major difference is the concept of resolution: Actual sensors are characterized by modulation transfer functions, while the theory uses the simplified concept of IFOV. To overcome the discrepancy, linear system theory [Castleman, 1979] allows the estimation of the image that would have been observed with an ideal sensor based on the actually observed image. The technique is implemented by a linear filter and is commonly applied in image restoration [Castleman, 1979]. An additional filtering step thus brings the remotely sensed image in a form that is compatible with the IFOV-based resolution concept.

A second difference is evident in the "mixtures" that cannot directly be observed by actual sensors. The solution are image classification methods that can handle "mixed pixels" [Goodchild, 1993] and thus yield reasonable estimations of "mixtures". One unresolved problem is the incapability of the proposed theory to model uncertainty introduced by the confusion of different objects (or classes), for example, due to an insignificant difference between their spatial signatures.

An image corrected for the non-ideal sensor resolution and classified with a mixed-pixel-conscious method is a good approximation of a "mixture field". It can be abstracted to a "feature partition" with the method provided by the theory.

A difference to common practice is the creation of "transition zone pixels" during feature abstraction. Also a feature partition is created in a two-step procedure (consisting of mixed pixel conscious classification and feature abstraction) rather than by a single image classification that assumes pixels to be homogeneous.

7 Comparison to Scaling in Cartography

The concept of scale used in today's GISs has its origin in (manual) cartography. Here, cartographic abstraction is the process that reduces detail of geographic reality according to the limitations of a graphic media at a given scale [Robinson, 1984].

Cartography is often considered an art more so than a science and, consequently, has a long tradition of subjective procedures. The reconciliation of cartography with formal theories is therefore a difficult undertaking, as is evident in the many problems of digital cartography. The following discussion reasons that a reconciliation requires a strict separation of objective and subjective aspects of cartography. With this approach, the proposed spatial theory is a good approximation of the objective aspects of cartographic practice, and a basis for the separately implemented subjective aspects.

Traditionally, cartography has the sole purpose of communicating spatial information to humans. For humans, subjectivity in the cartographic process is not only acceptable, but in most cases it is intentionally used to optimize communication for human interpretation.

With the introduction of GISs, maps increasingly serve as (analog) storage media for spatial information; and cartographic abstraction defines the representation model that is used for machine reasoning in GISs. While human interpreters of maps use location solely to visually extract relations among features of a single map, GISs use location in a much more precise way to extract quantitative properties and/or evaluate relations among features of different maps (i.e., data sets).

From this point of view, it is not surprising, that the more precise philosophy of GIS is incompatible with subjective cartographic practices such as displacement, exaggeration, and certain kinds of symbolization. While these practices facilitate human interpretation, they cause misunderstandings and errors when GISs interpret the data on the basis of precise locations.

It is proposed that a two stage approach can be used to overcome the diverse requirements of human and machine interpretation: a first stage of *formal spatial modeling* represents and manipulates spatial information in an objective, machine interpretable manner; and a second stage of *visualization* of formal representation is optimized for human consumption.

The proposed spatial theory is based on this two stage approach but focuses primarily on the stage of formal spatial modeling. The following properties of its scaling concept document that it formalizes many concepts of cartographic practice (see [Bruegger, 1994] for detail):

- it guarantees displayability within the limits of graphical media.
- it supports arbitrary steps of scale reduction
- it models the simplification of shapes
- it models absorption of small inhomogeneities (such as islands) and changes of topology
- the theory includes an equivalent to cartographic dimension change[2], a method to preserve features to scales where they would normally be lost.

Not surprisingly, the theory also has similarities to Perkal's attempt to formalizing cartographic generalization [Perkal, 1966], (see [Bruegger, 1994] for a detailed discussion).

While the presented work predominantly focuses on formal spatial modeling, some visualization methods have also been proposed [Bruegger, 1994]. Compared to cartographic tradition, they are rather simplistic since they lack optimization for human interpretation. While some of the proposed methods may be adequate for use in interactive GISs, future research must develop more sophisticated visualization methods that have their roots in cartographic expertise.

8 Discussion

The paper has described some fundamental problems of spatial data integration. It has demonstrated, how a spatial theory that is based on the concepts of format-independent modeling and consistency can overcome these problems. A concept of scale was proposed as a major component of such a theory. It was used to discuss practical implications of the stricter definition of concepts and procedures in the formalization.

While the proposed spatial theory promises far-reaching automation of the targeted data integration problems, these benefits can only be harvested if the theory is actually implemented in GIS packages. The following discussion focuses on problems, possible strategies, and benefits of such an introduction.

The majority of problems originate in a lack of complete downward compatibility with respect of current GIS implementations. The following lists of necessary modifications and additions illustrates this:

- Data structures (only vector) must be extended by the proposed uncertainty model. The locally varying uncertainty is captures by a kind of band along boundaries. A highly suitable vector structure [Gold, 1992] has already been developed for other purposes [Bruegger, 1994]; raster data structures do not require any extensions.
- Meta data standards have to be extended by a formally precise notion of scaling effects, namely "resolution" and "homogeneity".

---

2 For example, cartographic dimension change maps an areal feature such as a city (that would become unperceivably small with normal scaling) to a point symbol.

- GIS operations have to be extended to propagate meta data and uncertainty (see [Bruegger, 1994] for examples of "overlay" and "reclassification").
- Format conversion routines have to be extended to propagate uncertainty.
- Scale conversions have to be replaced by those of the proposed spatial theory.
- Automatic fitness for use assessment has to be added.
- Format-independent user interface has to be added in the form of a query optimizer.
- Query optimizer for the automatic integration of scales has to be added.

A complete implementation of the proposed theory is obviously a significant undertaking, although a high percentage of existing code is expected to be re-usable or easily adaptable. Considering the significant investment in the current theory, a complete adaptation of the proposed theory is unlikely to happen in a single step.

The most realistic strategy for the introduction of the proposed theory is a step-by-step procedure. The theory is then only used within limited modules. These could, for example, handle scale change, or the integration of a group of data sets. While such modules can internally take advantage of the formal theory, they communicate with the external world through a "theory translator".

This translator has to convert between the formally precise concepts of the proposed theory, and the partly subjective or vague concepts used in current practice. Since this is an ill-defined problem, theory translations will have to implement certain pragmatic assumptions: For example, a cartographic scale of 1:25,000 could be translated to a certain "resolution" and "homogeneity"; and "uncertainty" could be assumed to be homogeneous over space and of a magnitude associated with the cartographic scale.

A step-by-step introduction could quickly take advantage of certain benefits with a relatively small implementation effort and without loss of investment. Gradually, the whole theory with all its benefits could be implemented.

The proposed theory promises GISs with a significantly improved support of the targeted data integration problems since it opens the way to machine reasoning about scale and format issues. GISs can thus automate or support users in the following tasks:
- conversions between formats
- necessary scale-reductions
- propagation of uncertainty and meta data
- fitness for use assessment
- optimization of queries (e.g., for response-time or data quality)

The proposed theory thus solves some long-standing problems thanks to its formal approach. Formalization necessarily causes changes in current GIS practice which makes use of some subjective or vaguely defined concepts. While it is possible to find different solutions to data integration problems, I believe that they will necessarily be formal and thus will have a comparable impact on practice. The considerable benefits of an introduction in practice may make it well worth the also considerable cost.

The practical benefit of the proposed theory is mostly its potential for automation of data integration processes. From an academic point of view, the theory formalizes certain aspects of geography and consequently allows the use of spatial information in a less subjective manner. In this respect, the proposed theory contributes to a spatial information science [Goodchild, 1992b].

References

Abel, D.J. and Wilson, M.A., 1990: A System Approach to Integration of Raster and Vector Data and Operations. In: *Proceedings, 4th International Symposium on Spatial Data Handling,* Brassel, K. and Kishimoto, H. (eds.), Zürich, Switzerland, July 23-27, pp. 559-566.

Beard, M.K., 1991: Theory of the Cartographic Line Revisited / Implications for Automated Generalization. *Cartographica, 28,* 4, pp. 32-58.

Bruegger, B.P., August 1994: *Spatial Theory for the Integration of Resolution-Limited Data.* Ph.D. dissertation, University of Maine, Orono, Internet: ftp://grouse.umesve.maine.edu/pub/SurveyEng/Thesis/Phd/Bruegger1994.PS.Z

Buttenfield, B.P., DeLotto, J.S., and McKinney, J.V., 1989: *Multiple Representations: A Bibliography.* NCGIA, Technical Paper no. 89-11, SUNY, Buffalo, NY, December.

Castleman, K.R., 1979: *Digital Image Processing.* Prentice-Hall, Englewood Cliffs, NJ, Signal Processing Series.

Couclelis, H., 1992: People Manipulate Objects (but Cultivate Fields): Beyond the Raster-Vector Debate in GIS. In: *Proceedings, International Conference, GIS--From Space to Territory: Theories and Methods of Spatio-Temporal Reasoning in Geographic Space,* Springer-Verlag, Pisa, Italy, September 21-23, pp. 65-77.

Davis, F.W. and Simonett, D.S., 1991: GIS and Remote Sensing. In: *Geographical Information Systems: Principles and Applications.* Maguire, D.J., Goodchild, M.F., and Rhind, D.W. (eds.), ch. 14, pp. 191-213, Longman, London.

Douglas, D.H. and Peucker, T.K., 1973: Algorithms for the Reduction of the Number of Points Required to Represent a Line or its Caricature. *The Canadian Cartographer, 10,* 2, pp. 112-123.

Flowerdew, R., 1991: Spatial Data Integration. In: *Geographical Information Systems: Principles and Applications.* Maguire, D.J., Goodchild, M.F., and Rhind, D.W. (eds.), ch. 24, pp. 375-387, Longman, London.

Gold, C.M., 1992: An Object-Based Dynamic Spatial Model, and its Application in the Development of a User-Friendly Digitizing System. In: *Proceedings, 5th International Symposium on Spatial Data Handling,* Bresnahan, P., Corwin, E., and Cowen, D. (eds.), Charleston, SC, August 3-7, pp. 495-504.

Goodchild, M.F., 1990: A Geographical Perspective on Spatial Data Models. In: *GIS Design Models and Functionality Conference,* Leicester. U.K., March 21-22, pp. 1-16.

Goodchild, M.F., 1992a: *Hard Challenges and Opportunities in GIS-Related Research.,* GIS-L, summarized by Tom Poiker, November 27.

Goodchild, M.F., 1992b: Geographical Information Science. *International Journal of Geographical Information Systems, 6,* 1, pp. 31-45.

Goodchild, M.F., 1993: Fundamentals of an Integrated GIS. *Journal of Vegetation Science,* July.

van der Knaap, W.G.M., 1992: The Vector to Raster Conversion: (Mis)use in Geographical Information Systems. *International Journal of Geographical Information Systems, 6,* 2, pp. 159-170.

Maguire, D.J., Kimber, B., and Chick, J., 1991: Integrated GIS: The Importance of Raster. In: *Proceedings of the ACSM-ASPRS Annual Convention,* Baltimore, MD, pp. 107-116.

Monmonier, M.S., 1983: Raster-Mode Area Generalization for Land Use and Land Cover Maps. *Cartographica, 20,* 4, pp. 65-91.

Perkal, J., 1966: An Attempt at Objective Generalization. *Community of Mathematical Geographers Discussion Papers,* 10, pp. 1-18.

Piwowar, J.M. and LeDrew, E.F., 1990: Integrating Spatial Data: A User's Perspective. *Photogrammetric Engineering & Remote Sensing, 56,* 11, November, pp. 1497-1502.

Rhind, D.W., Green, N.P.A., Mounsey, H.M, and Wiggins, J.C., 1984: The Integration of Geographical Data. In: *Proceedings of Austra Carto Perth*, Australian Cartographic Association, pp. 273-293.

Robinson, A.H., Sale, R.D., Morrison, J.L., and Muehrcke, P.C., 1984: *Elements of Cartography*. John Wiley & Sons, New York.

Shepherd, I.D.H., 1991: Information Integration and GIS. In: *Geographical Information Systems: Principles and Applications*. Maguire, D.J., Goodchild, M.F., and Rhind, D.W. (eds.), ch. 22, pp. 337-360, Longman, London.

Smith, T.R., 1992: Towards a Logic-Based language for Modeling and Database Support in Spatio-Temporal Domains. In: *Proceedings, 5th International Symposium on Spatial Data Handling*, Bresnahan, P., Corwin, E., and Cowen, D. (eds.), Charleston, SC, August 3-7, pp. 592-601.

Stephan, E.M., Vckovski, A., and Bucher, F., 1993: Virtual Data Set--An Approach for the Integration of Incompatible Data. In: *Proceedings, Auto-Carto 11*, McMaster, R.B. and Armstrong, M.P. (eds.), Minneapolis, MN, October 30-Novermber 1, pp. 93-102.

Strahler, A.H., Woodcock, C.E., and Smith, J.A., 1986: On the Nature of Models in Remote Sensing. *Remote Sensing of the Environment, 20*, pp. 121-139.

A Hierarchical Representation of Qualitative Shape based on Connection and Convexity

A G Cohn

Division of Artificial Intelligence,
School of Computer Studies,
University of Leeds,Leeds LS2 JJT, UK.
Telephone: +44 113 2335482.
Email: agc@scs.leeds.ac.uk.

Abstract. In this paper we consider the problem of representing the shape of a region, qualitatively, within a logical theory of space. Using just two primitive notions, that of two regions connecting, and the convex hull of a region, a wide variety of concave shapes can be distinguished. Moreover, by applying the technique recursively to the inside of a region (i.e. that part of the convex hull not occupied by the region itself), a hierarchical representation at varying levels of granularity can be obtained.

1 Introduction

In this paper we consider the problem of representing the shape of a region, qualitatively, within a logical theory of space, known as RCC theory (Randell and Cohn 1989, Randell, Cui and Cohn 1992, Cohn, Randell and Cui 1994). This logic originally had just one primitive notion, $C(x, y)$, that of two regions x and y connecting. However, to make an initial attack on this problem, the notion of the convex hull of a region, $\text{conv}(x)$ was introduced (see figure 1); this allowed relations such as topologically-inside[1], scattered-inside and geometrically-inside to be defined – see figure 2. This latter concept was then refined to distinguish tunnel-inside and containable-inside (see figure 3). In 2D these subdivisions of inside reduce to just three, as illustrated in figure 2.

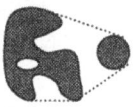

Fig. 1. The convex hull of a region

---

[1] By the 'inside' of a region, we mean that part of the convex hull not occupied by the region itself.

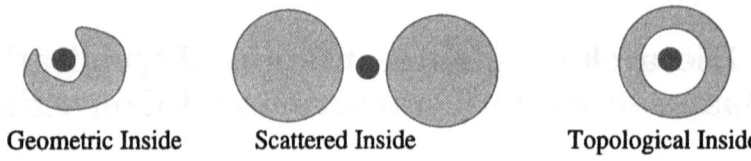

Geometric Inside   Scattered Inside   Topological Inside

Fig. 2. The three kinds of inside possible in 2D; in each subfigure the darker region is inside the lighter one and the lighter one is outside the darker.

Fig. 3. Two kinds of geometrical inside: tunnel-inside, containable-inside. The mug has a containable inside (where the liquid is normally put) and a tunnel inside (the handle). Actually it also has a third kind of inside since the sum of the mug and its containable and tunnel insides does not comprise the entire convex hull of the mug: there remains that part of the convex hull which one would pass through when inserting ones finger into the handle.

Clearly if one region has another inside it, then it is concave, and depending on the specific kind of inside relation, we can say something about the nature of the concavity. Moreover, it turns out that even without the additional conv primitive, a great deal can be said about the qualitative shape of a region as (Gotts 1994) has shown: the configurations in figure 4 can all be distinguished by first order formulae whose only non logical constant is C.

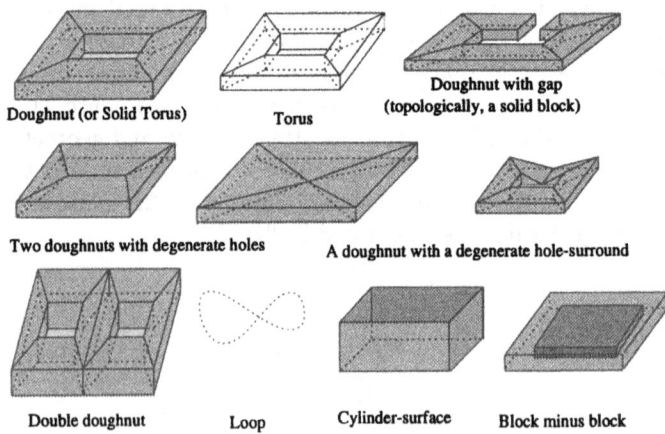

Fig. 4. Gotts can distinguish all these shapes using C alone.

The main concern of this paper however, is to develop techniques for distinguishing shapes such as those shown in figure 5 using C and conv. In the

Fig. 5. The proposed approach will distinguish all these shapes

next section we will present a brief overview of RCC theory, before the main development of the new theory.

2 An Overview of RCC Theory

The focus of research on spatial representation and reasoning at Leeds has been to evaluate, extend and implement a theory[2] of space and time based upon Clarke's (1981, 1985) 'calculus of individuals based on connection', and expressed in the many sorted logic LLAMA (Cohn 1987). Our revised and extended theory, now known as 'RCC-theory' has been developed in a series of papers, including (Randell and Cohn 1989), (Randell 1991), (Randell and Cohn 1992), (Randell, Cohn and Cui 1992b), (Cohn et al. 1994), (Bennett 1994), and (Gotts 1994). The most distinctive feature of Clarke's 'calculus of individuals', and of our work, is that extended *regions* rather than points are taken as fundamental. Our formal theory supports regions having either a spatial or a temporal interpretation (temporal 'regions' are periods of time with a non-zero duration, as opposed to temporal 'points', or instants). Informally, these regions may be thought of as infinite in number, and 'connection' may be any relation from external contact (touching without overlapping) to spatial or temporal identity. Spatial regions may have one, two, three, or even more than three dimensions, but in any particular model of the formal theory, all regions are of the same dimensionality. Thus, if we are concerned with a two-dimensional model, such as one in which regions are areas of land, the boundary lines and points at which these regions meet are not themselves considered regions.

. The basic part of the formal theory assumes a primitive dyadic relation: $C(x, y)$, read as 'x connects with y' (where x and y are regions). Two axioms are used to specify that C is reflexive and symmetric. C can be given a topological interpretation in terms of points incident in regions. In this interpretation, $C(x, y)$ holds when the topological *closures* of regions x and y share at least one point.

Using the relation C, further dyadic relations are defined. These relations are DC (is disconnected from), P (is a part of), PP (is a proper part of), EQ or = (is spatiotemporally identical with), O (overlaps), DR (is discrete from),

---

[2] We use the word 'theory' in its logical/mathematical sense, meaning a set of formal axioms which specify the properties and relations of a collection of entities, not in the natural scientist's sense of an empirically testable explanation of observed regularities.

PO (partially overlaps), EC (is externally connected with), TPP (is a tangential proper part of), and NTPP (is a nontangential proper part of). The relations P, PP, TPP and NTPP have inverses (here symbolised Pi, PPi, TPPi and NTPPi — a slight change in terminology from some earlier papers).

The complement (compl) of a region, and the sum (sum or +), product or intersection (prod or *) and difference (diff or −) of a pair of regions are also axiomatised in terms of C. The product of two regions may not always be a region: if the two do not overlap, this product is an object of the sort NULL (spatial and temporal regions belong to distinct subsorts of the sort REGION). Similarly, if two regions are EQ, then diff(x, y) is of sort NULL, as is the complement of the universal region, u. If DC regions are summed, then clearly a multi-piece region will be formed. We will often need to test whether a given region is one-piece or not, which we will denote CON(x).

$$\text{CON}(x) \equiv_{def} \neg \exists(y, z)[x = y + z \wedge \text{DC}(y, z)]$$

A 'nonatomic' axiom can be added, ensuring that every region has at least one NTPP:

$$\forall x \exists y[\text{NTPP}(y, x)].$$

This has the consequence that space is indefinitely divisible. There are therefore no 'atoms' — regions which cannot be subdivided — in this version of the theory (though see (Randell, Cui and Cohn 1992) for how this might be achieved). Also see (Bennett 1995) for a discussion of other existential axioms that might be added to the theory.

The relations defined in terms of C can be embedded in a relational lattice with the top element interpreted as tautology, and the bottom element as contradiction. This relational lattice is shown as figure 6, together with pictorial representations of the eight 'base relations' (known as RCC-8, which make up the layer immediately above the bottom element:) {DC, EC, PO, TPP, NTPP, EQ, TPPi, NTPPi}. These form a jointly exhaustive and pairwise disjoint (JEPD) set: one and only one of these relations will hold between a pair of regions.

In (Randell and Cohn 1989) we first defined the notion of a quasi-manifold, i.e. an n-dimensional region which is well connected in the sense that it cannot be divided into two regions of dimension n, which only connect along a boundary of dimension n-2 or less. Thus the region in figure 7 is not a quasi-manifold. Various definitions (some making use of conv in their formulation) can be found in previous papers from the Leeds group; (Gotts 1994) explores this notion of well connected regions much more fully. Here, we will use Manifold(x) to predicate that a region has this property.

Manifold$(x) \equiv_{def}$

$$\forall(x_1 x_2)[x = x_1 + x_2 \rightarrow \exists z[\text{O}(z, x_1) \wedge \text{O}(z, x_2) \wedge \text{DC}(z, \text{compl}(x)) \wedge \text{CON}(z)]]$$

Another concept, defined in (Cohn et al. 1994) is the idea of a maximal connected (i.e. one piece) subpart of a region (known as a 'component' in topology). In this paper we will want to insist that each maximal connected subpart of a region is also a quasi-manifold, so we will use the following slightly revised definition:

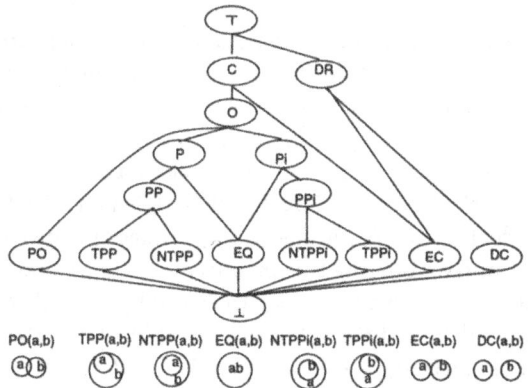

Fig. 6. A lattice defining the subsumption hierarchy of dyadic relations defined in terms of C

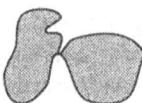

Fig. 7. This region is not a quasi-manifold since it can be divided into two point connected regions: there is no one piece connected region overlapping both parts which is DC to the compl of the region.

$$\text{MAX\_P}(x, y) \equiv_{def} \text{Manifold}(x) \wedge \text{P}(x, y) \wedge \neg \exists z[\text{PP}(x, z) \wedge \text{P}(z, y) \wedge \text{Manifold}(z)]$$

Several axiomatisations of conv have been given in various papers (e.g. (Bennett 1995), (Cohn et al. 1994), (Bennett 1994)). The most recent of these axioms are all repeated below for convenience along with some suggested by Antony Galton (personal communication).

$\forall x[\text{conv}(\text{conv}(x)) = \text{conv}(x)]$

$\forall x[\neg x = \text{conv}(x) \rightarrow \text{TPP}(x, \text{conv}(x))$

$\forall x \forall y[\text{P}(x, y) \rightarrow \text{P}(\text{conv}(x), \text{conv}(y))]$

$\forall x \forall y \text{P}((\text{conv}(x) + \text{conv}(y), \text{conv}(x + y))$

$\forall x \forall y[\text{conv}(x) = \text{conv}(y) \rightarrow \text{C}(x, y)]$ [3]

$\forall x \forall y[\text{conv}(x) * \text{conv}(y) = \text{conv}(\text{conv}(x) * \text{conv}(y))]$

$\forall x \forall y[\text{DC}(x, y) \rightarrow \neg\text{CONV}(x + y)]$

$\forall x \forall y[\text{NTPP}(x, y) \rightarrow \neg\text{CONV}(y - x)]$

$\forall x \forall y[[\text{CONV}(x) \wedge \text{CONV}(y)] \rightarrow \text{CONV}(x * y)]$

$\forall x \forall y \forall z[[\text{EC}(x, y) \wedge \text{CONV}(x + y) \wedge \text{EC}(y, z) \wedge \text{CONV}(y + z) \wedge \text{DC}(x, z)] \rightarrow \text{CONV}(y)]$

where the predicate $\text{CONV}(x)$ is defined in terms of the primitive function $\text{conv}(x)$ thus:

---

[3] Actually this is not necessarily true for infinite regions.

$\mathsf{CONV}(x) \equiv_{def} x = \mathrm{conv}(x)$

Notice that the above axiomatisation also defines a predicate $\mathsf{CONV}(x)$ which is true for convex regions. A major open question is whether these axioms fully capture the intended meaning of conv and which if any of these axioms are redundant. One possible line of attack would be to introduce an alternative primitive, 'region y is between regions x and z' (cf Tarski's axiomatisation of Geometry (Balbiani, Dugat, del Cerro and Lopez 1994), (Bennett 1995), (Tarski 1929), (Quaife 1989)), which uses a point based betweenness primitive, and define conv in terms of this primitive. Linking this primitive to Tarski's point based betweenness relation may provide a way to verify the completeness of the axiomatisation.

We use conv to define three relations: '$\mathsf{INSIDE}(x,y)$' ('x is inside y'), 'P-$\mathsf{INSIDE}(x,y)$' ('x is partially inside y') and '$\mathsf{OUTSIDE}(x,y)$' ('x is outside y'), each of which also has an inverse. Two functions[4] capturing the concept of the inside and the outside of a particular region are also definable: $\mathrm{inside}(x)$ and $\mathrm{outside}(x)$. This particular set of relations refines $\mathsf{DR}(x,y)$ in the basic theory. In (Randell, Cui and Cohn 1992, Randell, Cohn and Cui 1992a) we generated a JEPD set of relations by taking the relations given above, their inverses, and the set of relations that result from non-empty intersections. The set of base relations for this particular set were then finally generated by defining an EC and DC variant for each of these relations. A new set of base relations (using the relations defined immediately above) is constructed according to the following schema:

$$\alpha\_\beta\_\gamma(x,y) \equiv_{def} \alpha(x,y) \wedge \beta(x,y) \wedge \gamma(x,y)$$

where: $\alpha \in \{\mathsf{INSIDE}, \mathsf{P\text{-}INSIDE}, \mathsf{OUTSIDE}\}$, $\beta \in \{\mathsf{INSIDEi}, \mathsf{P\text{-}INSIDEi}, \mathsf{OUTSIDEi}\}$, and $\gamma \in \{\mathsf{EC}, \mathsf{DC}\}$ excepting where $\alpha = \mathsf{INSIDE}$, $\beta = \mathsf{INSIDEi}$ and $\gamma = \mathsf{DC}$. E.g. $\mathsf{INSIDE\_P\text{-}INSIDEi\_DC}(x, y)$ is true when x is inside y, y is partially x and the two regions are DC. The reason that we exclude the case of two regions being mutually inside each other and DC is that such a configuration of regions is only possible when certain kinds of infinite regions are allowed in one's ontology – which we do not wish to permit (see the fifth axiom in the list above): see (Cohn et al. 1994) for further discussion. This gives a total of 23 base relations instead of the original 8. Further distinctions in INSIDE can be made, as mentioned above which we will not consider further here (Cohn et al. 1994).

3 Using the Convex Hull Operator to Describe Qualitative Shapes

For the present, we will just consider the shapes of 2D regions which are quasi-manifolds. We will also only consider geometric insides (so excluding regions which have interior holes (i.e. have 'topological insides') – this includes the case

---

[4] Note that it does not really make much sense to define a functional analogue of P-INSIDE as this would simply be the sum of the inside and the outside, i.e. the complement of x!

of an interior hole which is point connected with the rest of the exterior (e.g. a region which meets itself at a point). The technique we will describe is based on identifying each maximal connected part (MAX_P) of the inside of the region and describing the relationships between the maximal parts of the inside. We will define a predicate Concavity(x, y) which is true when x is a maximal connected part of the inside of y:

Concavity(x, y) $\equiv_{def}$ MAX_P(x, inside(y))

Thus in figure 8 we can distinguish eight different parts of the inside, named $I_1...I_8$.

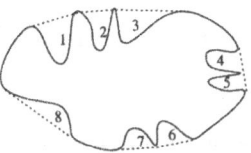

Fig. 8. This shape has 8 maximal subparts of the geometric inside (concavities).

We can now start to describe the relationship between the I_j:

$$EC(I_1, I_2) \wedge EC(I_2, I_3) \wedge EC(I_4, I_5) \wedge EC(I_6, I_7) \wedge \bigwedge_{\langle i,j \rangle \in \gamma} DC(I_i, I_j)$$

$$where \ \gamma = \{\langle i, j \rangle : \ i < j\} - \{\langle 1, 2 \rangle, \langle 2, 3 \rangle, \langle 4, 5 \rangle, \langle 6, 7 \rangle\}$$

However this does not distinguish between the shape of figure 8 and that in figure 9 which is different because the single inside (I_8) is between the pair of double insides in figure 9 whilst it is between the triple and the double in figure 8.

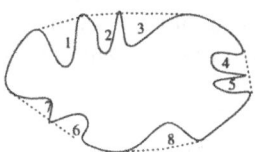

Fig. 9. This shape differs from the previous one in the permutation of the EC-groups of 'insides'.

In order to distinguish these two situations (and other similar ones in general[5]), we may divide the region (whose convex hull we have formed) into n+1 separate parts, where n is the number of concavities, such that n of them all EC the n+1st

---

[5] In fact the analysis below only works for regions with four or more concavities.

part, and the others each externally connect with the complement of the convex hull just once and with exactly two other of the n parts. The n+1st part is an NTPP of the shape. Suitable partitions for figures 8 and 9 are displayed in figure 10.

Fig. 10. Partitioning can distinguish these two shapes.

Thus $p_1, ...p_n, q$ is a suitable partition of a region p if
$$\forall(1 \leq i \leq n)\exists(j,k)[DC(p_j, p_k)\wedge EC(p_i, p_j)\wedge EC(p_i, p_k)\wedge EC1(p_i, compl(conv(x)))\wedge$$
$$NTPP(q,p) \wedge q = (p_1 + ... + p_n) \wedge EC(p_i, q)]$$
where
$$EC1(x,y) \equiv_{def} EC(x,y) \wedge \exists z[CONV(z) \wedge \neg Manifold(x - z + y)]$$

We can now distinguish the two figures since in figure 10(a), there is a p_i (shown shaded) which ECs both the triple-inside and the single-inside, whereas this is not the case in figure 10(b). It is worth commenting briefly of the definition of $EC1(x,y)$: the predicate is intended to be true when the 'arm' x and the complement of the shape, y, EC just once (i.e. they have only one boundary in common). This is the case if we can remove part of one of the regions (x) and 'separate' the two regions. The part which is removed must be convex since otherwise one could easily remove the tips of two touching fingers with a single non convex U shaped region. Moreover the separation condition cannot be DC since x may be locally non convex where it diverges from y. Thus we insist that the sum of the two regions minus the part removed is not a quasi-manifold.

However note that we cannot distinguish the shapes in figure 11 which differ purely by virtue of the ordering of their various insides around their perimeters. This is not surprising since if one of them is viewed 'from the other side of the

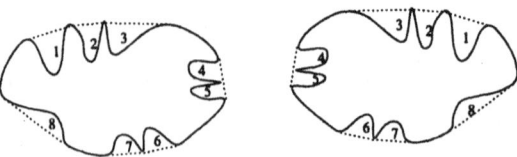

Fig. 11. These two shapes are indistinguishable with our technique.

paper', they are identical. We could imagine introducing orientation information (c.f. (Freksa 1992b)), to distinguish such situations, but will not do this here.

The partitioning technique described above could be used to define a predicate Adjacent(x, y) which is true when x and y are adjacent (round the perimeter) concavities, whether or not they are DC or EC. In terms of the analysis above it would mean that a single p_i is EC to both. However, we can give a direct definition which does not make use of the above partitioning technique.

Adjacent$(i1, i2, x) \equiv_{def}$ Concavity$(i1, x) \wedge$ Concavity$(i2, x) \wedge$
$\exists z[\text{EC}(z, i1) \wedge \text{EC}(z, i2) \wedge \text{PP}(z, x) \wedge \text{CON}(z) \wedge \text{DC}(z, \text{inside}(x) - i1 - i2) \wedge$
$\forall(i3, i4)[[\text{Concavity}(i3, x) \wedge \text{Concavity}(i4, x) \wedge \text{DC}(i1 + i2, i3 + i4)] \rightarrow$
$\exists w[\text{CON}(w) \wedge \text{PP}(w, x) \wedge \text{EC}(w, i3) \wedge \text{EC}(w, i4) \wedge \text{DR}(w, z)]]]$

According to this definition, two concavities are Adjacent iff there is a one-piece part of x that ECs both, but no other concavity, (i.e if the two concavities can be linked together) and every other pair of concavities of x can be similarly linked without any of these links crossing the first link.

There is another distinction we can make; consider the two shapes in figure 12. Both have two indentations, but will have identical qualitative shape descriptions in our formalisation thus far. However we can distinguish them: in the right hand

Fig. 12. We want to distinguish these two shapes.

shape we can remove a region (the middle 'finger') and we will still be left with one concavity composed in part of each of the original concavities. In the left hand shape there is no sub-region which can be removed (leaving the region in one piece as a quasi-manifold) and leave just one concavity with the same properties. We can define a predicate to detect the fact that the two insides $i1$ and $i2$ in the right hand shape(x) in figure 12 are on the 'same side' thus:

SameSide$(i1, i2, x) \equiv_{def}$ Concavity$(i1, x) \wedge$ Concavity$(i2, x) \wedge$
$\exists z[\text{P}(z, x) \wedge \text{Manifold}(i1 + i2 + z) \wedge \text{Manifold}(x - z) \wedge$
$\text{O}(i1, \text{conv}(x - z)) \wedge \text{O}(i2, \text{conv}(x - z))]$

Thus if $i1$ and $i2$ are concavities of x then they are on the same side if there is some part of x which when added to $i1$ and $i2$ forms a quasi-manifold whilst leaving the rest of x as a quasi-manifold as well. Moreover, at least part of each concavity must still be a concavity of the remainder of x. Notice that this definition means that the two concavities in each of the shapes in figure 13 are SameSide. If we wanted to distinguish these two cases, then we could modify the final two conjuncts of the definition of SameSide to yield the following predicate which is true of the left hand shape but not the right hand one. In this case there is a point/part of each the three 'arms' which are colinear (a straight edge can EC all three arms): hence we term this predicate SsColinear. Notice that whereas SsColinear is transitive (on its first two arguments) SsNotColinear is not (see figure 14).

$\text{SsColinear}(i1, i2, x) \equiv_{def} \text{Concavity}(i1, x) \wedge \text{Concavity}(i2, x) \wedge$
$\quad \exists z[\text{P}(z, x) \wedge \text{Manifold}(i1 + i2 + z) \wedge \text{Manifold}(x - z) \wedge$
$\quad\quad \text{P}(i1 + i2 + z, \text{inside}(x - z))$
$\text{SsNotColinear}(i1, i2, x) \equiv_{def} \text{SameSide}(i1, i2, x) \wedge \neg\text{SsColinear}(i1, i2, x)$

Fig. 13. Can we distinguish these two kinds of same sideness?

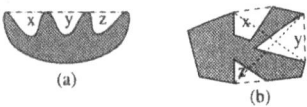

Fig. 14. Examples to demonstrate the transtivity of SsColinear and the lack of transitivity of SsNotColinear. In (a) x, y, z are all SsColinear pairwise. However in (b), although x, y and y, z are SsNotColinear, x, z are not.

It is worth summarising the shape language we have introduced so far. The shape of an object is described by its maximal one-piece inside (concavities); the relationship between any two concavities can be described using four attributes:

– Are they DC or EC?
– Are they SsColinear, SsNotColinear or ¬SameSide?
– Are they Adjacent or not?

Thus there are 12 possible descriptions for each pair of concavities – except that EC implies Adjacent, so DC and Adjacent is impossible, yielding a total of 11 actual JEPD descriptions. Also note that it requires at least four concavities before a pair can be ¬Adjacent.

4 A Hierarchical Representation

We can turn to more fine grained distinctions. One might like to distinguish the two regions depicted in figure 15. The technique described thus far will not distinguish them, but this is easily achieved if we simply apply the technique recursively to each maximal inside of the original shape. Figure 16 depicts the two maximal insides from figure 15 and it is clear the two original shapes are now distinguished. Of course this technique can be applied as often as necessary; consider figure 17.

Fig. 15. A hierarchical representation will distinguish these two shapes.

Fig. 16. The insides of the two previous shapes, which are distinguishable by virtue of the shape of *their* insides.

The technique would seem to be ideally suited to describing objects such as coastal regions with many bays – the Norwegian fjords for example.

We can use this shape description technique to test whether two regions will possibly 'plug together'. For instance which of the 'pieces' on the right hand side of figure 18 might 'plug into' the shape depicted on the left hand side of figure 18? We can answer this by determining if a part of one of the 'pieces' has the same qualitative shape description as an inside of the left hand shape.

This might be useful for applications just as (qualitative) jigsaws, locking together of mechanical parts or describing the shape made by a rigid body on a compliant one or, in a GIS context, when reasoning about the original configuration of separated landmasses prior to geological movement.

5   Other Aspects of Qualitative Shape

There are other distinctions in shape which are expressable in our formalism. For example, we can define predicates for convex polygons (i.e. convex closed straight sided contours) of arbitrary degree. The definition for a triangle is as follows:

$$\text{Triangle}(x) \equiv_{def} \text{CON}(x) \wedge \text{CONV}(x) \wedge \exists (u, v, w)[\text{compl}(x) = (u + v + w) \wedge$$
$$\text{DC}(u, v) \wedge \text{DC}(v, w) \wedge \text{DC}(u, w) \wedge \text{CONV}(u) \wedge \text{CONV}(v) \wedge \text{CONV}(w)]$$

This definition states that a triangle is a connected convex region, whose complement can be partitioned into three disjoint convex regions. Similar definitions can of course be written for quadralaterals, pentagons, hexagons etc.

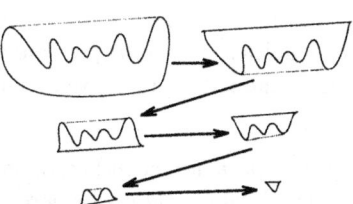

Fig. 17. The technique can be applied recursively as often as necessary to distinguish shapes.

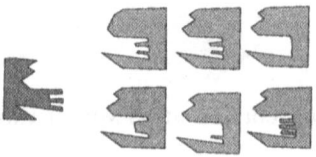

Fig. 18. Which of these shapes might 'plug' into the leftmost one?.

Given the definition for a triangle, it is of course possible to test for a region being an arbitrary polygon using the well known property that any polygon can be represented as the sum of triangles.

Polygon$(x) \equiv_{def}$ Triangle$(x) \vee$
$\quad$ [CON$(x) \wedge \exists y$[PP$(y, x) \wedge$ Triangle$(y) \wedge$ Polygon$(x - y)$]]]

Convex polygons can of course be tested for separately with CONV(x).

Although symmetry is of course not definable quantitatively within this system, we could define a notion of qualitative symmetry (i.e. any shape which is qualitatively symmetrical would be continuously deformable to one which was quantiatively symmetrical and which has the same qualitative description). For example, a 180 degree qualitative rotational symmetry exists if a region x can be partitioned into two DC and CON regions $x1$ and $x2$ each of which has the same description in our language of shape. We must ensure that the partition between these two regions takes the form of a straight line, i.e. that
CONV(conv(x) − conv$(x1)$) $\wedge$ CONV(conv(x) − conv$(x2)$)

6 Changes in Shape

In earlier papers (Cohn, Randell, Cui and Bennett 1993, Cohn et al. 1994) we presented the transition networks[6] for our sets of relations. These networks tell us that if a particular relation holds between a pair of regions at some particular time, then, assuming continuous motion and deformation[7], what the 'next' possible relationship that might hold could be. For example figure 19 diagrams the situation for RCC8. Here, we are not interested in relationships between regions, but rather, a shape description of a single region. Since the number of possible shape descriptions is rather large (in fact unbounded since a region may have arbitrarily many concavities), it is not as easy to draw a similar diagram for the shape descriptions defined in this paper. So in figure 20 we give a diagram which shows the possible continuous transformations for regions (without holes) with up to two top level concavities – *assuming a single local deforming process only*[8] (otherwise, for example, the left hand convex shape could be transformed directly to any of the others).

---

[6] (Freksa 1992a) terms these conceptual neighbourhoods, and (Egenhofer and Al-Taha 1992) have a very closely related notion: closest topological distance.

[7] In the GIS context, processes such as erosion and flooding would yield these kind of changes.

[8] In the sense of (Leyton 1988), e.g. making an indentation.

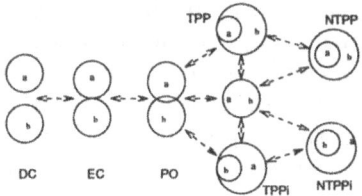

Fig. 19. The continuous transitions (conceptual neighbourhood) for RCC8

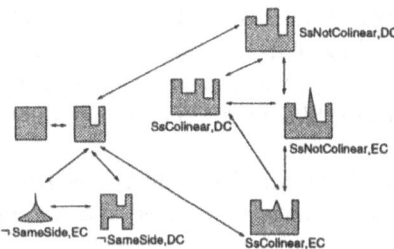

Fig. 20. A diagram showing the continuous transformations possible between regions with up to 2 concavities.

7 Related Work

Perhaps the most closely related work to the underlying RCC8 calculus and the description of holes, is that of (Egenhofer, Clementini and Di Felice 1994, Egenhofer 1994). This work is not based on a logical framework and the notion of connection and convexity as ours is, but rather on regions defined in terms of sets of points and relations are defined using matrices which compare the set intersections of the interior and boundary of two regions. In the first of these two papers an analysis of 2D regions with holes is presented whilst in the latter the model is refined to consider not just whether the intersection is empty or not but its dimensionality; moreover regions are broken into components and complex relationships are built by composing the relationships between the sequence of the components of two regions.

The well known work of (Leyton 1988) is highly relevant to this research. Leyton describes the shape of 2D[9] regions by noting the sequence of four different kinds of curvature extrema along the perimeter of a region. He also associates a process with each of these kinds of curvature extrema: protusion, indentation, squashing and internal resistance. He gives a set of 6 rewrite rules which apply to a single curvature extrema which provide a set of transformation rules for evolving shapes over time (e.g. 'squashing continues till it indents', 'internal resistance continues till it protrudes'). These give a notion somewhat similar to the notion of 'conceptual neighbourhood' described above. He also shows how these rules can be applied to give the possible process paths between two given

---

[9] He does in fact consider the 3D case as well, but not at the same level of detail.

shapes, analogously to our work on qualitative spatial simulation (Cui, Cohn and Randell 1992). In many ways the work presented here is very similar to Leyton's. One crucial difference is that he has no notion of convex hull and so cannot distinguish between **SsColinear** and **SsNotColinear**; our approach will however distinguish these two shapes. His approach does make distinctions our approach cannot though: for example all convex objects have identical shape descriptions in our approach whereas there are three different kinds of one piece convex object having a maximum of eight different curvature extrema.

Another approach to the representation of shape is to use voronoi regions rather than convex hulls. (Edwards 1993) outlines some of the possibilities.

It turns out that the idea of describing shape using convex hulls and a hierarchical decomposition of the concavities has already been briefly explored in the computer vision literature (Sklansky 1972), under the terminology of *concavity trees*.

Finally, the work of (Casati and Varzi 1994) is worth mentioning; they present an excellent treatise on the nature of holes, and an appendix starts to give a formal axiomatisation of this which is continued in (Varzi 1993).

8 Final Comments

This paper has presented a relatively informal description of a possible approach to qualitative shape description using a conceptually very simple formalism with just two primitive concepts, both of which are well known, natural, and easy to compute if necessary.

It should be clear that the technique outlined above could also be used to describe the shape of holes in a region: for example consider the holes in the region displayed in figure 21. Equally, the extension to 3D and non manifolds should be possible though the details still need to be worked out.

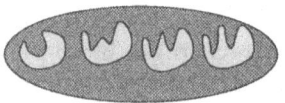

Fig. 21. The technique can easily be extended to differentiate these holes.

Future work will concentrate on the continued formalisation of these ideas, on theoretical aspects such as complexity analysis and on their practical application. Possible refinements of 'scattered-inside', i.e. the shape of multipiece objects should also be investigated. The conceptual neighbourhood diagram presented in section 6 needs to be placed on a sound theoretical footing by building in proper theory of spatial processes and their effect on the shape of regions.

Finally, a proper evaluation of the utility of the formalism should be carried out: can it really be used to describe and distinguish between the shapes of a wide variety of real objects? Certain objects clearly have a natural and obvious description in the formalism: consider, for example, the teeth of a tenon saw – these are a series of concavities which are all SsColinear and EC. How well does the formalism describe other commonplace objects?

9 Acknowledgements

The support of the EPSRC under grant GR/H/78955 and also the CEC under the Basic Research Action MEDLAR 2, Project 6471 is gratefully acknowledged as are the comments of Longin Latecki, Nick Gotts, John Gooday and Brandon Bennett and three anonymous referees.

References

Balbiani, B., Dugat, V., del Cerro, L. F. and Lopez, A.: 1994, *Eléments de géométrie mécanique*, Editions Hermes.

Bennett, B.: 1994, Spatial reasoning with propositional logics, *in* J. Doyle, E. Sandewall and P. Torasso (eds), *Principles of Knowledge Representation and Reasoning: Proceedings of the 4th International Conference (KR94)*, Morgan Kaufmann, San Francisco, CA.

Bennett, B.: 1995, Modal logics for qualitative spatial reasoning, *Bulletin of the Interest Group on Propositional and Predicate Logics (IGPL)*. To appear.

Casati, R. and Varzi, A.: 1994, *Holes and Other Superficialities*, MIT Press, Cambridge, MA.

Clarke, B. L.: 1981, A calculus of individuals based on 'connection', *Notre Dame Journal of Formal Logic* **23**(3), 204–218.

Clarke, B. L.: 1985, Individuals and points, *Notre Dame Journal of Formal Logic* **26**(1), 61–75.

Cohn, A. G.: 1987, A more expressive formulation of many sorted logic, *Journal of Automated Reasoning* **3**, 113–200.

Cohn, A. G., Randell, D. A. and Cui, Z.: 1994, Taxonomies of logically defined qualitative spatial relations, *in* N. Guarino and R. Poli (eds), *Formal Ontology in Conceptual Analysis and Knowledge Representation*, Kluwer. To appear.

Cohn, A. G., Randell, D. A., Cui, Z. and Bennett, B.: 1993, Qualitative spatial reasoning and representation, *in* N. P. Carreté and M. G. Singh (eds), *Qualitative Reasoning and Decision Technologies*, CIMNE, Barcelona, pp. 513–522.

Cui, Z., Cohn, A. G. and Randell, D. A.: 1992, Qualitative simulation based on a logical formalism of space and time, *Proceedings AAAI-92*, AAAI Press, Menlo Park, California, pp. 679–684.

Edwards, G.: 1993, The voronoi model and cultural space: Applications to the social sciences and humanities, *in* A. U. Frank and I. Campari (eds), *Spatial Information Theory: A Theoretical Basis for GIS*, Vol. 716 of *Lecture Notes in Computer Science*, Springer Verlag, Berlin, pp. 202–214.

Egenhofer, M.: 1994, Topological similarity, Proc FISI Workshop on the Topological Foundations of Cognitive Science.

Egenhofer, M. J. and Al-Taha, K. K.: 1992, Reasoning about gradual changes of topological relationships, *in* A. U. Frank, I. Campari and U. Formentini (eds), *Theories and Methods of Spatio-temporal Reasoning in Geographic Space*, Vol. 639 of *Lecture Notes in Computer Science*, Springer-Verlag, Berlin, pp. 196–219.

Egenhofer, M. J., Clementini, E. and Di Felice, P.: 1994, Toplogical relations between regions with holes, *Int. Journal of Geographical Information Systems* 8(2), 129–144.

Freksa, C.: 1992a, Temporal reasoning based on semi-intervals, *Artificial Intelligence* 54, 199–227.

Freksa, C.: 1992b, Using orientation information for qualitative spatial reasoning, *in* A. U. Frank, I. Campari and U. Formentini (eds), *Proc. Int. Conf. on Theories and Methods of Spatio-Temporal Reasoning in Geographic Space*, Springer-verlag, Berlin.

Gotts, N. M.: 1994, How far can we 'C'? defining a 'doughnut' using connection alone, *in* J. Doyle, E. Sandewall and P. Torasso (eds), *Principles of Knowledge Representation and Reasoning: Proceedings of the 4th International Conference (KR94)*, Morgan Kaufmann.

Leyton, M.: 1988, A process grammar for shape, *Artificial Intelligence* p. 34.

Quaife, A.: 1989, Automated development of tarski's geometry, *J. Automated Reasoning* 5(1), 97–118.

Randell, D. A.: 1991, *Analysing the Familiar: Reasoning About Space and Time in the Everyday World*, PhD thesis, University of Warwick.

Randell, D. A. and Cohn, A. G.: 1992, Exploiting lattices in a theory of space and time, *Computers and Mathematics with Applications* 23(6-9), 459–476. Also appears in "Semantic Networks", ed. F. Lehmann, Pergamon Press, Oxford, pp. 459-476, 1992.

Randell, D. A., Cohn, A. G. and Cui, Z.: 1992a, Computing transitivity tables: A challenge for automated theorem provers, *Proceedings CADE 11*, Springer Verlag, Berlin.

Randell, D. A., Cohn, A. G. and Cui, Z.: 1992b, Naive topology: Modelling the force pump, *in* P. Struss and B. Faltings (eds), *Advances in Qualitative Physics*, MIT Press, pp. 177–192.

Randell, D. A., Cui, Z. and Cohn, A. G.: 1992, A spatial logic based on regions and connection, *Proc. 3rd Int. Conf. on Knowledge Representation and Reasoning*, Morgan Kaufmann, San Mateo, pp. 165–176.

Randell, D. and Cohn, A.: 1989, Modelling topological and metrical properties of physical processes, *in* R. Brachman, H. Levesque and R. Reiter (eds), *Proceedings 1st International Conference on the Principles of Knowledge Representation and Reasoning*, Morgan Kaufmann, Los Altos, pp. 55–66.

Sklansky, J.: 1972, Measuring concavity on a rectangular mosaic, *IEEE Trans. on Computers* C-21(12), 1355–1364.

Tarski, A.: 1929, Les fondaments de la géométrie des corps, *Ksiega Pamiatkowa Pierwszego Polskiego Zjazdu Matematycznego* pp. 29–33. A suplement to *Annales de la Société Polonaise de Mathématique*. English translation, 'Foundations of the Geometry of Solids', in A. Tarski, *Logic, Semantics, Metamathematics*, Oxford Clarendon Press, 1956.

Varzi, A. C.: 1993, Spatial reasonng in a holey world, *Proceedings of the Spatial and Temporal Reasoning workshop, IJCAI-93*, pp. 47–59.

Representational Structures for Cognitive Space: Trees, Ordered Trees and Semi-Lattices

Stephen C. Hirtle
University of Pittsburgh
hirtle+@pitt.edu

During the past twenty years, numerous researchers and papers have discussed inclusion of a hierarchical component to the cognitive representation of spatial knowledge. However, such discussion has occurred without serious consideration of alternative representations. This paper examines the nature of hierarchical representation for spatial representations, in detail, and considers several alternative representational schemes, including ordered trees and semi-lattices. In addition, differences between these representations are demonstrated by mapping two sample datasets onto a variety of representations.

1 Introduction

The nature of the mental representation of space remains an important question for the future development of spatial information systems. The ability to present and to interpret spatial data in a method that is consistent with the internal cognitive map of the user would lead to systems that are more flexible and will provide greater functionality in terms of cognitive spatial tasks (Hirtle, & Heidorn, 1993; Medyckyj-Scott, & Blades, 1992).

Many early studies of the underlying representation of space focused on the ability to generalize beyond traveled pathways and acquire what is now referred to as survey knowledge of the world (e.g., Tolman, 1948). This view was supported by work on visual images (Kosslyn, 1978; Shepard, 1978) and lead to a holistic, continuous, image-like view of the spatial representation of spatial information. In sharp contrast, more recent studies have focused on the structural components of cognitive space and have examined a series of organizing principles, such as the use of reference points (Sadalla, & Staplin, 1980), the use of rotational and alignment heuristics (Tversky, 1981), the effect of orientation (Holyoak, & Mah, 1982), and other related principles (see Tversky, 1991, and McNamara, 1992, for reviews).

A common conclusion that has emerged from the research on the structure of cognitive mapping is that spatial memory is organized hierarchically, which results in processing biases and errors in judgments (Hirtle, & Jonides, 1985; McNamara, & Hirtle, 1989; Stevens, & Coupe, 1978). Hierarchical structuring is incorporated into the anchor-point hypothesis of spatial cognition (Couclelis, & Tobler, 1987; Golledge, 1992). Car and Frank (1994) have applied hierarchical modeling to a route planning task on a large network, while numerous researchers have explored the use of hierarchical storage and display of geographic information (Ballard, 1981; Frank, & Timpf, 1994; Samet, 1989). Spatial choice models often include a hierarchical component (Fotheringham, 1983; Fotheringham & Curtis, 1992). Poucet (1993) has incorporated a hierarchical component into new models of the spatial cognitive maps of animals, while Yoshino (1991) has shown that a hierarchical representation can be highly efficient at storing and accessing spatial information.

Often the claim of a hierarchical representation is modified in the discussion, without providing an explicit alternative. For example, this author and his colleagues have argued that their data is consistent with a "partially hierarchical model" (McNamara, et al, 1989) and have warned against the conclusion that only structure in a cognitive map is of a hierarchical nature (Hirtle and Jonides, 1985). While such qualifications are intriguing, they are stated without proposing an explicit alternative. That is, the argument is often presented as one of hierarchical structure versus non-hierarchical structure, rather than the more subtle argument of what is the appropriate hierarchical structure for modeling cognitive space.

In this paper, I look carefully at the assumption of hierarchical processing, by first examining what is meant by hierarchical representations and then describing various ways in which the term has been interpreted in the literature. Next, two alternative, closely-related representations, that of ordered trees and semi-lattices are discussed. Finally, two sample datasets are mapped to each of the representational classes to further highlight critical differences.

As will become clear, the focus in this paper is on hierarchical *representations*. The issue of hierarchical *processing*, while related to the issues discussed here, cannot be adequately addressed within the limits of this paper.

2 Hierarchies

The term hierarchy has been used in the literature in at least three related ways. First, the term may refer to implicit or explicit tree structure. For example, Stevens and Coupe (1978) showed how people consistently misjudged certain directions, such as assuming that Reno, Nevada is northeast of San Diego, California. To account for such effects, Stevens and Coupe (1978) presented a nested, propositional model, with San Diego as part of California, Reno as part of Nevada, and California to the west of Nevada. Here, the reasoning processes occur on a hierarchical tree structure, which contains cities nested within states.

A second use of the term hierarchy consists of ordered levels of specificity. For example, Kuipers and Levitt (1988), discuss a semantic hierarchy for navigation systems, consisting of sensorimotor, procedural, topological, and metrical levels. Timpf, Volta, Pollock, & Egenhofer (1992), present a model for wayfinding that is based on three ordered levels of information: a driver level, a highway level, and a network level.

A third use of the term hierarchy relates to the granularity of the representation. For example, Hernández (1994) discusses a model of qualitative reasoning that uses a hierarchical structure to reason at different levels of coarseness by making fewer distinctions at higher levels.

Note that these interpretations are not mutually distinct, but rather suggest overlapping characteristics. A reasoning model that includes a global level and recursively finer details of local information may fit all three of the definitions above. However, it is important to make the distinctions above to distinguish those uses in the literature of the term hierarchy without regard to any possible tree structure.

The strong claim is made here that only the first interpretation, that of a tree, is appropriate for a hierarchical *representation*. Thus, a hierarchy is assumed to be formally equivalent to a rooted tree, in a graph-theoretic form. It can be defined as a collection of sets such that for any two sets in the collection, either one set is contained in the other or the two sets are disjoint (Alexander, 1965). As an example, consider the collection of sets {NH VT}, {ME NH VT}, {CN MA RI}, and {CN MA ME NH RI VT}. This collection is a tree and can be represented by the drawing in Figure 1.

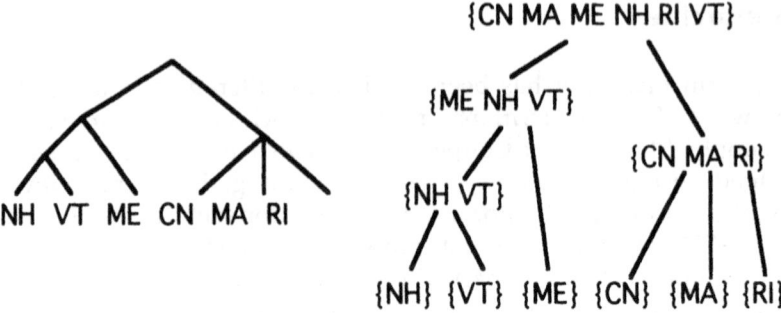

Fig. 1. Tree and lattice diagram for hierarchical tree

A multi-level task or a task of differential granularity is only hierarchical under the tree definition above, if the information from one level or granularity is passed on the other levels during the reasoning process, as occurred in the Stevens and Coupe (1978) example cited above.

Many real-world phenomena can be represented by a tree, such as cities within states, states within countries, countries within continents, and so on. However, the stress in the last sentence is on the word "can." Whether it is correct to do so is a separate issue. In fact, in close examination, we find that most attempts to force a hierarchy onto anything other than artificial examples usually fail. Gary, Indiana, in terms of influences, transportation, and even time zones, is more closely associated with Chicago than with the rest of Indiana. Lake Tahoe represents a single geographical "neighborhood" that lies in both California and Nevada. Such examples might be considered noise in the data to be ignored. However, in discussing the structure of cities, Alexander (1965) has argued that a natural city is by nature not hierarchical, but contains overlapping clusters that are better represented in semi-lattice. I believe this claim is true, not only for cities, but for other large-scale geographical spaces, as well. Before discussing a semi-lattice, I turn to another data structure, that of ordered trees.

3 Ordered Trees

A technique that has proven useful for uncovering hierarchical structure in cognitive maps has been that of the ordered tree algorithm for free-recall data (Hirtle, & Jonides, 1985; McNamara, & Hirtle, 1989). An ordered tree is a rooted tree where the children of a node, at any level, may be ordered, as a unidirectional or bidirectional node, or unordered, as a nondirectional node. Ordered trees, as discussed here, were first introduced by Reitman and Rueter (1980) and differ from two other uses in the literature of the term. Aho, Hopcroft, and Ullman (1974) define an ordered tree as one in which all children are strictly ordered from left to right. In a third use of the term,

Barthelemy, Leclerc, & Monjardet (1986) define an ordered tree as a rooted tree where the nodes are ordered by the height of the nodes. In this paper, the discussion is restricted to the first use of the term, as defined by Reitman and Rueter (1980).

An ordered tree is built by examining the regularities in a set of recalls over a fixed set of items. In fact, an ordered tree is a generalization that allows for some overlapping structure. As an example, the collection of sets {NH VT}, {ME NH VT}, {CN MA}, {MA RI}, {CN MA RI}, and {CN MA ME NH RI VT} can not be represented by a tree, since the sets {CN MA} and {MA RI} are overlapping and violate the tree definition, given above. However, this collection can be represented by the ordered tree, as seen in Figure 2.

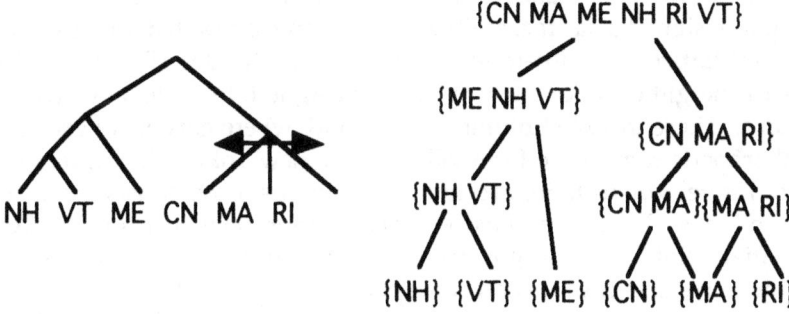

Fig. 2. Tree and lattice diagram for ordered tree.

4 Semi-lattices

A semi-lattice is a generalization of an ordered tree. It is defined formally as a collection of sets, such that for any two overlapping sets in the collection, the intersection of the sets is also in the collection (Alexander, 1965). Therefore, if the sets {A B C D E F} and {B C E G H} are in the collection, then the set {B C E} must be in the collection, as well. As an example, consider the collection of sets {NH VT}, {CN MA}, {ME NH VT}, {CN MA VT}, {CN MA RI}, and {CN MA ME NH RI VT}. Such a collection cannot be represented as either a tree or an ordered tree, but can be represented as a semi-lattice. The sets {ME NH VT}, {CN MA VT} and {CN MA RI} are overlapping and thus violate the definition of an ordered tree, given above. However, this collection can be represented by the graph structure shown in Figure 3.

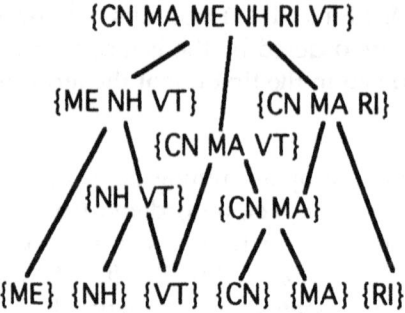

Fig. 3. Diagram for semi-lattice structure.

Alexander (1965) makes a distinction between artificial cities, which are often designed as a tree, and natural, organic cities that are best viewed as a semi-lattice. To illustrate the differences, he describes several new cities or neighborhoods, which were designed to follow a strict tree structure. These include Columbia, Maryland, where clusters of exactly five neighborhoods combine to form villages, or a new Tokyo plan developed by Kenzo Tange, where there are four major loops, each consisting of three medium loops. The medium loops contain minor loops that consist of either residential neighborhoods, government offices, transportation, or industrial buildings. He also cites several conceptions of natural cities that try to force a hierarchical tree onto the spatial structure. For example, a 1943 plan of Greater London by Abercrombie and Forshaw argues for a "large number of communities, each separated from all adjacent communities," which are further subdivided into neighborhoods, each with their own shops and schools.

Alexander (1965) goes on to argue that a natural, living city, despite the ill-advised wishes of the urban planner, does not conform to the hierarchical structure of a tree. Instead, the areas served by the post office, a local school, a social club, or water authority are of different sizes and scope. The resulting structure is better conceptualized to be that of a semi-lattice.

5 Mapping Data to Structures

5.1 Data and Trees

To demonstrate further the differences among these three representation schemes, I turn to two small datasets collected on the recall for countries in Europe. During the academic year of 1994-1995, students on two different campuses in two different European countries were asked to make an ordered list, from memory, of either all the countries in Europe, or all the capitals in

Europe. No other instructions were given to the subjects. To equate the two samples, the capitals were converted into the country name for those receiving the capital task. It is further acknowledged that the capital task was harder and that exclusions might occur, not from forgetting the country, but because the subject does not know or is unsure of the name of the capital. However, these two datasets are considered only to highlight the differences between the representations discussed in this paper, and not to generalize about specific regional understanding of European geography. Furthermore, the purpose of this exercise was to explore the possible clustering that exists among countries and not the set inclusion principle of aggregating a capital to its host.

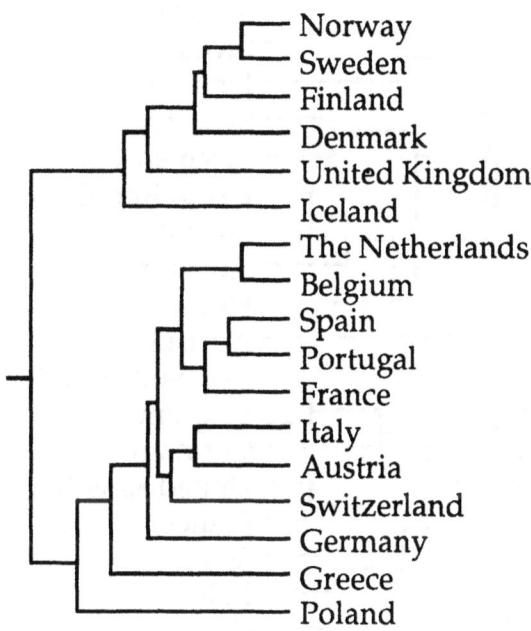

Fig. 4. Hierarchical tree of European countries generated by the Norwegian subjects

A group of 18 subjects in Norway, who were asked to recall countries of Europe, produced a total of 48 distinct entries, even after collapsing obvious synonyms, such as "The Netherlands" and "Holland." A smaller set of 17 countries was recalled by at least 14 out of 15 subjects. Three subjects had far fewer countries recalled and were deleted from further consideration. The ordered list from the remaining 15 subjects were clustered into a strict hierarchical tree, using a standard, average-link clustering algorithm with the city-block metric. This metric is equivalent to stating that the distance between any two countries is proportional to the total number of intervening items between them across all the ordered lists. For example, the pair

Norway and Sweden has a distance of 17.00, which indicates that in all but one of the 15 recall patterns, the pair of items were contiguous, with two additional intervening countries in the remaining recall pattern (17 = 14 * 1 + 1 *3). The resulting tree is shown in Figure 4.

A group of 12 subjects in Austria, who were asked to list all the capital cities of Europe, produced a total of 35 distinct entries. A smaller set of 18 countries was recalled by at least 7 out of 12 subjects. The ordered lists of these 18 countries were clustered into a strict hierarchical tree, also using the average-link clustering algorithm with the city-block metric. The resulting tree is shown in Figure 5.

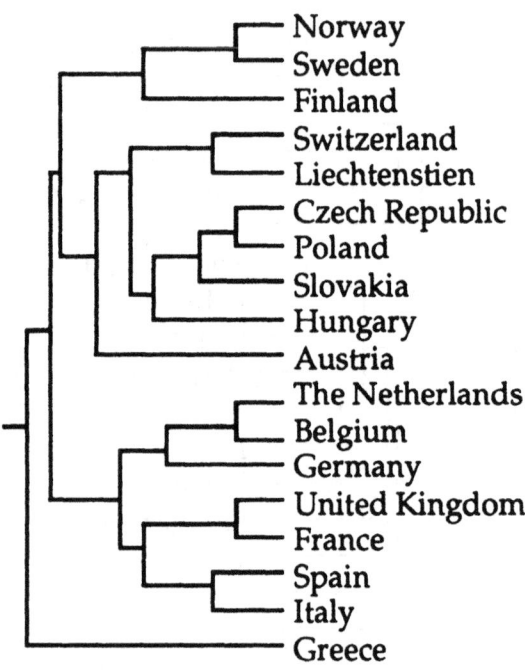

Fig. 5. Hierarchical tree of the country names for the European capitals generated by the Austrian subjects

In examining these trees, the limitation of a hierarchical tree becomes obvious. Each country is placed uniquely in a single cluster within the tree, by the very definition of a tree. The multiple relationships, which Alexander (1965) argued convincingly in favor of, are not able to be incorporated in the representational structure. In Figure 4, the United Kingdom is placed, albeit weakly, in a cluster with other Nordic countries. A representation that would allow the United Kingdom to cluster in part with France and Belgium, and in part with Norway and the other Nordic countries, might capture closer the essence of the true representational features of the UK, but is impossible to incorporate into a single tree.

5.2 Ordered Trees

An ordered tree might allow some overlapping relationships to emerge. Unfortunately, an immediate application of the existing ordered tree algorithm of Reitman and Rueter (1980) is not possible. The algorithm was developed to account for the strong representational structures within a single subject for a domain of interest and not to build an average structure across many subjects. Thus, the algorithm is deterministic and produces clusters that exist across all recall patterns. Within the Norwegian sample, there was not a single cluster that was common to all subjects, whereas in the Austrian sample, only the single cluster of {Norway Sweden} existed for all the subjects.

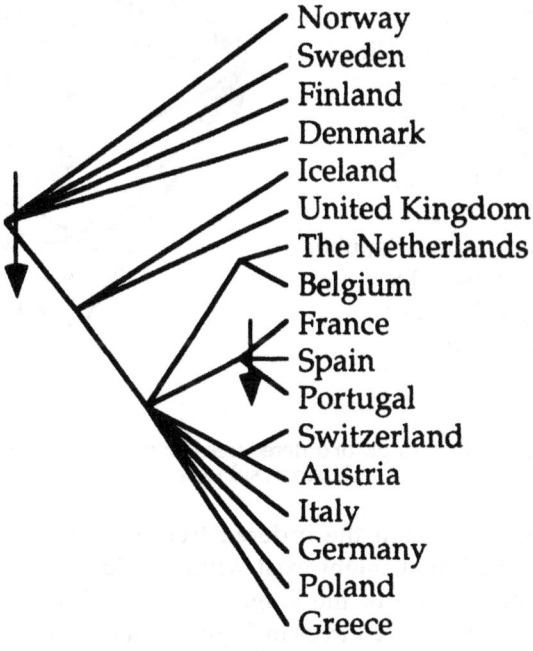

Fig. 6. An example of an ordered tree for a subgroup of
Norwegian subjects

However, by examining subgroups of subjects within each sample, one can identify small groups of subjects with common strategies, for which one can calculate non-trivial ordered trees. Figure 6 shows one tree from a subset of the Norwegian subjects and Figure 7 shows two trees from a subset of the Austrian subjects. It is interesting to note the predominance of the home country, as expected, in each sample. In addition, the two ordered trees in Figure 7, from the Austrian sample, indicate two very different strategies, one that is geographically oriented (Figure 7a), and another that is ordered by prominence (Figure 7b). The former strategy resulted in Austria clustered with Switzerland and Liechtenstein, whereas the latter strategy resulted in Austria being followed by France and United Kingdom.

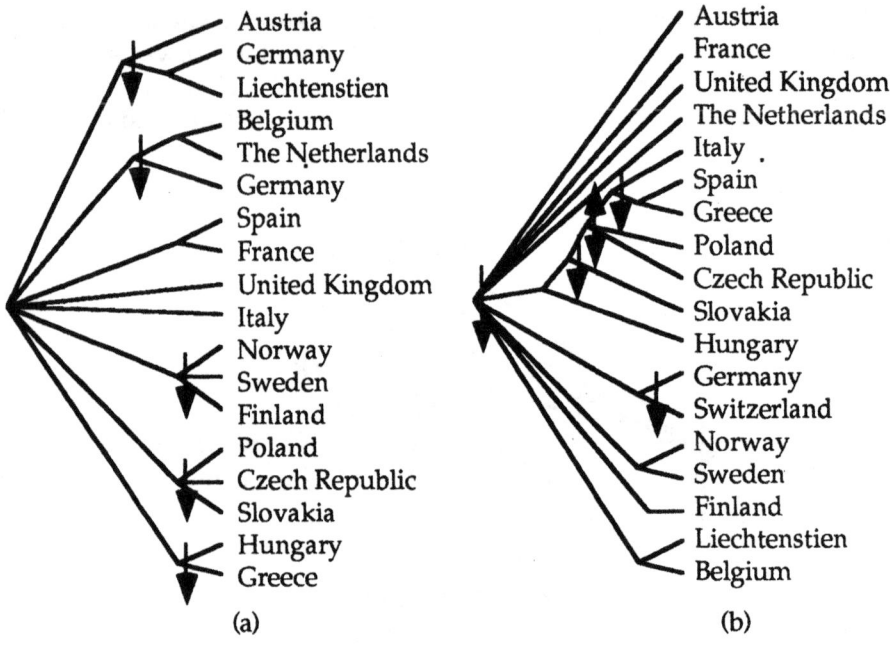

Fig. 7. An example of ordered trees for two subgroups of Austrian subjects

Two benefits arise from the ordered tree over the strict hierarchical tree. First, any order that might exist within a cluster is preserved. This can be seen by dominance of the home countries of Norway and Austria within their respective ordered trees in Figures 6 and 7. The ordered clusters also provide examples of implicit overlapping internal clusters. For example, in Figure 6, the ordered cluster of {France, Spain, Portugal} is created by the two underlying, overlapping clusters of {France, Spain} and {Spain, Portugal} both of which have strong surface validity, while the excluded relationship {France, Portugal} has much weaker surface validity. A strict hierarchical model would imply that every pair of items in a cluster would be associated at the same level.

5.3 Semi-lattices

The final representational scheme of a semi-lattice lacks any direct method to produce, which may account for why the representation of a semi-lattice for cognitive maps has not been considered to the extent of the previous two representations. One solution would be to use the MAPCLUS model (Shepard, & Arabie, 1979) of overlapping clusters to provide a seed set of potential clusters to build a semi-lattice upon. An initial application of this algorithm to the data from the Austrian subjects resulted in three large overlapping clusters. One cluster was a mixture of eastern and northern European countries, from Scandinavia to Greece. The second cluster was northern and western countries, from Scandinavia to Spain, whereas the third cluster consisted of the larger, prominent central European countries of the United Kingdom, France, Austria, Germany, Spain, Norway, Italy and Sweden. While such an analysis is promising, it is clear that any implementation of semi-lattice models will require the additional development of appropriate algorithms.

6 Conclusions and Summary

In summary, a tree structure is one realization for a hierarchical structure for the representation of space. It is easily constructed and understood, but it is also a rigid structure that does not allow for overlap. Ordered trees provide an extension that allows for some degree of overlap, whereas a semi-lattice is an even richer structure that appears to be consistent with many aspects of cognitive space (Alexander, 1965). There are many other possibilities for representing spatial clusters, including additive trees (Sattath & Tversky, 1977), pseudo-hierarchies or pyramids (Diday, 1986), pathfinder associative networks (Schvaneveldt, 1990), extended trees (Carroll, & Corter, 1995), and hybrid scaling models (Carroll, & Pruzansky, 1980). A survey and review of the mathematical properties of many of these representations can be found in Cutsem (1994). The methods discussed here were chosen to illustrate the limitations of a strict, hierarchical representation and because they have been used in the past to model cognitive spaces.

It is important to realize that each of the representational structures discussed in this paper are success refinements of one another. As such, a hierarchical tree remains a good, first pass for modeling the representation of space. Research that is built on the assumption of a hierarchical tree should not be discarded, but rather interpreted with care, and some violations to a strict hierarchy should be expected.

As spatial information systems develop and evolve, the importance of considering alternative structures to strict hierarchical trees and the necessity of being explicit about the nature of the assumed representational structure will only increase. Finally, it is important to note that a consistent theme behind all of the representations discussed is that of a highly structured representation. To replace use of hierarchical trees with unstructured representation, such as an undifferentiated network, would be a serious mistake. Rather, the goal of future research should be to clarify the exact nature of the underlying, structured representation.

7 Acknowledgments

This paper was prepared while the author was on sabbatical at the Department of Computer Science, Molde College, in Molde, Norway. Their support is gratefully appreciated. The author wishes to thank Adrijana Car, Kai Olsen and an anonymous reviewer for their comments concerning the issues presented in this paper.

8 References

Aho, A. V., Hopcroft, J. E., & Ullman, J. D. (1974). The design and analysis of computer programs. Reading, MA: Addison-Wesley.

Alexander (1965). A city is not a tree. *Design*, 46-55.

Ballard, D. H. (1981). Strip trees: A hierarchical representation for curves. *ACM Communications, 24* (5), 310-321.

Barthelemy, J. P., Leclerc, B., & Monjardet, B. (1986). On the use of ordered sets in problems and consensus of classification. *Journal of Classification, 3*, 187-224.

Carroll, J. D., & Corter, J. E. (1995). A graph-theoretic method for organizing overlapping clusters into trees, multiple trees, or extended trees. *Journal of Classification*, in press.

Carroll, J. D., & Pruzansky, S. (1980). Discrete and hybrid scaling models. In Feger, E. D. L. &. H. (Eds.), *Similarity and Choice, Bern: Hans Huber*.

Car, A., & Frank, A. (1994). Modeling of the hierarchy of space applied to large road networks. . Paper presented at IGIS'94, Monte Verita, Ascona, Switzerland.

Couclelis, H., Golledge, R. G., Gale, N., & Tobler, W. (1987). Exploring the anchor-point hypothesis of spatial cognition. *Journal of Environmental Psychology, 7*, 99-122.

Diday, E. (1986). Orders and overlapping clusters in pyramids. In de Leeuw, J., Heiser, W., Meulman, J., & Critchley, F. (Eds.), *Multidimensional data analysis* (pp. 201-234). Leiden: DSWO Press.

Fotheringham, A. S. (1983). A new set of spatial interaction models: The theory of competing destinations. *Environment and Planning A, 15*, 15-36.

Fotheringham, A. S., & Curtis, A. (1992). Encoding spatial information: The evidence for hierarchical processing. In Frank, A. U., Campari, I., & Formentini, U. (Eds.), *Theories and methods of spatio-temporal reasoning in geographic space* Lecture Notes in Computer Science, 639 (pp. 269-287). Berlin: Springer-Verlag.

Frank, A. U., & Timpf, S. (1994). Multiple representations for cartographic objects in a multi-scale tree - An intelligent graphical zoom. *Computers & Graphics, 18*, 823-829.

Golledge, R. G. (1992). Place recognition and wayfinding: Making sense of space. *Geoforum, 23*, 199-214.

Hernández, D. (1994). *Qualitative representation of spatial knowledge.* Lecture Notes in Artificial Intelligence, 804. Berlin: Springer-Verlag.

Hirtle, S. C., & Heidorn, P. B. (1993). The structure of cognitive maps: Representations and processes. In Gärling, T., & Golledge, R. G. (Eds.), *Behavior and environment: Psychological and geographical approaches* (pp. 170-192). Amsterdam: North-Holland.

Hirtle, S. C., & Jonides, J. (1985). Evidence of hierarchies in cognitive maps. *Memory and Cognition, 3*, 208-217.

Holyoak, K. J., & Mah, W. A. (1982). Cognitive reference points in judgments of symbolic magnitude. *Cognitive Psychology, 14*, 328-352.

Kosslyn, S. M., Ball, T. M., Reiser, B. J. (1978). Visual images preserve metric spatial information: Evidence from studies of image scanning. *Journal of Experimental Psychology: Human Perception and Performance, 40*, 47-60.

Kuipers, B. J., & Levitt, T. S. (1988). Navigation and mapping in large-scale space. *AI Magazine, 9(2)*, 25-43.

McNamara, T. (1992). Spatial representation. *Geoforum, 2*, 139-150.

McNamara, T. P., Hardy, J. K., & Hirtle, S. C. (1989). Subjective hierarchies in spatial memory. *Journal of Experimental Psychology: Learning, Memory, and Cognition, 15*, 211-227.

Medyckyj-Scott, D. J., & Blades, M. (1992). Human spatial cognition. *Geoforum, 2*, 215-226.

Poucet, B. (1993). Spatial cognitive maps in animals: New hypotheses on their structure and neural mechanisms. *Psychological Review, 100*, 163-182.

Reitman, J. S., & Rueter, H. R. (1980). Organization revealed by recall orders and confirmed by pauses. *Cognitive Psychology, 12*, 554-581.

Sadalla, E. K., Burroughs, W. J., & Staplin, L. J. (1980). Reference points in spatial cognition. *Journal of Experimental Psychology: Human Learning and Memory, 5*, 516-528.

Samet, H. (1989). *The design and analysis of spatial data structures.* Reading, MA: Addison-Wesley.

Sattath, S., & Tversky, A. (1977). Additive similarity trees. *Psychometrika, 42,* 319-345.

Schvaneveldt, R. W. (Ed.). (1990). *Pathfinder associative networks: Studies in knowledge organization.* Norwood, NJ: Ablex.

Shepard, R. N. (1978). The mental image. *American Psychologist, 33,* 125-137.

Shepard, R. N., & Arabie, P. (1979). Additive clustering: Representation of similarities as combinations of discrete overlapping properties. *Psychological Review, 86,* 87-123.

Stevens, A., & Coupe, P. (1978). Distortions in judged spatial relations. *Cognitive Psychology, 10,* 422-437.

Timpf, S., Volta, G. S., Pollock, D. W., & Egenhofer, M. J. (1992). A conceptual model of wayfinding using multiple levels of abstraction. In Frank, A. U., Campari, I., & Formentini, U. (Eds.), *Theories and methods of spatio-temporal reasoning in geographic space* Lecture Notes in Computer Science, 639 (pp. 348-367). Berlin: Springer-Verlag.

Tolman, E. C. (1948). Cognitive maps in rats and men. *Psychological Review, 55,* 189-208.

Tversky, B. (1981). Distortions in memory for maps. *Cognitive Psychology, 13,* 407-433.

Tversky, B. (1991). Distortions in memory for visual displays. In Ellis, S. R., Kaiser, M. K., & Grunwald, A. (Eds.), *Pictorial communication in virtual and real environments* (pp. 61-75). London: Taylor and Francis.

Van Cutsem, B. (Ed.). (1994). *Classification and dissimilarity analysis,* Lecture Notes in Statistics, No. 93. New York: Springer-Verlag.

Yoshino, R. (1991). A note on cognitive maps: An optimal spatial knowledge representation. *Journal of Mathematical Psychology, 35,* 371-393.

Reasoning About Ordering

Christoph Schlieder

University of Freiburg, Institute of Computer Science and Social Research,
Center for Cognitive Science, Friedrichstraße 50, D-79098 Freiburg
email: cs@cognition.iig.uni-freiburg.de, fax: (+49)-761-203-4938

Ordering information is a special type of spatial information that derives from the linear, planar or spatial ordering of points. A definition of ordering information in terms of the orientation of simplexes is used in this paper to introduce a system of line segment relations which generalizes Allen's system of interval relations to two dimensions. It shows that this generalization differs in interesting properties from the generalizations based on topological relations which have been proposed so far. The conceptual neighborhood structure of the line segment relations provides the foundation of ordering information reasoning. This is illustrated with an example from motion planning. Finally, the problem of representing ordering information is addressed. In that context the cell complex representation of Frank and Kuhn is compared with the approach presented here.

1 Introduction

The linear ordering of a finite number of points on a directed line constitutes the simplest example of what we will define below as *ordering information*. Even though this information abstracts from metric relationships (i.e. distances between points), it allows us to answer some questions concerning a point's position. We can, for instance, infer which is the nearest neighbor of a given point in the direction of the line's orientation. One-dimensional ordering information reasoning needs not to be restricted to points, also intervals can be handled since they are definable in terms of points. Algorithmic and representational issues of reasoning with points and intervals on a line have been studied thoroughly in temporal logic. Some of this work has encountered considerable interest in the spatial reasoning community. The most successful example for an import from the temporal to the spatial domain is probably Allen's (1983) system of interval relations. Explicit use of the interval relations is made for example by Guesgen (1989) and by Hernández (1992) who both devise spatial reasoning formalisms that are more general than Allen's calculus in that they allow two-dimensional extended objects to be reasoned about.

This paper also proposes a two-dimensional generalization of Allen's interval relations. The core idea consists of using the fact that linear ordering is the one-dimensional specialization of a general n-dimensional concept of point ordering. In section 2 it is shown that just as the linear ordering of points gives rise to the system of interval relations, the two-dimensional point ordering induces a system of relations describing the relative position of two line segments in the plane. This system of *line segment relations* is defined in section 3. Reasoning with these relations exploits their conceptual neighborhood structure. Section 4 gives the conceptual neighborhoods for the line segment relations and shows how to use them in reasoning. The problem of efficiently representing two-dimensional ordering information is dealt with in section 5 where a representation which encodes ordering information together with topological information, the cell complex representation of Frank & Kuhn (1986), is discussed.

2 Two-dimensional ordering information

The ordering of points in the plane and in space has been studied extensively in discrete and computational geometry. Especially relevant to our discussion is a series of papers by Jacob Goodman and Richard Pollack which treat combinatorial and algorithmic problems of two-dimensional point ordering (for a survey see Goodman & Pollack, 1993). The concept of two-dimensional ordering that is defined in this paper corresponds to what they call the order type of a set of points.

Ordering and orientation are closely related. This becomes clear by examining the one-dimensional case. The orientation of a line induces an orientation on all segments on that line: the segment $p_i p_j$ is oriented positively (negatively) iff the path from p_i to p_j follows the positive (negative) direction of the line. From the oriented segments one easily obtains the ordering relation on the points because $p_i < p_j$ is only another way of expressing that $[p_i p_j] = +$, i.e. the positive orientation of the segment. In the following, we will use the functional notation instead of the relational and thus consider the *one-dimensional ordering* of a set of points $P = \{p_1, ..., p_n\}$ to be the function $[\ ]: P \times P \to \{-, 0, +\}$ which maps a segment $p_i p_j$ onto its orientation $[p_i p_j]$, where $[p_i p_j] = 0$ for $p_i = p_j$. The connection between ordering and orientation also holds in higher dimensions and we use the orientation of a plane to define the orientation of a triangle in that plane.

Definition 1

Let p_i, p_j and p_k be points in a plane Π with counterclockwise positive orientation. The *orientation of a triangle* $p_i p_j p_k$ is defined as

$[p_i p_j p_k] = +$ iff the path $p_i p_j p_k$ follows the positive orientation

$[p_i p_j p_k] = 0$ iff the path $p_i p_j p_k$ is rectilinear

$[p_i p_j p_k] = -$ iff the path $p_i p_j p_k$ follows the negative orientation

In other words, $[p_i p_j p_k] = +$ iff the point p_k lies on the left side of the directed line through p_i and p_j, $[p_i p_j p_k] = -$ iff p_k lies on the right side of that line and $[p_i p_j p_k] = 0$ iff p_k falls on the line. Clearly, the orientation depends on the order in which the points are enumerated.

Observation 1

$[p_1 p_2 p_3] = + \Leftrightarrow [p_1 p_3 p_2] = - \Leftrightarrow [p_2 p_1 p_3] = + \Leftrightarrow$
$[p_2 p_3 p_1] = + \Leftrightarrow [p_3 p_2 p_1] = - \Leftrightarrow [p_3 p_1 p_2] = +$

We continue to proceed in analogy to the one-dimensional case where the ordering of a set of points is a function that assigns orientations to the line segments over the points. The orientation of triangles plays the same role for defining the two-dimensional ordering of points as the orientation of segments did for one-dimensional ordering.

Definition 2

The *two-dimensional ordering* of a set of points $P = \{p_1, ..., p_n\}$ is the function $[\ ]: P \times P \times P \to \{-, 0, +\}$ which maps a triangle $p_i p_j p_k$ onto its orientation $[p_i p_j p_k]$ as specified by definition 1

For a configuration of four points such as the one depicted in Fig. 1 four different triangles exist. The two-dimensional ordering of the configurations is determined by the values of the orientation function. According to observation 1 it is sufficient to give one orientation for each triangle:

$$[ABC] = + \quad [ABD] = - \quad [ACD] = + \quad [BCD] = +$$

As can be seen from the righthand diagram in Fig. 1 the point D may move inside the shaded area without any of the four triangles changing its orientation. Generally speaking, if a point is moved such that it does not cross any line defined by other points of the configuration, the two-dimensional ordering of the configuration is not changed. Note that in drawing the configuration we introduce additional metric information. The two-dimensional ordering itself does not put metric constraints on the positions of the points.

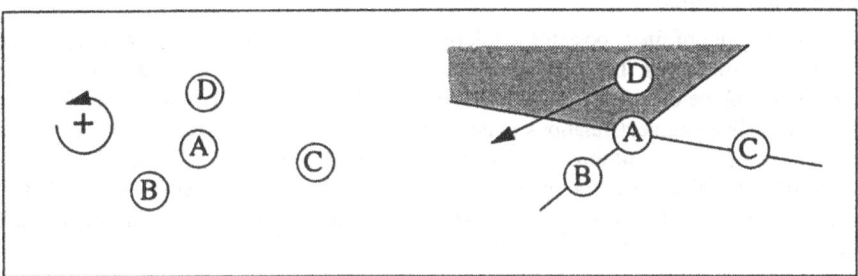

Figure 1: Ordering information about a configuration of four points

We are now in a position to make the notion of ordering information more precise.

Definition 3

A relation encodes one-dimensional (two-dimensional) *ordering information* about a set of points iff it is definable in terms of the one-dimensional (two-dimensional) orientation function.

Obviously, ordering information is less constraining than metric information. It may be less clear though, that ordering information yields stronger constraints than topological information. Consider a convex polygon. The polygon's convexity is generally not preserved by topological transformations, but it is definable in terms of triangle orientations of the vertex points. To learn that a polygon is convex provides one with more than topological information, namely ordering information in the sense of definition 3.

Let us briefly sketch how to generalize definition 3 to higher dimensions. A n-dimensional simplex is the convex hull of *n* points in general position: segments and triangles are the simplexes of dimension 1 and 2, the tetrahedron is the simplex of dimension 3. In n-dimensional space the oriented n-dimensional simplexes can be used to define point ordering proceeding similarly as in definition 2 and 3.

3 The line segment relations

Allen's (1983) system of interval relations is probably the best-known example of a spatial representation formalism dealing with ordering information. Each of the 13 relations describes a class of the relative positions of two line segments AB and CD on a directed line. The interval relations are defined in terms of the one-dimensional ordering of the points A, B, C, D. All situations for instance, in which the segments meet in the points B and C (i.e. $A < B = C < D$) are described by the interval relation *meets*, written AB *m* CD. For the *meets* relation the orientation function takes the following values

$[AB] = +$ $[AC] = +$ $[AD] = +$ $[BC] = 0$ $[BD] = +$ $[CD] = +$

By systematically assigning all combinatorially possible orientations and eliminating the assignments that are not geometrically realizable one obtains exactly the system of 13 interval relations. An assignment of orientations is geometrically realizable iff it is possible to find coordinates for A, B, C, D that realize the orientations. For example, any orientation function with $[AB] = +$, $[AC] = -$ and $[BC] = +$ is not geometrically realizable because it violates the transitivity of ordering. To summarize, the interval relations describe exactly those positions of the two intervals that can be distinguished on the grounds of one-dimensional ordering information.

This principle of finest possible resolution with respect to ordering information generalizes to two dimensions. In a preparatory step, we determine those positions of two line segments in the plane that are distinguishable by two-dimensional ordering information. Then, a system of line segment relations is defined in such a way that each relation describes a class of distinguishable positions. There are four triangles to consider for obtaining ordering information about the position of two intervals AB and CD in the plane. See Fig. 2 for an example. The two interval positions shown are distinguishable since they differ in the orientation of the triangle ACD.

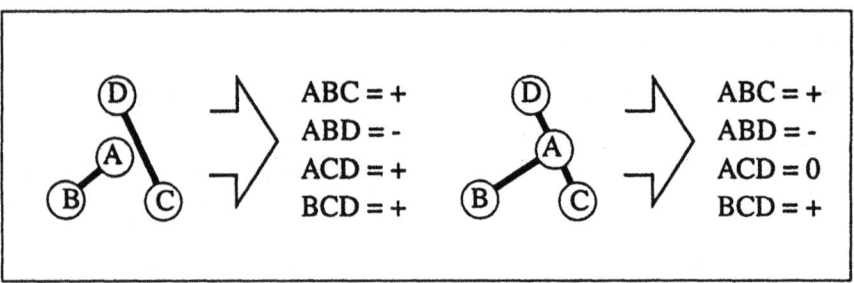

Figure 2: Line segment positions distinguishable by ordering information

If the points A, B, C and D are collinear then the orientation of all four triangles is zero – this is the one-dimensional case described by Allen's interval relations. Zero orientations appear also when three, but not all four points are collinear. In other words, all orientations are non-zero iff the points are in general position. We will restrict the following overview of distinguishable interval positions to cases where the points are in general position. The reason for this restriction is simply that it permits us to present the result in a table of manageable size. Following the restriction all orientations are either + or - which means that $2^4 = 16$ combinatorially possible orientation assignments exist. Enumerating the triangles in lexicographic order allows us to encode the orientation pattern by a number, e.g. $([ABC]\ [ABD]\ [ACD]\ [BCD]) = (- + - +) = 0101_{bin} = 5_{dec}$. We will denote by $o_{ABC}(n)$ the orientation of the triangle ABC defined by the encoding n and in the same sense use the notations $o_{ABD}(n)$, $o_{ACD}(n)$ and $o_{BCD}(n)$. It turns out that two orientation assignments, namely those with encodings 5 and 10 are not geometrically realizable.

Observation 2

Let AB and CD be two line segments in the plane. If A, B, C and D are in general position then the 14 line segment positions depicted in Fig. 3 can be distinguished by two-dimensional ordering information.

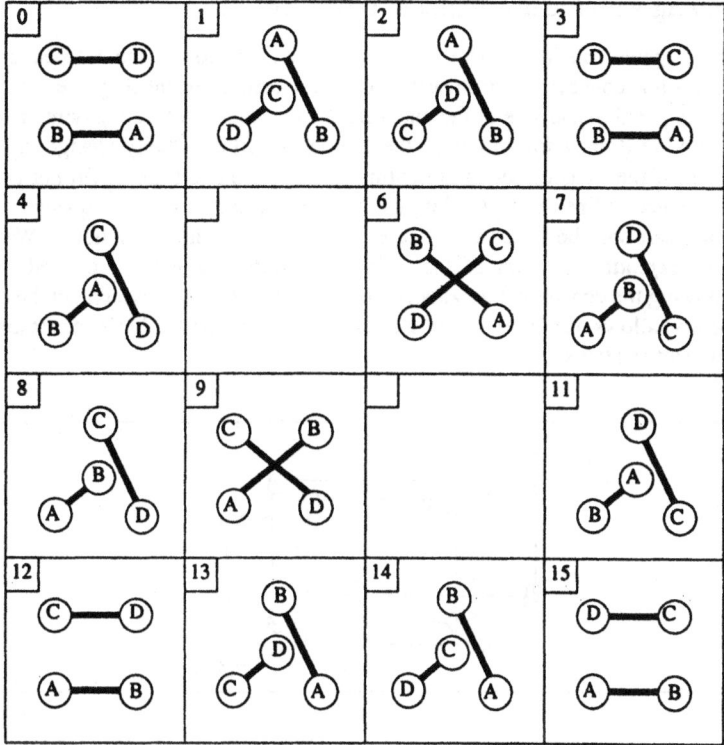

Figure 3: The 14 line segment relations

We may now define 14 relations over the set of the line segments in the plane, each of which describes a class of interval positions that is equivalent with respect to two-dimensional ordering information.

Definition 4

For $0 \leq i \leq 15 \land i \notin \{5,10\}$ the *line segment relation* r_i (AB, CD) holds iff

[ABC]=$o_{ABC}(i)$, [ABD]=$o_{ABD}(i)$, [ACD]=$o_{ACD}(i)$, [BCD]=$o_{BCD}(i)$.

How many more line segment relations are obtained when the restriction on the point's position is removed? For points in special position two cases must be distinguished: (1) all four points lie on a line, (2) only three of them are collinear. As has already been mentioned, the first case leads to Allen's 13 interval relations. Note that since all triangle orientations are zero the relations must be distinguished by one-dimensional ordering information. In the second case, 36 positions can be distinguished on the basis of two-dimensional ordering information alone. That is, the complete relation system for representing the position of two intervals in the plane consists of $13 + 36 + 14 = 63$ relations. By *line segment relations in the broad sense* we mean this complete relation system.

4 Reasoning with the line segment relations

The line segment relations abstract from the exact location of the segment's extremal points and do not convey metric information. To a certain extent they are insensitive to changes in the position and size of the intervals. Nevertheless, deforming (stretching, shortening) or moving (translating, rotating) the segments will eventually change the line segment relation. If the transformation is continuous in space and time it will not produce an arbitrary sequence of line segment relations. It turns out that in such a sequence a line segment relation can only be followed by certain other line segment relations. We will call them the *conceptual neighbors* of that relation, adopting a term introduced by Freksa (1992). Knowing the conceptual neighborhood structure (see Fig. 4 which will be explained in detail below) allows one to predict changes and thus provides the basis for reasoning with the line segment relations.

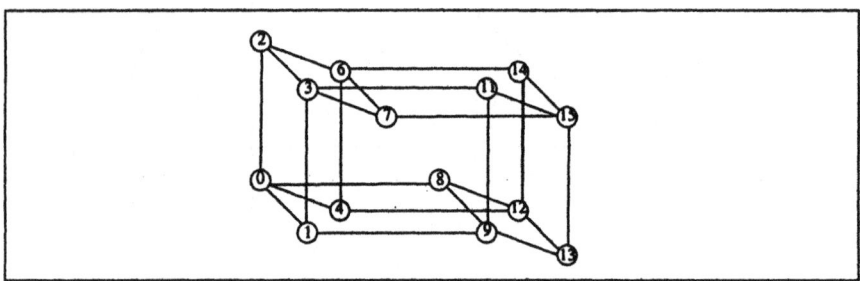

Figure 4: Conceptual neighborhood structure for the line segment relations

Consider as an example the line segment relation r_4. The corresponding orientation function has the following values: $[ABC] = -, [ABD] = +, [ACD] = -, [BCD] = -$. A transformation continuous in space and time that moves just a single point will alter the triangle orientations one after another, but never simultaneously. Starting with the orientation pattern of r_4, that is (- + - -), and inverting the orientation of a single triangle, four different patterns may be reached: (+ + - -), (- - - -), (- + + -) and (- + - +). The first three correspond to the relations r_{12}, r_0 and r_6, the last pattern with the encoding 5 is not geometrically realizable. We thus find that r_4 has three conceptual neighbors, namely r_0, r_6 and r_{12}. In a similar way the conceptual neighbors of the other relations can be determined. The result is summarized by the graph in Fig. 4 whose nodes are line segment relations specified by their encoding and whose edges indicate conceptual neighborhood. Remember that the conceptual neighborhood graph reflects a special kind of transformation: movements of single points. The reason for making this restriction is that the 14 line segment relations are closed under these transformations. Obviously, the richest conceptual neighborhood structure will be found for the complete system of 63 line segment relations and arbitrary continuous transformations.

Nevertheless, interesting inferences are also possible with the small neighborhood graph over 14 relations. A motion planning problem will illustrate this. Let $r_4(AB,CD)$ specify the segment's start position and $r_7(AB,CD)$ the goal position. The problem consists in finding how to move the segment AB such that the start position is transformed into the goal position without a collision of the segments in-between. In order to solve the problem we have to translate it into the line segment formalism. Moving a segment corresponds to a sequence of line segment relations in which every two consecutive relations are conceptual neighbors. To avoid a collision means that the relations r_6 and r_9 may not appear in the sequence. The shortest path in the conceptual neighborhood graph leads to a collision as shown in Fig. 5. But the two next shortest paths provide solutions: AB may move around CD in a left turn (4-12-14-15-7) or a right turn (4-0-2-3-7).

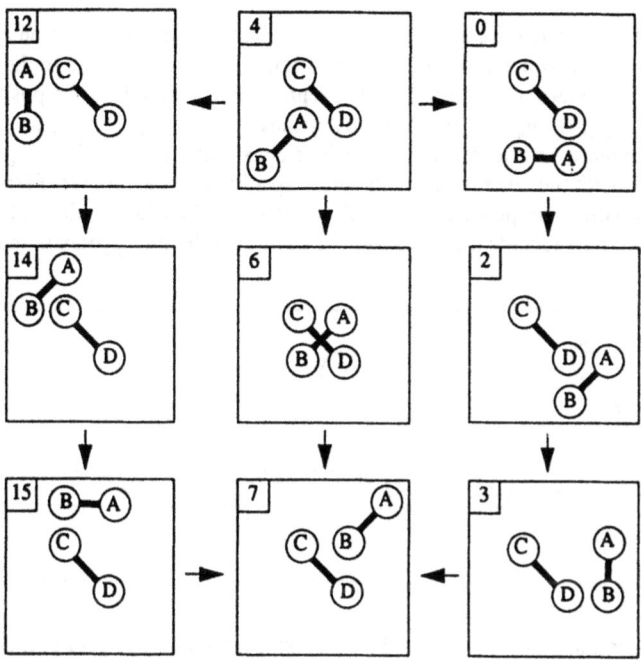

Figure 5: Possible transformations from r_4 to r_7.

5 Representing ordering information

Ordering information has rarely been recognized as such in spatial reasoning and consequently there is no self-contained treatment of the topic – for the beginnings of such a study see Schlieder (1993, 1995). However, a considerable amount of research has been devoted to the development of topological representation formalisms (an overview is given by Freksa & Röhrig, 1993). Since the threefold distinction between metrical information, ordering information and topological information is not generally made in the spatial reasoning community, we find that some representational formalisms that describe themselves as topological in fact encode topological information as well as ordering information.

Thus, there are on the one hand the purely topological approaches to spatial reasoning. Two prominent examples are the c-relation calculus of Randell & Cohn (1989) and the 4-intersection-model of Egenhofer & Franzosa (1991) which have both undergone a number of refinements and extensions. These and similar approaches encode only topological invariants (connectedness, topological closure). A genuine ordering information property such as convexity can not be represented – the c-relation calculus for example was extended by Randell, Cui & Cohn (1992) to overcome this limitation. On the other hand, there are approaches in which topological information is encoded together with ordering information. Among them the cell complex representation of Frank & Kuhn (1986) is particularly interesting because it makes use of oriented simplexes and thus bears some resemblance to the approach proposed in this paper. We will therefore devote the rest of this section to discuss in what way the cell complex representation encodes ordering information and how it differs from the concept of two-dimensional ordering that was introduced in definition 2.

In order to facilitate the comparison we will refer to a simplified version of the cell complex representation which is accurate to the original only with respect to the representation of ordering information. The cell complex representation describes the position of a set of points $P = \{p_1, ..., p_n\}$ in the plane by means of a triangulation. The triangulation is build incrementally as illustrated in Fig. 6. The construction starts with the points p_1, p_2, p_3 and proceeds with the other points in the order specified by the indexes. Fixing the ordering in which the points are processed uniquely determines the triangulation. The cell complex representation now records the orientation of each triangle that appears in the triangulation.

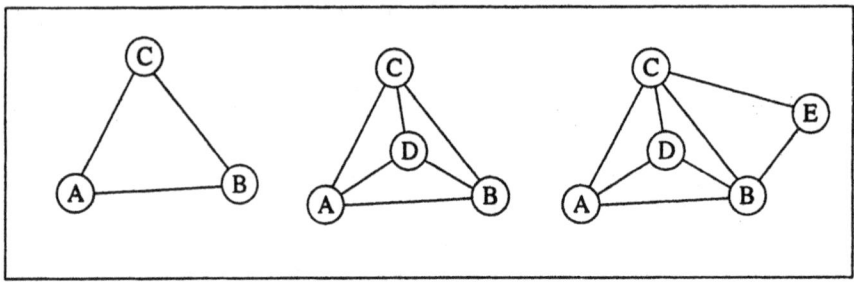

Figure 6: Triangulation arising from the successive integration of points A, B, C, D, E.

The cell complex representation obviously encodes ordering information in the sense of definition 3. However, it does not use all triangles for that purpose. Orientations are only recorded for the triangles appearing in the triangulation. We can quantify this difference to the concept of two-dimensional ordering of definition 2. The number of triangles of any triangulation of n points is $O(n)$ whereas the number of all triangles is $O(n^3)$. This difference is reflected by a difference in expressive power. There are configurations of points that can be distinguished on the grounds of two-dimensional ordering information but have the same cell complex representation (see Fig. 7 for an example). The two configurations of points differ only in a single triangle orientation: [ADE] is - for the left and + for the right configuration. Since the configurations lead to the same triangulation and the triangle ADE does not appear in it, they are indistinguishable by their cell complex representation.

We have seen that the cell complex representation encodes part of the two-dimensional ordering information of a set of points in the plane. However, the advantage of being more economical in storage requirements is paid for by less expressive power. It seems interesting though to investigate other schemes that allow to select from the set of all $\binom{n}{3}$ triangles a subset that is relevant for the spatial reasoning problem under consideration. Some insight for future work in that direction may be gained by looking at cognitive processes that deal with ordering information (e.g. spatial problem solving, mental mapping).

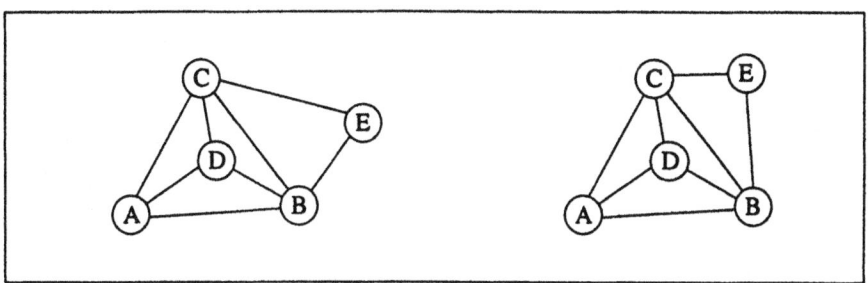

Figure 7: Configurations that are equivalent for the cell complex representation

To sum up, it has been shown in this paper that ordering information constitutes a well-defined type of spatial information deriving from a general concept of n-dimensional point ordering. Allen's interval relations which are often characterized as "topological relations" are more adequately described as relations representing ordering information. The reason for this is simply that these relations distinguish between left and right, a distinction which can not be defined in terms of topological invariants. Once that the interval relations are recognized as encoding ordering information their generalization to two dimensions is straightforward. The line segment relations described in section 3 provide this generalization. In two dimensions the difference in expressiveness between topological and ordering information becomes very clear. Most of the interval positions described by the line segment relations can not be discriminated on the grounds of topological invariants. Topological information only allows to distinguish between intersecting line-segments (i.e. r_6, r_9) and non-intersecting segments (all other relations). Both types of spatial information should therefore be clearly distinguished. Such a differenciated perspective is needed for designing representational formalisms which deal with the indeterminacy of spatial information.

References

Allen, J. (1983). Maintaining knowledge about temporal intervals. *Comm. of the ACM, 26,* 832-843.

Egenhofer, M. & Franzosa, R. (1991). Point-set topological spatial relations, *International Journal of Geographical Information Systems, 5,* 161-174.

Frank, A. & Kuhn, W. (1986). Cell graphs: A provable correct method for the storage of geometry. In Proc. *2nd International Symposium on Spatial Data Handling,* Seattle, WA, 411-436

Freksa, C. (1992). Temporal reasoning based on semi-intervals. *Artificial Intelligence, 54,* 199-227.

Freksa, C., & Röhrig, R. (1993). Dimensions of qualitative spatial reasoning. In N. Piera Carreté & M. Singh (Eds.), *Qualitative reasoning and decision technologies* (pp. 483-492). Barcelona: CIMNE.

Güsgen, H. (1989). Spatial reasoning based on Allen's temporal logic. *TR-89-049,* ISCI, Berkeley, CA

Hernandez, D. (1992). *Qualitative representation of spatial knowledge,* Ph.D. thesis, TU München.

Goodman, J., & Pollack, R. (1993). Allowable sequences and order types in discrete and computational geometry. In J. Pach (Ed.), *New trends in discrete and computational geometry* (pp. 103-134)

Randell, D. & Cohn, T. (1989). Modelling topological and metrical properties in physical processes, In R. Brachman et al. (eds.) Principles of Knowledge Representation and Reasoning

Randell, D., Cui, Z. and Cohn, A. (1992). A Spatial Logic based on Regions and Connection, In *Proc 3rd Int. Conf on Knowledge Representation and Reasoning,* Boston

Schlieder, C. (1993). Representing visible locations for qualitative navigation. In N. Piera Carrete & M. Singh (Eds.), *Qualitative Reasoning and Decision Technologies* (pp. 523-532). Barcelona: CIMNE.

Schlieder, C. (1995). Qualitative shape representation In A. Frank (ed.) Spatial conceptual models for geographic objects with undetermined boundaries, London: Taylor & Francis.

Improving the Selection of Appropriate Spatial Interpolation Methods

Felix Bucher and Andrej Včkovski

University of Zürich
Winterthurerstr. 190, CH-8057 Zürich, Switzerland
{bucher,vckovski}@gis.geogr.unizh.ch

Abstract. In many natural sciences as well as in spatial data processing data describing *continuous fields* are a major information type. One of the main challenges when using such field representations in analysis and modelling arises from restrictions caused by the discretization that occurs when sampling the field. As a consequence, interpolation is often required before using the data in a particular application. In this paper the family of *kriging methods* is used as an example to illustrate the decision-making process when selecting appropriate interpolation methods. The analysis of the properties of various kriging methods shows that the decision-making process should be based on a considerable body of information, including implicit and external knowledge, i.e., information not related and derivable from the data.

The acquisition of the information needed and the examination of the relevant data characteristics needs can be a demanding procedure, including many statistical tests and other means of exploratory data analysis. *Extended Exploratory Data Analysis* is a strategy to support the decision process when selecting an interpolation method. This standardized procedure supports the derivation of implicit information and navigates users through the decision process. In combination with *Virtual Data Sets* Extended Exploratory Data Analysis results in more reliable estimated field values.

1 Introduction

Data describing continuous fields are a major information type used in many natural sciences. As a rudimentary rule we can state that as soon as "independent variables" and "dependent variables" are involved (and the "independent variables" are *continuous*), the information type dealt with are *fields*. Whenever data sets describing fields are used, one has to deal with some consequences of *discretization* that occurs when sampling the field. Discretization primarily affects the (continuous) domain of the "independent variables", i.e., the *support* of the field. A field has to be described with a *finite* set of samples. Each of these samples, i.e., the *field values* ("dependent variables"), are somehow related to an element (or subset) of the support ("independent variable") which we call *index* for a particular value. Consider a data set consisting of annual mean surface air temperature over Switzerland sampled at irregularly distributed sites. The *support* then might be the surface of Switzerland, the *indices* are sample site coordinates and sample time (i.e., year), the field values are each site's mean temperature value averaged from quasi-continuous temperature measurements.

The use of data describing fields of natural phenomena usually has two basic problems (Včkovski, 1995):

Global property: Field values are needed for indices that are not available in the given set of indices, that is, "values at unsampled locations". This is a global issue of the data set, i.e., it depends on the entire structure of the set of indices and the support.

Local property: Since the field values are describing a natural phenomenon they are never accurate, i.e., they are always affected with uncertainty to some degree. This is a local property since it affects mainly the *field value* at a certain index and not the index itself[1].

The latter of these two problems was presented in (Včkovski, 1995). In this contribution we will discuss issues concerning discretization, and present strategies to overcome some of the effects of discretization of a field's support. The discussion will be restricted to fields where the support is both a part of the earth's surface and two-dimensional. First, a short overview of the processes generating field data is given, in order to introduce the terminology and concepts used. The section following will focus on one consequence of discretization which is the need for subsequent *interpolation* to estimate field values at unsampled locations. Using *kriging methods* as an example the epistemic way towards those estimated field values is illustrated. The decisions, potential problems and steps that have to be taken are analyzed. Discussion of the procedure finally leads to proposition of two strategies that could help reduce problems found within this procedure.

2 Sampling and Analysis of Continuous Fields

2.1 Sampling of Fields

A *field* on a given support S is a function $z(\cdot)$ assigning every element s of the support a corresponding field value $z(s)$ from the value domain $\mathbb{F}$:

$$s \xrightarrow{z} z(s), \quad s \in S \subset \mathbb{R}^N, \quad z(s) \in \mathbb{F} \tag{1}$$

For the sake of simplicity we will assume the value domain $\mathbb{F}$ real and one-dimensional, i.e., $\mathbb{F} \in \mathbb{R}$.

The notion of *continuity* of the support S implies that there has to be a *metric* $\|\cdot\|$ defined on S, to express "distance" $d(x,y)$ between two elements of the support:

---

[1] Uncertainties affecting the indices can be mapped to uncertainties of the field values, at least if a probabilistic point of view is used: Using the joint distribution of both index and field value one can use its projection along the index-axis, i.e. the marginal distribution and assume the index to be a single, fixed value.

$$d(x,y) = \|x - y\|, \quad x, y \in \mathbb{S} \tag{2}$$

Therefore, the support $\mathbb{S}$ is a (dense and compact) subset of a metric space[2].

If the field describes a physical property (e.g., surface air temperature) a *measurement model* for this property is needed, i.e., a way to describe the property with a number $z(s)$. We assume this model is defined and that we know how to measure a property to get the associated number $z(s)$. Therefore, the sampling of a field consists of a set of measurements of the property at different indices s_i, and, using the measurement model, a *selection* of field values "at" indices s_i. This selection can be written as a projection of the field onto the measured value, using a selection function $\phi_i(s)$ corresponding to the index s_i:

$$z(s_i) = \int_{\mathbb{S}} \phi_i(s) z(s) ds \tag{3}$$

The selection function is normalized:

$$\|\phi_i\| = \int_{\mathbb{S}} \phi_i(s) ds = 1 \tag{4}$$

That is, the sampled field value $z(s_i)$ is a weighted average of the field $z(s)$ "around" the index s_i, where weights are given by the selection function ϕ_i, which in turn depends on the index s_i. If s_i is a point and the measurement/selection retrieves the field value exactly at that point, then $\phi_i(s)$ is a Dirac-distribution centered at s_i:

$$\phi_i(s) = \delta(s_i - s) \tag{5}$$

In many cases (e.g., remote sensing images) the index is a proper subset of $\mathbb{S}$, e.g. a square pixel. Then the selection function ϕ_i yields a corresponding aggregation of the field $z(s)$. A simple example is the mean value over s_i with a selection function given by:

$$\phi_i(s) = \begin{cases} \frac{1}{V(s_i)} & : \quad s \in s_i \\ 0 & : \quad s \notin s_i \end{cases} \tag{6}$$

where the volume $V(s_i)$ of s_i is

$$V(s_i) = \int_{s_i} ds \tag{7}$$

A typical sampling of a field consists of a set of (estimates of) indices $\hat{s}_i$ and corresponding measured field values $\hat{z}_i$. In order to understand these data it has to be figured out

– how the $\hat{s}_i$ relate to a corresponding selection function ϕ_i, and

---

[2] It is also necessary to define a *measure* $\mu(\cdot)$ on the support $\mathbb{S}$. This allows the introduction of the concept of a *volume* of a subset of $\mathbb{S}$. If $\mathbb{S} \subset \mathbb{R}^N$ then this is typically a Riemann-measure.

- how the field value estimates $\hat{z}_i$ relate to the field values projected out through ϕ_i.

If the samplings are *sparse*, i.e., there are "few" $\hat{s}_i$ within the whole support $\mathbb{S}$, and the $\hat{s}_i$ can be approximated by points, then these two mappings are usually simple, since the selection function just yields the field value at $\hat{s}_j$ and can therefore be modelled as given by (5). For field samplings covering most or all of $\mathbb{S}$ these mappings are typically more difficult. Consider a remotely sensed image of surface air temperature distribution. Each measured field value represents an aggregation of field values within and near the pixel $\hat{s}_i$. If the variation of the field $z(s)$ is small within each cell, then a simple, constant selection function as given by (6) can be assumed. If the variations of the field over a few cells are not insignificant, then more realistic selection functions have to be modelled, typically using as much knowledge of the measurement process as possible and the charactersitcs of the instruments involved (e.g., satellite sensor).

To allow for uncertainties, the field values have to be modelled as entities that characterize uncertainty. A useful and commonly used way is to model field values as *random variables*, i.e., applying probabilistic concepts on the field value's uncertainty. The mapping (1) is enhanced so that every s is mapped to a random variable $Z(s)$ of an appropriate probability space. $Z(s)$ is sometimes called a *random function* (Isaaks & Srivastava, 1989). The subsequent selection of certain field values is fully analogous to the "undisturbed" case discussed above, except that the selection function maps to a random variable instead of a real number.

The objective of this section was to show how the samplings of a field are related to the field itself. This understanding is mandatory to the analysis of continuous fields asking the simple question: "Given a set of samplings $\{\hat{s}_i, \hat{z}_i\}$, what does the field $z(s)$ look like?". In the next subsection we will cover certain specific issues concerning the analysis of continuous fields, namely *interpolation* or prediction of values at unsampled locations.

2.2 Analysis of Continuous Fields

The primary objective of the analysis of fields is the estimation of a field value $z(s_0)$ at a subset s_0 of $\mathbb{S}$. Depending on the distribution of the indices s_i and the set s_0, one might distinguish:

Interpolation: The objective of interpolation is to estimate values "in between" sampled values, i.e., to increase the resolution of a currently available information source. $z(s_0)$ is found by interpolation, if s_0 is within the convex hull of all indices s_i and if s_0 and all s_i are disjunctive, i.e., $s_0 \cap s_i = \emptyset$ for all s_i.

Extrapolation: Extrapolation is similar to interpolation except that the location s_0 is *not* with the indices' convex hull.

Aggregation: $z(s_0)$ is determined by aggregation if s_0 and the indices s_i are *not* disjunctive, i.e., $s_0 \cap s_i \neq \emptyset$ for some i.

In the following discussion we will focus on *interpolation*, and more explicitely on *spatial* interpolation.

3 Spatial Interpolation

3.1 Selecting an Appropriate Interpolation Method

Spatial interpolation has occupied scientists from various disciplines for a very long time, and still commands a great deal of attention. Over the past decades various interpolation methods have been developed for a wide range of applications[3].

It is a truism that no one method will ever exist that will be able to reliably account for all potential applications of spatial interpolation. As in shown in (Englund, 1990), selecting an interpolation method that is appropriate to a particular situation is at the very heart of reliable use of interpolated data in subsequent applications, such as visualizing a field or using the data as inputs to a particular spatial analysis. In order to select an appropriate interpolation method, the decision-making process needs to be based on several criteria. These criteria depend on the characteristics of the field under consideration, characteristics of the sampled field values available, and their mutual relationship (e.g., the selection functions ϕ_i). Furthermore, properties of the various interpolation methods, expertise of the researchers, purpose of the interpolation and many other factors influence an appropriate selection, parametrization and application of an interpolation method. Taking all these dependencies into account leads to a very demanding interpolation procedure. For various reasons[4] the interpolation procedure often is not properly performed. Ignoring the complexity of this task may significantly affect the quality of interpolated field values.

In the remainder of this contribution it is the objective to propose strategies to manage this complexity, i.e., standardized procedures for appropriate selection and parameterization of an interpolation method. The following discussion of a subset of interpolation methods – the family of kriging methods – is intended to motivate these strategies.

3.2 Criteria for Selecting an Appropriate Kriging Method

Kriging is a collection of methods that covers in its totality a considerable range of potential spatial interpolation situations. These methods all have in common that they are generalized linear regression techniques for minimizing an estimation variance defined from (a) prior model(s) for the degree of spatial dependence of nearby sample values, i.e., covariance(s). Over the past years, some kriging methods have become most favourable interpolation methods particularly in

---

[3] Major reviews on spatial interpolation methods are given in (Lam, 1983).

[4] These reasons include: Researchers do not have sufficient expertise or insufficient knowledge on the data properties and origin, software provides only some few and simple methods.

environmental sciences. The reason for this evolution is that kriging is known as a BLUE estimator, i.e., best linear unbiased estimator. Despite its popularity, it remains not very well known that kriging is a collection of several dozen methods. Each method responds to a particular situation, such as purpose of interpolation (what are the interpolated values used for), real-world behaviour of the phenomenon being represented, and the representativity of data to the real-world behaviour of the phenomenon. Beyond that, each method has some specific requirements for the data being used. Tab. 1 gives an overview of kriging methods, their properties, and some of the key criteria used for selecting a method. This table is far from being comprehensive, both with respect to available kriging methods, and criteria to be considered, respectively, but it should give an impression of the complexity of the decision-making process when selecting an appropriate method. The criteria listed in Tab.
refKriging are described in more detail below.

Stationarity required for the mean value: Stationarity of the random function $Z(s)$ to be kriged means invariance of its multivariate *cumulative distribution function* (cdf) under any translation of any coordinate vectors within the support S. Invariance of the multivariate cdf results by definition in invariance of any lower order cdf, including invariance of all their moments, such as mean value and covariances. Strictly speaking, statistical inference – such as performed with kriging – requires stationarity of the random function to be kriged. In most empirical situations, however, the decision of stationarity is critical to the representativity and reliability of the interpolation method being used. As an example, kriging performed under strict stationarity does not adapt to local trends since it relies on the mean value, assumed to be known and constant throughout the support. As a consequence, in most kriging methods, the assumption for stationarity in the mean value is relaxed by some appropriate means, such as re-estimating the local mean value within moving search neighbourhoods, or by restricting the stationarity assumption to a residual component of the random function. In some situations, however, it is strictly recommended to apply the stationarity assumption, such as for the normal score transformation used for Multi Gaussian kriging method. The hypothesis of having a random function $Z(s)$ being stationary with respect to its mean value can be proven by calculating moving window statistics, i.e., moving a window over the area of investigation and calculating the mean value of each individual window. The size of the window, however, is a critical parameter and thus has to be carefully determined.

Explicitly modelling global spatial trend: Some of the more simple kriging methods consider the random function stationary in the mean value, or provide means for slight deviations from stationarity. For some of the phenomena being kriged, this random function model might not be appropriate, since the phenomenon has a significant global spatial trend. This trend results from the physics of processes that determine the spatio-temporal behaviour of the phenomenon, such as impacts of climate, topography, and soil

Table 1. Criteria and properties of various kriging methods (•: yes, o: no, n/a: not applicable)

	stationarity required for the mean value of the RF	global spatial trends explicitly modelled	modelled trend specified by the physics of the problem	use of secondary variables	secondary variable much denser sampled	resulting estimation (mean, ccdf)	index type (area / point)	reflection of uncertainty	accounts for soft / fuzzy data	accounts for categorical data	smoothing effect
simple kriging	•	o	n/a	o	n/a	mean	point	σ_k	n/a	n/a	•
ordinary kriging	o	o	n/a	o	n/a	mean	point	σ_k	n/a	n/a	•
universal kriging	only for residual	•	o	o	n/a	mean	point	σ_k	n/a	n/a	•
kriging with an external drift	o	•	•	•	o	mean	point	σ_k	n/a	n/a	•
co-kriging	o	o	n/a	•	•	mean	point	σ_k	n/a	n/a	•
multi gaussian kriging	•	o	n/a	o	n/a	ccdf	point	univariate ccdf	o	o	•
simple indicator kriging	•	o	n/a	o	n/a	ccdf	point	univariate ccdf	•	•	•
indicator co-kriging	o	o	n/a	•	n/a	ccdf	point	univariate ccdf	•	•	•
block kriging	o	o	n/a	o	n/a	mean	area	σ_k	o	o	•
gaussian related stochastic simulation	rather	o	n/a	o	n/a	joint ccdf	point	joint ccdf	o	o	o
indicator simulation algorithms	•	o	n/a	o	n/a	joint ccdf	point	joint ccdf	•	•	o

to vegetation. In order to reliably krige the data, such global trends must be adequately taken into account. Examining the presence of a global spatial trend may be performed by testing the degree of fitness of a low order polynomial function of the coordinates.

Trend model specified by the physics of the processes: Modelling an existing global spatial trend is ideally based on an (partial) understanding of the genesis of the phenomenon being kriged. In the absence of knowledge about the physics of underlying processes, or in the absence of appropriate data, the trend most often is modelled as a low order polynomial of the coordinates. If the physics of the processes are partially known, the global spatial trend may be modelled with either an appropriate function, e.g., considering a sine function if spatial variability has some periodic component, or by using secondary variables that reflect some statistical coherence to the kriged data and which makes physical sense as well.

Use of secondary data: Secondary data is data that is not kriged, but rather incorporated, in order to reduce the estimation variance of kriged data. Basically, secondary data may be used for both explicitely modelling a global spatial trend (e.g., in kriging with an external trend) or to improve adaptation to local trends (e.g., in some co-kriging methods). The basic idea behind these methods is that the secondary data is spatially cross-correlated with the data being kriged, and thus potentially contain useful information about the kriged data. This condition can be examined both visually with cross h-scatterplots (Isaaks & Srivastava, 1989, p. 61), and analytically by means of cross-correlation, cross-covariance, or cross-variogram functions, respectively. Additionally, these methods require the secondary data varying smoothly in space, and the statistical coherence between the data being reasonable with respect to the physics of the problem.

Sampling density of secondary data: Some of the methods incorporating secondary data are extremely demanding, such as co-kriging methods, that require covariance functions to be modelled between each of the variables used. In these cases, thus, it is recommended to carefully examine whether the incorporation of secondary data might result in a significant reduction of estimation variances. Usually, this is only guaranteed in cases where the kriged data is undersampled with respect to secondary data. Examining this condition, however, is not as trivial as it might sound at first. This condition is not only made for the number of samples, but also for the spatial configuration of sample sites, since both areas with spatially clustered data, and areas with data sparsity may create problems when performing a kriging. Tests for the spatial configuration might be done visually or by calculating some measure of local spatial concentration.

Accounts for categorical data: Most of the kriging methods require variables to be continuous. Some of the methods, however, are based on a mechanism where the kriging process is repeated for a series of cutoff values. Consequently, these methods require the continuous variables to be discretized within their interval of variability.

Accounts for soft data: Some of the methods, such as indicator kriging, are based on a Bayesian update of the local prior cdf into a posterior cdf using information supplied by neighbouring local prior cdf's. One of the major advantages of this mechanism is its ability to account for *soft data*. I.e., as long as some sort of uncertainty information, e.g., information about measurement errors, can be coded into prior local probability values, these methods can be used to integrate the uncertainty information into a posterior probability value. Actually, the method that yields the conditional cdf (ccdf) estimates can be seen as a repeated co-kriging for a series of cutoff values that pools hard data and the soft prior probabilities.

Smoothing effect vs. emphazising local maxima: Kriging was initially developed as a collection of methods that have in common estimation of data values as weighted moving averages of the sampled field values. Kriging, thus, works as a low-pass filter that tends to smooth out details and extreme values of the original data. More recently developed kriging methods, however, allow the construction of probabilistic models of uncertainty for values being estimated. These methods can be used to estimate a series of posterior conditional probability distributions from which "unsmoothed images of the attribute spatial distribution can be drawn" (Deutsch & Journel, 1992, p. 62). The decision whether to krige the data with a method that emphasizes smoothing, or a method that is more sensitive to patterns of local variability, respectively, is heavily dependent on the intended use. For example, *mapping* the kriged data would prefer a smoothing method, whereas data used as input in a *simulation* should be kriged with a method that is sensitive to patterns of local variability.

Estimation type, reflection of uncertainty: Most kriging methods provide a best linear unbiased estimate (BLUE) for unsampled attribute values, with the estimation (kriging) variance σ_k being used to define Gaussian-type confidence intervals:

$$\mathcal{P}\left(Z(s) \in [z_k(s) \pm 2\sigma_k(s)]\right) \approx \alpha_c \quad \alpha_c \in \{0.95, 0.99, \ldots\} \tag{8}$$

The kriging variance σ_k is often considered, and communicated, as uncertainty information. It is important to note that kriging variances are independent of the data values, and thus usually do not provide measures of local estimation accuracy. Additionally, kriging estimators are "best" only in the least-squared error sense for a given model of spatial correlation, e.g., a variogram model. Minimizing an expected squared error does not need to be the most relevant estimation criterion. Rather it would be preferable to have a method that minimizes the impact of the resulting error (Deutsch & Journel, 1992, p. 15), but this requires the estimation of the ccdf for the random variable being estimated. Actually, the kriging algorithm has two basic properties that facilitate the estimation of the posterior ccdf, namely:

- If the random function model is multivariate Gaussian then the kriging estimate and variance identify the mean and variance of the posterior ccdf. Moreover, the ccdf is fully determined by these two parameters. However, since the sample distribution usually is not normal, a normal score transformation needs to be performed before kriging the data.

- Instead of the random variable $Z(s)$, one might krige its binary indicator transform $I(s; z)$. This provides an estimate that is also the best linear squared estimate of the conditional expectation of $I(s; z)$. Moreover, the conditional expectation of $I(s; z)$ is itself equal to the ccdf of $Z(s)$. Thus, kriging applied to binary indicator transforms provides least-squared estimates of the ccdf.

Kriging methods that are based on these two basic properties are not aimed at estimating the unsampled value $z(s)$ or its transform, but at providing a ccdf model of uncertainty about $z(s)$.

These ccdf models, however, are univariate, thus only providing an indication of the uncertainty for a given index s. Any indication about the joint spatial uncertainty, e.g., the probability that all values $Z(s_i)$ simultaneously exceed a given threshold value, cannot be derived from a set of univariate ccdfs. Some of the recently developed methods known as conditional simulations, however, allow the derivation of information about the joint spatial uncertainty. This is provided by the differences among a series of alternative, equally probable simulations of a model of the spatial distribution of the random variable $z(s)$. Finally, some kriging methods allow direct estimation of attribute values that are averaged over a prescribed area.

4 Decision-making Processes behind the Selection of Interpolation Methods

4.1 Acquisition of Information

The kriging example shown in the previous section hints at the complexity of the decision-making process that is crucial to selection of an appropriate interpolation method. The decision-making process must be based on a considerable body of information. This information can be divided into three groups:

Explicit information: The set of basic information as provided by the data, i.e., the set of indices $\hat{s}_i$, data values $\hat{z}_i$, and possibly metadata (e.g., description of the sampling, selection functions used, etc.).

Implicit information: Knowledge derivable from the explicit information. The extraction of implicit knowledge is largely based on the examination of various data characteristics, such as tests of the degree of spatial autocorrelation, or tests of distributional properties of data.

Data-unrelated information: A reliable selection of an appropriate interpolation method should be based on knowledge that often is neither explicitly nor implicitly represented in the data. This knowlegde largely consists of information on the specifications made about the sampling of the data. These specifications determine the degree of representativity of data according to the real-world behaviour of the phenomenon (this is sometimes referred to as *first abstraction* (Mackaness & Beard, 1993)). Most often, this knowledge is merely present outside the data producer's domain[5].

4.2 Problems with Information Acquisition

There are various problems associated with information acquisition, particularly for implicit and data-unrelated knowledge. The tests to derive implicit information may cause various problems, including:

- Reliably examining data characteristics requires a considerable statistical and mathematical background, e.g., knowledge about the criteria to be considered, and test procedures that may yield relevant information for the selection of an appropriate interpolation method.
- The interpretation of test results is often not simple and clear.
- Most tests are not robust, i.e., they fail under certain circumstances.
- One is often not fully aware of the necessity and usefulness of implicit information.

Data-unrelated information, that is, information that cannot be reconstructed through examination of any data characteristics, could possibly be made available at the data producer, but often in a quite intuitive form, not to mention digitally. This problem, in fact, has been accentuated over the recent years since data users now increasingly process data not sampled by themselves. Recently, the demand has increased for digital provision of such data semantics with the data set, i.e., an enhancement and standardization of metadata (e.g., (FGDC, 1994, Bucher *et al.*, 1994)). This is, even on a conceptual level, far away from being realized, particularly the issue on how to formalize such data semantics.

5 Strategies for Reliable Spatial Interpolation

The previous sections have shown, that the procedure for spatial interpolation involves many decisions and requires a lot of both implicit and data-unrelated information. The acquistion of the necessary information of both types is demanding, cumbersome and tricky. It is also a non-trivial step to derive an appropriate interpolation method using this information due to the variety of methods and their specific characteristics. Therefore, it is necessary to formalize and structure the whole process of decision-making to overcome the problems when selecting an appropriate interpolation method. The following proposes two strategies, not mutually exclusive, which support data users (and data producers) in the analysis of continuous fields.

---

[5] Data-unrelated information includes knowledge about measurement errors, preprocessing and other information that *should* be included in the metadata but often is missing.

5.1 Decision Support Based on Extended Exploratory Data Analysis

A set of tests for examining data characteristics, i.e., implicit information, for reliably using the data in a particular task, such as interpolation, is often referred to as *Exploratory Data Analysis* (EDA) (Tukey, 1977, Getis, 1993). However, an EDA that provides reliable decision support is very demanding and requires a considerable statistical background. Such statistical background – and the awareness of the importance of EDA – is often lacking within the community of data users. Hence, a structured specification of EDA steps needed at the various stages of the selection process would be helpful. This specification may serve as a conceptual framework for both a decision support system, and a digital tutorial for user training.

Basically, the specification is based on a decision tree. At each node of the decision tree, the user is provided with both a set of relevant data characteristics to be examined, and corresponding test procedures. Moreover, information is provided for a straightforward interpretation of test results that support the user's correct navigation through the decision tree. On the path through the decision tree the user is required to examine all relevant data characteristics in a logical order, ending up with a proposition for an appropriate interpolation method.

5.2 Incorporating Data-unrelated Information

The use of extended EDA might not always yield appropriate choice of an interpolation method due to missing data-unrelated information needed in the decision-making. It was previously shown that the data producer usually is best informed about these issues. Consequently, it would be reasonable to charge the data producer with the responsibility to select the appropriate interpolation method. This task ideally would include the selection of several interpolation methods, each for a specific potential purpose to be performed with the data. Additionally, it should be the responsibility of the data producer to consistently communicate this selection[6].

Virtual Data Sets (VDS) (Stephan *et al.*, 1993, Včkovski, 1995) provide a powerful and simple vehicle to digitally communicate the data producer's selection of interpolation methods and other relevant meta-inforation. This is established in VDS by enhancing the sampled field values with procedural information, e.g., appropriate interpolation methods. A VDS represents a field $\hat{z}(s)$ instead of a set of values $\{\hat{z}_i, \hat{s}_i\}$ via rules (interpolation methods) that allow the VDS to calculate $\hat{z}(s_0)$ at every s_0 in the support $\mathbb{S}$ out of the sampled field

---

[6] The claim to charge the data producer with the responsibility to select the downscaling method has a similar approach in the domain of data quality. There, the so-called "truth in labelling"- principle (Lanter & Veregin, 1992) makes the data producer responsible to report the quality of data, whereas it is the user who must interpret this information in order to evaluate the fitness-of-use of data for a particular application.

values $\{\hat{z}_i, \hat{s}_i\}$. In order to avoid the "black-box"-effect these rules also provide information about the quality of rule-based derived (virtual) data.

Generating VDS (by the data producer) requires all of the extended EDA steps mentioned above in order to reliably select and define the rules and methods within the VDS.

6 Conclusions and Future Research

Spatial interpolation is a basic requirement for various usages of (spatial) field samplings. Using the familiy of kriging methods as an example has shown the *complexity* of the decision-making process when selecting an approriate interpolation method. Properties of each available method, characteristics of the data and metadata all influence the *fitness-of-use* of a method for the particular interpolation purpose in mind. The discussion of the acquistion of this necessary information revealed many potential problems and risks within the whole selection process.

These problems and risks are met by an emphasis on the information aquistion process, which is proposed to be supported by an *extended exploratory data analysis* (EEDA) scheme. This is realized by a decision tree guiding users through the selection process, asking "the right questions at the right time". Combined with *Virtual Data Sets* (VDS), the EEDA promises better and more sound selection of interpolation methods and therefore yields more reliable field estimates.

Future research will focus on refinement and implementation of the conceptual framework based on EEDA and VDS. In a first stage, the decision tree will be restricted to the familiy of kriging methods and be implemented into an interactive decision support system. A gradual extension of the decision support system with other families of interpolation methods, such as splines and stochastic simulation, will improve the generation of VDS.

References

Bucher, Felix, Stephan, Eva-Maria, & Včkovski, Andrej. 1994. Integrated Analysis and Standardization in GIS. *In: Proceedings of the EGIS'94 Conference.*

Deutsch, Clayton V., & Journel, André G. 1992. *GSLIB: Geostatistical Software Library and User's Guide.* Oxford University Press, New York.

Englund, Evan J. 1990. A Variance of Geostatisticians. *Mathematical Geology,* **22**(4), 417–455.

FGDC. 1994 (June 8). *Content standards for digital geospatial metadata.* Federal Geographic Data Committee.

Getis, Art. 1993. GIS and Modeling Prerequisites. *Pages 322–340 of: Proceedings of European Conference on Spatial Information Theory (COSIT).*

Isaaks, E.H., & Srivastava, M.R. 1989. *An Introduction to Applied Geostatistics.* Oxford University Press, New York.

Lam, N. S. 1983. Spatial Interpolation Methods: A Review. *American Cartographer,* **10**, 129–149.

Lanter, D. L., & Veregin, H. 1992. A reserach paradigm for propagating error in layer based GIS. *Photogrammetric Engineering and Remote Sensing*, **58**, 825–833.

Mackaness, William, & Beard, Kate. 1993. Visualization of interpolation accuracy. *Pages 228–237 of: Proceedings of the AUTOCARTO 11 Conference.*

Stephan, Eva-Maria, Včkovski, Andrej, & Bucher, Felix. 1993. Virtual Data Set: An Approach for the Integration of Incompatible Data. *Pages 93–102 of: Proceedings of the AUTOCARTO 11 Conference.*

Tukey, J. W. 1977. *Exploratory Data Analysis.* Addison Wesley.

Včkovski, Andrej. 1995. Representation of Continuous Fields. *Pages 127–136 of: Proceedings of the AUTOCARTO 12 Conference.*

Spatial Pattern and Spatial Autocorrelation

Yue-Hong Chou

Department of Earth Sciences
University of California
Riverside, CA 92521-0423, USA
hong@ucrac1.ucr.edu

Abstract. The spatial pattern of a distribution is defined by the arrangement of individual entities in space and the geographic relationships among them. The capability of evaluating spatial patterns is a prerequisite to understanding the complicated spatial processes underlying the distribution of a phenomenon. Spatial autocorrelation indicates the extent to which the occurrence of one feature is influenced by similar features in the adjacent area. As such, statistics of spatial autocorrelation provide a useful indicator of spatial patterns. This study shows that quadrat analysis of spatial autocorrelation is not suitable for evaluating point patterns. The evaluation of spatial patterns for area features, using Moran's I coefficient, must take into consideration the log-linear relationship between map resolution and spatial autocorrelation. The topological structure of a complex spatial pattern can be revealed by the correlograms constructed based on higher-order spatial relationships. The spatial pattern can be characterized by the behavior of the correlogram's wavelength and amplitude within a specific range of spatial orders.

1. Evaluation of Spatial Patterns

The spatial pattern manifested in the distribution of a phenomenon is determined by the way individual entities are arranged in space and the geographical relationships among such entities. In general, each pattern is the reflection of the underlying spatial process at a specific stage of time and it also influences the process for the next stage. Accordingly, the capability of generalizing and quantifying spatial patterns is a prerequisite to understanding the complicated processes governing the distribution of spatial phenomena. For instance, in a complex ecosystem the distribution of a specific species is affected by several interrelated factors. Any attempt to modeling the distribution and formulating the underlying process requires the specification of the fundamental structure of its spatial pattern.

Spatial autocorrelation indicates the extent to which the occurrence of one feature is influenced by the distribution of similar features in the adjacent area. As such, statistics of spatial autocorrelation provide a useful measure of spatial patterns. If attraction among entities acts as the driving force in the distribution, i.e., the existence of one feature attracts similar features to occur in its neighborhood, the spatial autocorrelation is positive and the distribution would be characterized by clusters of similar entities. When competition among features dominates the spatial process, i.e., the existence of one entity tends to expel similar entities from the neighborhood, the distribution would illustrate a scattered pattern associated with negative spatial autocorrelation. If neither attraction nor repulsion dominates the spatial process, the spatial pattern of distribution would be random and no significant spatial autocorrelation exists.

Commonly used statistics of spatial autocorrelation include the joint-count statistics, Moran's I coefficient [20, 21], and Geary's c coefficient [13]. Their statistical properties and procedures for significant testing are treated thoroughly in [9, 10]. Recently Getis and Ord proposed the G statistics [14, 15]. Issues that are important to the evaluation of spatial autocorrelation include the measure of contiguity as weighting functions [11], the identification of extreme values [18], the generalized procedures for evaluating spatial autocorrelation [17], the assessment of correlograms [22], and the extension of spatial autocorrelation into multivariate spatial correlation [29]. General discussions on spatial autocorrelation are available in [12, 16, 23, 24, 25, 26, 28]. Among the available statistics, Moran's I coefficient is employed in this study because of its wide adoption in the literature.

2. Moran's I Coefficient of Spatial Autocorrelation

Formally, Moran's I coefficient is defined as:

$$I = \frac{n \Sigma_i \Sigma_j \delta_{ij}(x_i - \bar{x})(x_j - \bar{x})}{S_o \Sigma_i (x_i - \bar{x})^2}$$

where n is the number of geographic units; δ_{ij} denotes the spatial relationship between the I-th and j-th units; x_i denotes the frequency of the spatial phenomenon in question; $S_o = \Sigma_i \Sigma_j \delta_{ij}$ is the total number of pairs that hold the spatial relationship specified by δ_{ij}.

In evaluating spatial patterns, the significance of the I coefficient can be tested under the assumption of normality specified in [10], such that,

$$E(I) = - (n - 1)^{-1}$$

and

$$Var(I) = \frac{n^2 S_1 - n S_2 + 3 S_0^2}{S_0^2 (n^2 - 1)}$$

where $E(I)$ and $Var(I)$ are the mean and variance of the I Coefficient, respectively, and

$$S_1 = (\tfrac{1}{2}) \Sigma_i \Sigma_j (\delta_{ij} + \delta_{ji})^2$$

$$S_2 = \Sigma_i (\Sigma_i \delta_{ij} + \Sigma_j \delta_{ji})^2$$

In the simplest form, the spatial relationship (δ_{ij}) is defined by contiguity, i.e., whether or not two geographic units are adjacent to each other. In this case, δ_{ij} equals 1 if the I-th and j-th units share a common border. Appropriate spatial weighting functions such as distance, area, the length of common border, and their combinations can been defined and examined [3, 6].

The range of the value of the I coefficient lies between -1 and 1 although theoretical extremes can be identified [18]. A larger positive value implies a clustered pattern while a negative value significantly different from 0 is associated with a scattered pattern. When the I coefficient is not significantly different from 0, there is no spatial autocorrelation and the spatial pattern is random. A typical example of the relationship between spatial patterns and the I coefficient can be found in [4].

In empirical applications using spatial autocorrelation statistics, several critical issues must be considered [7]. First, an interval- or ratio-scaled measurement of the study variable is preferred over nominal-scaled measurements. In case the study variable is a binary response denoting the presence or absence of a phenomenon, efforts should be taken to convert the binomial classification into a more reasonable measurement if possible.

Second, geographic units are best defined by one or more selected variables that are meaningful to the phenomenon in question. Also, in delineating boundaries of the geographic units, polygon features derived from the selected variables must be non-overlapping and form complete coverage of the study area. Patches of unclassified surfaces must be corrected or reclassified. Furthermore, the geographic units defined by such boundaries should not produce a clear spatial pattern by themselves.

Third, the problem of topological invariance discussed in [11] must be taken into consideration, i.e., once the measure of spatial relationship (δ_{ij}) is determined, the size and shape of geographic units, and the relative strength of their relationships would be ignored. A possible way to alleviate this problem is by specifying and examining spatial weighting functions such as contiguity, area, boundary length, distance, and their combinations [6]. Correlograms based on higher-order spatial relationships should be constructed in order to detect the effects of each weighting function.

3. Quadrat Analysis of Point Features

Evaluating spatial autocorrelation of point features usually takes the form of quadrat analysis where a grid is overlain onto the point pattern and the frequency of points in each cell is counted. Fig. 1 shows a hypothetical spatial pattern where each dot represents the occurrence of a point feature.

Fig. 1-A shows a relatively clustered point pattern in which the point features are located in either the northeast quadrant or the southwest quadrant. In this case, the I coefficient is -1 and this negative spatial autocorrelation implies a perfect scattered pattern. However, when the map resolution is doubled with the area covered by a grid of 16 cells (Fig. 1-B), the calculated I coefficient (0.33) becomes positive and implies a clustered pattern. Further doubling the map resolution makes the area covered with 64 cells (Fig. 1-C), and changes the spatial autocorrelation back to negative.

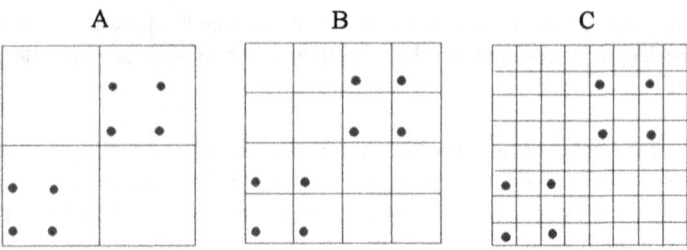

Fig. 1. Three quadrat configuration of a hypothetical point pattern.

The effects of varying map resolution on the measure of spatial autocorrelation can be further examined from three hypothetical patterns, a clustered pattern (Fig. 2-A), a random pattern (Fig. 2-B), and a uniform/scattered pattern (Fig. 2-C). This experiment is conducted such that a series of five quadrat configurations of varying resolution are overlain onto each pattern and the I coefficient is computed for each configuration [19].

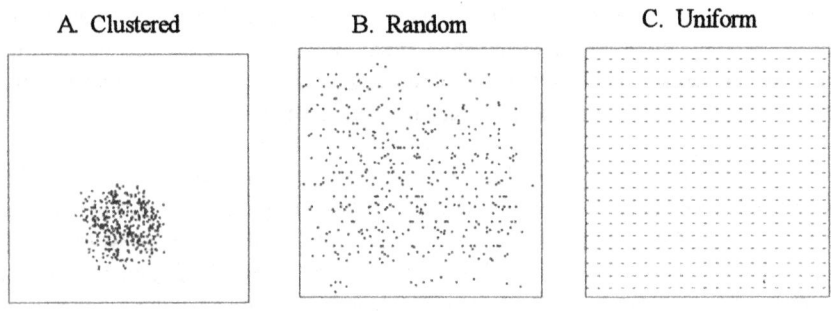

Fig. 2. Three hypothetical point patterns for evaluating spatial autocorrelation.

At the lowest level of resolution, the area is covered by a 2-by-2 grid consisting of four cells. As the resolution is systematically increased, the same map area is covered by a grid of more cells. In this experiment, the configurations of the next levels of resolution are arbitrarily selected at 5-by-5 (25 cells), 10-by-10 (100 cells), 20-by-20 (400 cells), and 40-by-40 (1600 cells). The computed I coefficients for these configurations are listed in Table 1. For convenience, map resolution is defined as the number of cells in a configuration and a higher level of resolution is associated with a grid of higher density [4].

Table 1. Moran's *I* coefficients of designed configurations

Pattern	Map Resolution				
	4	25	100	400	1600
1. Clustered	-.29	.36	.72	.89	.40
2. Random	.25	-.08	-.01	.03	.36
3. Scattered	*	*	*	*	-.12

*: Spatial autocorrelation undefined

For the clustered pattern, the *I* coefficient is -.29 at the lowest level of resolution. Although the point pattern is apparently clustered, the computed negative *I* coefficient implies a scattered pattern. This is because the only unit with an extraordinarily high frequency count (the southeast quadrant) is surrounded by units of very low frequency. At the second level of resolution, the computed *I* coefficient (0.36) indicates a slightly clustering pattern. This trend goes on for the next two levels of map resolution, i.e., the increase in resolution leads to a greater positive *I* coefficient which implies a more clustered pattern. However, the trend is reversed at the highest level of map resolution when the map area is covered by 1600 cells. At this stage, the computed *I* coefficient drops to .40.

The above example demonstrates that, in evaluating clustered patterns of point features, Moran's *I* coefficient has a tendency to increase with an increased level of map resolution. However, this tendency holds only within a certain range beyond which the coefficient may drop.

For the random point pattern (Fig. 2-B), at the lowest level of resolution the low positive *I* coefficient suggests a slightly clustered pattern. It drops gradually for the next two resolution levels. The trend is then reversed and the *I* coefficient increases to the positive domain again. Between the second and the fourth level of map resolution, all the standard normal deviates indicate insignificant autocorrelation. Therefore, at these levels the *I* coefficient correctly indicates spatial randomness. This example illustrates that evaluating point patterns by spatial autocorrelation is valid only within a specific range of resolution.

Scattered point patterns cannot be evaluated by spatial autocorrelation under the quadrat configuration. For Fig. 2-C, the *I* coefficient is undefined for the first four levels of map resolution because the frequency count for each grid cell is identical to the mean. When the level of resolution becomes high enough, (i.e., 1600 in this case,) the *I* coefficient becomes meaningful and represents a slightly scattered pattern.

The analysis shows that the use of spatial autocorrelation for evaluating spatial patterns of point features is unreliable and could be erroneous. The *I* coefficient is unstable over the configuration of the quadrat system and its accuracy depends on map resolution.

4. Quadrat Analysis of Area Features

Unlike point features, the spatial autocorrelation for area patterns varies in a consistent and predictable manner. Fig. 3 shows a series of quadrat configurations for a hypothetical area pattern similar to Fig.1. In Fig. 3-A, the area is divided into four quadrants where the feature occurs in both the northeast and southwest quadrants. The I coefficient is -1 which indicates a perfect scattered pattern of negative spatial autocorrelation. As the map resolution is doubled, the area is covered by 16 cells (Fig. 3-B). The I coefficient changes into 0.333 - a positive spatial autocorrelation associated with a clustered pattern. Further increase in map resolution always increase the I coefficient further (Fig. 3-C).

Regarding the relationship between map resolution and spatial autocorrelation, the difference between point features and area features is that points are discrete and dimensionless while area makes two dimensional, continuous patterns. The consistent change in spatial autocorrelation over the increased map resolution enables one to evaluate the spatial patterns of area features using statistics of spatial autocorrelation.

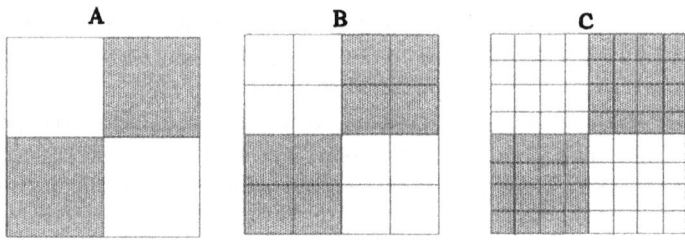

Fig. 3. Three quadrat configurations of an area pattern.

The evaluation of spatial autocorrelation for area features need not be limited to quadrat configurations [3, 5, 8]. A reated study of the effect of resolution on autocorrelation is given in [1]. In [4], the effects of map resolution on spatial autocorrelation were examined using the empirical data of the distribution of wildfires in the Idyllwild quadrangle of California. In that study, geographic units are polygons delineated by the natural boundaries of vegetation and soils. Fig. 4 shows a curve representing the generalized relationship between map resolution and spatial autocorrelation for area features.

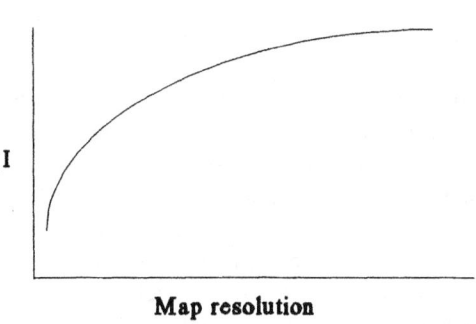

I

Map resolution

Fig. 4. Generalized relationship between map resolution and the I coefficient.

In [4], four different cases were investigated: a hypothetical clustered pattern, a hypo thetical scattered pattern, an empirical pattern evaluated by nominal scale, and the same empirical pattern evaluated by interval scale. The results show that spatial autocorrelation increases systematically with the increase in map resolution. The consistent relationship exists because an increase in map resolution always enhances the clustering pattern due to the continuous, two-dimensional property associated with area features. In general, this relationship can be expressed by a log-linear function, such that,

$$I = \beta_0 + \beta_1 Log_2 RL$$

where I is moran's I coefficient; RL is the level of map resolution; β_0 and β_1 are parameters; Log_2 is the logarithm of base 2.

In summary, statistics of spatial autocorrelation provide a valid measure for evaluating spatial patterns of area features as long as the relationship between map resolution and spatial autocorrelation is taken into consideration. However, the spatial relationship dealt with so far is limited to the direct connection between geographic units. In reality, spatial relationships between geographic units tend to go beyond the immediate neighbors and thus certain variations in spatial patterns may not be detected by statistics derived from the direct spatial relationship alone. To evaluate spatial patterns using statistics of spatial autocorrela-tion, it is necessary to consider indirect neighborhood relationships.

5. Evaluating Spatial Patterns with Spatial Correlograms

Fig. 5 shows six patterns of area features organized in the ascending order of spatial concentration. Pattern A is a perfect scattered system in which every shaded cell is sur-rounded by blank cells while every blank cell is surrounded by shaded cells. Pattern B is similar to A except that two vertical cells form a cluster of identical shade. In pattern C, clusters are formed by groups of four cells. Pattern D combines two vertical four-cell clusters into a eight-cell cluster. Pattern E consists of four sixteen-cell square clusters. Pattern F is a complete clustered configuration where all shaded cells are clustered together while all blank cells are clustered together. In all six patterns, the number of shaded cells is equal to that of blank cells thus autocorrelation is solely affected by spatial arrangement.

Effects of spatial configuration on the measure of spatial autocorrelation must be examined from spatial correlograms. A correlogram shows the variations of the I coefficient over a higher order spatial relationship (also known as spatial lag elsewhere). Naturally, the first order spatial relationship is defined by contiguity, i.e., two cells are called the first-order neighbors if they share a common border. The second order relationship is defined by indirect contact through a common first-order neighbor, i.e., two cells are called the second-order neighbors if they share a common first-order neighbor instead of a common border. Accordingly, spatial relationship can extend from the first order to the order in which the two cells that are farthest apart in the configuration can reach each other.

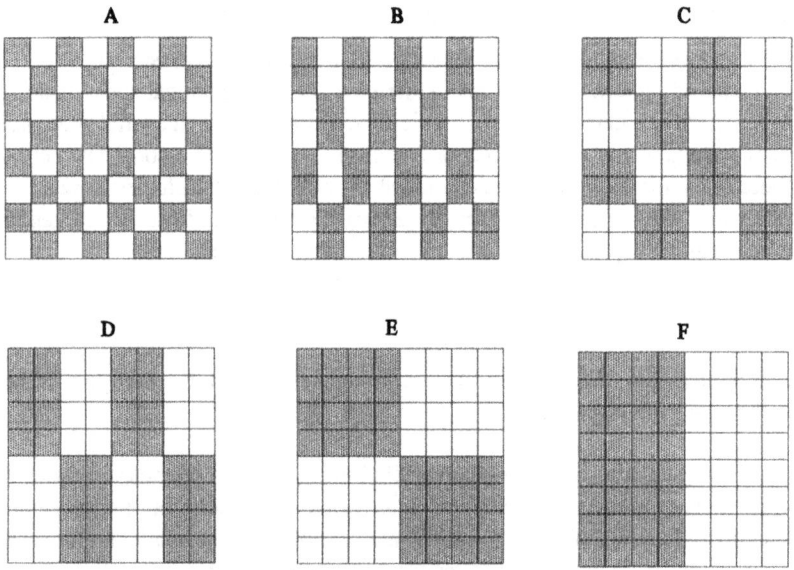

Fig. 5. Six patterns designed for analysis of correlograms.

The maximum order of meaningful spatial relationship is equivalent to the "diameter" in terms of the graph theory. Practically, the maximum order is less than the order associated with the diameter, depending on spatial configuration, because in most cases at the order of the diameter the number of qualified neighboring pairs is too small to be statistically meaningful.

Moran's I coefficient for the second-order neighbors is evaluated in such a way that the spatial weight, δ_{ij}, is equal to one if the i-th and j-th units are second-order neighbors, otherwise δ_{ij} equals zero. The plot of all the calculated I coefficients over the increasing spatial order makes a correlogram. Fig. 6 shows the correlograms of the six patterns.

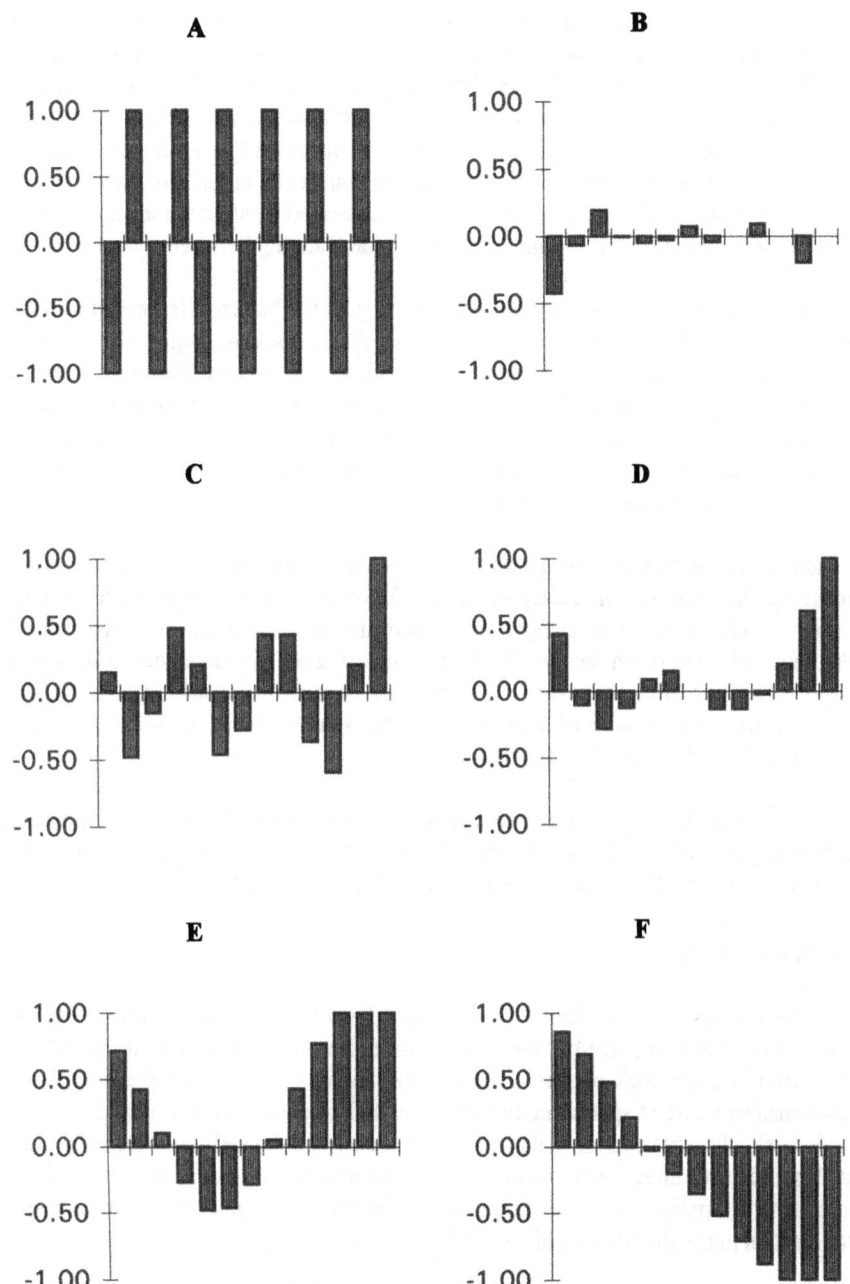

Fig. 6. Correlograms of the six designed patterns. Vertical axis
denotes I coefficient. Horizontal axis denotes spatial order.

For pattern A, Moran's I for the first-order relationship is -1. For the second-order relationship, every shaded cell can reach only shaded cells and every blank cell can only reach blank cells, thus the I coefficient becomes 1. For the third order relationship, again, every shaded cell can only reach blank cells and every blank cell can only reach shaded cell, thus the I coefficient returns to -1. The I coefficient becomes 1 again for the fourth order relationship. The same pattern continues to the thirteenth order, the maximum order incorporated in this study. The correlogram is characterized by short wavelength (covering about two spatial orders, or, 2^1) and large amplitude (ranging between -1 and +1).

Pattern B can be considered a transition from A to C. Pattern C is formed by squared clusters of single cells in pattern A, i.e., every four cells forms an aggregate shaded or blank block. In this case, the correlogram shows a pattern characterized by a wavelength extending about four orders (2^2) with amplitude lower than in pattern A. Then, we consider pattern D as a transition from C to E and compare the arrangement of pattern E with that of pattern C. The same systematic change can be identified, i.e., the wavelength now covers about eight orders (2^3) with amplitude even lower.

Pattern F is the most clustered pattern in this system. Within the fourth order, most cells can only reach cells of the same category. At the fifth order, most cells can reach about equal number of cells of the same category as cells of the opposite category. Beyond the fifth order, most cells can reach fewer cells of the same category and more cells of the opposite category. The resulting correlogram illustrates a system of constant decline in spatial autocorrelation with a wavelength covering the entire span of spatial orders. The magnitude varies roughly within the +1 and -1 range.

This analysis shows that variations in spatial patterns can be detected from the I coefficients of higher-order spatial relationships. Once a spatial correlogram is constructed, its wavelength and amplitude can be used for analyzing spatial patterns.

6. Conclusions

Spatial autocorrelation, equivalent to Tobler's "first law of geography" [27] that "everything is related to everything else but near things are more related than distant things", is an important building block of spatial information theories because it represents the most fundamental structure of spatial relationships. Spatial autocorrelation is a valid measure of what is implicitly assumed in spatial interaction models (e.g., the well-known gravity model) that closer features affect each other more than features that are farther apart. As such, the evaluation of spatial autocorrelation may reveal the topological structure and spatial relationships in the distribution of a spatial phenomenon.

Previous efforts in spatial pattern recognition emphasized the geometric properties of map features. Modeling spatial processes were technically difficult due to the complexity in their spatial relationships. Recent technological advances in geographic information systems (GIS) have made it possible to manipulate large amounts of geographic data and construct the topological structure underlying complicated spatial phenomena [2]. Consequently, spatial autocorrelation can now be effectively evaluated even for higher-order, indirect spatial relationships.

To evaluate spatial patterns based on statistics of spatial autocorrelation, one must consider carefully how the study variable is measured, how geographic units are delineated, and whether or not spatial weighting functions should be incorporated. In addition, this study shows that the quadrat analysis of spatial autocorrelation is not suitable for evaluating the spatial patterns of point features. For area features, the evaluation of spatial autocorrelation must take into account the log-linear relationship between map resolution and spatial autocorrelation.

Certain variations in complicated spatial patterns are not directly discernible and require investigation into the spatial relationships beyond the immediate neighborhood. In this case, correlograms can be constructed for revealing higher-order spatial relationships and detecting systematic variations. The wavelength and amplitude of a correlogram, identified from the cyclic pattern of fluctuation in the I coefficients, provide a useful indicator of the topological structure underlying the spatial relationships in the distribution of the study phenomenon.

Acknowledgments

This study is sponsored by the U.S. Forest Service under grant PSW930022CA. The statements may or may not represent the official opinions of the U.S. Forest Service.

References

1. P.A. Burrough, Sampling designs for quantifying map unit composition, in Mausbach and Wilding (eds), *Spatial Variability of Soils and Landforms*, Soil Science Soc. America Special Pub. #28, 89-125 (1991).
2. Y.H. Chou: Analyzing the spatial autocorrelation of polygonal data in GIS, *Proceedings of Urban and Regional Information Systems Association*, IV:138-148 (1989).
3. Y.H. Chou, R.A. Minnich, L.A. Salazar, J.D. Power, and R.J. Dezzani: Spatial auto correlation of wildfire distribution in the Idyllwild Quadrangle, San Jacinto Mountain, California, *Photogrammetric Engineering and Remote Sensing*, 56:1507-1513 (1990).
4. Y.H. Chou: Map resolution and spatial autocorrelation. *Geographical Analysis*, 23, 228-246 (1991).
5. Y.H. Chou: Management of wildfires with a geographical information system. *International Journal of Geographic Information Systems*, 6, 123-140 (1992).
6. Y.H. Chou: Spatial autocorrelation and weighting functions in the distribution of wildland fires. *International Journal of Wildland Fire*, 2, 169-176 (1992).
7. Y.H. Chou: Critical issues in the evaluation of spatial autocorrelation, *Lecture Notes in Computer Science*, 716: 421-433, Springer-Verlag (1993).
8. Y.H. Chou, R.A. Minnich, and R.A. Chase: Mapping probability of fire occurrence in the San Jacinto Mountains, California. *Environmental Management*, 17, 129-140 (1993).
9. A.D. Cliff and J.K. Ord: 1973, *Spatial Autocorrelation*, Pion (1973).

10. A.D. Cliff and J.K. Ord: *Spatial Processes: Models and Applications*, London: Pion (1981).

11. M.F. Dacey: A review of measures of contiguity for two and K-color maps, in B.J.L. Barry and D.F. Marble (editors), *Spatial Analyses: A Reader in Statistical Geography*, Englewood Cliffs, New Jersey: Prentice-Hall (1965).

12. P.J. Diggle: *Statistical Analysis of Spatial Point Patterns*, Academic Press (1983).

13. R.C. Geary: The contiguity ratio and statistical mapping, *The Incorporated Statistician*, 5, 115-145 (1954).

14. A. Getis: Spatial interaction and spatial autocorrelation: a cross-product approach, *En vironment and Planning A*, 23, 1269-1277 (1991).

15. A. Getis and J.K. Ord: The analysis of spatial association by use of distance statistics, *Geographical Analysis*, 24, 189-206 (1992).

16. D.A. Griffith: *Spatial Autocorrelation: A Primer*, Association of American Geogra phers, Washington, D.C (1987).

17. L.J. Hubert, R.G. Golledge, and C.M. Costanzo: Generalized procedures for evaluating spatial autocorrelation, *Geographical Analysis*, 13: 224-33 (1981).

18. P. de Jong, C. Sprenger, and F. van Veen: On extreme values of Moran's I and Geary's C, Geographical Analysis, 16, 17-24 (1984).

19. P.S. Liu and Y.H. Chou: Quadrat size and spatial autocorrelation in point pattern analysis, *Proceedings of 1994 ESRI User Conference*, Palm Springs, California, 1371-1378 (1994).

20. P.A.P. Moran: The interpretation of statistical maps, *Journal of the Royal Statistical Society*, Series B, 37, 243-51 (1948).

21. P.A.P. Moran: Notes on continuous stochastic phenomena, *Biometrika*, 37, 17-23 (1950)

22. N.L. Oden: Assessing the significance of a spatial correlogram, *Geographical Analysis*, 16: 1-16 (1984).

23. J. Odland: *Spatial Autocorrelation*, Sage (1988).

24. B.D. Ripley: *Spatial Statistics*, Wiley & Sons (1981).

25. R.R. Sokal and N.L. Oden: Spatial autocorrelation in biology 1: methodology, *Biological Journal of the Linnean Society*, 10: 199-228 (1978a).

26. R.R. Sokal and N.L. Oden: Spatial autocorrelation in biology 2: some biological applic ations of evolutionary and ecological interest, *Biological Journal of the Linnean Soci ety*, 10: 229-49 (1978b).

27. W.R. Tobler: A computer movie simulating urban growth in the Detroit region, *Economic Geography*, 46, 234-240 (1970).

28. G.L. Upton and B. Fingleton: *Spatial Data Analysis by Examples, Vol. 1: Point Pattern and Quantitative Data*. Wiley (1985).

29. D. Watenberg: Multivariate spatial correlation: a method for exploratory geographical analysis. *Geographical Analysis* 17: 263-83 (1985).

Towards a Qualitative Theory of Movement

Antony Galton
Department of Computer Science
University of Exeter
Exeter EX4 4PT, UK
email: antony@dcs.exeter.ac.uk

Abstract

The phenomenon of movement arises whenever the same object occupies different positions in space at different times. Therefore a theory of movement must contain theories of time, space, objects, and position. We provide a theoretical basis for describing movement events in terms of the conditions for their occurrence, which refer to the holding or not holding of various positional fluents at different times. For this we need to bring together a formal model of time with a formal model of space. By attending closely to the constraints imposed by continuity on the temporal behaviour of different fluents we develop theory of *dominance*, which enables us to generate *ab initio* the perturbation relation on the full set of positional relations.

1   Introduction

The phenomenon of movement arises whenever the same object occupies different positions in space at different times. This bare definition already suggests the main ingredients of a theory of movement:

1. *A theory of time*, comprising (a) a theory of *times*, i.e., a fundamental set of temporal entities, usually either instants or intervals, which act as loci in the temporal dimension for things to happen in, and (b) a theory of *temporal order* by which the fundamental set of times is endowed with an ordering relation, representing temporal sequence.

2. *A theory of space*, analogous to the theory of time but with points or regions as the fundamental entities, and a more complex theory of spatial ordering.

3. *A theory of objects*, which allows objects to be rigid or non-rigid (the latter being capable of a wider variety of modes of motion than the former), and to have parts, and which can include objects which are not discrete individuals (consider the water in a river—which certainly moves), or which are not even concrete substances at all (e.g., holes and shadows, both capable of movement).

4. *A theory of position*, which brings together the theory of objects with the theory of space, the idea being that, at any one time, each object occupies a certain part of space, which is called its position at that time. The position of an object can be specified as the total region of space occupied by it at a time. An object's position will thus be a region of space precisely congruent, in the geometrical sense, to the body itself.

Our aim is to provide a framework for formalising our "common-sense" knowledge of the world. This is a different enterprise from that of physics, which seeks to go beyond our common-sense view, correcting it where it is in error, and providing a unified explanatory framework. But we should not ignore the fact that a well-developed physico-mathematical theory of motion exists; and ideally our common-sense view should be reconcilable with the physico-mathematical view. A common-sense theory is typically qualitative in nature, whereas the physico-mathematical picture is quantitative. We shall be particularly interested in examining how phenomena such as continuity, which find a natural expression in the quantitative theory, can be expressed when we move over to a qualitative theory. Throughout, it should be borne in mind that all theories, whether qualitative or quantitative, naive or scientific, are idealisations and as such cannot be regarded as absolutely true to the complete underlying reality.

2 Theory of Time

We shall use an instant-based model $T = (T, <)$, which posits a set T of entities called *instants* to be the fundamental set of temporal loci. They are endowed with a relation $<$ of *temporal succession*. We use abbreviations as follows:

$$t \leq u \quad \overset{\Delta}{=} \quad t < u \lor t = u$$
$$t < u < v \quad \overset{\Delta}{=} \quad t < u \land u < v$$

Properties typically ascribed to this relation (cf. (van Benthem 1983)) are as follows (we use the capitalised parts of the names as labels for the axioms):

1. IRReflexivity. No time precedes itself:

$$\forall t \neg (t < t).$$

2. TRAnsitivity. If time t precedes time u which in turn precedes time v, then t also precedes v:

$$\forall t, u, v(t < u < v \rightarrow t < v).$$

3. LINearity. Of any two distinct times, one precedes the other:

$$\forall t, u(t \neq u \rightarrow t < u \lor u < t).$$

4. UNBoundedness. Every time has a time preceding it, and a time which it precedes (so there are no first and last times):

$$\forall t \exists u, v(u < t < v).$$

5. DENsity. Between any two times there is a third (and hence, given IRR and TRANS, infinitely many):

$$\forall t, u(t<u\rightarrow\exists v(t<v<u)).$$

An *interval* is defined by specifying, for each instant, whether it precedes, begins, divides, ends, or follows the interval. It is sufficient to specify just two instants, namely the unique instants at which the interval begins and ends. We could thus define an interval as an ordered pair $\langle t, u \rangle$ of instants, where $t < u$. We call this the *pair model* of an interval. The instants which precede $\langle t, u \rangle$ are precisely the instants which precede t, and the instants which follow $\langle t, u \rangle$ are the ones which follow u.

If $i = \langle t, u \rangle$, we write $Beg(i) = t$ and $End(i) = u$. This notation allows us to refer to the instants marking the beginning and end of any interval we can refer to. We write $Lim(t, i)$ whenever either $t = Beg(i)$ or $t = End(i)$, and say that t *limits* i in this case. We write $t \varepsilon i$, and say that t *divides* i, as an abbreviation for $Beg(i) < t < End(i)$. The motivation for this term is simply that t divides i into two contiguous subintervals $\langle Beg(i), t \rangle$ and $\langle t, End(i) \rangle$.

An interval is often identified with a range of instants, namely all those instants which divide it. For the interval $\langle t, u \rangle$ these are the instants v such that $t < v < u$. The instants in the range are thus members of the set $\{v \mid t < v < u\}$, and it is customary to *identify* the interval with this set. We call this the 'set model' of an interval. Although it is widely used in mathematical modelling, the set model of an interval has nothing to offer over and above the pair model. In particular, the distinction between open and closed intervals which is suggested as a natural extension of the set model does not seem to have any application either to physical time or to any of our common-sense conceptions of time.

There are three distinct temporal successions relation on intervals, and any of them can be taken as fundamental. Their definitions are

- *immediate succession*, in which the first interval 'meets' the second at an instant, without any intervening interval:

$$\langle t, u \rangle | \langle v, w \rangle \quad \overset{\Delta}{=} \quad u = v.$$

The instant at which the intervals meet in this case is u ($= v$).

- *delayed succession*, in which the first interval is separated from the second by an intervening interval:

$$\langle t, u \rangle < \langle v, w \rangle \quad \overset{\Delta}{=} \quad u < v.$$

The intervening interval in this case is $\langle u, v \rangle$.

- *general sucession*, which covers the preceding two cases:

$$\langle t, u \rangle \lhd \langle v, w \rangle \quad \overset{\Delta}{=} \quad u \leq v.$$

General succession is the disjunction of immediate and delayed succession, which are themselves mutually incompatible. We write things like $i|j|k$ and $i < j < k$ with the obvious meanings. We write $i]|j$ to denote the instant at which i meets j (so if $i = \langle t, u \rangle$ and $j = \langle u, v \rangle$ then $i]|j = u$). Note the 'polymorphic' character of the symbol '$<$', which we use both as a relation on instants and as a relation on intervals; it is natural to think of these as the 'same' relation (since with instants, succession is always delayed succesion). Note, however, that our axioms for $<$ are intended to apply only to the relation on instants (in particular, LIN does not hold when the terms of the relation are intervals).

Using our axioms for $<$ we can prove a number of important properties of $|$, as follows (the proofs are easy and are omitted here—but we indicate which axioms we need to use; where no axioms are cited, nothing more is needed than the definitions of $|$ and $<$). In M2, '$\oplus$' is the exclusive 'or' connective.

(M1) $\forall i, j, k, l(i|k \wedge i|l \wedge j|k \to j|l)$ [No axioms]

(M2) $\forall i, j, k, l(i|k \wedge j|l \to i < l \oplus i|l \oplus j < k)$ [LIN]

(M3) $\forall i \exists j, k(j|i|k)$ [UNB]

(M4) $\forall i, j, k, l(i|j|l \wedge i|k|l \to j = k)$ [No axioms]

(M5) $\forall i, j[i|j \to \exists k \forall l[(l|k \leftrightarrow l|i) \wedge (k|l \leftrightarrow j|l)]]$ [No axioms]

(M6) $\forall i \exists j, k[j|k \wedge \forall l[(l|i \leftrightarrow l|j) \wedge (i|l \leftrightarrow k|l)]]$ [DEN]

Note that M4 and M5 together imply that if i meets j then there is a unique interval which meets whatever j meets and is met by whatever i is met by (existence is given by M5, uniqueness by M4). We denote this interval $i + j$; it is the interval which begins when i begins and ends when j ends, and hence spans the entire time taken up by i and j together.

We can define further relations on intervals as follows. We use infix notation after the fashion of (Allen 1983):

$$iOj \quad \triangleq \quad \exists i', j', k(i = i' + k \wedge j = k + j')$$

$$iSj \quad \triangleq \quad \exists k[j = i + k]$$

$$iDj \quad \triangleq \quad \exists k, l[j = k + i + l]$$

$$iFj \quad \triangleq \quad \exists k[j = k + i]$$

$$i \sqsubset j \quad \triangleq \quad iSj \vee iDj \vee iFj$$

$$i \sqsubseteq j \quad \triangleq \quad i \sqsubset j \vee i = j$$

(O, S, D, F are read as 'overlaps', 'starts', 'is during', and 'finishes', respectively.) The relations $\sqsubseteq$ and $\sqsubset$ are analogous to the set-theoretic relations of subset ($\subseteq$) and proper subset ($\subset$), and indeed could be defined to identical to them if the set model of an interval were to be adopted.

Following Allen, it is often regarded as more satisfactory to base one's temporal model on intervals as the fundamental temporal units rather than instants. Allen and Hayes (1985) do this, using equivalents of (M1)–(M6) as their axioms for immediate succession. They also show how instants can be defined in their system; the resulting system satisfies our axioms for instants. It follows that it does not matter whether we base our temporal model on instants or intervals. By choosing appropriate definitions we will end up with equivalent systems starting from either choice. It is certainly convenient to be able to use both instant and interval notations!

The model of time presented here is dense, because of axiom M6. The obvious mathematical model for dense time, satisfying all our axioms, is to represent instants by real numbers, with temporal succession represented by the 'less than' relation. We could either choose the set of all real numbers to represent T, or any dense subset of them, such as the rational numbers. Dense time implies that there is no lower limit to the length of an interval, and while this is a satisfactory idealisation for many purposes, sometimes there are good reasons not to accept it. For example, in a context in which intervals are only known through observation and measurement, there is an effective lower limit to the length of an interval, corresponding to our chronometrical resolving power; or again, one may be concerned with physical processes which always require a certain minimum duration in which to occur.

For these reasons, an alternative *discrete* model of time is often preferred: mathematically, this is tantamount to representing instants as integers rather than real numbers. In our axiomatisation, it is necessary to replace the axiom DEN by

6. DISCreteness. If time t precedes time u then there is an earliest time v which t precedes and a latest time v' which precedes u:

$$\forall t, u(t < u \rightarrow \exists v \forall w(t < w \leftrightarrow v \leq w) \wedge \exists v' \forall w(w < u \leftrightarrow w \leq v)).$$

Note how we say that v is the earliest time which t precedes: an arbitrary time w is preceded by t if and only if it is either equal to or preceded by v, so that the only times which t precedes are v and anything which v precedes; and analogously for the latest time which precedes t.

In discrete time, the elements of T are *atomic intervals*, or *moments*. A general interval in discrete time is the concatenation of one or more consecutive moments, a natural measure of its duration being the number of moments involved. This is quite different from dense time, where it makes no sense to speak of 'consecutive instants', and duration has to be introduced as an additional primitive, not derivable from the temporal order alone. As will be seen below, we still need the notion of 'instant' in discrete time: we shall need to speak of the instant at which two consecutive moments meet. The reason for this will become apparent when we consider the occurrence conditions for an instantaneous event in the next section.

In the rest of this paper we shall mostly confine our attention to dense rather than discrete time. None the less, much of what we say about movement in the next section can be adapted fairly straightforwardly to the latter case. On the other hand, the later material on continuity and the theory of dominance applies specifically to dense time, and does not make much sense in the discrete case.

3   Movements and their Occurrence Conditions

We use the RCC-8 system (Randell, Cui and Cohn 1992) for specifying relations between regions. There are eight basic relations, as follows:

DC	A is disconnected from B
EC	A is externally connected to B
PO	A partially overlaps B
EQ	A is equal to B
TPP	A is a tangential proper part of B
NTPP	A is a non-tangential proper part of B
TPPI	A has B as a tangential proper part
NTPPI	A has B as a non-tangential proper part

These eight relations correspond closely to the eight relations determined by Egenhofer's 4-intersection method (Egenhofer 1991). Regions will be mainly of interest to us as possible positions for movable bodies. The position of a body can be given, with greater or less precision, by the RCC-8 relation which it bears to some known region, such as the position of another body. To be able to talk about motion, we need only relativise this to time.

We take as our fundamental notion for the analysis of change the idea of a *fluent*. A fluent can take different values at different times. If f is a fluent and a is a value it can take, then $f = a$ is a proposition that can be true or false at different times, in other words a *Boolean fluent*, or *state*. Likewise, if f_1 and f_2 are fluents, and R is a relation which may hold between values that they can take, then $R(f_1, f_2)$ is also a state.

We write $Holds\text{-}at(S, t)$ to indicate that state S holds at instant t, and $Holds(S, i)$ to indicate that S holds throughout the interval i (Allen 1984, Galton 1990). These two notations are connected by the rule

$$Holds(S, i) \leftrightarrow \forall t{\in}i\, Holds\text{-}at(S, t),$$

which says that a state holds throughout an interval if and only if it holds at every instant which divides the interval. This could be taken as a definition of $Holds$ in terms of $Holds\text{-}at$, if desired. An immediate consequence is Allen's rule

$$Holds(S, i) \wedge i \sqsubseteq j \rightarrow Holds(S, j),$$

which says that a state holds throughout every subinterval of any interval throughout which it holds.

We write $S \sqcap S'$ to refer to the state which holds when and only when both S and S' hold (*state-conjunction*), and $-S$ to refer to the state which holds when and only when S fails to hold (*state-negation*). Formally, they obey the rules

$$
\begin{aligned}
Holds\text{-}at(S \sqcap S', t) &\leftrightarrow Holds\text{-}at(S, t) \wedge Holds\text{-}at(S', t), \\
Holds\text{-}at(-S, t) &\leftrightarrow \neg Holds\text{-}at(S, t).
\end{aligned}
$$

We write $pos(a)$ to denote the position of body a. This is a region having exactly the same shape and size as a, so that a can fit into it with no space left over. Since the position of a can change over time, $pos(a)$ is a fluent. We write

$$Holds(R(pos(a), r), i)$$

to indicate that throughout interval i, the position of a bears the RCC relation R to region r.

To handle movement we need a formalism for referring to events, since a movement is an event, not a state. We write $Occurs(e, i)$ to indicate that an event of type e occurs over interval i. Allen's rule for $Occurs$ is

$$Occurs(e, i) \wedge j \sqsubset i \rightarrow \neg Occurs(e, j),$$

which says that an event does not occur over any proper subinterval of an interval over which it occurs. (So events can be described as *unitary*, in contrast to states, which are *homogeneous*.) This rule can be regarded as a constraint on the allowable event-types. We shall also have cause to talk about *instantaneous events*. For these we write $Occurs\text{-}at(e, t)$ to indicate that an event of type e occurs at the instant t (see Galton (1994) for a detailed treatment of instantaneous events).

Our paradigm for analysing movement will be to specify a movement event e in terms of its *occurrence conditions*, that is in terms of a formula of one of the forms

$$Occurs(e, i) \quad \overset{\Delta}{=} \quad \cdots$$
$$Occurs\text{-}at(e, t) \quad \overset{\Delta}{=} \quad \cdots$$

where the right-hand side is a formula not containing e (Galton 1993, Galton 1994). We shall consider a number of examples.

Suppose we wish to define what it is for a to *move* from position r_1 to position r_2 over the interval i. A natural first attempt might be to stipulate that a must be at r_1 throughout some interval j which meets i, and at r_2 throughout some interval which i meets. In order to ensure that the event occurs over the whole interval i, and not some proper subinterval (in accordance with Allen's rule for $Occurs$), we should add that a is not at either r_1 or r_2 at any time during i itself. This gives us the definition:

$$Occurs(move(a, r_1, r_2), i) \quad \overset{\Delta}{=}$$
$$\exists j, k(j|i|k \wedge Holds(pos(a) = r_1, j) \wedge Holds(pos(a) = r_2, k)) \wedge$$
$$Holds(pos(a) \neq r_1, i) \wedge Holds(pos(a) \neq r_2, i).$$

This definition is adequate if we do not allow states to be said to hold at instants, for example if our model of time is discrete; but it will not do in general. For suppose a moves from position r_0 to r_3, passing through positions r_1 and r_2, in that order, but without stopping at either of them. Then there is no interval throughout which a is at either r_1 or r_2, yet a still moves from r_1 to r_2 over the interval between the times at which it is at these positions. We must replace our definition of *move* by:

$$Occurs(move(a, r_1, r_2), i) \quad \overset{\Delta}{=}$$
$$Holds\text{-}at(pos(a) = r_1, Beg(i)) \wedge Holds\text{-}at(pos(a) = r_2, End(i)) \wedge$$
$$Holds(pos(a) \neq r_1, i) \wedge Holds(pos(a) \neq r_2, i).$$

Note that this definition subsumes the previous one, since if a is at position r_1 over an interval meeting i, then by continuity it must be at r_1 at the beginning of i, and likewise with r_2 at the end.

Suppose next that we wish to say that a *enters* region r over interval i. A natural first attempt would be to postulate that a must be just outside (EC) r throughout some interval j which meets i, and just inside (TPP) r throughout some interval k which i meets; during i itself, a must be partly inside and partly outside (PO) r:

$$Occurs(enter(a,r),i) \triangleq$$
$$\exists j, k (j|i|k \wedge Holds(EC(pos(a),r),j)) \wedge$$
$$Holds(PO(pos(a),r),i) \wedge Holds(TPP(pos(a),r),k)).$$

As before, we can argue that a does not need to be EC or TPP to r for more than an instant: consider the case where a approaches r from a distance and enters it without pausing in either the EC or the TPP positions. A more general definition is therefore:

$$Occurs(enter(a,r),i) \triangleq$$
$$Holds\text{-}at(EC(pos(a),r),Beg(i)) \wedge$$
$$Holds(PO(pos(a),r),i) \wedge Holds\text{-}at(TPP(pos(a),r),End(i))).$$

Suppose finally we wish to characterise the event of two objects' coming into *contact*; this is an instantaneous event. It can happen at the meeting point of two intervals i, j, such that a is separated from b throughout i and touching b throughout j:

$$Holds(DC(pos(a),pos(b)),i) \wedge Holds(EC(pos(a),pos(b)),j).$$

The event itself occurs at the instant $i][j$. However, this does not cover the case where a moves towards b and then enters it. Then a first makes contact with b at $i][j$, where

$$Holds(DC(pos(a),pos(b)),i) \wedge Holds(PO(pos(a),pos(b)),j)).$$

Here we can infer, by continuity, that a is EC to region r at the instant t. Even this does not cover the case where the objects move apart as soon as they have touched. If the moment of touching is $i][j$, then we have

$$Holds(DC(pos(a),pos(b)),i) \wedge Holds(DC(pos(a),pos(b)),j)$$

but this does not tell us that they ever touched. Hence we have to bring in explicit reference to the state of affairs holding at the instant $i][j$ itself. Our fully general definition, subsuming the others, will therefore be

$$Occurs\text{-}at(connect(a,r),t) \triangleq$$
$$\exists i(t = End(i) \wedge Holds(DC(pos(a),r),i) \wedge Holds\text{-}at(EC(pos(a),r),t))$$

This says that a is touching b at the instant which ends an interval throughout which a is separated from b: that is the instant at which a makes contact with b. Nothing need be said about what state holds after that instant.

Note that in discrete time, the same analysis will apply, but it requires us to locate the instantaneous event of making contact at the instant where two intervals meet. This will always be expressible as the meeting point of two atomic intervals (moments). This example shows why it is necessary to have instants as well as moments in a discrete model; but they do not have to be postulated separately, since the existence of the instants follows of necessity from the fact of each moment's immediately preceding the next.

4 Continuity

We have mentioned continuity several times; in this section we examine more closely what it entails. Two different notions of continuity are suggested by our experience of the physical world. On the one hand there is the *continuity of space and time*, which we are accustomed to regard as "seamless" continua admitting arbitrarily fine subdivision and no "gaps". On the other hand, there is the *continuity of change*, according to which measurable physical magnitudes such as the position, velocity, acceleration, or temperature of a body vary smoothly in time, again presenting an appearance of seamlessness, with no instantaneous jumps.

Notice here that we only refer to the *appearance* of seamlessness. Of course, many phenomena which appear continuous at one scale are seen to be discrete when observed more closely. An obvious example is afforded by the atomic structure of matter. Another example, with a temporal dimension, is provided by the illusion of continuity produced by the rapid succession of frames in a cine-film. As mentioned in the introduction, we are working with idealisations of the world we experience, and our purpose in this section is to examine closely the relationship between those idealisations in which phenomena are naturally represented as continuous (e.g., using the real number system) and those—such as the qualitative system of RCC-8—in which continuity does not find an obvious or natural expression.

Continuity of space and time is represented mathematically by modelling space and time in terms of the ordered set $(\mathbf{R}, <)$ of real numbers. The time dimension is represented by $\mathbf{R}$ itself, each number corresponding to a temporal instant; space is represented by the Cartesian product $\mathbf{R}^3$, each triple of real numbers corresponding to a single spatial point. The special features of $(\mathbf{R}, <)$ which suit it for this role are

- *Density*: between any two real numbers there is a third, and hence, by iteration, infinitely many. (Cf. our axiom DEN.)

- *Dedekind completeness*: if $\mathbf{R}$ is partitioned into two disjoint subsets L and R such that every member of L is less than every member of R, then either L has a greatest member, or R has a least member, but not both.

Note that neither the integers nor the rational numbers possess both these properties. The integers fail on both counts, the rationals on only the second. In the absence of Dedekind completeness, the temporal sequence admits 'gaps', at which the totality of instants can be divided into two parts L and R without there being a unique instant to mark the point of division; this is felt to be incompatible with continuity.

Temporal intervals and spatial regions, on this picture, must be specified in terms of the relationships they bear to the instants or points that have been identified with real numbers or triples thereof. From a physical point of view, we require each point (or instant) P to bear exactly one of the following relations to each region (or interval) R:

- P is *inside R*;

- P is *on the boundary of R*;

- P is *outside R*.

We shall regard an interval/region as entirely determined once it is known for every instant/point whether it lies inside, outside or on the boundary of the interval/region.

Note that this criterion of identity for intervals and regions is blind to the open/closed issue. There is no physical significance to the idea of a membership relation that is separate from the inside/outside/boundary trichotomy. This trichotomy *excludes* the idea that a region can 'contain its own boundary', there being no separate notion of containment apart from 'inside'. Nothing is gained by identifying a region either with the set of its interior points or with the set of its interior points plus boundary points, since a set is an abstract notion such that the relation it bears to its members is *sui generis* and not to be confused with the relation between a whole and its parts or between a region and the points of its interior or boundary.

The second kind of continuity, continuity of change, is modelled mathematically by representing measurable magnitudes as functions (of time) that are continuous in the special mathematical sense, namely:

> The function $f : \mathbf{R} \to \mathbf{R}$ is continuous at the point $x_0 \in \mathbf{R}$ so long as, for every real number $\epsilon > 0$, there exists a real number $\delta > 0$ such that for every $x \in \mathbf{R}$, if $x_0 - \delta < x < x_0 + \delta$ then $f(x_0) - \epsilon < f(x) < f(x_0) + \epsilon$.

The intuitive notion of continuity demands that a function should be continuous at *every* point: so there are 'no jumps'. This excludes certain well-known 'pathological' cases, e.g., a function which is continuous at irrational points but not at rational ones.

Granted that the mathematical model is well suited to modelling continuous change, there is now a problem about modelling *discontinuous* change. This is the classical 'dividing instant' problem, which goes back to Plato and Aristotle. It may be phrased as follows: Let S be a state such that $Holds(-S, i) \wedge i|j \wedge Holds(S, j)$; can we determine whether $Holds\text{-}at(S, i][j)$?

An easy instance of this problem is the following: let S be the state $pos(a) = r$. We are supposing that an interval throughout which the position of a is different from r meets an interval throughout which a is in position r:

$$Holds(pos(a) \neq r, i) \wedge i|j \wedge Holds(pos(a) = r, j).$$

For $pos(a) \neq r$, it is enough that the space occupied by some part of a is disjoint from r—in particular, a does not have to be right outside r to count as not being at r. If the motion of a is continuous, then for a to move to r from any position r' distinct from r it must pass through a range of intermediate positions forming a path from r' to r.

Suppose, then, that a is not at r at the instant $t = i][j$. Let u be any instant dividing j: since a is at r thoughout j, it is at r at u. So a moves from r' at t to r at u. Hence, by continuity, it must occupy positions along a path joining r' to r at some times in the interval $\langle t, u \rangle$. But this is a subinterval of j. Hence there are times during j when a is not at r—contradicting our assumption that a is at r throughout j. It follows that a must be at r at instant t as well. There is an asymmetry between the states represented by 'a is at r' and 'a is not at r', which could be expressed by saying that the former must be true on closed sets of instants, the latter on open sets (compare the *continuity rule* of (Williams 1990)). In the somewhat infelicitous terminology of Galton (1990), 'a is at r' is a *state of position* whereas 'a is not at r' is a *state of motion*.

In general, if the state S describes the state of the world with respect to some continuously variable property (such as the position of a body), then the above type of argument can be used to determine a solution to the dividing instant problem. The problem becomes more vicious when S is essentially discontinuous, i.e., when there are no intermediate states between a state of the world described by S and a state of the world described by $-S$.

An example which is often cited in this connection concerns a lamp which may be on or off. Suppose the lamp is off over the interval $i = \langle t_1, t_2 \rangle$ and on over $j = \langle t_2, t_3 \rangle$. The problem is whether the lamp is on or off at the instant $t_2 = i][j$. There are a number of different responses one might make here:

1. The lamp is neither on nor off at t_2. This response comes in two flavours:

    (a) The proposition 'The lamp is on at t_2' is ill-formed—propositions have truth-values over intervals, not at instants.

    (b) The proposition does have a truth-value—but it is neither true nor false. Instead we use a three-valued logic, and assign the third truth value to this case.

2. The lamp is both on and off at t_2. This is bizarre, but I have heard it seriously suggested.

3. The lamp is off at t_2. To justify this, we treat the state of the lamp as continuously variable: we postulate a real variable l, taking values in the range $[0, L]$ and representing the amount of illumination from the lamp. Then 'The lamp is off' means $l = 0$, while 'The lamp is on' means $l > 0$ (this case is exactly parallel to that of our body a moving *away* from the position r).

4. The lamp is on at t_2. This can be similarly justified by means of a different interpretation of 'on' and 'off': this time say that 'The lamp is off' means $l < L$, and 'The lamp is on' means $l = L$. This case is parallel to a moving *to* position r, exactly as in our first example.

5. Finally, one might reject the premises, and deny that it is possible for an interval over which the lamp is off to be immediately followed by an interval over which it is on. To justify this, one must deny that 'The lamp is off' is the negation of 'The lamp is on'. We can do this by defining 'The lamp is off' to mean $l = 0$,

and 'The lamp is on' to mean $l = L$. In order for the lamp to change from being off to being on, it has to pass over the range of values for which $0 < l < L$, and this must take time. What happens is that the lamp is off over some interval $\langle t_1, t_2 \rangle$, it is on over an interval $\langle t_3, t_4 \rangle$, and neither on nor off over the interval $\langle t_2, t_3 \rangle$, which may be of extremely short duration. Of course the dividing instant problem arises again with respect to the instants t_2 and t_3, but this time it is of the relatively harmless continuous variety that we have dealt with already: the lamp is off (i.e., $l = 0$) at t_2, and on (i.e., $l = L$) at t_3.

Response 1(a) is interesting because it has become prominent in AI, largely owing to the influence of Allen. Allen's work was prefigured by that of Hamblin in philosophy (Hamblin 1969, Hamblin 1971). Hamblin was very much motivated by the dividing instant problem. Here we highlight the fact that there is a radical incompatibility between the view of the world espoused by Allen and Hamblin and that implicit in the standard mathematical view which models continuity using the real numbers.

This incompatibility does not only arise from the dividing instant problem. A more serious problem concerns what I call *instantaneous tenure* (Galton 1994). By *tenure* of a state I mean an event which consists of that state's holding for a certain time, flanked by times at which it does not hold. For example, if the lamp is off over interval $\langle t_1, t_2 \rangle$, on over $\langle t_2, t_3 \rangle$, and off again over $\langle t_3, t_4 \rangle$, then an event of tenure of the state of the lamp's being on occurs on the interval $\langle t_2, t_3 \rangle$. Hamblin explicitly denied the possibility that a tenure event could be instantaneous, referring to the impossibility of a red book turning green just at midnight and then immediately becoming red again, so that there is only a single instant at which it is green. We may grant Hamblin his example, but what of the ball tossed up into the air: surely there is an instant, and only an instant, at which it is moving neither up nor down? On the standard mathematical view, this is inescapable. Aristotle would have said that in this situation what actually happens is that the ball comes to rest at the highest point of its trajectory, *stays there for a short interval*, then begins descending. If this is correct, then there is a serious mismatch between what actually happens and the mathematical apparatus standardly used by physicists to describe what happens in cases like this. (Note that this issue cannot, in principle, be settled by observation or measurement, since the length of the supposed interval might always be below the threshold of discrimination; once again, the matter at stake is one of finding a workable idealisation.)

Even more seriously, consider the case of a body a moving uniformly along a line from position p to position r, passing through position q on the way. This is the example used to criticise Allen in (Galton 1990). Suppose the whole movement takes up interval $\langle t_1, t_3 \rangle$. The part of the movement during which a moves from p to q must occupy some initial interval of this, say $\langle t_1, t_2 \rangle$. The remainder of the interval, namely $\langle t_2, t_3 \rangle$, is taken up by the movement from q to r. It is natural to say that a is at q at instant t_2. On the other hand, it is not at q at any other time during $\langle t_1, t_3 \rangle$. We thus have instantaneous tenure of the state of a's being at q.

We cannot dismiss this by claiming that it does not matter what we say is the case at t_2, for if we are not prepared to say that a is at q at t_2, we must deny that a is at q at all during the interval $\langle t_1, t_3 \rangle$, despite the explicit supposition that a passes through q. We should have to allow that a body can *pass through* a position without *being* there.

The alternative is to say that if a passes through q then it must spend some time there; but this destroys continuity. For there is not enough time for a to spend a positive duration at each of a nondenumerable set of positions along the path from p to r, so either we have to deny that the path contains nondenumerably many positions (which entails denying Dedekind completeness, and hence the continuity of space) or we must deny that a occupies every position along the path at some time during its move (which entails denying continuity of movement).

If we are committed to the beliefs that (a) all motion must be continuous, and (b) a body cannot occupy a position without spending some positive length of time there, then the only way out is to deny that motion is possible at all. Perhaps something of this sort was what motivated Zeno's arrow paradox, though of course Zeno was working with a conception of continuity that had not yet been formulated in terms of the Dedekind property.

5 The Theory of Dominance

When giving the occurrence conditions for *connect*, we noted that if we have DC over i and PO over j, where i meets j, then we must, by continuity, have EC at the instant $i][j$. In this section we explore further the role that continuity has to play in arguments of this kind. Essentially, we shall look at the structure of the state-space consisting of the RCC-8 relations from the point of view of its being a qualitative projection of an underlying continuous space.

We say that a state S' is a *perturbation* of state S if and only if one of these states can hold at an instant which limits an interval throughout which the other state holds, i.e., at least one of the following situations can occur:

$$Holds(S, i) \land Lim(t, i) \land Holds\text{-}at(S', t)$$
$$Holds(S', i) \land Lim(t, i) \land Holds\text{-}at(S, t)$$

If only the first of the above situations can occur, then we shall say that state S' *dominates* state S, written $S' \succ S$, whereas if only the second can occur, S' is dominated by S, written $S' \prec S$. The motivation for the term 'dominance' is as follows: suppose that S holds throughout i and that S' holds throughout j, where i meets j. Then we can think of S and S' as being in competition as to which of them, if either, should hold at the instant $i][j$. It is the dominant state which wins.

To illustrate these ideas, we divide the state-space for a single real variable into the three qualitative states positive, negative and zero, abbreviated P, N, and Z respectively. Assuming continuous variation, we have

$$Holds(Z, i) \land Lim(t, i) \rightarrow Holds\text{-}at(Z, t).$$

This is because if the value of the variable is non-zero at t, then by continuity it must assume values arbitrarily close to that non-zero value, and hence non-zero themselves, at all times sufficiently close to t, and this is incompatible with its being zero throughout an interval beginning or ending at t. This means that neither of the situations

$$Holds(Z, i) \land Lim(t, i) \land Holds\text{-}at(P, t)$$
$$Holds(Z, i) \land Lim(t, i) \land Holds\text{-}at(N, t)$$

can occur. On the other hand, if our variable increases uniformly over the interval $\langle -1, 1 \rangle$ so that its value is 0 at time 0, then we have

$$Holds(N, \langle -1, 0 \rangle) \wedge Lim(0, \langle -1, 0 \rangle) \wedge Holds\text{-}at(Z, 0)$$
$$Holds(P, \langle 0, 1 \rangle) \wedge Lim(0, \langle 0, 1 \rangle) \wedge Holds\text{-}at(Z, 0).$$

It follows that Z dominates both N and P. Moreover, N and P are not perturbations of each other (and hence neither can dominate the other), since it is not possible for the value of the variable to change from N to P or vice versa without passing through Z. The complete dominance relation on the set $\{N, Z, P\}$ is therefore given by $Z \succ N \wedge Z \succ P$.

While it is quite possible two specify two states which are mutual perturbations although neither dominates the other, we shall only be concerned with state spaces in which all perturbation relations involve dominance in one direction or other, as in the example above. The general definition is:

A *dominance space* is a pair $(S, \succ)$, where

- S is a finite set of *states*
- $\succ$ is an irreflexive, asymmetric relation on S, where $S' \succ S$ is read "S' dominates S",

and the following *temporal incidence rule* holds

$$\forall S, S' \in S(Holds(S, i) \wedge Lim(t, i) \wedge Holds\text{-}at(S', t) \rightarrow S' \succeq S)$$

(where $S' \succeq S$ abbreviates $(S' \succ S) \vee (S' = S)$).

The temporal incidence rule ensures that if a state S holds throughout an interval i then the only states apart from S itself which can hold at the limits of i are ones which dominate S.

The key fact about dominance spaces is that a set of such spaces can be combined into a composite dominance space, as shown by the following theorem.

Theorem 1. Let $(S_1, \succ_1), (S_2, \succ_2), \ldots, (S_n, \succ_n)$ be dominance spaces. Then

$$(S_1 \times S_2 \times \cdots \times S_n, \succ)$$

is also a dominance space, where $\succ$ is defined by the rule

$$\vec{S} \succ \vec{S'} \text{ if and only if } S_i \succeq S'_i \text{ for } i = 1, \ldots n \text{ and } \vec{S} \neq \vec{S'},$$

where $\vec{S}$ denotes the ordered n-tuple $(S_1, S_2, \ldots, S_n)$, understood as representing the state-conjunction $S_1 \sqcap S_2 \sqcap \cdots \sqcap S_n$.

Proof. First, since $S_1, S_2, \ldots, S_n$ are all finite, so is $S_1 \times S_2 \times \cdots \times S_n$. Next, we must check the properties of $\succ$. That $\succ$ is irreflexive follows immediately from the condition $\vec{S} \neq \vec{S'}$ appearing in the definition. To show that $\succ$ is asymmetric, suppose that both $\vec{S} \succeq \vec{S'}$ and $\vec{S'} \succeq \vec{S}$. Then for $i = 1, 2, \ldots, n$, both $S_i \succeq S'_i$ and $S'_i \succeq S_i$, so by asymmetry of $\succ_i$, $S_i = S_i'$. It follows that $\vec{S} = \vec{S'}$. Hence if we have $\vec{S} \succ \vec{S'}$ then we do *not* have $\vec{S'} \succ \vec{S}$.

Finally, we must check the temporal incidence rule. Suppose that

$$Holds(\vec{S}, i) \wedge Lim(t, i) \wedge Holds\text{-}at(\vec{S'}, t).$$

Then for $k = 1, 2, \ldots, n$ we have

$$Holds(S_k, i) \wedge Lim(t, i) \wedge Holds\text{-}at(S_k', t),$$

which by the kth temporal incidence rule implies that $S_k' \succeq_k S_k$. We thus have

$$(S_1' \succeq_1 S_1) \wedge (S_2' \succeq_2 S_2) \wedge \cdots \wedge (S_n' \succeq_n S_n),$$

i.e., $\vec{S'} \succ \vec{S}$ as required. □

The importance of this theorem is that it enables us systematically to build complex dominance spaces from simpler ones. We can start with very simple spaces where the dominance relations are easy to verify 'by hand', and then build up to more complicated cases where these relations are less straightforward to determine. Since we are dealing with dominance spaces, information regarding dominance also provides complete information about perturbation as well. In the next section we illustrate this with two examples of particular relevance to the present study. Further examples can be found in (Galton 1995).

6 Applications to Spatial Change

We can construct RCC-8 as a dominance space by noting that the RCC-8 relations are uniquely determined by knowing (a) whether all, some, or none of A is inside B, (b) whether all, some, or none of B is inside A, and (c) whether A and B share any boundary points. (Here we are using the word 'some' in its exclusive sense, i.e., some but not all). This representation is related, but not identical, to Egenhofer's 4-intersection. The three factors can be represented numerically as follows:

(a) Let

$$\alpha = \frac{\text{area of A inside B}}{\text{area of A}}.$$

Then the three states are $\alpha = 0$ (none of A is inside B), $0 < \alpha < 1$ (some of A is inside B), and $\alpha = 1$ (all of A is inside B). The state 'some' is dominated by both 'none' and 'all'.

(b) As with (a), but with A and B swapped around:

$$\beta = \frac{\text{area of B inside A}}{\text{area of B}}.$$

(c) Let γ be the minimum distance between a boundary point of A and a boundary point of B. Then we have two states $\gamma = 0$ (i.e., A and B share at least one boundary point) and $\gamma > 0$ (A and B do not share any boundary point). The state $\gamma = 0$ dominates the state $\gamma > 0$.

PROPORTION OF B INSIDE A

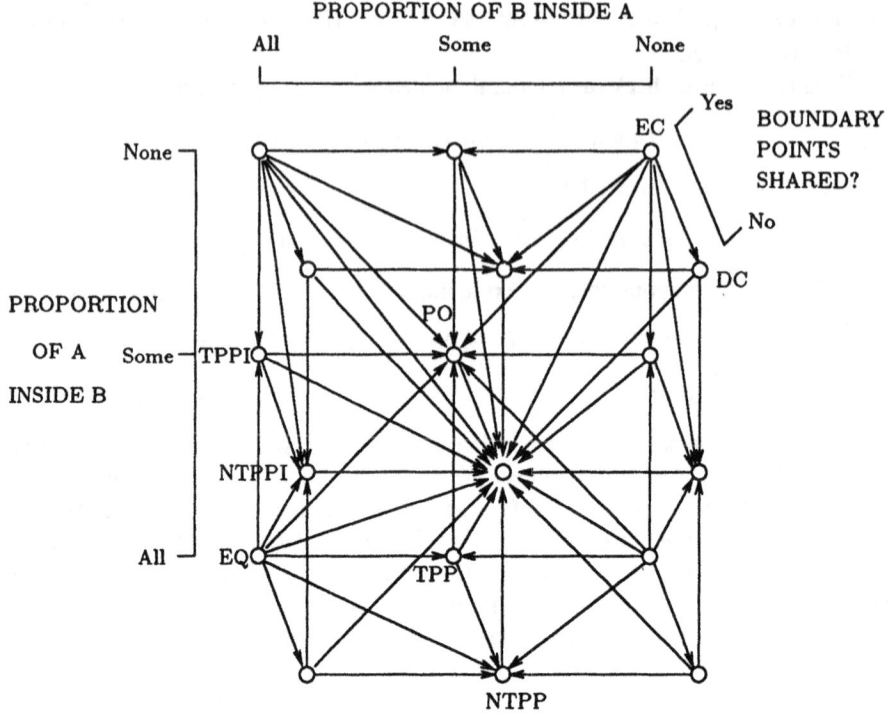

Figure 1: *Generation of the dominance space for the RCC-8 spatial relations.*

These three little dominance spaces would, if they were mutually independent, combine to give a space with $3 \times 3 \times 2 = 18$ elements. The dominance relations are computed using Theorem 1, as shown in Figure 1, where the arrows indicate the direction of dominance.

In fact the three spaces are not independent; all but 8 of the elements are impossible (e.g., if none of A is inside B then none of B can be inside A either). The eight possible combinations correspond exactly with the spatial relations in RCC-8, as follows[1]:

	B in A	A in B	Share boundary
DC	none	none	no
EC	none	none	yes
PO	some	some	yes
TPP	some	all	yes
NTPP	some	all	no
TPPI	all	some	yes
NTPPI	all	some	no
EQ	all	all	yes

[1]Note that this analysis assumes that each region consists of a single connected component.

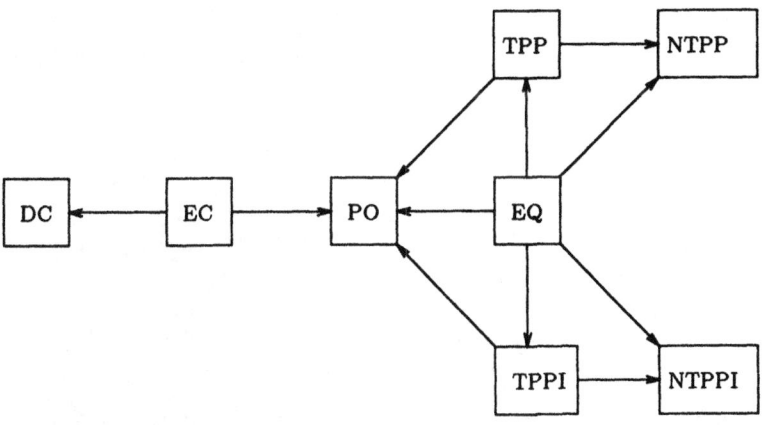

Figure 2: *The dominance space for the RCC-8 spatial relations.*

In Figure 1 the nodes corresponding to these eight possibilities are labelled with the appropriate RCC-8 designations. A clearer view of the dominance space can be obtained by deleting all the 'impossible' nodes and rearranging those that remain as in Figure 2. It is reassuring to note that this diagram is identical, apart from the addition of the dominance arrows, to that of (Randell et al. 1992)!

From this figure we can read off, for example, that a transition between DC and PO must involve passing through EC, though since this dominates both its neighbours the intermediate state need only hold for an instant. We can also see that the state TPP can hold for an single instant in the context of a transition between PO and NTPP (since it dominates both of these states), but if it holds in the context of a transition between EQ and NTPP then it must do so for an interval (since it is dominated by EQ). These conclusions are in conformity with the demands of continuity.

For a more complicated example, we consider the position of a non-rigid body in relation to two fixed disconnected regions. There are eight relations which the position of the body can stand in with respect to either region individually. If its position with respect to one region were free to vary independently of its position with respect to the other, this would give us a state space containing 64 product relations. In fact only 31 elements of the full Cartesian product can be realised. In particular, if the position of the body in relation to one region is EQ, TPP, or NTPP, then it can only be DC with respect to the other. This is because the two regions are themselves DC. This gives us the six composite relations (EQ,DC), (TPP,DC), (NTPP,DC), (DC,EQ), (DC,TPP), and (DC,NTPP). The remaining five simple relations are genuinely independent for each of the two positions, giving a further 25 composite relations. All 31 relations are portrayed pictorially in Figure 3, with the dominance—and hence perturbation—relations as determined from Theorem 1.

Using this figure, we can observe that, for example, if a body is EC to region 1 and DC to region 2, then in order to become DE to 1 and EC to 2 it must first become either

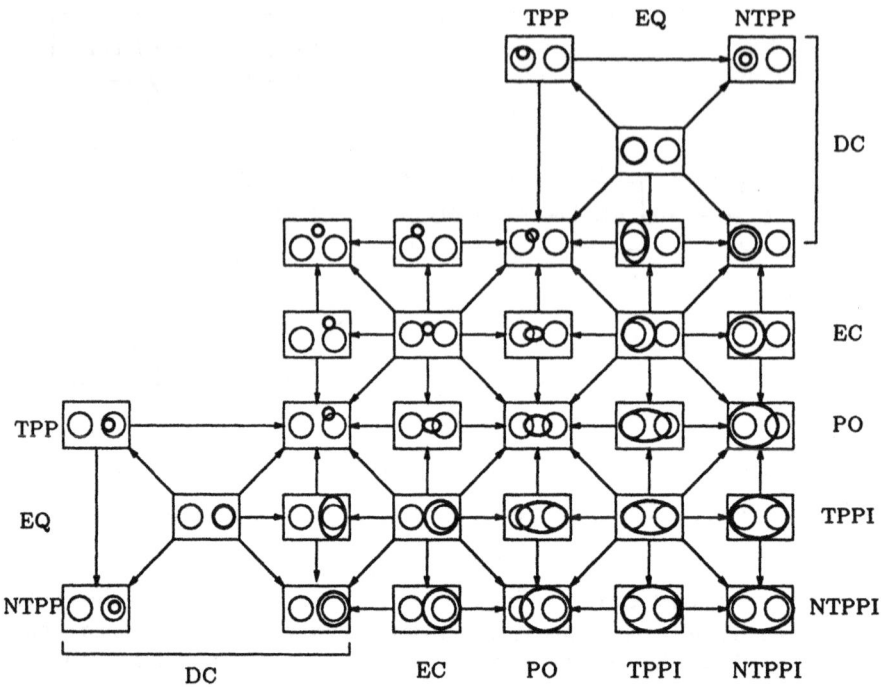

Figure 3: *Position of a non-rigid body in relation to two fixed regions.*

DC to both or EC to both—the latter case being possible for a single instant, the former requiring an interval.

7 Concluding Remarks

We have provided a theoretical basis for describing movement events in terms of the conditions for their occurrence, which consist of the holding or not holding of various positional fluents at different times. To do this we have brought together a formal model of time, based on a set of instants endowed with a total ordering, with a formal model of space based on regions. By attending closely to the constraints imposed by continuity on the temporal behaviour of different fluents we developed a theory of *dominance* by which we are able to generate from first principles the neighbourhood relation on the RCC-8 set of qualitative positional relations, as well as on more complicated state-spaces such as the possible positions of a non-rigid body in relation to two fixed regions.

We advocate the use of dominance as providing a systematic tool for investigating the structure of qualitative state-spaces derived from an underlying space that is conceptualised as continuous. As such, the theory of dominance is not only of interest in the spatial domain, although that does provide a highly appropriate testing ground, with perhaps the richest set of particular examples, for the theory. There is considerable scope for further research in investigation of the dominance structure of other qualitative spaces in the literature on spatial reasoning, as for example the orientation-based system of (Freksa 1992).

With regard to the work presented in this paper, an intriguing open question remains. Our theory of dominance depends on our being able to speak meaningfully of a state holding at an instant. We used the theory to help us generate the perturbation and dominance relations for complex state-spaces. The resulting diagrams can be read as supplying information about perturbation only, ignoring the dominance information. As such, they can be recognised as valid *even if one does not accept the premiss that states can be said to hold at instants.* The utility of dominance in deriving results relating to perturbation seems to indicate that dominance expresses deep-seated characteristics of the spaces in question. But as explicitly presented here, it is incompatible with the widespread view (as expressed for example by Allen and Hayes (1985)) that it does not make sense to speak of a state holding at an instant, but only over an interval. The question facing us is therefore whether there exists an alternative formulation of the dominance theory which can play the same role as that theory in uncovering the perturbation structure of complex state-spaces, without presupposing the contentious association of states with instants.

References

Allen, J. F. (1983). Maintaining knowledge about temporal intervals, *Communications of the ACM* **26**: 832–843.

Allen, J. F. (1984). Towards a general theory of action and time, *Artificial Intelligence* **23**: 123–154.

Allen, J. F. and Hayes, P. J. (1985). A common-sense theory of time, *Proceedings of the 9th International Joint Conference on Artificial Intelligence*, pp. 528–531.

Egenhofer, M. J. (1991). Reasoning about binary topological relations, *in* O. Günther and H.-J. Schek (eds), *Advances in Spatial Databases*, Springer-Verlag, pp. 143–160.

Freksa, C. (1992). Using orientation information for qualitative spatial reasoning, *in* A. U. Frank, I. Campanari and U. Formentini (eds), *Theories and Methods of Spatio-Temporal Reasoning in Geographic Space*, Springer, Berlin, pp. 162–178. Lecture Notes in Computer Science, Volume 639.

Galton, A. P. (1990). A critical examination of Allen's theory of action and time, *Artificial Intelligence* **42**: 159–188.

Galton, A. P. (1993). Towards an integrated logic of space, time, and motion, *Proceedings of the Thirteenth International Joint Conference on Artificial Intelligence*.

Galton, A. P. (1994). Instantaneous events, *in* Hans Jürgen Ohlbach (ed.), *Temporal Logic: Proceedings of the ICTL Workshop*, Max-Planck-Institut für Informatik, Saarbrücken.

Galton, A. P. (1995). A qualitative approach to continuity, *Time, Space, and Movement: Meaning and Knowledge in the Sensible World*. Working papers of TSM'95, Château de Bonas, 23–27 June 1995, Groupe 'Langue, Raisonnement, Calcul', Toulouse.

Hamblin, C. L. (1969). Starting and stopping, *The Monist* **53**: 410–425.

Hamblin, C. L. (1971). Instants and intervals, *Studium Generale* **24**: 127–134.

Randell, D. A., Cui, Z. and Cohn, A. G. (1992). A spatial logic based on regions and connection, *Proceedings of the Third International Conference on Knowledge Representation and Reasoning*, pp. 165–176. Cambridge, Massachusetts, October 1992.

van Benthem, J. F. A. K. (1983). *The Logic of Time*, Kluwer, Dordrecht. Second Edition, 1991.

Williams, B. C. (1990). Temporal qualitative analysis: Explaining how physical systems work, *in* D. S. Weld and J. de Kleer (eds), *Qualitative Reasoning about Physical Systems*, Morgan Kaufmann, San Mateo, California, pp. 133–177.

Qualitative Causal Modeling in
Temporal GIS

Eric Allen[1,2], Geoffrey Edwards[1,2] and Yvan Bédard[2]

[1] Chaire industrielle en géomatique appliquée à la foresterie

[2] Centre de recherche en géomatique
Université Laval, Pavillon Casault
Sainte-Foy, Québec, G1K 7P4
e-mail: edwardsg@vm1.ulaval.ca

ABSTRACT: A generic model for explicitly representing causal links within a spatio-temporal GIS is presented. The current litterature on causal modeling within the fields of geomatics, computer science and geography, is surveyed. The contributions of several major philosophers to the theory of causality are also briefly summarised. Based on the lessons extracted from such earlier work, the elements of the new model are presented. The model consists of a small number of elements which are presented via a conceptual data model using an extended Entity-Relationship formalism. These elements include the following entities — Objects and their States, Events, Agents, and Conditions — as well as the relations Produces, Is Part Of, Conditions. Both objects and events may have spatial representations, and all entities may have temporal representations. The model is simple but hierarchical. A specific means of modeling uncertainty has been included in the development of the model. An example of the model proposed applied to data related to environmental impact assessment is shown, using a formal visual language designed to help understand the linkages implicit in the model. Furthermore, the relative strengths and weaknesses of the model are outlined and its relationship to previous work is also discussed.

KEY WORDS: causality, temporal GIS, events, databases, models, simulation, object-oriented design, uncertainty

1 Introduction

Temporal GIS has attracted considerable interest as a research topic over the past few years (Langran 1992). A wide variety of studies have been undertaken, some more conceptual, some concerned with implementation of temporal capabilities within existing systems, some oriented towards developing entirely new functionality within GIS.

In an early study of the subject, Al-Taha and Barrera (1990) indicated that temporal GIS would permit a number of new capabilities. Their list included: the prediction of future behaviour (scenarisation); better planning, based on knowledge of past history (planning); the capability to observe or infer the rules which cause certain changes (causal inference); the capability to analyse past histories (historical databases); and the ability to explain the existing status of data (interpretation). Studies of temporal GIS, to date, have focussed on planning, scenarisation and the development and use of historical databases. Much less work has been done on causal inference and on the interpretation of data based on knowledge about causal links in the data.

This paper addresses these last two issues. Because so little work has been done on the role of causality within (temporal) GIS, it has been necessary to return to the fundamental principles of causality as studied by the philosophers, in order to extract a useful framework for the generic modeling of causality within GIS. However, other disciplines are more advanced in the use of concepts of causality. This is particularly true of many areas of computing science. Hence, for example, the issue of causality has been addressed within the study of temporal logics, non-spatial information systems, and object-oriented systems design. Other areas where causality has been addressed include the development of mathematical models for a variety of natural resource modeling applications and research in the administrative sciences concerned with the nature of organisations and with business process modeling.

For this paper, we shall focus on a brief overview of work within computing science, geography and geomatics as it relates to our problem. This will be carried out in the second section of the paper. We shall not address issues of temporal GIS, except as they related directly to the concept of causality. Section 3 will present an overview of issues raised by the natural philosophers which are relevant. Section 4 will present the model we have adopted, and section 5 will deal with potential analysis tools and applications of the model we have developed, with some worked out examples. In section 6, we shall summarise the results and present our conclusions.

2 Causality and GIS

2.1 Temporal GIS

As mentionned earlier, very little work has focussed explicitly on the question of causal modeling within the framework of temporal GIS. One or two cases are worth presenting, however. Elmes and Cai (1994) have developed a method for understanding how uncertainty and error propagates in a complex interconnected set of modules within an expert system using a causal model. They attempted to identify the variables responsible for different kinds of error that arise in such complex models. These variables were then organised in a dependency hierarchy, which the authors label a "causal hierarchy". This approach allows them to prioritize variables in terms of the importance of their influence on the propagation of errors within the model. They found that the causal error model they developed helped in untangling the interelationships between the different variables in the models they were studying, but that the causal error analysis handled poorly the more quantitative aspects of error propagation.

In quite a different approach, Beller (1991) studied the problem of implementing events, states and evidence within a GIS which includes remotely sensed imagery. Beller defined an event as a change in a measured variable. He observed that an event may have a spatial location and a geometry, and hence be represented as a polygon within a GIS. In this way, Beller introduces events as spatial objects which can be managed in a manner similar to any other geometric object in a database. Beller does not address specifically the question of causality, but he indicates that once one has a means of modeling events as objects in the database, it then becomes possible to infer causes from an analysis of the objects.

Whigham (1993) discussed the problem of designing a spatial-temporal database using a hierarchical order. In particular, Whigham posits a hierarchical structure for temporal events, so that more complex events are made up of simpler events in a clearly defined structure. Although he did not discuss in detail causal aspects of the problem, he indicated that causal links could be modelled between events, allowing an explanation of one event in terms of another. Whigham, however, does not precisely define what is meant by an event within his formulation, nor what is meant by a causal connection. Hence the causal link he discussed simply consisted of the system developer or user identifying a link between two events and a reason for the link.

Gagnon (1993) introduced a rich data structure for temporal GIS which allows the management of the existence, presence/absence, activation/disactivation, descriptive and geometric evolution of spatial objects as well as of temporal topology. Some of these concepts were successfully tested for forestry and urban databases and included in a new formalism for spatio-temporal database design (Caron *et al.* 1993; Bédard *et al.* 1994). This formalism, called MODUL-R, and some of Gagnon's concepts, were used for the present research. Edwards *et al.* (1992) presented a very brief overview of some of these issues as well as an introduction to the issue of causality.

2.2 Computing science developments

Although the development of traditional (non-spatial) temporal information systems would seem to be a useful source of insight into the causal modeling problem, much of the work on temporal information systems has been carried out within the domain of temporal logics (Clifford and Warren 1983). Possible applications of the latter are much wider, and indeed overlap heavily with the six goals identified by Al-Taha and Barrera for temporal GIS. Temporal logics were developed to support temporal reasoning, which consists of analysing sets of statements about events in order to deduce information about temporal relationships. Causal relationships are often implicit in the temporal relationships which are studied.

In an overview paper, Long (1989) describes three classes of temporal logics — first-order logics, modal logics and reified logics. However, these are all monotonic logics. Shoham (1988) points out that a *causal* logic must be non-monotonic. This means that the addition of new information may invalidate existing information - this is not usually a property of logic. He further describes several problems when one wishes to reason about causality. One problem, called the *qualification* problem, is the fact that any causal relationship may be associated with an endless series of conditions. Another, related problem is called the *frame* problem, the term deriving from the use of frames in knowledge representation within Artificial Intelligence. It consists of the need to manage large numbers of default propositions in order to be able to reason about causality.

Other temporal logics of interest include Allen's interval calculus (1984) and Kowalski and Sergot's event calculus (1986). In the latter, events are considered to be the primitive temporal objects, and time emerges from the ordering of these events. The event calculus was developed specifically for temporal database applications. It represents events as structureless primitives. Recent extensions to this approach allow continuous processes and also the existence of an event hierarchy. Also, probabilistic models of causality have been developed by, among others, Mackey (1974), Lewis

(1973) and Suppes (1970), as well as Shoham (1989). Shoham's work overcame many of the difficulties associated with the other logics.

Another area of research within which causal modeling plays a part is the development of object-oriented systems design methodologies (Rumbaugh *et al.* 1991). Hence, for example, the work of Embley *et al* (1992) focusses on a number of formal tools which can be used to model different aspects of temporal systems which involve controlling and/or synchronizing different activities within an operating system. There are useful lessons to be learned from such design methodologies, especially if we desire to borrow parts of the design methodology which might be conveniently structured in our model of causality. The systems nature of these models is congruent with the kind of phenomena we wish to model.

2.3 Spatio-temporal mathematical models

Finally, mathematical models have been developed within a systems perspective (Roberts *et al* 1983) to understand and, sometimes, predict a wide variety of natural phenomena. Such models almost always involve changes over time; more recently, many such models have been extended to include the spatial dimensions. Cause and effect are modelled via functional relationships between variables, and these models also usually involve accumulator variables (reservoirs) and flows as well as purely functional dependencies. They require a quantitative approach to modeling, although they are sometimes used for qualitative inference about system behaviour. Hence, for example, the user must not only specify which variables affect each other, but also the exact functional relationship between the two variables.

In these models, causal connections between variables are explicitly modelled via mathematical formulae. Changing one variable 'causes' another variable to change also. However, as we shall see later on, this is related to a particular definition of causality, and not necessarily the only one we would like to model within GIS. Furthermore, as was the case in much of the work done in object-oriented systems design, the models used overspecify the information required to model the existence of causal relations. Hence, for example, testing and validating these models tends, necessarily, to focus on an evaluation of the forms of the functional dependencies, and much less on whether or not a variable should be included in the model. The latter is usually a decision which must be made by the model developer (Huggett 1980). We believe it is appropriate to develop more qualitative tools for causal modeling. Such tools may help the systems modeler to better define the variables for which causal relations are sought. They may also help when systems become too complex to model quantitatively.

2.4 GIS constraints

The goal of the development of a generic causal modeling capability within GIS means that several constraints on the nature of the causal model must be met for the results to be useful. These constraints include, in general terms, at least the following:

    (i)   a data structure which is sufficiently expressive to cover most of the needs identified by Al-Taha and Barrerra;

    (ii)   a data structure which is accessible and lends itself to analysis;

(iii)   a means of spatially referencing causes and changes;

(iv)   the extension of conceptual modeling tools such as the conceptual data model (CDM) to handle causal entities;

(v)   a means of modeling uncertainty; and

(vi)   a data structure compatible with existing GIS.

3   The Concept of Causality in Natural Philosophy

Arguably the most influential early writer on the subject, Aristotle classified causes into four kinds: (i) the formal cause, what an object represents; (ii) the material cause, or what an object is made of; (iii) the efficient cause, or the maker of the object; and (iv) the final cause, or the goal of the object. In Aristotle's scheme, no clear distinction is made between an object, an event, or a 'thing' in general. However, he does observe that causes are associated with change - without change, there is no need to have recourse to a causal explanation.

In the 18th century, Hume (1947) emphasised that causality is the major factor which allows humans to make sense of their experience, but that it is impossible to 'prove' that a given causal relation exists (and hence that causality itself exists). Futhermore, the inference of a given causal relation is a choice on the part of the human, and not a given, not a fact which can be determined independently of the human perceiver. To attempt to model the causality present in the 'real world' in a purely objective manner is therefore impossible.

Kant (1944) went on to establish that causality is not a property of 'things', but rather a property of relations. Like Hume, Kant saw causality as the fabric, the 'understanding' by which we are able to make sense out of the phenomena that we perceive. According to Kant, however, this understanding is not derived from our perceptions, as Hume thought, but is what allows us to organise them. From this, we may retain the idea that causality should be a system which explains the connections between events.

Kuhn (in Bunge *et al.* 1971) notes that modern science ascribes two distinct meanings to the concept of causality. The first, narrow sense consists of the notion of production, whereby a given cause produces an effect (close in concept to Aristotle's efficient cause). The second, broader sense is that of an explanation. Davidson (1993) retains the idea of events as the central idea in the study of causal logic, but his focus is more exclusively concerned with the roles of agents with respect to causality. In particular, he ascribes the actions of agents to a reason, which, combined with the means of production, may be interpreted as the cause. Hence, Davidson's use of the term "causality" generally refers more to the explanatory role of causality than to its genetic role.

The philosopher whose approach has been the most useful to the present study is Bunge (1966). Bunge makes clear the relationship between determination and causality. In Bunge's view, causality is a particular class of determination, which is a much more general principle. Hence, for example, the functional relationship between two variables, as was described in the context of spatio-temporal mathematical modeling, is not truly a causal relationship although it is a deterministic one. Bunge makes the distinction that causality is determination plus production. Indeed, for Bunge, determination is a lawful relationship between two 'things', whereas causalism is a lawful and *genetic* relationship.

The definition of determination, according to Bunge, is "a constant and unique connection". Bunge identified many different kinds of determination, including what he called quantitative self-determination, causal determination, interaction, mechanical determination, statistical determination, structural determination, and qualitative self-determination. Quantitative self-determination exists where one condition (the consequent) is determined uniquely by a previous condition (the antecedant). Causal determination occurs when there is a productive or genetic link between two conditions. Bunge considers causal determination to involve explicit, external causes.

Interaction is defined as a form of mutual or reciprocal causation. Mechanical determination is similar to quantitative self-determination, except that the consequent is determined by an antecedant in combination with other factors, such as efficient (external) causes. Statistical determination occurs when a consequent is determined by the joint action of several independent antecedants. Structural determination refers to the determination of parts by the whole. Dielectical determination (or qualitative self-determination) occurs when the inner struggle of essential and opposite components of a single system result in a sythesis.

Furthermore, Bunge discusses a variety of causal principles, rejecting many on philosophical grounds (i.e. because of logical inconsistencies or undesirable properties). Bunge considers also the effect of including conditions in the causality principle, such as "If C happens under the same conditions, then and only then E is always produced by it". He notes that the cause, in this case, must be interpreted as a "trigger" of a process which cannot occur without certain external or internal conditions, and not at a necessary and sufficient producer of the effect.

4 The Model Proposed

Let us return now to the criteria we listed for building a generic causal model within GIS. The first requirement was for expressive power. We believe that the formulation of Bunge, which distinguishes between determination and (productive) causality, provides the kind of expressive power we seek to establish. The reason for this is that Bunge's concept of causality is quite simple, while his concept of determination is fairly complete and much more sophisticated. For GIS applications, it may not be necessary to handle all the forms of determination described by Bunge. Several are useful, however. Furthermore, the work on the event calculus (Kowalski and Sergot 1986) as well as the hierarchical event structure discussed by Whigham (1993) suggests that a hierarchical approach to modeling causality will enhance significantly the expressive power of the model.

Secondly, our requirement that the data structure be accessible for efficient analysis means that causal relations must be explicitly represented within the database itself. Our survey of earlier work indicates that events are the most useful and most easily modeled entities within the context of causality. Representing events explicitly in the database is not difficult - they can be treated as can any other object. As noted by

Beller, events may be treated as being both spatially and temporally referenced, although there are nuances in how this might be modelled. Bunge, on the other hand, is careful not to say what the antecedant and consequant are in his formulation of the problem. Hence we have adapted Bunge's theory to a theory of events.

Third, the conceptual data model is a powerful tool for helping elucidate the relations between entities in the database, and hence can be used to advantage as a tool for reflection. Furthermore, the inclusion of causal information (not to speak of temporal information) is likely to increase manyfold the complexity of the databases within GIS. Without the development, in parallel with the concepts themselves, of appropriate data models, it will be difficult to carry out implementation of the concepts beyond very rudimentary prototypes.

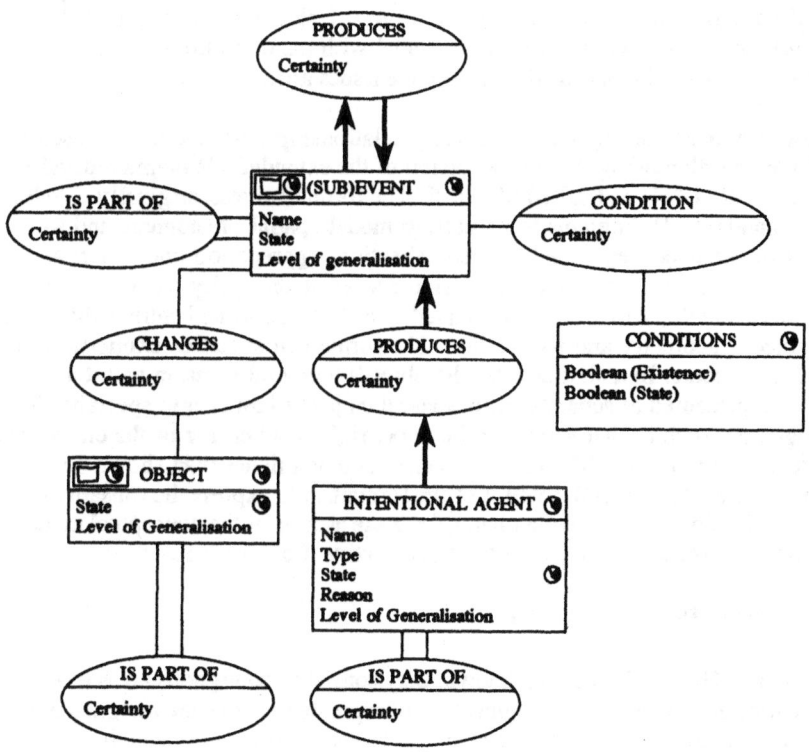

Fig.1: Conceptual data model showing the proposed generic causal model in a slightly modified form of the MODUL-R ER formalism. Boxes represent entities and ellipses represent relations. Cardinalities are not shown. The "pie" symbols (circle with pie shaped piece) indicate that the variable in question evolves over time as either a point or an interval. When present in the title bar, on the right, it indicates that the existence of the entity may evolve. The boxed symbol on the title bar, which contains a representation of a polygon as well as the pie symbol, indicates the entity may have geometry, and that the geometry may evolve over time. The heavy arrows indicate the causal relations in the model. Other connections are symmetrical, following standard ER practice. Cardinalities are not shown.

For our purposes, conditioned causality is strictly necessary in order to make the modeling of the causal link tractable. This is because, by separating causes from conditions, we can simplify the structural requirements on the former. With regard to the question of an individual versus a class formulation of the causality principle, we have adopted an individual formulation, because we are interested in modeling specific causal links. This does not, however, exclude the possibility of inferring collective or class causal links by analysis, and indeed will probably be one of the goals of causal analysis. A further restriction on the causal principle which we apply is to note that not all events are necessarily causally related to other events. From a philosophical perspective, this restriction is probably abhorent, but from a practical perspective the GIS designer must be able to choose to present some causal links and not others, according to his or her perception of what is pertinent.

In the following sections, the generic model we developed according to these specifications is laid out. Furthermore, Figure 1 shows the conceptual data model (CDM) which has been developed in parallel with the definitions. Each aspect of the CDM is discussed in turn, as the appropriate issues arise.

A few comments on the extended entity-relationship (ER) formalism used in this CDM are worth making. We use a version of the extended ER diagram developed by Bédard *et al* (1994) and called Modul-R. The formalism was originally developed to allow standard ER diagrams to be used to model spatial phenomena and to include object-oriented concepts such as generalisation, aggregation and inheritance. The spatial pictograms in the upper left-hand corner of the entity labelled "Object" are indicative that this entity can be expanded to include associated entities that represent the object's geometry and its location. The formalism specifies explicitly how this unfolding occurs. In a similar way, the object's temporal location and duration may also be represented as separate entities and then packed away into specially designed pictograms - such as that shown in the upper right hand corner of the entity labelled "Object". Although the Modul-R formalism is much richer than shown here, we did not need to exploit the full richness of the formalism to express the causality concepts we have developed. On the other hand, for some aspects of the conceptual modeling of causality, the formalism was found to be restrictive if not inadequate.

4.1 Objects and States

Objects in a GIS world are well known: they consist of contiguous areas (polygons or collections of grid cells), line segments and/or points to which are assigned attributes, or of non-cartographic objects such as contracts, persons, etc., which are stored in tables not connected to graphic features. Usually, it is possible to define a topology between objects which intersect, indicating how they are related to each other spatially. Although objects (especially in the object-oriented sense) may also include a variety of other entities, we shall restrict our use of the word to entities which may be localised in space (although they need not be so localized) or which can be localised in time (although they need not be so localised). Hence objects may or may not persist over time. States refer to the attribute values of each object, as well as to each object's geometry and also to an existence referent (i.e. the object exists in the database at a given moment in time or it does not). States may also be associated with the presence of an agent. In Figure 1, objects are represented as entities which may have both spatial and temporal location, as well as attribute information (which has been indicated in the diagram by a single variable, called a state).

4.2   Events

Events, as mentionned earlier, are considered to be changes of state in objects. They are the central entity of our model. This is consistent with the viewpoint of the philosophers, but differs somewhat from common usage. Hence the Concise Oxford Dictionary, for example, defines an event as a "happening" or as an "outcome". If a happening is understood as referring to a change, the two definitions converge. An outcome, however, is associated with the arrival state of the change, and this meaning of event is not retained in our model.

It is clear that changes of state may involve changes in existence or changes in the geometry associated with objects, as well as changes in attribute values. It should also be clear from the above discussion that we do not consider one state to have been "caused" by another, but rather one change of state in an object to have been "caused" by another change of state of either the same object or of a different object.

Events will be associated with temporal intervals and with spatial locations to the extent that the objects with which they are associated are localised temporally or spatially. However, events may have additional or alternative localisations than those which are derived from the objects, and hence the conceptual data model also shows an explicit pictogram for temporal or spatial location for events.

Finally, it should be noted that Events may also be called "Sub-Events" (see Figure 2). This is an important aspect of the proposed model. At some level, any event may always be decomposed into a sequence of sub-events, which may or may not be causally linked. We expect that, given the cost and difficulty of acquiring data, one would always wish to model causal links between events at the most general level consistent with the targetted task. Furthermore, all hierarchical levels of generalisation in the causal relationships between events will be maintained within the database, so that queries may be answered at whatever level desired, consistent with the level at which causal data has been entered into the system. The attribute "Level of generalisation" listed inside the Event/Sub-Event entity in Figure 1 refers to the level of generalisation in the causal hierarchy.

Each of the other entities shown in Figure 1 also may be grouped into entities of the same kind at other levels of generalisation, hence supporting the generalisation of events. Thus, for example, in Figure 3 the institution (intentional agent) responsible for building the road (change of state of road) at one level of generalisation may contain the crew (intentional agent) which causes a road trunk to be build (change of state of road trunk) at a lower level of generalisation.

4.3   Conditions

As stated earlier, conditions play an important role in our model. Conditions may arise due to the presence of any other entity within the model, including other conditions, any pre-existing state, any previous event occurence, or any agent's state (to be described below). They are expressed as boolean statements (e.g. State A < Value I). Hence a condition which consists of the presence of a given object within the database would appear as "Presence(Object A) = True". Conditions are assumed to be necessary to the caused occurence of a given event and are associated with the outcome event of a causal pair of events (although they may also be associated with an event for which no cause is assigned). Hence conditions serve a restrictive role in the model. They are likely to play a very important role during causal analysis, but until we develop further the operations which can be performed on our model, we will be unable to evaluate the full consequences of the existence of such conditions in the model.

4.4   Intentional Agents

From Davidson, we retain the notion of a causal agent motivated by a reason. Davidson discussed the role of agents and of actions in his causal model. Davidson's definition of an agent implies intentionality. In our model, intentional agents are included because not all events are caused by another event. Sometimes events are caused by the action of an intentional agent — and some events may occur spontaneously. Note that we define an action as an event combined with a causal agent. This is somewhat different fron the definitions used by Davidson. Our justification for modifying slightly these definitions is related to the use of causal models in the context of land management. Indeed, several agents may operate within the same territory and there is need to track events to different causal agents.   The agent may be one of several different types: a person, an animal, or an organisation. It may even be a biological entity, such as a virus. An agent will also usually have a reason for acting. Hence these properties of intentional agents are included in the attribute lists shown in Figure 1.

The respective roles of agents and events are not symmetrical, in our model. This is the reason the directionality of causal links are explicitly indicated in the conceptual data model (Figure 1). Agents may cause events, and events may cause other events, but events may not cause the actions of agents, nor may agents themselves cause the actions of other agents. This constitutes an explicit recognition that intentional agents are autonomous, and that their actions arise from internal choices. The action of agents may be conditionned (this is already present in the model, since the condition is associated with the effect event) and they may also be determined in any of several ways (because they are part of another agent, because they are constrained to act in a certain way, and so forth), but they may not be caused. Hence, analysis which involves agents will tend to focus on determination relations and conditions rather than strictly causal relations. We believe this is a more appropriate model in the context of land management than one which permits the actions of agents to be "caused" directly.

4.5   Uncertainty and causality

One of the requirements for a GIS-compatible model of causality that we identified in the first section of this paper was a means of modeling uncertainty. There are several reasons for this. Extensive work is under way to better characterise and represent uncertainty, spatial, temporal and attribute uncertainty, within the context of GIS and new capabilities which might be introduced into GIS need to address the same issues. Secondly, causality can be used (Elmes and Cai 1992) to try to understand where uncertainty arises. Thirdly, our study of the natural philosophers indicated that what one chooses to represent as a causal relation is more a question of human perception than of real world properties. Hence an ability to indicate the level of credence which should be associated with a given causal relation will be important. Finally, the choice of conditions which are associated with any particular causal relation will never be complete. This is the qualification problem outlined by Shoham (1988). Hence, although a cause is defined as a necessary and sufficient condition for producing a given effect, and hence the set of conditions are necessary, sufficiency is very difficult to guarantee. For this reason, the modeling of causal links will always occur in the presence of incomplete information.

We have handled this issue by introducing a "certainty" attribute in each of the relations which are part of the causal model. The certainty attribute may be modelled using any of the current approaches (fuzzy set theory, certainty factors or the theory of evidence). We do not believe this attribute needs to be finely subdivided - three or four major labelled levels of certainty are probably adequate. For example, "highly certain, moderately certain, low certainty" may well be good enough. An alternative approach to handling uncertainty might be to use a causal logic such as the one proposed by Shoham (1988) to infer the causal consequences of a given event. Shoham's logic, however, deals only with primitive events — there has been no published extension for handling hierarchical event structures. Furthermore, none of the logics described in the first section of this paper deal with spatial reasoning. For use in a GIS environment, spatial reasoning capabilities is a *sine qua non*.

5 An Example Application

In the previous section, a data structure (and associated conceptual data model) for representing causal relationships was proposed. This model meets all of the criteria that were laid out for a causal model which could be integrated with GIS. The data structure is expressive; it is accessible and lends itself to analysis; causes, changes and events may be spatially referenced; a generic conceptual model for causality has been developed; and uncertainty has been explicitly included in the structure. In addition, the data structure has been designed to be fully compatible with temporal and spatial data structures which exist now or which are being developed. Furthermore, the data structure is modular, in the sense that a system designed to permit causal modeling does not require a user to exploit this functionality of the system. The causal modeling functionality may be grafted onto existing capabilities.

In this section, we will present a limited example designed to illustrate how this structure might be implemented. Space prohibits us from examining a more fully worked out example. In particular, we present a visual language or formalism which helps one to visualize the causal relationships between entities and to track individual case histories. We then go on to discuss briefly possible analyses which might be performed on the data, and we conclude with a presentation of possible applications which have been identified.

Figure 2 shows a template of the formal language developed to track causal links. Each type of entity (event, condition, agent, or object) has been assigned a different container symbol, and each type of relation (causal/production, generalisation determination, object change, conditioning and agent association) has been assigned to a particular connector symbol. Figure 3 shows an example of the use of the visual language applied to a geographical data set. This example used consists of part of an environmental impact study for a dam construction project by a utility company. The

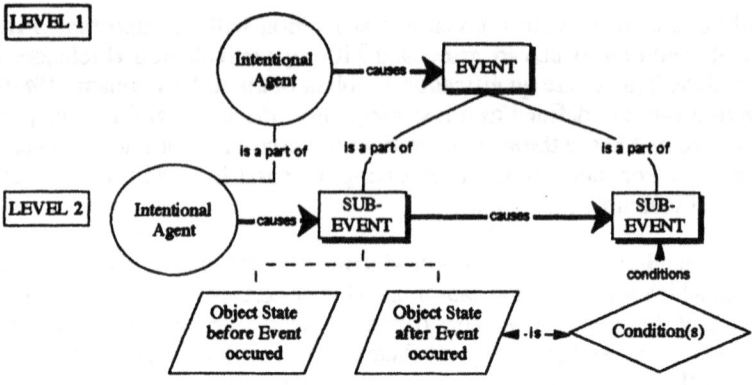

Fig. 2: Template of the visual language showing the different types of entities as different container symbols, and the different types of relations as different types of connector lines. Also shown are different hierarchical levels or levels of generalisation.

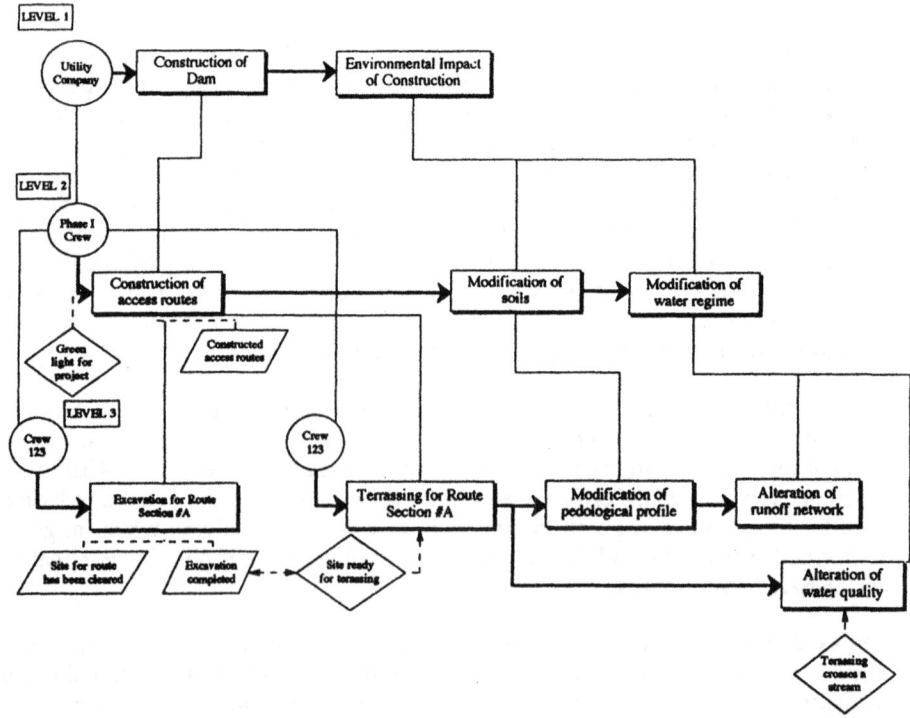

Fig. 3: Example of a hierarchical causal model for a dam construction project, showing three levels of generalisation and the hierarchical relationships between them.

event data have been modelled in a series of generalisation levels, as described in section 4.4, of which three levels are shown.

There are several interesting items to note with respect to this diagram. First of all, the generalisation hierarchy can be seen more clearly. This three level breakdown of the structure is somewhat oversimplified - it is easy to come up with as many as five or six levels. Secondly, the direct actions of the utility company, or of parts of the utility company, are easily visualized. The consequences of these actions are also easy to see.

It should be noted that the breakdown of an event into several sub-events requires dependency links between all sub-events, but does not require strictly causal links.

Hence, for example, "excavation" and "terrassing" are both sub-events within the event "Construction of access routes", but they are related by contiguity and not causality (temporal contiguity may be viewed as an additional form of determination - Kant 1944). Specifically, the second event cannot occur until the first has been completed. Hence they are part of a dependency group but not necessarily a causal group.

We may define causal chains, however, as groups of causally linked events and agents. Hence "Phase I Crew" causes "Construction of access routes" which causes "Modification of soils and water properties of the site" are part of a causal chain. When the effect events from a given causal event multiply, we may call the resulting causal chain a causal "cascade". Both a chain and a cascade will always contain at most one intentional agent, at the head of the chain, although they may contain no intentional agent.

Some of the analysis (causal queries) which might be carried out include the following

(i)   What events are caused directly by a given intentional agent and all its components?
(ii) Which effects result from the action of a given intentional agent?
(iii) What conditions mediate the effects of a given intentional agent?
(iv) What is the geometrical extent of the region affected by a given event and its subsequent effects?
(v)   What is the spatial extent of the effects of a given event in the five years following its occurence?
(vi) What was the state of a given object when a given event occured?

Undoubtedly, the reader will be able to imagine other queries which such a system could be used to model. The potential power of such a system may be quite extensive. There remains, however, the challenge of transforming the theory into a practical system. This will be a non-trivial task, since, to some extent, it is dependent upon new developments in the area of temporal GIS. However, we believe a prototype system could be developed for a small region.

The approach developed is being tested on an environmental impact assessment study, and such a study is clearly one application of this kind of system. However, the system is also likely to be useful in any area where the relationships among data are complex, especially spatial-temporal data, or where there are formal requirements for accountability. Furthermore, it should be possible to develop specific tools which could be used to extract or infer causal relationships from data which has been recorded without causal information, for hypothesis testing or scenarisation of potential future effects of present day actions. The analysis part of the project is just in its early stages, but there appear to be many advantages to the approach we have developed.

6 Discussion and Conclusions

The causal model we have developed should be consistent with other kinds of temporal logic, although we have not worked out the specifics of how this might be done. The advantage to our model, however, is that it explicitly deals with spatial phenomena as well as temporal phenomena. No existing temporal logic yet deals adequately with spatial relations. The specifics of our model also render it consistent with the dynamic systems models used in natural resource modeling. The latter, however, require specifying all functional dependencies in the model, and hence our model is necessarily under-specified. Nonetheless, it should be possible to construct a "translator" which converts our causal models to an appropriate framework for dynamic systems modeling. In this way, our model might be used to explore interactively and qualitatively the expected causal links between objects in order to decide on a subset of variables to be used in a more quantitative spatial model.

Although we have not explored extensively the use of our model for planning and scenarisation of possible actions among possible worlds, our model can be usefully applied to the latter problem, provided suitable modifications are made.

Our model is based on certain choices, and hence is limited by these choices. Hence, for example, we have explicitly excluded causal connections between agents, or from events to agents. Other approaches permit such causal links to be modelled (e.g. Davidson 1993). There is also some ambiguity in our treatment of causality versus dependence. It would be possible, using our framework, to handle every event as being caused by an agent, with the occurence of a previous event as a condition. All the "causality" in such a system would be expressed via dependencies. Hence there is some choice as to how much of a given scenario is represented via productive causal links and how much via conditioned dependencies.

Furthermore, we have not addressed the issue of obtaining or inferring causal connections between events. This is a controversial topic which bears much reflection, and caution should be exercised before venturing too quickly along such lines. Schreuder and Thomas (1991) discuss many of the pitfalls involved in attempting to infer causal relationships between variables in the context of forestry.

Our model deals with some, but not all, of the determination types enumerated by Bunge. In particular, statistical determination is not formally addressed by the current version of the model, except insofar as a certainty factor is included. More work will be required to extend the model correctly for handling more statistical aspects.

In conclusion, we began by defining the requirements for a generic causal model which would be compatible with existing (and projected) GIS functionality. We then examined both existing work on causal modeling in several disciplines, and a number of important philosophical theories of causality. This earlier work was then used to construct the broad outlines of a causal theory appropriate to our requirements. We then constructed a conceptual data model using an entity-relation formalism (Modul-R), which defines the specific elements which make up our causal theory and the relationships between them. Definitions were provided for each element. Following this, we developed a visual formalism for expressing the causal relationships of our model for specific case histories and showed a simple example of such a case history.

The resulting model provides directly, in principle, tools for analysing past histories and for understanding the current status of a region as a function of causal dependencies in the past. Furthermore, provided appropriate analysis tools are developed, the model should facilitate the development of tools for infering causes and for predicting changes in the future, at least in a qualitative sense. In fact, we believe that qualitative modeling may play a much larger role in spatial data analysis than it presently does, if appropriate tools were available to carry out such modeling activities. The causal model proposed here is an example of one kind of tool which might be used to perform qualitative analysis of data and the relationships between data.

7 Acknowledgements

This research was supported by an FCAR team grant and via the funding to be found in the Industrial Research Chair in Geomatics applied to forestry, which was financed by the Association des industries forestières du Québec (AIFQ) and the Canadian Natural Sciences and Engineering Research Council (NSERC).

8 References

Al-Taha, K., and Barrera, K.R. Temporal data and GIS: An overview. *Proceedings of GIS/LIS '90*. November 1990, 244-254.

Allen, J.F. Towards a general theory of action and time. *Artificial Intelligence*. Volume 23(2), 1984, 123-154.

Bédard, Y., Vallière, D., and Gagnon, P. MODUL-R 2.0: A New formalism for spatio-temporal database design. *Technical document*. Centre de recherche en géomatique, Université Laval, Québec, 140 pp.

Beller, A. Spatial/Temporal Events in a GIS. *Proceedings of GIS/LIS '91*. Volume 2, 1991, 766-775.

Bunge, M. *Causality*. Dover Publications: New York, 1966.

Bunge, M., Halbwachs, F., Kuhn, Th. S., Piaget, J., and Rosenfeld, L. *Les théories de la causalité*. Presses Universitaires de France: Paris, 1971, 210 pp.

Caron, C., Bédard, Y., and P. Gagnon. MODUL-R: un formalisme individuel adapté pour les SIRS. *Revue de géomatique*. Volume 3(3), 1993, 283-306.

Clifford, J., and Warren, D.S. Formal Semantics for Time in Databases. *ACM Transactions on Database Systems*, Vol. 8(2), 1983, pp. 214-254.

Davidson, D. *Actions et événements*. P. Engel (transl.). Presses Universitaires de France: Paris, 1993, 402 pp.

Edwards, G., Gagnon, P., and Bédard, Y. Spatial-Temporal Topology and Causal Mechanisms in Time-Integrated GIS: From Conceptual Model to Implementation Strategies. *Proceedings of the Canadian Conference on GIS '93*. Ottawa, Canada, 1993, 842-857.

Elmes, G.A., and Cai, G. Structural reasoning for spatial database accuracy assessment. *International Symposium on the Spatial Accuracy of Natural Resource Data Bases*. ASPRS: Williamsburg, Virginia, 1994, 141-149.

Embley, D.W., Kurtz, B.D., and Woodfield, S.N. *Object-Oriented Systems Analysis: A Model-Driven Approach*. Prentice-Hall: Englewood Cliffs, New Jersey, 1992, 302 pp.

Gagnon, P. Concepts fondamentaux de la gestion du temps dans les SIG. *Mémoire de maîtrise*. Université Laval, Québec, 1993, 160 pp.

Huggett, R.J. *Systems Analysis in Geography*. Clarendon Press: Oxford, 1980, 208 pp.

Hume, D. *Enquête sur l'entendement humain*. A. Leroy (transl.). Éditions Montaigne: Paris, 1947, 224 pp.

Kant, E. *Critique de la raison pure*. A. Tremesaygues and B. Pacaud (transl.) Presses Universitaires de France: Paris, 1944, 584 pp.

Kowalski, R.A., and Sergot, M.J. A logic-based calculus of events. *New Generation Computing*. Volume 4, 1986, 67-95.

Langran, G. *Time in Geographic Information Systems*, Taylor and Francis: New York, 1992, 189 pp.

Lewis, D. Causation. *Journal of Philosophy*. Volume 70, 1973, 556-567.

Long, D. A review of temporal logics. *The Knowledge Engineering Review*. Vol. 4(2), 1989, pp. 141-162.

Mackey, J.L. *The Cement of the Universe: a Study of Causation*. Oxford University Press, 1974.

Roberts, N., Andersen, D., Deal, R., Garet, M. and Shaffer, W. *Introduction to Computer Simulation: A System Dynamics Modeling Approach*. Addison-Wesley: Don Mills, Ontario, 1983, 562 pp.

Rumbaugh, J., Blaha, M., Premerlani, W., Eddy, F., and W. Lorensen. *Object-Oriented Modeling and Design*. Prentice Hall: Englewood Cliffs, New Jersey, 1991, 500 pp.

Schreuder, H.T., and Thomas, C.E. Establishing Cause-Effect Relationships Using Forest Survey Data. *Forest Science*. Volume 37(6), 1991, 1497-1512.

Shoham, Y. *Reasoning about change*. MIT Press: Massachusetts, 1988, 201 pp.

Suppes, P. *A Probabilistic Theory of Causation*. North Holland, 1970.

Whigham, P.A. Hierarchies of Space and Time. *Lecture Notes in Computer Science*, Vol. 716, Springer-Verlag: Berlin, 1993, pp. 190-201.

A Design Support Environment for Spatio-Temporal Database Applications

P. A. Story and M. F. Worboys

Department of Computer Science, Keele University
Staffordshire, ST5 5BG, UK
Email: phil@cs.kl.ac.uk, michael@cs.kl.ac.uk
Fax: + 1782 713082

Abstract. Much work has been undertaken in recent years to provide temporal support for geographic information systems (GIS). Attention has focussed primarily on uniform temporal modelling, where the temporal type (for example, interval, bitemporal element) is constant throughout the model, as is the level at which it is integrated. This approach has only a limited range of application. This paper proposes an alternative strategy which allows an application developer, within a non-prescriptive design environment, to develop a spatio-temporal model specific to the particular application requirements. A set of temporal classes are given which aim to capture sufficient temporal semantics for a wide range of application domains. Additionally, a model is defined which provides a framework for the integration of the spatial, temporal and attribute components of an object. A prototype CASE tool has been developed to support the method.

1 Introduction

Many application domains require information systems which enable the storage and retrieval of past and/or future data, in addition to that which is current. This is certainly the case for many applications which involve spatial data. This paper describes our approach to supporting the development of such systems with particular emphasis on spatio-temporal (ST) database design.

Let us consider a spatio-temporal database design and implementation based on a specific ST-model. The system is likely to provide an accurate representation for those application domains which fit the model. However, it may subsequently be found that for other application domains a different ST-model would have been more appropriate, either in the way the model represents the developer's view of the application or in terms of the database storage and query efficiency of the temporal component. A preferable scenario is one in which a developer is able to choose the most appropriate ST-model for the specific application and generate the database design accordingly. This paper investigates how far it is possible to pursue this ideal in practice by providing a design support environment for ST-application developers.

A prototype has been implemented by extending a non-temporal CASE (Computer Aided Software Engineering) tool of a proprietary GIS. The temporal CASE tool developed enables a variety of designs to be applied and tested. The development of the tool required as pre-requisites:

- a well defined temporal model containing sufficient temporal classes to cater for the majority of temporal applications;
- an integration model which permits the full range of timestamping options and allows for temporal constraints between the various components of an object and between interacting objects.

The temporal model is described in section 3, which contains a brief summary of a model given in [23]. The issues relating to the integration of time with data are discussed in section 4 including the definition of an ST-application class and the treatment of temporal constraints. The design and implementation of the CASE tool is given in section 5 followed by an example of its use from the application developer's perspective.

2 Background

There has been much research on developing database models which incorporate time [20]. At a basic level, temporal and spatio-temporal modelling alike involve the association of a data value or collection of values (either attribute[1] or spatial) with a time value. The nature of the time value depends on the particular time model but is commonly either an interval [24] or a temporal element[2] [7] and can also occur in more than one temporal dimension (for example, bitemporal element [19]). This attachment of time values to data values is often known as *timestamping*. One of the major issues concerned with timestamping is the level at which the timestamp is applied and a variety of approaches have been tried. For example, Snodgrass [19] timestamps at the tuple level using either a single valid timestamp, an interval or a bitemporal element. In other relational temporal extensions the timestamp is simultaneously applied at more than one level. For example, Clifford and Croker [2] introduce lifespans at the tuple and attribute level. Elmasri et al. [6] temporally extend the EER model by timestamping at the entity and attribute level. In the area of spatio-temporal modelling Worboys [28] attaches a bitemporal element to a simplex (the simplest component of a generic topological model). A bibliography of temporal database research as applied to spatial objects is given in [1].

As most computational spatial data models contain objects which are aggregates of simpler spatial objects, the level at which to attach the timestamps within the aggregate object is an important decision in spatial modelling. Another option with spatial data, originating from research in CAD/CAM systems, is the versioning of spatial objects. This approach has often been used within GIS

---

[1] The term attribute is used here to refer to any non-spatial or non-temporal data associated with an object.

[2] The concept of a temporal element was introduced by Gadia and Vaishnav [7] but other terms are also used including lifespan [3], temporal set [25].

for the management of multi-user transactions rather than for recording temporal changes to spatial objects [11]. For the purpose of spatio-temporal versioning it would be necessary to timestamp each version. To avoid data duplication between versions it is common practice to record only changes to a completed 'base' state for each version. A whole object can then be reconstructed for a given time by combining all the changes from the base state. Various strategies are possible: which to choose is a balance between the number of steps required to reform a completed object at a given level and reducing storage overheads. Recently, more advanced strategies have been investigated that attempt to further reduce storage [10]. Further complexity arises with spatial applications over whether to timestamp topology or geometry [14]. Often such design decisions are dependent upon the frequency of the updates and the nature of potential queries.

Some alternative approaches have been developed in recent years by extending existing object-oriented models to incorporate time [13],[16]. Of particular interest is the work by Wuu and Dayal [29],[4] who temporally extend the OODAPLEX functional data model. The temporal type is introduced as a supertype, point with other types defined as subtypes of point. The model supports attribute and object level versioning and allows both together, as in Clifford and Croker [2].

Although current models provide a good basis for the study of temporal systems they are often limited. Many are restricted to one temporal dimension [2], [8], often there is no treatment of future time [25], [6] and nearly all are based on a totally ordered (linear) time line. As many applications require temporal support beyond these restrictions it is desirable to have a model which incorporates a complete range of temporal features.

The problem of finding a generic temporal model seems particularly acute in the realm of spatial applications due to their inherent complexity. An alternative approach to developing spatio-temporal systems may be the way forward. This should incorporate many of the concepts of the current temporal models but provide a coherent application development environment, emphasising a non-prescriptive model with a mechanism to permit the application of alternative integration strategies. This is the approach adopted here, set in an object-oriented framework because of the proven advantages that this provides for spatial domains [27],[26],[15],[5].

3 Temporal Class Model

The temporal model is based on a discrete representation of time and consists of a set of temporal classes which are used as the basis for the design of the temporal CASE tool. It includes a class to represent alternative object states at a given time and uncertainty of object states in the past or future. The model aims to capture sufficient temporal semantics for the majority of applications involving time.

Two main levels of classes are defined: application definition layer classes required to define the nature of a particular temporal domain and object definition layer classes, instances of which can be associated with individual application objects. Where appropriate, standard terms and definitions from the glossary of Jensen et al. [9] have been used.

3.1 Application Definition Layer

The application definition layer contains information on the temporal bounds of the application, the system (temporal) granularity (see section 3.2) and the aspect of time required for the application, namely, valid, transaction or bitemporal dimensions. This information is instantiated for a particular application via the temporal domain class, each instance of which defines a separate temporal domain. The temporal dimension(s) chosen affects the structure of the classes of the object definition layer.

3.2 Object Definition Layer

This layer may be subdivided into four groups of classes: temporal, event modelling, parallel state and auxiliary classes.

Temporal Classes. These may be considered 'pure' time classes as they are only concerned with temporal information. This layer contains the standard temporal classes chronon, interval[3] and temporal element. An instance of chronon is a single unit of time at the granularity of the application's temporal domain. It is therefore the shortest duration of time supported by the application and as such cannot be decomposed into any sub-unit. An instance of interval is a continuous and connected set of chronons and may be represented by a pair of chronons $< t_s, t_f >$, such that $t_f > t_s$, where t_s is the chronon representing the start of the interval and t_f is the chronon representing the finish of the interval. An instance of temporal element is a finite disjoint union of intervals.

There is also an additional class extended chronon which is a specialization of interval. The extended chronon is an interval that is considered, for the purposes of the application, to be a chronon and is treated as such by the system. This enables chronons to be defined which are of greater duration than that specified by the granularity of the system. For example, if the granularity of the system is hour but we wish to timestamp certain objects using years then we can introduce the extended chronon year. We can then compare the times of objects stamped with this temporal object in terms of chronons rather than intervals. Dealing with extended chronons explicitly in the model, rather than intervals, enables the modeller more closely to model his/her perception of time in the application domain. Allowing time chronons at different granularities leads to the problem of how to compare such chronons [18]. However, as an extended chronon can easily be reconverted to an interval the standard chronon/interval comparisons can be used.

---

[3] The term interval is becoming standard [9] but has also been known as time period by some workers, for example, Sarda [17]

Event Modelling Classes. An *event* acts on one or more objects to change their states. Events are often considered to occur at specific points in time and are therefore often referenced by chronons. However, it may sometimes be useful to model events which have a duration. In such cases events can be associated with intervals. Which method to use depends on the needs of the application and its granularity, but time intervals are necessary where it is desirable for an event to be decomposed in order to provide more information about the event. We therefore specialize a general **event** class into the classes **simple event** and **complex event** corresponding to the above methods. Both inherit an event identifier and event descriptor but the **simple event** class has a chronon as its time reference whereas the **complex event** class is associated with an interval. Additionally, the **complex event** class definition includes an *event list* of timestamped elements. Each element consists of a spatial reference and a set of dynamic attribute values.

For each event class, an instance is associated with two consecutive application object states to form a *state change* which identifies the state of the object before the event, the time of the event, and the state of the object following the event. The following syntax is used for event modelling:

- simple event ::= ⟨event_id, event_description, chronon⟩
- complex event ::= ⟨event_id, event_description, interval, event_list⟩
- event_list ::= ⟨spatial_object_id, attribute_list, chronon⟩
- state change ::= ⟨application_object_id, event_id, application_object_id⟩

The example below illustrates the use of the **complex event** class. Let us consider modelling the effect of a storm on the composition of a coastal area. The map (Fig.1) shows the state of the coastal area before and after the event, and is represented by $\langle O_1, E_1, O_2 \rangle$ where O_1 and O_2 are application object identifiers and E_1 is an event identifier. If the details of the storm are not important then the event may be marked using a **simple event** instance. If, however, more information about the storm is required a **complex event** instance is used. If we model the event using a **simple event** instance, then we have:

$$\langle E_1, D_1,\ 11\text{-}06\text{-}83\text{:}1400 \rangle$$

where D_1 is the storm description identifier which references, for example, 'a force 9 gale moving from East to West across the Island.' and 11-06-83:1400 is a chronon (at granularity minute). If, on the other hand we use a **complex event** instance we get:

$$\langle E_1, D_1,\ 11\text{-}06\text{-}83\text{:}1400\ -\ 12\text{-}06\text{-}83\text{:}1630,\ L_1 \rangle$$

where 11-06-83:1400 - 12-06-83:1630 is an interval and L_1 is an event list. The required storm details at different locations at various times can then be recorded as follows:

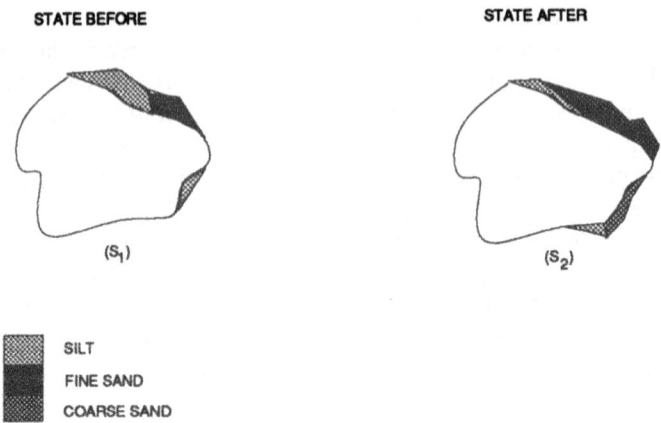

Fig. 1. State of a coastal region before and after a storm.

L_1:

$$
\begin{array}{ccc}
\textbf{spatial\_object\_id} & \textbf{windspeed} & \textbf{time} \\
124 & 54 & 11-06-83:1731 \\
125 & 62 & 11-06-83:2309 \\
126 & 59 & 12-06-83:0630 \\
\vdots & \vdots & \vdots
\end{array}
\tag{1}
$$

The time component of the list is constrained such that each chronon is within the bounds of the interval used to timestamp the complex event instance. An event list object must be tailored for each application event class being modelled, in this case storm.

Parallel State Class. The types of classes described so far are used to model objects which have single states at a given valid time. The state of an object at a particular time is represented, in a spatio-temporal application, by its spatial reference and attribute values at that time. Most ST-models do not permit an object to have more than one state at a particular valid time[4,5]. We term this

---

[4] Many versioning models permit multiple representation of an object for a single transaction time. These models have generally been developed to allow concurrent multi-user updating of an object to handle the problem of long transactions in spatial applications. However, in these models an object still has only one state at a particular valid time.

[5] Note the work of Stonebraker and Keller [22] and Stonebraker [21] on hypothetical relational databases.

the *spatio-temporal exclusivity* constraint. For many applications this is precisely the constraint we require. For example, if we would like to know the boundary of the county of Staffordshire in 1987, we only expect to retrieve one boundary for that time. On the other hand, we may also want to know about the proposed boundaries for a county. Let us consider a situation where a change is proposed for the county boundary of Staffordshire and several possible options are considered. We assume the proposed change is made in 1986 and is due to take effect in 1987. In the database we therefore need to represent and be able to distinguish between several representations for the boundary of Staffordshire in 1987. We also need to be able to record which option was chosen to be the actual boundary and maintain sufficient information so that in the future we can query proposed and actual boundaries. If we model the proposed boundaries in the same way as the actual boundary chosen, then we have a problem in distinguishing between the multiple representations of the boundary for a given valid time. To overcome this problem we identify two types of object state. We define *actual* states as those states which actually exist in terms of the application domain (for example, the actual boundary of Staffordshire at a particular time) and *hypothetical* states as those states which are potential (for example, the proposed boundaries of Staffordshire at a particular time). We also relax the spatio-temporal exclusivity constraint to allow multiple states of an object to exist co-temporally, providing that only one state is an actual state. Each co-temporal state of an object is known as an *alternative*. Each alternative may be either in a hypothetical, actual or additionally a transition state (see below). An individual alternative may change its state over time.

An instance of the **parallel state** class is used to represent the multiple alternatives of an application object and is structured such that the set of hypothetical alternatives as well as the actual alternative at a particular time may be yielded. This allows for analysis of proposed schemes against the actual scheme chosen and the recording of periods of uncertainty in an object's history or future. The **parallel state** class will be further explained by means of an example. We will consider a proposed rail link between Dover and London, illustrated in Fig.2. Let us assume six stages in the construction of the rail link, as shown below, with $t_1 - t_6$ representing intervals.

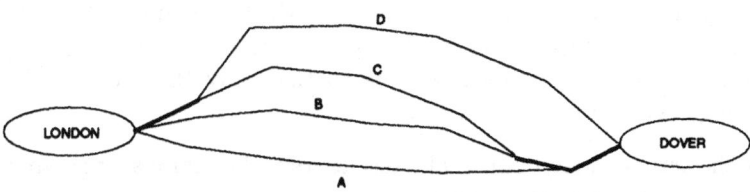

Fig. 2. Proposed Channel Tunnel Rail Links

t_1 Options A,B,C proposed.
t_2 An additional option D proposed.
t_3 Option B rejected.
t_4 A chosen as prefered option.
t_5 Construction begins.
t_6 A completed.

From the above it can be seen that t_1 to t_4 represent a period of uncertainty with several alternatives possible during each interval, t_5 is the interval during which the object is in transition and t_6 is the interval when the object has an actual state. We simply represent this as a **parallel state** instance, which is a list. Each member of the list is a set of alternatives of the object together with an interval. During the transition period an alternative may consist of an actual/hypothetical composite. Each composite within the transition period is in a *transition* state. Depending on the amount of information required several transition states of the alternative over time are possible during the transition period. The example above may be represented by:

Attribute-set	Time
(A,B,C)	t_1
(A,B,C,D)	t_2
(A,C,D)	t_3
(A)	t_4
(A)	t_5 (Transition state).
(A)	t_6 (Actual state).

It should be noted that the temporal references t_1 to t_6 could be bitemporal if transaction time support was required. This would enable the information to be analysed on the basis of what was known by the database at a particular time. In our model a transaction time interval can be associated with each attribute-set in addition to the valid-time interval. The data can then be queried on the basis of either or both temporal dimensions.

Each element (alternative) of the attribute-set consists of an attribute component plus a spatial mapping to an appropriate spatial object. In practice, the implementation requires versioning strategies to reduce the storage of duplicate geometry:

- firstly, between alternatives of one set which contain intersecting spatial sections for example, links marked with a broad line on Fig.2;
- secondly, between consecutive sets so that only those alternatives which change between sets are recorded.

While the actual and hypothetical alternatives each require a single spatial reference, that of transition alternatives is more complex and requires a pair $\langle S_a, S_h \rangle$ representing the actual and hypothetical parts of the spatial element of the alternative respectively. In fact the spatial representation of the alternative may be split topologically where the actual and hypothetical components meet, as shown in Fig.3.

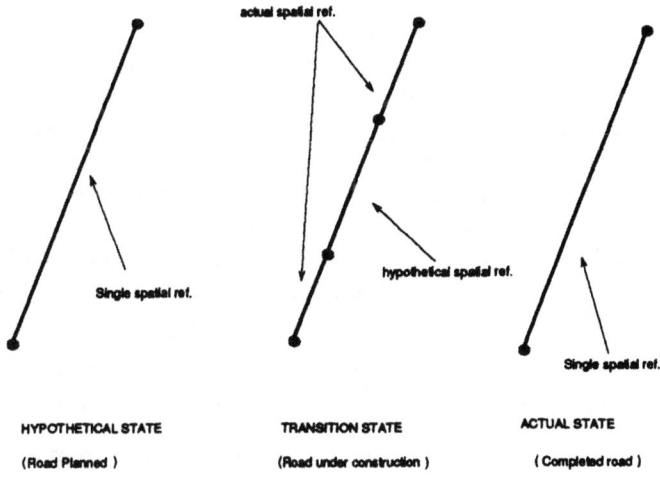

Fig. 3. The spatial representation of an alternative in various states.

Auxiliary Classes. In addition to the above three groups, auxiliary classes such as granularity are required. An instance of **granularity** is associated with an instance of temporal domain to define the granularity for the system (that is to say, day, month, decade, etc.). We introduce the constant chronon *future*, for the purpose of modelling events which occur at some future unspecified date. For example, a mapping agency may be given information regarding a boundary change to take place at the next general election, the date of which is not known at the time. Therefore the relevant information may be entered and stamped with future. An actual time value may then be substituted for future when the date of the election is known.

4 Integration

This section looks at the issues related to the integration of the various data components of a spatio-temporal (ST)-application. If an ST object's spatial or attribute components are non-atomic then the integration of time within it may take place at several levels. It is apparent therefore, that there are many ways in which a particular ST-application class may be specified. The definition below provides a framework within which such a specification can take place and a simple example is given to illustrate its use.

4.1 Definition: ST- application scheme

An ST-application scheme is defined as $\langle \mathcal{C}, \mathcal{R} \rangle$, where:

– $\mathcal{C}$ is the set of all ST-application classes defined for the application.

– $\mathcal{R}$ is the set of all relationships among the elements in $\mathcal{C}$.

The elements of $\mathcal{C}$ will be explained in terms of their constituents, such as their spatial and attribute components and their associated temporal classes. A class $C \in \mathcal{C}$, may be written as $C = \langle C_s, C_a \rangle$, where C_s and C_a are the spatio-temporal and the attribute-temporal components of C respectively. The relationships in $\mathcal{R}$ are considered to be of the usual kind found in any object-oriented model, for example, inheritance, and are not considered further. The behavioural aspect of the class is handled in the normal way through the definition of methods on the class. We therefore proceed with a more detailed definition of the structural component of C.

Definition: structural component of C. A class consists of a tree with C_s and C_a forming the two children of the root node: C_s and C_a are defined below.

– The spatio-temporal component, C_s, is a tree formed from elements of the spatial model, S, of the application, with or without an associated temporal class. More formally, if T is the set of temporal classes, then each spatial node of the tree n_s, is written as

$$n_s = \begin{cases} s_l(t) & \text{with associated temporal class} \\ s_l & \text{without associated temporal class} \end{cases}$$

where $t \in T$, $s \in S$ and l represents the level of the node within the tree. For convenience the node may also be represented by $s_l(t)$ where t may be null. The levelling in the tree represents the aggregation (is-a-part-of) relationship between the elements. The form of the tree is constrained such that the arrangement of the nodes within the tree must not violate the aggregation relationships present in the underlying spatial model.

– Similarly, the attribute-temporal component, C_a, consists of attribute classes arranged into a tree. The attributes are members of the set of all the attribute classes of the scheme (A). Each node of the tree is written as:

$$n_a = \begin{cases} a_l(t) & \text{with associated temporal class} \\ a_l & \text{without associated temporal class} \end{cases}$$

where $t \in T$, $a \in A$ and l represents the level of the node within the tree. Again for convenience the node may be represented by $a_l(t)$ where t may be null.

The diagram Fig.4 illustrates the structure of an ST-application class. The definition allows spatial or attribute aggregates to be represented within the class through the levelling within the tree. Additionally, a temporal object class may be introduced at any level within the spatial or non-spatial component of an ST-application class. In practice it is likely that only one or two spatial object classes will be associated with an ST-application class at level one within the C_s sub-tree. A derived relationship between a node of the spatial and attribute trees is possible where an attribute's value is a function of a spatial or spatio-temporal object. A dotted line may be used to represent this on the diagram.

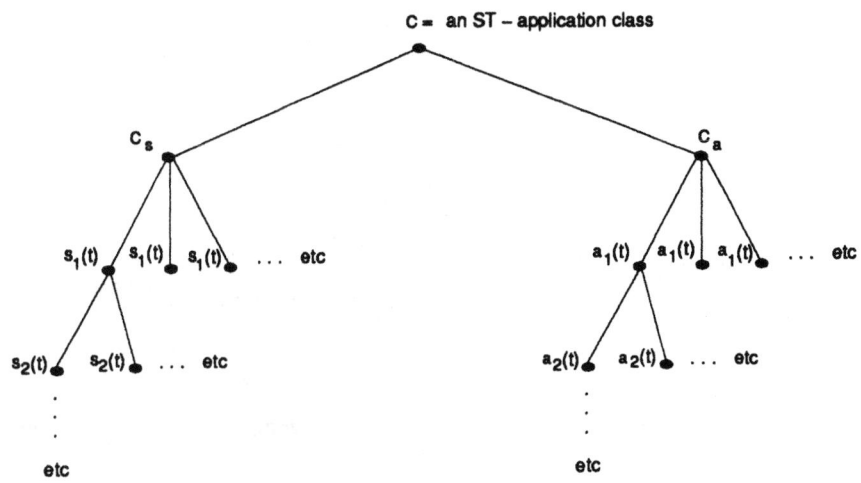

Fig. 4. Representation of an ST-application class.

Example. Figure 5 shows a diagrammatic representation of the ST-application object: house. We assume one atomic attribute, owner, a composite attribute address and one spatial object polygon to represent the spatial extent of the house. We have an additional attribute area derived from the area of the polygon. Let us assume that we use the spatial model of ARC INFO. In this model a polygon is a complex spatial object composed of arcs. In this example the designer has decided to reference at the polygon level using interval and the owner attribute with temporal element. Such decisions would in practice be based on the requirements for the particular application.

Constraints. Temporal constraints are required between each parent and child node of the tree to ensure the validity of the structure for class instances. A range of constraints are needed depending upon the nature of the component of the aggregate, namely, dependent[6] or independent, and the type of associated temporal object. Furthermore, spatio-temporal constraints are needed for aggregate spatial objects to prevent violation of the spatial topology of the underlying spatial model at any time.

It is also necessary to consider relationships between objects or classes in a temporal application as only objects which exist at the same time can participate in a relationship. Many temporal models include temporal constraints, for example Wuu and Dayal [29] and Petrounias and Loucopoulos [12]. These have generally been concerned with classes/entities time-stamped with a uniform temporal object. For our model it is also necessary to define constraints

---

[6] A dependent component object can only exist as part of the aggregate object whereas an independent object can exist in its own right regardless of the presence of the aggregate.

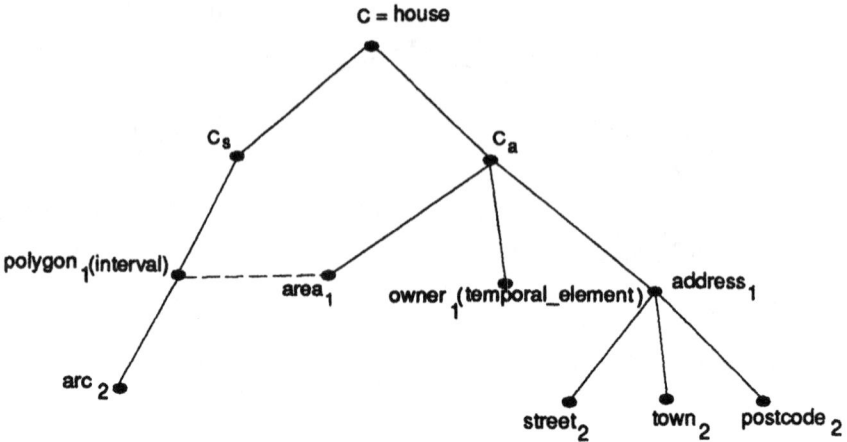

Fig. 5. ST-application class - house.

between objects which are associated with different temporal classes, for example, interval, temporal element, and consider two temporal dimensions. These constraints are formally defined in [23].

5 Tool Design

The aim of the CASE tool is to facilitate the design of spatio-temporal applications by enabling a comparison of various design strategies through rapid prototyping. In order to achieve this, a flexible interface is required which caters for various possible strategies to be tried, applied and if necessary altered and retried. The tool should enable a developer to:

- define those features of the temporal domain applicable to the entire application;
- select from a variety of temporal classes;
- associate a temporal class with an application class at any required level within its structure;
- test out the model at the application user level in terms of benchmarks and usability.

In addition, the tool should generate appropriate temporal constraints for the user model and select the appropriate temporal editor for the particular application class.

5.1 Temporal System Development

Before proceeding with the description of the implementation it is first necessary to outline some general points relating to temporal systems development. These will be considered in terms of the 'levels of interaction' within an object-oriented GIS. We describe a three level architecture and outline the pre-requisite temporal development required at each level.

System designer level: this consists of the design of standard object classes (for example, standard object editors) and system tables (for example, tables to store geometry) used by or available for use by all applications.

Temporal requirements: the introduction of standard temporal classes, system tables and editors, and procedures to integrate them with the non-temporal system. The definition of standard temporal constraints (as discussed in section 3) which may be applied to the instances of ST-application classes which take part in a relationship. Additionally, operations should be defined between objects of different temporal classes.

Application designer level: the application classes (for example, house), functionality and class relationships for specific applications are constructed.

Temporal requirements: enhancement of data manipulation language (DML) commands to access the temporal system level, in conjunction with standard GIS class libraries and system tables, to enable the creation of application classes with temporal support.

User level: class instances (for example, '22 Acacia Avenue') are created and are inserted to system and application level tables via the manipulation of editors defined at the system level.

Temporal requirements: the user will interact with the standard temporal editor appropriate for the class.

5.2 Implementation

The implementation is based on the temporal class model of section 2, the ST-application definition of section 3 and has been undertaken using the Smallworld GIS and CASE tool. The Smallworld system will be outlined in terms of the architecture shown in Fig.6.

In the Smallworld GIS the application level data modelling is facilitated by the CASE tool. The developer can define application classes and their relationships with other classes graphically. Details such as fields (physical, logical or geometric) and the manifolds to which the object belongs are specified using a set of editors. Interaction with the system level is established when the model is applied to the database.

Fig.6(C) shows the architecture of our temporal implementation. Limitations were applied to simplify the initial prototype.

– Since our research is primarily concerned with spatio-temporal integration, attribute data was considered time invariant.

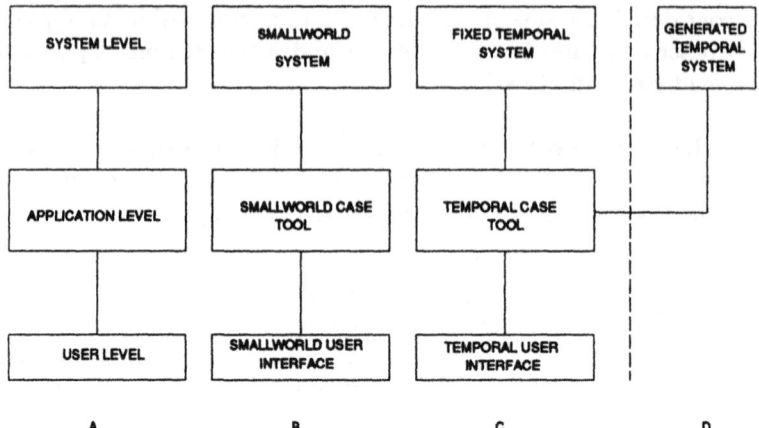

Fig. 6. System Architecture

- The temporal dimension is permitted to be valid or bitemporal but not transaction time alone.
- Only a single spatial object is allowed at level one within the tree of each ST-application class (see section 4.1).

The extensions required for the fixed temporal system level have been described above. However, instead of having a static all encompassing system level temporal model the required temporal system tables are generated dynamically from the CASE tool so that only those tables required to support the application data model are created (Fig.6(D)).

At the application (CASE tool) level the current window system has been extended by including temporal CASE editors and redefining the methods which access the system level tables and classes. Due to the different types of temporal modelling being supported it has not been possible for all temporal classes to be handled in a similar way to the standard temporal classes. This is particularly true for the **parallel state** class which has required different treatment at the CASE level and more complex user editors.

One of the aims of the implementation was to work within the existing structure of the Smallworld CASE tool so that it could still function as a non-temporal spatial CASE tool for non-temporal applications. As the treatment of the spatial component of application classes requiring temporal support was substantially different than for standard Smallworld CASE objects[7], it was necessary to develop additional CASE editors to handle the geometry of temporal application classes.

Developer Walk-through. The following section provides an overview of the system, at the two levels of temporal design described in the model of section

---

[7] Smallworld uses the term CASE objects to refer to application classes.

2. It is viewed from the application developer's perspective and is illustrated by way of two simple examples. Firstly, the definition of granularity at the application definition level and secondly the selection of a temporal class at the object definition level.

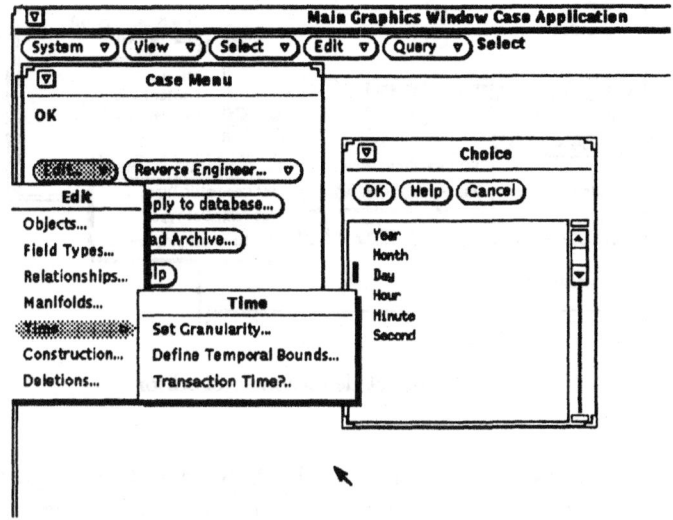

Fig. 7. Application definition level editor

Figure 7 shows the time option button at the application definition level plus the pull-down menu from which various options may be selected by the developer. In the example shown the *Set Granularity* option has been selected. The developer is presented with a choice of granularities from which to make a selection. The granularity chosen then becomes the default granularity for the application which effects among other things the storage type used for time. The *Transaction Time?* option defines whether transaction time support is required for the application. As the default is valid time only, selecting this option means that all temporal classes defined for the application will be bitemporal. The transaction time support is provided by the system associating a transaction time interval with each application class defined by the developer. The remaining option is used to define the valid time bounds of the application. The time bounds are used by various constraints, for example, to ensure that a time value inserted is within the bounds allowed by the application.

Figure 8 shows the definition of an application class. The time menu has two options, *Set Local Granularity* and *Set Temporal Geometry* which is again a pull-down menu. The granularity option at this level enables the developer to override the system granularity for the particular application class. Only larger granularities than system granularity are permitted, for example, default system granularity day, class granularity year. This is in fact the method via which ex-

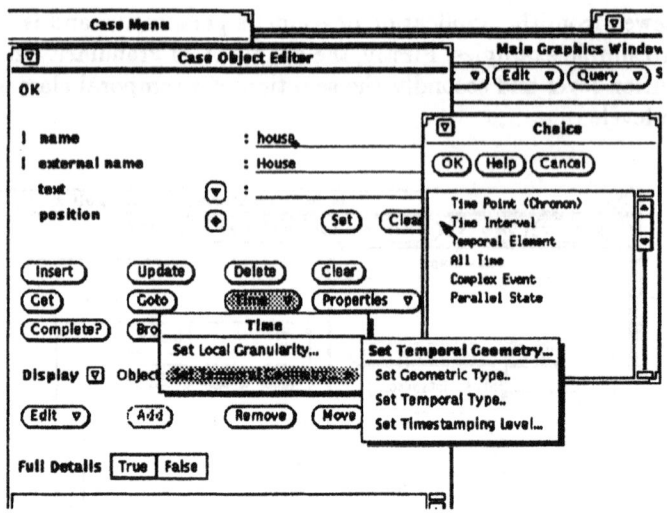

Fig. 8. Object definition level editor

tended chronon is associated with an application class. The temporal geometry option is used to specify geometry for temporal application classes to avoid effecting the standard Smallworld treatment of geometry for non-temporal CASE objects. The *Set Geometric Type* option is used to select from a list of spatial types from the Smallworld spatial model and name the temporal geometric attribute. For example, polygon may be selected and given the name Location. The type selected (polygon) is then associated with a particular temporal class from the next menu option. This is the example shown in the screen dump with the *Set Temporal Type* option chosen. The developer selects a temporal type from the list provided. The spatial component of 'house' will be associated with the temporal class, in this case interval. The developer can then select the level at which the timestamp is to be applied by selecting the timestamping option. The choice offered here depends on the geometric type selected, for example, if the geometric type was point it can only be timestamped at one level.

Once the definition of the application class(es) is complete the data model is then applied to the GIS as usual for the Smallworld CASE tool. This generates the required temporal support for the classes defined in the CASE tool. The user is then able to edit and create individual objects of the type(s) defined in the GIS, using the temporal editor appropriate for the particular temporal class associated with the application class. It should be noted that the appropriate editor is automatically assigned from the CASE tool definition.

Future Work. Currently the temporal editors provide support for queries based on the operations defined on the classes in the temporal and integration models. Although not the current goal, it is clear that the generated system would be greatly enhanced by the development of a spatio-temporal query language either:

- specific to this implementation: based on the integration of the temporal class operations and the existing Smallworld spatial query facility (to be effectively used in practice this would also require much work on visualization techniques to adequately represent the result of queries);
- at a more general level: requiring an extension of the temporal and spatio-temporal class operations into a full query language.

Additionally, the implementation may be extended by allowing the temporal modelling of attribute data, providing benchmarking guidelines for the temporal integration options and enhancing the graphical CASE interface to indicate which CASE objects and relationships are temporal.

Acknowledgements. One of the authors is jointly funded by the Ordnance Survey of the United Kingdom and the United Kingdom EPSRC (ref: 92566298). The authors would also like to thank Asoka Karunananda for his useful comments.

References

1. K. Al-Taha, R. T. Snodgrass, and M. D. Soo. Bibliography on spatio-temporal databases. *SIGMOD Record*, 22:59–67, 1994.
2. J. Clifford and A. Croker. The historical relational data model HRDM and algebra based on lifespans. In *Proceedings of the Third International Conference on Data Engineering*, pages 528–537, 1987.
3. J. Clifford and A.U. Tansel. On an algebra for historical relational databases: Two views. In S. Navathe, editor, *Proceedings of ACM SIGMOD International Conference on Management of Data*, pages 247–265, Austin, Texas, USA, May 1985.
4. U. Dayal and T. Wuu. Extending existing DBMSs to manage temporal data: an object-oriented approach. *Proceedings of the International Workshop on an Infrastructure for Temporal Databases*, 1993.
5. M.J. Egenhofer and A. Frank. Object-oriented modelling for GIS. *Journal of the Urban and Regional Information Systems Association*, 4:3–19, 1992.
6. R Elmasri et al. A temporal model and query language for EER databases. In A. Tansel et al., editors, *Temporal Databases: Theory, Design and Implementation*, chapter 5, pages 213–229. Benjamin Cummings, 1993.
7. S Gadia and J Vaishnav. A query language for a homogeneous temporal database. In *Proceedings of the ACM Symposium on Principles of Database Systems*, pages 51–56, 1985.
8. S.K. Gadia. A homogeneous relational model and query language for temporal databases. *ACM Transactions on Database Systems*, 13(4):418–444, December 1988.
9. C.S. Jensen et al. A consensus glossary of temporal database concepts. *SIGMOD Record*, 23:52–64, 1994.
10. T. Kampke. Storing and retrieving changes in a sequence of polygons. *International Journal of GIS*, 8(6):493–513, 1994.

11. R. Newell and M. Easterfield. Version management - the problem of the long transaction. Technical Report Technical Paper 4, Smallworld Systems, 1990.

12. I. Petrounias and P. Loucopoulos. Time dimension in a fact based model. *1st International Conference on Object Role Modelling*, 1, July 1994.

13. N Pissinou and K Makki. A unified model and methodology for temporal object databases. *International Journal of Intelligent and Cooperative Information Systems*, 2(2):201–23, June 1993.

14. L. Rackham. Development of a system for the management and supply of data on administrative areas and public boundaries, February 1992. Paper to CERCO Working Group IX , Updating of Digital Maps and Topographic Databases. Sweden March 1992.

15. S.A. Roberts, M.N. Gahegan, J. Hogg, and B. Hoyle. Application of object-oriented databases to geographic information systems. *Information and System Technology*, 33(1):38–45, 1991.

16. J. Rose and A. Segev. A temporal object-oriented data model with temporal constraints. *Proceedings of the 10th International Conference on the Entity-Relationship Approach*, pages 205–229, 1991.

17. N. L. Sarda. Algebra and query language for a historical data model. *The Computer Journal*, 33(1):11 – 18, 1987.

18. N.L. Sarda. HSQL: A historical query language. In A. Tansel et al., editors, *Temporal Databases: Theory, Design and Implementation*, chapter 5, pages 110–140. Benjamin Cummings, 1993.

19. R. Snodgrass. The temporal language TQuel. *ACM Transactions on Database Systems*, 12(2):247–298, 1987.

20. M. D. Soo. Bibliography on temporal databases. *SIGMOD Record*, 20:14–23, 1991.

21. M. Stonebraker. Hypothetical data bases as views. In *Proceedings of SIGMOD Conference*, pages 224–229. ACM, 1981.

22. M. Stonebraker and K Keller. Embedding expert knowledge and hypothetical data bases into a data base system. In *Proceedings of SIGMOD Conference*, pages 58–66. ACM, 1980.

23. P. A. Story and M. F. Worboys. An object-oriented model of time. Technical Report TR95-03, Department of Computer Science, Keele University, Keele, Staffordshire, UK., 1995.

24. A. U. Tansel. Adding the time dimension to relational model and extending relational algebra. *Information Systems*, 11(4):343–355, 1986.

25. A.U. Tansel. A generalized relational framework for modeling temporal data. In A. Tansel et al., editors, *Temporal Databases: Theory, Design and Implementation*, chapter 5, pages 110–140. Benjamin Cummings, 1993.

26. M.F. Worboys. Object-oriented approaches to geo-referenced information. *International Journal of Geographical Information Systems*, 8:385–399, 1994.

27. M.F. Worboys, H.M. Hearnshaw, and D.J. Maguire. Object-oriented modeling for spatial databases. *International Journal of Geographical Information Systems*, 4:369–383, 1990.

28. Michael F. Worboys. A uniform model for spatial and temporal information. *The Computer Journal*, 37(1):26 – 34, 1994.

29. T Wuu and U. Dayal. A uniform model for temporal and versioned object-oriented databases. In A. Tansel et al., editors, *Temporal Databases*, chapter 10, pages 230 – 247. Benjamin Cummings, 1993.

Internal vs. External Spatial Information and Cultural Emergence in a Self-Organizing City.

Itzhak Benenson, Juval Portugali
Department of Geography, Tel-Aviv University, Tel-Aviv 69978, Israel.
Tel: (+3) 6423619, Fax: (+3) 6414148, E-Mail: bennya@ccsg.tau.ac.il

Modern, or rather postmodern, cities are characterized by spatial, social and cultural pluralism: the city can be described as a spatial mosaic of coexisting cultural and social groups, some of which are the product of "old' ethnic groups who emigrated to the city as already established cultural groups, while others, and this is the more recent phenomenon, are the dialectical product of the city itself. In a series of previous studies on the city as a self-organizing system we have examined various facets of the city's cultural and social dynamics. This was done by means of a family of models we have specifically designed for this purpose. The first model, *City*, reffered to already established cultural groups. The second model, *City-1*, was a planning oriented cell-space model which introduced, in addition to socio-cultural properties of individuals, also their economical status, as well as the changing land value surface of the city. With the third model, *City-2*, we have examined the possibility that the city dynamics can generate the emergence of a new cultural spatial entity. the present parer further elaborates on the question of social-spatial emergence and investigates the conditions by which changes in individuals' spatial mobility in conjunction with spatial cognitive dissonance entail structural changes in the spatio-cultural composition of the city.

Introduction.

Much of the city dynamics is created by individuals' intra- and inter-urban migration movements. That is to say, by inhabitants who change their residential location in the city and by immigrants who enter the city and try to find a home in it. The spatial decisions individuals take in this process are based on two forms of information: internal and external. *Internal information* is derived from the spatial consequences of the individual's cultural identity and socio-economic status. For example, a person whose cultural identity is *Blue* might prefer to spatially locate among Blue neighbors, whereas his/her socio-economic status might constraint his/her choice to certain areas in the city.

External information refers to actual spatio-cultural and spatio-economic configurations in the city at a certain moment. For example, a Green individual might find him/her self at a certain time living among Blue neighbors, or in an area which is beyond his/her economic status; or a Blue immigrant might find that at a certain time all vacant houses in the city are surrounded by Green neighbors. The location decision and migration behavior of the individual in the city thus reflect the dialectical tension between the internal and external information.

The external information which is usually available to individuals in the city comes in two spatial scales: local-scale information, concerning a house and its immediate neighbors, and mezo-scale information concerning the vacant houses in the city at a certain time. The global-scale information, concerning the cultural-socio-economic spatial configuration and structure of the city as a whole, as it evolves in time, is usually not available to the individual nor can it be: The global spatial structure of the city as a whole is constantly evolving and changing, to a large extent as a consequence of the movement of millions of individuals, in the city and outside it, whose decision to move is based on their internal information and on the local- and mezo-scale information available to them at a certain moment. The city as a whole is therefor an open, complex and consequently self-organizing system associated with phenomena of nonlinearity, bifurcation and phase transition.

One of the most prominent expressions to the city as a self-organizing system concerns its spatial cultural and social fabrics. Modern, or rather postmodern, cities are characterized by spatial social and cultural pluralism: the city can be described as a spatial mosaic of coexisting cultural and social groups, some of which, such as of Little-Italy or China Town, are the product of "old" ethnic groups who emigrated to the city as already established cultural groups, while others, and this is the more recent phenomenon, are the dialectic product of the city itself (Yappies, Gays, Lesbians, etc.).

In a series of previous studies on the city as a self-organizing system we have examined various facets of the city's cultural and social dynamics. This was done by means of a family of models we have specifically designed for this purpose.

The first model, termed *City*, is a probabilistic Cellular Automata simulation model and we have used it to examine the evolution of socio-spatial relations between large cultural groups in a city, as they evolve out of the city dynamics and the local- and mezo- level information available to individuals (Portugali, Benenson and Omer, 1994). This model referred to cultural groups whose cultural identity was determined and established long before they came to the city. The second model, *City-1*, is a planning-oriented cell-space model which introduced, in addition to socio-cultural properties of individuals, also their economic status, as well as the changing land-value surface of the city (Portugali and Benenson, forthcoming). Here too we focused on already established cultural groups (the specific case-study being the recent migration wave from ex-USSR to Israel). With the third model, *City-2*, we examined the possibility that the city dynamics can generate the emergence of, or give birth to, a new cultural spatial entity. The basic idea here is that the above noted dialectical tension between internal and external spatial information available to individuals, might entail cognitive dissonance on the part of individuals and consequently a change in their spatio-cultural identity. The model City-2 which we have designed for this purpose, can be described as a two layers model composed of a migration sub-model, describing the inter- and intra-city migration movements, superimposed on a Cellular Automata sub-model describing the dynamics of the urban landscape itself (Portugali, Benenson and Omer, forthcoming). The present paper further elaborates on the question of spatio-cultural emergence: whereas in the previous paper our main aim was to theoretically justify the notion of spatial cognitive dissonance and to operationally establish the possibility to model the process, here our aim is to investigate the conditions by which spatial cognitive dissonance emerges and entails a structural change in the spatio-cultural composition of the city, that is to say, the birth of a new cultural group.

The Model.

The model is built, as noted, of two layers: a migration layer, superimposed on an infrastructure layer, describing the city. For a detailed description of the model we refer the reader to Portugali, Benenson and Omer (forthcoming). Here we present the features essential for the discussion that follows.

The infrastructure of the city is a "cellular automata" model of a rectangular lattice of cells (houses). Each house is characterized by *a value* $V_{ij}(t)$ $(0 <= V_{ij} <= 1)$, which changes in time depending on the values of neighboring houses and their occupants. We define the (local) neighborhood of house P_{ij} as a 5x5 square with P_{ij} in the center.

$$U(P_{ij}) = \{P_{kl}, |k-i| < 3, |j-1| < 3\} \qquad (1)$$

This local neighborhood constitutes the external, local-scale, spatial information field, available to an individual decision-maker.

On top of this infrastructure we superimpose a layer of "individuals" which describes the various migration movements in the city. Each individual inhabitant in the city occupies a house P_{ij} at the infrastructure layer (we allow for one inhabitant in a house only) and is characterized by two features. The first one is the individual's *cultural identity* $C_{ij}(t)$ and the second one is the individual's *economic status* $S_{ij}(t)$ $(0 <= C_{ij} <= 1, 0 <= S_{ij} <= 1)$.

The value of houses and the features of the city habitants change in time. The economical status of the individual changes in time in a most simple manner.

$$S_{ij}(t + 1) = S_{ij}(t) + \sigma \qquad (2)$$

where σ is a random variable, normally distributed with mean 0.02, variance 0.0001.

The evolution of the house value depends on local neighborhood information from both layers. Namely, at iteration $t+1$ the value V_{ij} of house P_{ij} is an average of the values of the vacant places in $U(P_{ij})$ and the status of individuals, occupying houses in $U(P_{ij})$ at iteration t:

$$V_{ij}(t+1) = (\sum\{V_{kl}(t)| P_{kl} \text{ unoccupied}\} + \sum\{S_{kl}(t)| P_{kl} \text{ occupied}\}) / 25 \qquad (3)$$
$$P_{kl} \varepsilon U(P_{ij}) \qquad\qquad P_{kl} \varepsilon U(P_{ij})$$

The driving forces of the City evolution are the migration processes defined by the *cultural* (ΔC) and the *economic* (ΔS) *dissonances* between the individual and his/her local neighborhood.

For an individual occupying a house P_{ij} we define ΔC_{ij} and ΔS_{ij} as the absolute value of a difference between the individual's internal spatial information and the average value of the external local spatial information regarding the individual's neighbors:

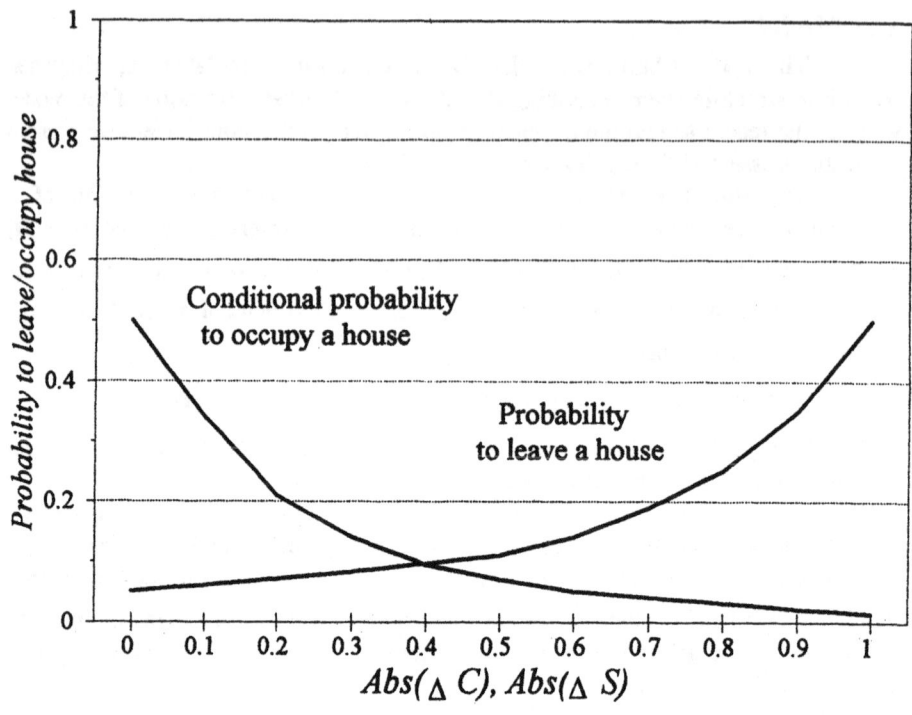

Fig. 1

$$\Delta C_{ij} = | \sum \{C_{kl}\} /N_{kl} - C_{ij} |, \, \Delta S_{ij} = | \sum \{S_{kl}\} /N_{kl} - S_{ij} | \qquad (4)$$

$$P_{kl} \, \varepsilon \, U(P_{ij}) \qquad\qquad P_{kl} \, \varepsilon \, U(P_{ij})$$
$$P_{kl} \text{ occupied} \qquad\qquad P_{kl} \text{ occupied}$$
$$P_{kl} \# P_{ij} \qquad\qquad P_{kl} \# P_{ij}$$

where N_{ij} is the number of occupied houses in $U(P_{ij})$.

The cultural and economic dissonances define the probability $\lambda(\Delta C_{ij}, \Delta S_{ij})$, for an individual situated at P_{ij} to leave the house, and the conditional probability $\kappa(\Delta C_{kl}, \Delta S_{kl})$ for a house searcher to settle at some empty house P_{kl} if it is the only possible choice.

An individual can occupy house P_{kl} when his/her current status is higher than the current value of the house:

$$S_{ij} > V_{ij} \qquad (5)$$

and, s/he chooses one from *all* houses P_{kl}, satisfying (5), based on the relative values of $\kappa(\Delta C_{kl}, \Delta S_{kl})$ for them. The qualitative patterns of dependencies of λ and κ on ΔC and ΔS are presented on Fig 1.

An internal migrant, who decided to leave his house, but could not realize his chance to occupy a new one, either stays in his previous house (with constant probability 1–φ) or leaves the City and joins the *Queue* (with probability φ). Additionally, at each iteration a constant number of L external immigrants try to enter the City. When doing so they first join the Queue. Each iteration all queuing individuals try to occupy a house in the City. Queuing individuals who did not succeed to occupy a house during some predetermined number of iterations T, leave the system.

To study the above question of spatio-cultural emergence we suppose that the initial city inhabitants, as well as external immigrants, belong to one of two cultural groups - Greens and Blues. We assign zero cultural identity to Greens and unit cultural identity to Blues. As a result, according the definitions of probabilities λ and κ, a Blue individual, living in a Green neighborhood will try to leave it and to settle in a Blue neighborhood, and vice versa. It was shown (Portugali, Benenson and Omer, 1994) that segregative behavior on the part of one of the groups only is sufficient to produce the spatial segregation of Blues *and* Greens at the mezo-scale (several clusters of Blues/Greens in the City). Furthermore, a process of spatial segregation takes time, and limitation (5) slows it down. As a result, some individuals might find themselves living in a foreign neighborhood, despite their will, under high cultural dissonance. In *City-2* an individual had two ways to resolve such a dissonance. One, is to leave the house or the city. The second, to change his/her cultural identity "in favor" of the neighborhood's cultural identity. We describe a change of identity of an individual who stays at house P_{ij} in the simplest way:

$$C_{ij}(t + 1) = C_{ij}(t) + \mu\Delta C_{ij} \qquad (6)$$

where μ is a random variable uniformly distributed on [a, b], $0 <= a < b <= 1$. Below, $a = 0$, $b = 0.04$.

As a result, if an individual stays in an undesirable neighborhood for a relatively long period of time, s/he gradually becomes indifferent to the cultural identity of the local neighbors, that is to say, this individual becomes *Neutral*.

Formally, we define an individual as a Neutral, when his or her cultural identity is between 1/3 and 2/3. When $0 <= C_{ij} <= 1/3$ or $2/3 <= C_{ij} <= 1$ we still call an individual Green or Blue. Note, that external immigrants can be either Blues or Greens, and that the Neutrals can only appear as a consequence of, and during, the City evolution.

The change in individuals' identity is a local process. When this process occurs (it does not always occur, because of the possibility to resolve the dissonance by leaving the city), two-fold development of the model City is plausible. First, Neutral, Blue and Green individuals can stay in mixed neighborhoods. The presence of Neutrals thus entails a relatively low cultural dissonances for each of the other neighbors. As a result, we can obtain a random spatial distribution at each scale. The second possibility is the spatial segregation of Blues, Greens, as well as of Neutrals. In our recent paper (Portugali, Benenson and Omer, forthcoming) we show that the latter situation is principally possible for some values of the model parameters. This implies that Neutrals have been self-organized by the system to live together with the consequence that they have now become an observable spatio-cultural entity in the city.

In the present paper we study at what extend the phenomena of socio-spatial emergence could be a common one.

Below we denote the fractions of Greens, Blues and Neutrals in the City as G(t), B(t) and N(t) respectively, and measure the spatial segregation of each group by means of the Lieberson isolation index (Lieberson, 1981). The index xP*y, measures the group spatial segregation relative the other groups taken together. Comparing the obtained values of the Lieberson index with the observed spatial distributions of the various cultural groups, we found that there exists a direct correspondence between the value of xP*y and the visual impression of the segregation pattern. On this basis we have chosen the value of xP*y = 0.4 as the boundary. This implies that if the Lieberson isolation index between the Neutrals and the two other groups is below 0.4, we consider the group of Neutrals as a newly emerging cultural entity.

To summarize:

(i) If an individual lives for a relatively long time in a state of cultural cognitive dissonance (between his/her internal and external spatial information) s/he will turn into a (Blue or Green) Neutral in terms of behavior.

(ii) If a Neutral lives for a relatively long period of time among Neutral neighbors, and the spatial distribution of Neutrals at the mezo-scale is spatially segregated, the neutral individual will acquire a new cultural identity. We define (i.e.paint) this newly emerging cultural group as Reds, and use the yellow color to mark the houses occupied by Neutrals.

Results, Discussion and Conclusions.

As stated above, we investigate the conditions by which a new cultural group can emerge in the City. In particular, we focus here on the way individual mobility φ is related to the emergence of a new cultural group - the Red people. Note, that the higher is individual mobility (φ) higher is also the individual's ability to respond to the incoming flow of external and internal information.

We describe the time evolution of the fraction of Neutrals in the City by the dependence of $\Delta N(t) = N(t+1) - N(t)$ on N(t), and the segregation phenomena by the value of xP*y.

As shown in fig. 2, for each value of φ the fraction of the Neutrals in the City converges to an equilibrium. The resulting fraction of the Neutrals remains high (above 35%) irrespective of change in the mobility from 0.1 to 0.9: at φ = 0.1 we have close to 100% Neutrals, at φ = 0.5 the fraction of neutrals stabilizes around 60 - 70%, whereas at φ = 0.9 the stable fractions of neutrals are between 35 - 50%. We do not discuss here the fact that there exist two accumulation points for φ = 0.5 and φ = 0.9 in fig. 2. High fraction of Neutrals is thus a necessary, but not sufficient condition for the emergence of a new social group. For the value of φ = 0.9 the fraction of neutrals is high, but they mostly occupy houses in culturally mixed neighborhoods. As a result, the value of Lieberson xP*y remains above the threshold level of 0.4 (fig. 3) and a new group does not emerge at the mezo-level (see fig. 4).

From fig.2 we can conclude, that there exists a boundary value of φ that defines the region of self-organization, and this value is above φ = 0.5. The numerical estimation gives φ approx. 0.6 (Fig. 5). For φ below this value the Neutrals self-organize into a new, spatially continuous, group (Reds).

In figures 2 - 5 we can see the spatial implications of changes in individual mobility. At φ = 0.1 the entire city becomes Red and the originally Blue and Green groups almost disappear. This is an extreme case of "melting pot" situation by which extreme restrictions on free movement reduces the cultural variability in the city. With time the city population becomes one homogeneous intermediate cultural group.

At φ = 0.9 the situation reverses. Now most, but not all, individuals can resolve cognitive dissonance by moving to a new location or by leaving the city. The city is now highly pluralistic in the sense that there are randomly appearing and disappearing local neighborhoods of specific cultural identity, while most of the individuals of the various groups live side by side in culturally mixed areas.

φ = 0.5 is typical of a culturally segregated city. The city has produced a new cultural group (the Reds) and this newly born group segregates in its own area. The whole city is highly segregated and its territory is spatially divided between the three main cultural groups - the previously existed Blues and Greens and the newly self-organized Reds. The Neutrals are now spatially distributed along the boundaries between the large segregated areas. From the above follows, that spatio-cultural emergence and segregation are inherent properties of the city dynamics.

Bibliography.

Lieberson S, (1981) An Asymmetrical Approach to Segregation, In: *Ethnic Segregation in Cities*, Peach C, Robinson V, and Smith S, eds, Croom Helm, London.

Portugali J, Benenson I, and Omer I, (1994), Socio-Spatial Residential Dynamics, Stability and Instability within a Self-Organizing City. *Geographical analysis*, 26, 4, 321-340

Portugali J, Benenson I, (forthcoming), Artificial Planning Experience By Means of a Heuristic Cell-Space Model: Simulating International Migration in the Urban Process. *Environment and Planning A*.

Portugali J, Benenson I, and Omer I, (forthcoming), Spatial Cognitive Dissonance and Socio-Spatial Emergence in a Self-Organizing City. *Environment and Planning B*.

Probability to leave the city = 0.1

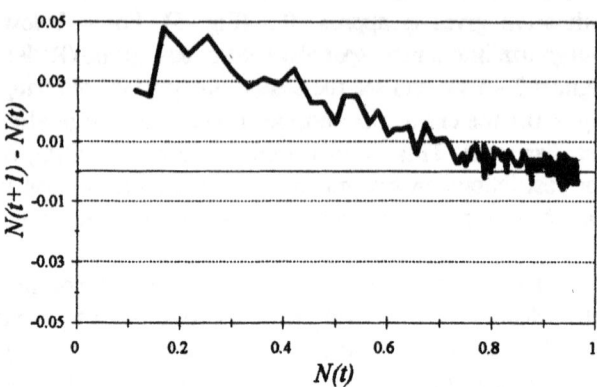

Probability to leave the city = 0.5

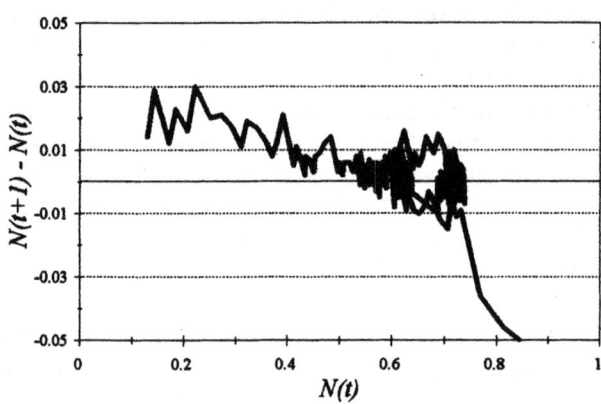

Probability to leave the city = 0.9

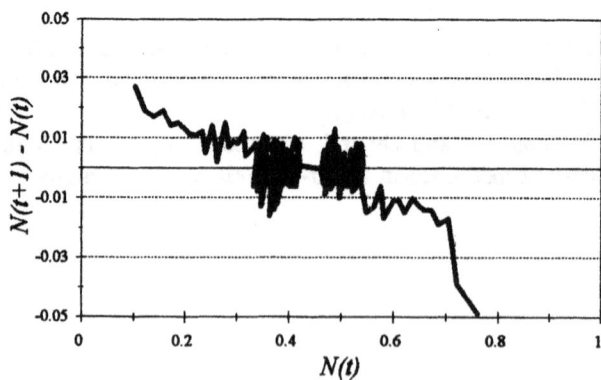

Fig. 2   Distribution of xP*y for the fraction of the Neutrals near equilibrium

Probability to leave the city = 0.1

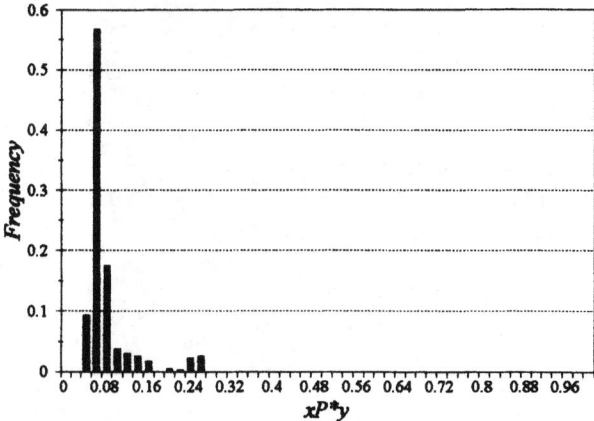

Probability to leave the city = 0.5

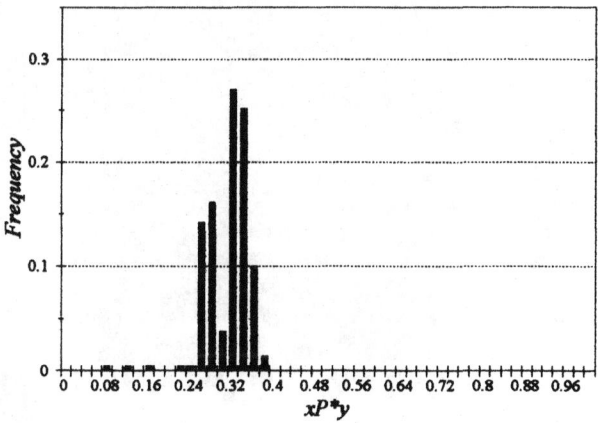

Probability to leave the city = 0.9

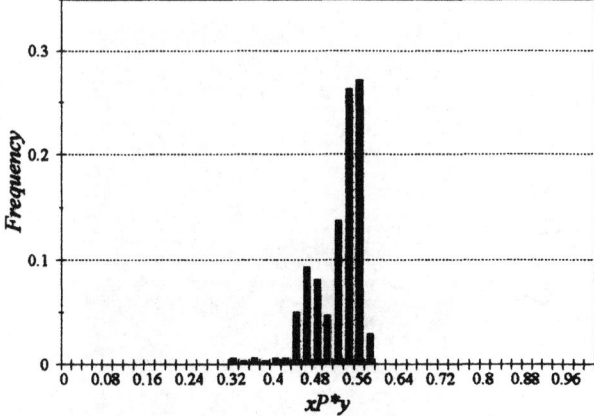

Fig. 3   Dependence of N(t+1) - N(t) on N(t)

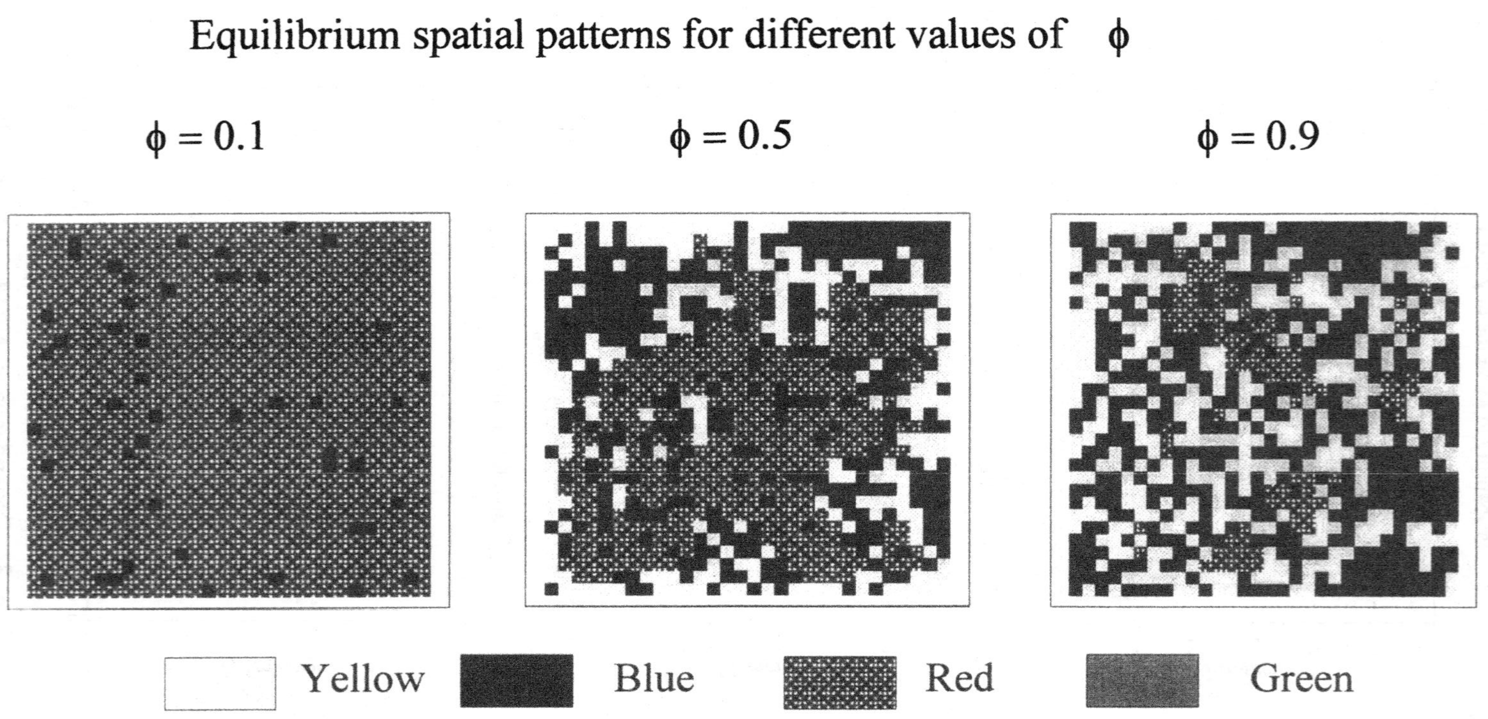

Equilibrium spatial patterns for different values of ϕ

$\phi = 0.1$ $\phi = 0.5$ $\phi = 0.9$

Yellow   Blue   Red   Green

Fig. 4.

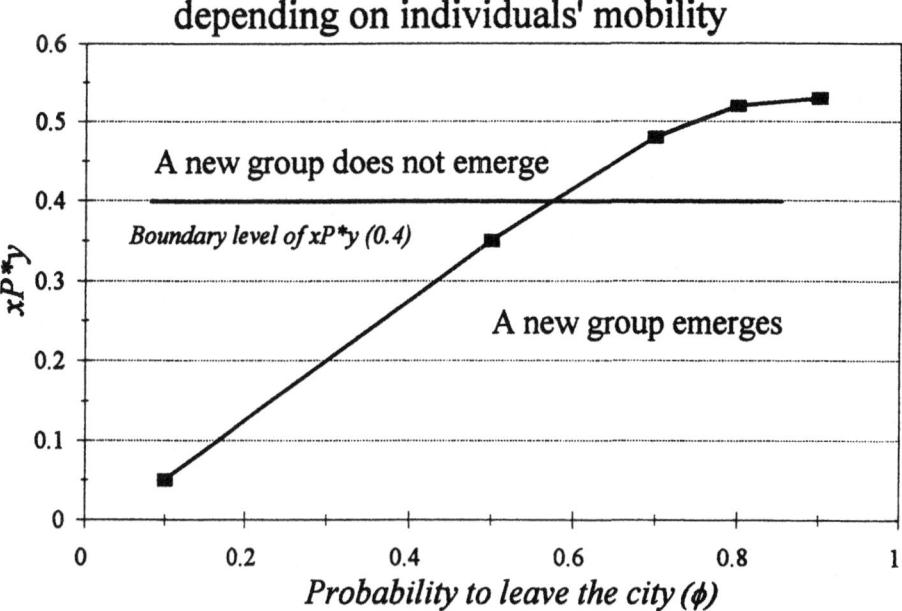

Fig. 5

The Social Perception of Space
Non-Spatial Determinants of the Use of Directionals in Tongan (Polynesia)[1]

Jürgen Broschart,
Dept. of Linguistics
University of Cologne
D-50923 Köln
e-mail: am004@rs1.rrz.uni-koeln.de

O. Abstract

In this paper it will be demonstrated that the use of so-called "directionals" in Tongan (*mai*, *atu*, *ange*, and *hake*) mainly depends on the social and communicative setting of the situation described, rather than on the positions or the movements of the people involved. Particular spatial interpretations seem to be a rather marginal aspect of a conceptual system created by routines of interaction between the observer and the observed.

1. Introduction

There are basically three models dealing with the relationship between space and social relations including linguistic communicative relations.

One of these models assumes that "concrete" spatial relations are primary and are metaphorized or extended to "abstract" social domains. This "space first" model is quite popular among "localists" such as Anderson (1971) etc., and is usually associated with Kant (1781/1956), who considers space an a priori form of "anschauung", which structures experience. Linguistic evidence in favour of this approach is the development of formerly local adpositions such as *with* (originally (spatially) 'against', cf. German *wider*) to comitative adpositions (*with* my friend). For details on this process of "grammaticalization" see Heine et al. 1991.

---

[1]This paper is based on a talk previously given at the Max-Planck-Institute for Psycholinguistics in Nijmegen, Holland. I wish to thank the Cognitive Anthropology Research Group for many helpful suggestions and for having provided me with their space kit so I could use it during my fieldwork in Tonga. As for my fieldwork, I am extremely grateful to the Helu and the Makaafi family, who made my stay in Tonga most enjoyable and profitable, as well as to Melenaite Taumoefolau (Univ. of Auckland), who gave many valuable comments on the use of the directionals in question. I am also grateful to the Sonderforschungsbereich 'Theorie des Lexikons' (Düsseldorf, Köln, Wuppertal) for funding my research on word classes in Tongan.

a) SPACE FIRST MODEL (Kant etc.)

metaphoric extension of space expressions to social relations

WITH ('against' (in space)) >>>>>>> WITH (my friend)

The second model, which we could call the "people first" model, maintains that society becomes spatialized in the sense that certain spatial positions will become symbols of social rank and social priviledge. Such an approach has been adopted, at least implicitly, by Bott (for discussion see Helu 1993) in her account of the kava ritual, where distance from the point of reference symbolizes a decrease of social power or prestige. In a similar vein, it has been observed that bodypart relations often become extended to spatial relations, so that e.g. HEART becomes to mean INSIDE (see Bowden 1992). In other words, according to this "anthropomorphic" model (von Foerster 1993a:77), the structure of space reflects the structure of a human body or a human society.

b) PEOPLE FIRST MODEL (Bott, Bowden)

extension of human body and human society to space relations

SOCIAL POWER symbolized by POSITIONING in ritual
HEART >>>>>>>> INSIDE ('heart of space')

The third model, i.e. the approach to be advocated here, would admit that to a certain extent both previous theories are right, but it might be argued that they fail to explain why extensions in either direction are possible. In order to be able to interpret the extensions mentioned there must be a common denominator for the different domains in question. Like the proponents of gestalt theory (see Smith 1988) and of constructivist action theories (see Mead 1938), I assume that domains of very different kind are structured according to the same general perceptual operations. According to the perceptual interaction model to be exemplified below, operational, interactive principles will structure space as well as society.

However, I will demonstrate that social relations will always be a more important trigger of these operations than spatial ones. Generally speaking, what counts are people, not frames, what counts are figures, not grounds. In other words, being localizable in space is not so much a matter of impersonal geometry but of routines of (social) interaction.

In particular, I shall demonstrate that the use of so-called "directionals" in Tongan (*mai, atu, ange,* and *hake*) mainly depends on the social and communicative setting of the situation described, rather than on the positions or the movements of the people involved. In other words, a native speaker of Tongan will be able to deduce far more from the use of these particles than just a description of a movement in space. Particular spatial interpretations seem to be a rather marginal aspect of a conceptual system created by routines of interaction between the observer and the observed[2].

---

[2]For discussion of the problem of the observer in the natural sciences and the humanities see Schmidt (ed.) 1993 and Köck 1986. According to Schmidt (1993:12), in "reflexive

If the history of a case marker etc. can be traced back to a spatial origin this only means that grammaticalized, "central" constructions are by default non-spatial, and that the expression of relations between people is often less explicit than the expression of spatial relations - which may require adpositions etc.. It is only through the process of grammaticalization that "peripheral" spatial expressions may enter constructions referring to more "interesting" relations between people, such as agent-patient-relations etc.

In other words, unlike most writers[3], I do not believe in the conceptual primacy of spatial relations, neither in terms of an a priori form of "anschauung" structuring experience, nor in the sense of a domain of primary interest to the observer. I assume that spatial relations are conceptually peripheral (cf. Tesnière's (1951) "circonstants") and a side-product of interactive principles structuring domains of any kind.

c)   PERCEPTUAL INTERACTION MODEL (Mead, Broschart)

<<<   same perceptual activity   >>>

High degree of grammaticalization

primary domain of interest	secondary domain of interest
SOCIAL	SPATIAL
unmarked expression	marked expression

Explicit marking of relation

2. The use of directionals in Tongan

In the following we shall take a look at the various uses of the Tongan directionals mentioned above, starting with a "standard" account of *mai, atu*, and *ange*. These so-called "directionals" have at least been reconstructed for Protopolynesian. *Mai* probably goes back to an action word 'to come', the lexical origin of *atu* and *ange* is unknown.

---

cybernetics [the] task of the observer includes the explanation of his own function and his development" (trl. JB).

[3]Compare Deane (1992) who advances "the thesis that the capacity to process syntactic structure is based upon cognitive structures and processes which apply in the first instance to physical objects" (p. 3). He argues "that human linguistic abilities depend upon the processing of linguistic information by brain structures whose primary function is the processing of spatial structure" (47). It is true that "stupid" neurons have nothing to operate on than physical stimuli, but as long as the organism does not manage to consider itself distinct from what it observes and is not able to "invent" such perceptual qualities as "feelings" and "self", there will not be any concept, be it spatial or non-spatial. For a survey of biosystems theory and cognition see Köck 1986.

Most commonly, *mai* is glossed as 'towards the speaker', *atu* as 'away from the speaker/towards the hearer', and *ange* as 'across to somebody other than speaker or hearer'. As we shall see later, especially the word *hake* '(moving) upward' is gradually intruding into this speech-act-related framework. Given the spatial set-up of diagram (1) below,

(1)

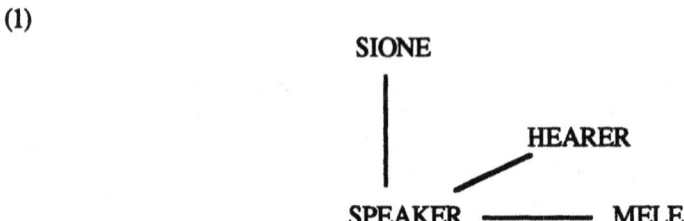

the following utterances are "good examples" for this standard "spatial" use:

(2a) 'Omai   ha('aku)   fo'i keke!
     give-MAI USP.POSS.1.SG   CLASS cake
     "Give me a piece of cake!"

(2b) 'E   'oatu   ha('o)   fo'i   keke?
     FUT give-ATU   USP.POSS.2.SG CLASS cake
     "Should a piece of cake be given to you?/Do you want a piece of cake?"

(2c) 'Oange   ha   fo'i keke kia Mele!
     give-ANGE USP.ART CLASS cake ALL Mele
     "Give a piece of cake to Mele!"

(2d) 'Alu hake   kia Sione!
       go HAKE   ALL Sione
     "Move up to Sione"

Diagram (3) illustrates these standard readings:

(3)

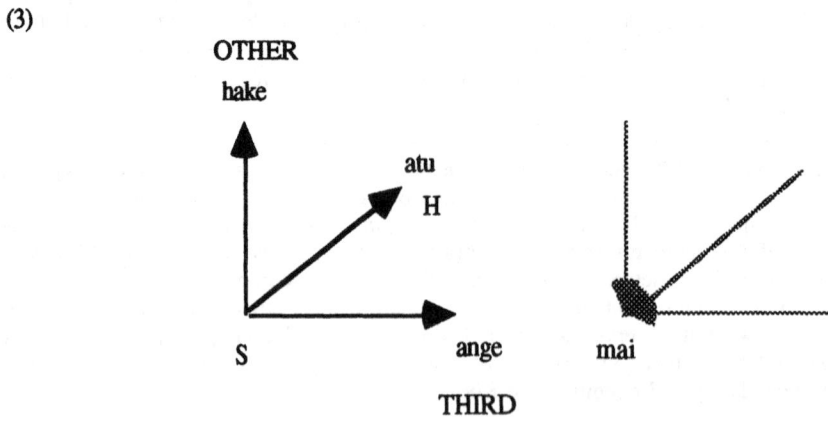

In this spatial model, the *mai* - direction is defined by the position of the speaker, regardless where the relation comes from. The *atu* - axis is defined by the positions of speaker and hearer. The *ange* - axis is defined by the positions of speaker and third, not including the position of the hearer, i.e. it is prototypically in a rectangular relation to the *atu* - axis. Note that if a third person is placed behind the hearer the relation would still be *atu* (just as it would still be *mai* if Third was located behind the speaker).

(4)

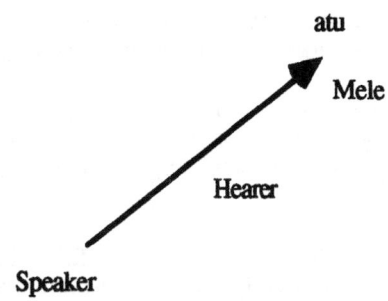

(5)　'Oatu　　ha　　fo'i　　keke　　kia Mele!
　　　give-ATU USP.ART CLASS　cake　　ALL Mele
　　　"Give a piece of cake to Mele (standing behind you)"

There are other spatial expressions based on the three spatial coordinates which are not or not necessarily speaker-oriented. E.g. there is an "absolute system" of cardinal directionals such as *hihifo* 'west' vs. *hahake* 'east', which are apparently based on *hifo* and *hake* in the sense of 'down' or 'setting' and 'up' or 'rising' (of the sun). North in Tongan is *tokelau*, and south is *tonga* (the Kingdom of Tonga being in the south of Polynesia).

(6)

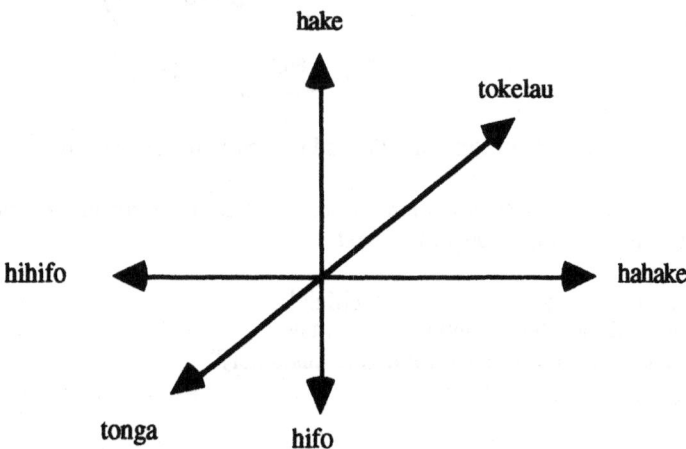

There are many other directionals of similar kind which shall not concern us here.

If we merely look at example (5) we might get the impression that the use of *ange* or *atu* is really just a matter of one's position in space at the time of the speech act.

However, as mentioned already by Tchekhoff (1990), the same spatial set-up allows for a completely different use of *mai*, *atu* and *ange*, and I do agree with her that this usage is dependent on social knowledge rather than purely "spatial" knowledge. However, quite a number of examples in Tchekhoff's provisional survey would have to be amended. So I will be using my own examples instead, which I collected during my fieldwork in Tonga. Consider, for instance, the following examples:

(7) (Na'e) 'ohovale 'a   Mele 'i ˙ ho'o 'oatu   ha'ane fo'i   keke?
     PAST   surprised ABS   Mele LOC your give:ATU her   CLASS cake
     "Was Mele (i.e. our neighbour) surprised when you presented a (piece of) cake
     to her?"

(8) Kuo ke 'osi   'oatu   ha   fo'i keke kia Mele?
     PERF 2.SG already give:ATU USP.ART CLASS cake ALL Mele
     "Have you already given a (piece of) cake to Mele?"

Recall that the standard form for giving something to somebody different from the hearer would be *'oange*, while *'oatu* usually points towards the hearer:

(9)

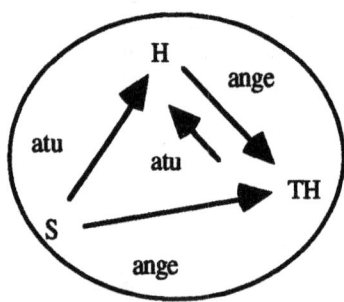

So why should it be that in examples (7) and (8) *'oatu* refers to third?

Still more, there are cases when *mai* refers to third, even when the spatial set-up is exactly the same as in the standard context:

(10) 'Omai   ha puha sipi   kia Mele!
     give-MAI a box mutton to Mele
     "Give a box of mutton to Mele (our customer)"

(11) 'Osi   'omai.
      Already   give-MAI
      "(I have) already given it to her"

In both types of examples the persons take up the same places as they did before. What has changed is the knowledge about the "social setting" of the act in question: As for *'oatu* being used with respect to a third person, the speaker considers himself part of the act of giving (e.g. speaker and hearer have agreed earlier on presenting Mele with a piece of cake, or speaker and hearer belong to the same household). We shall call this the *solidarity parameter* (in the sense of Brown/Gilman 1960), where speaker and hearer "act as one", and their "positions" (apparently defined by their activity) become conflated (in the *mai* - domain).

(12)

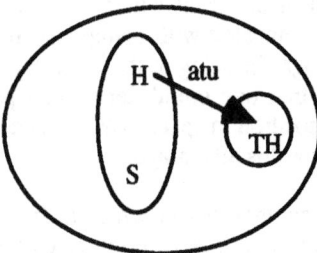

The use of *mai* in the selling situation must be explained differently. It is true that it might be argued that the client has been included into the *mai* - domain, (as in (13)),

(13)

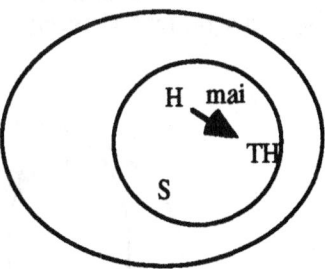

but it would be difficult to attribute this to the solidarity parameter. As other examples will tell, the use of *mai* in such a context is determined, above all, by politeness with respect to the one in power. We shall call this the *power* or *control parameter*. Consider, for instance, that in the kava ceremony the one who shouts the kava for a chief etc. will also usually say *'omai e ipú kia X* ('give the cup to X') rather than *'oange*. Conversely, *ange* in the form of its allomorph *angē* (**ange + ē* 'there') is used when someone asks someone in power for a favour (as in (14)):

(14) 'omai   angē   ha'aku fo'i   keke
     give-MAI ANGE:there my   CLASS cake
     "Please, give me a piece of cake"

If one wants to be extremely polite, one can even add the word *mu'a* 'front' as in *'omai angē mu'a ha'aku fo'i keke*, so literally this would be 'give towards me across there to the front a piece of cake for me' (which obviously does not make much sense in spatial terms).

Note that the front is institutionalized as a symbol for social priviledge in Tonga. For instance, there is a word *tulou* which one has to use if one passes in front of somebody, and noone is supposed to sit in front of the King of Tonga[4]. Of course, it is not the spatial idea of the front which is important, but the fact that there might be certain advantages connected with being first (for instance, at a meal. Of course, there may also be advantages in being at the back, especially in war). At any rate it would be nonsensical to say that one could derive this particular social meaning from space; rather, a particular position in space is associated with a social advantage and eventually becomes a symbol for this advantage.

Now let us turn to another aspect which at first sight seems to be strictly spatial, but which eventually turns out to be determined by non-spatial parameters:

Supposing you were to tell your Tongan friend that you will be sending a gift to him from Germany, you would have to say

(15) Te u   'omai (**'oatu) ha'o me'a 'ofa mei Siamane
     FUT 1.SG give-MAI   your thing love from Germany
     "I'll be sending you a gift from Germany"

(16)

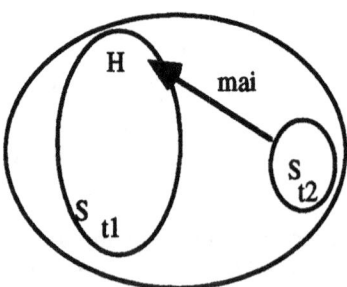

---

[4]It is still quite rare for the King of Tonga to sit in the rear of his limousine and quite common for His Majesty to sit right next to the driver.

In other words, rather than using 'oatu with reference to the hearer, one has to use *mai* as if speaker and hearer were at the same place. The reason seems to be simple: After, all, the speaker will be sending the parcel from a totally different country or different place, therefore Tonga, where speaker and hearer are now, counts as *mai*. This is true, but what really matters is something else. The speaker could easily 'omai something to the hearer from the neighbour's house or even from the next room, where under different circumstances one would use 'oatu. So what actually is "a different place" as opposed to the place where speaker and hearer are now? In fact, in this particular example it is the difference between one's current, present position and one's future position which determines the use of *mai* rather than *atu* . Apparently, the importance of a present activity overrides the importance of a non-present one. In other words, the interesting parameter is *time* , not space. And even more generally, it is not even time as such which is relevant, but the fact that the present situation provides a maximum of possible interaction.

The same *time parameter* will require *ange* instead of *mai* if the speaker leaves his friend and tells him to meet him later at the place where the speaker will be then (as in (17)):

(17) Te   u   'alu   ki   'api.   Te   ke   'alu ange 'apongipongi?
     FUT 1.SG go   to   home. FUT 2.SG go   ANGE tomorrow
     "I'm going home. Will you come to me tomorrow?"

(18)

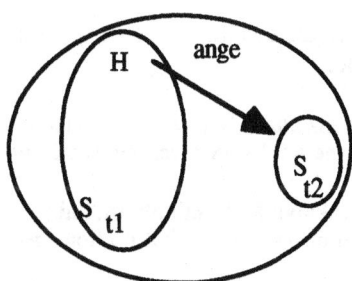

In an example like (17) one might assume that *ange* points to a place which is the speaker's habitual place of residence. But this is not true, as the following example reveals:

(19) Te   u   'alu   ki   he   kalapú   'aefiafi.   Kapau 'e   'a
     FUT 1.SG go   to   the club   tonight.   If   FUT wake.up

     'a   Maite pea   'alu ange mu'a   'o talaange   ke   u   ha'u.
     ABS Maite and.then go ANGE please (FRONT) and tell-ANGE SUBJ 1.SG come

     "I'll go to the kava club tonight. If Malte (my child who is asleep here) wakes
        up, please go there (ANGE to the club where I will be) and tell me so I will
        come back".

While habituality does not necessarily play a role with respect to the use of *ange* , it is decisive for another type of construction:

Supposing you decided to meet at a place which is neither your home nor the hearer's home, but which is different from the place where you are now, you would have to use *hake*:

(20) 'Alu hake mu'a   ki he mala'e vakapuná 'o tali mai (or *au* 1.SG) mei ai.
       go HAKE please (FRONT) to the field airplane   and wait MAI   from there
       "Please go (HAKE) to the airport and wait there for me"

(21)

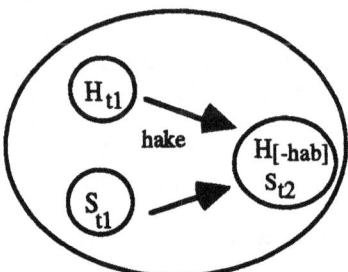

Hake need not refer to the meeting place of speaker and hearer, as is demonstrated by the following example:

(22)   'E folau vakapuna atu 'a 'Aiveni. 'Alu hake mu'a
       FUT travel airplane ATU ABS Alven. Go HAKE please (FRONT)

       ki he mala'e vakapuná 'o tali atu mei ai.
       to the field   airplane   and wait ATU from there

       "Alven is going to travel to you by plane. Please go (HAKE) to the airport and wait for him there".

Yet in every example *hake* implies that the meeting point is not the home of the hearer or the speaker, or does not count as the habitual domain of activity. We can assume, therefore, that the *habituality* or *familiarity parameter* is decisive in choosing what will count as a standard reference. Just as with the time parameter it is not habituality as such which is interesting, but the fact that a place you are familiar with is characterized by a maximum of interaction.

There is a very similar feature called "entrenchment" by Langacker (1987:59) and Deane (1992:35) which gives rise to the use of *mai* with reference to the topic of the discourse. In many Tongan folk tales (cf. Fanua (undated)) *mai* refers to the main person in the text (often identical with the person mentioned in the title of the story). The speaker as well as the speaker's position may not be known, and of course you do not really know the hero's spatial position either, but you do know that the hero is the information which is kept constant throughout the text, and therefore figures as a point of reference in the discourse. Entrenchment is generally dependent on familiarity or habituality of interaction.

The notion of a standard reference mentioned above implies that there is a non-standard reference, i.e. that there is an unmarked term as opposed to a marked one (in the sense of Jakobson 1985). In fact, the last example contains a very peculiar use of *atu* . Apart from the fact that the first *atu* in the sentence points towards Tonga (towards the hearer) and the second away from it (not to the hearer, but not to the speaker either, because the speaker does not go to Tonga), the second *atu* is used with respect to a third person. So why do we not use *ange*, especially since the third person does not do the same thing as either speaker or hearer? There is a very simple explanation: *Tali* is a two-place predicate, and it is only with three-place predicates like *'oange* 'to give' and *talaange* 'to tell' that *ange* refers to third. With predicates such as *tali* 'to wait', *a'u* 'to arrive at' or *'alu* 'to go' *atu* will usually be used as in the following example:

(23) na'á   ku   a'u   atu   (vs. *'oange 'a X kia TH* "give X to TH")
     PAST 1.SG   reach   ATU
     "I arrived there/at TH"

If *ange* was used it would refer to a speaker's non-present position:

(24) na'á   ke   a'u   ange?
     PAST   2.SG reach ANGE
     "did you arrive there (where I was then)?"

So if the predicate does not profile three arguments we shall normally use the form of the second person, even if we do not refer to the hearer. This suggests that *atu* is less marked than *ange*, and from the other examples we can deduce that *mai* is less marked than either *atu* or *ange*. *Ange*, in turn, seems to be less marked than the most marked expression, namely *hake*. The most unmarked point of reference is someone who is in control, present and familiar or a habitual point of contact (*mai*), the *atu* - reference is per default present and a habitual point of interaction but without control over the speech act (the hearer), and *ange* usually refers to someone who is neither controlling nor present in the speech act, but - unlike the *hake* - reference - may be familiar or a habitual point of interaction:

(25)

mai	<	atu	<	ange	<	hake
+control		-control		-control		-control
+pres		+pres		-pres		-pres
+hab		+hab		+hab		-hab

Thus, if the relations are fully unfolded as in figure (26a), *atu* differs from *mai* by the absence of control, *ange* by non-presence, and *hake* by non-habituality of reference, non-familiarity, non-topicality or non-entrenchment:

454

(26a)

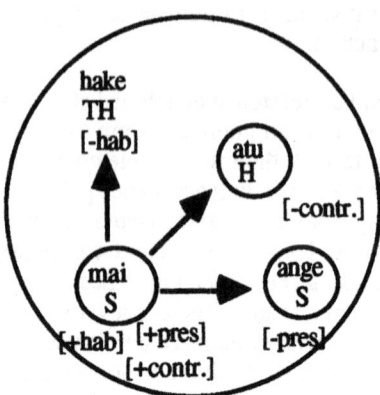

Under the conditions mentioned above, it may be that differentiations are conflated (remember the solidarity parameter). So apart from the maximum differentiation *mai*, *atu*, *ange* and *hake* (with *hake* referring to third and *ange* to non-present position of first) we also find *mai*, *atu*, *ange* (with *ange* referring to third), or just *mai* and *atu* (with *atu* referring to third (see (12)) and eventually just *mai* with *mai* potentially referring to third.

(26b)

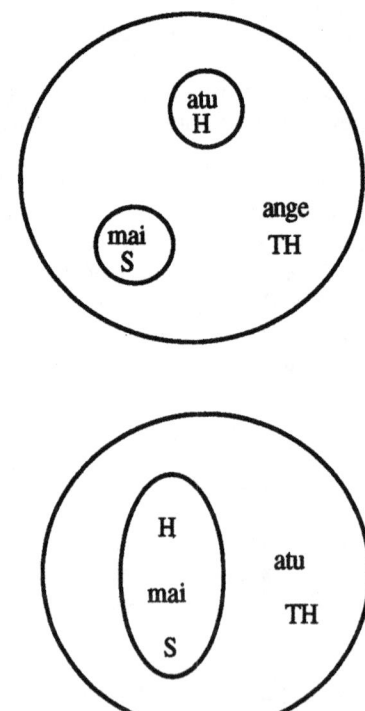

(26c)

(26d)

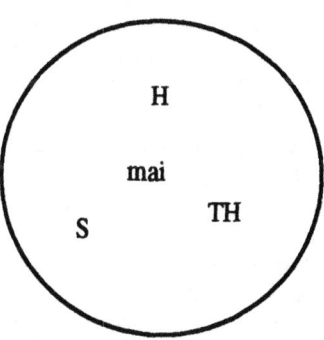

It goes without saying that it is not accidental that *hake*, which is the most marked term in this hierarchy, is also the most marked term in space (namely the third dimension).

Historically, though, *ange* might have been able to express a third dimension like *hake*. Thus, it functions as a comparative in sentences such as

(27) 'oku   lōloa   ange   'a   X   'ia   Y
      PRES   tall   ANGE   ABS   X   LOC   Y
      "X is taller (more or markedly tall) than Y"

Occasionally, *hake* is used in similar contexts.

Conversely, even if we are primarily interested in spatial motion, there are modern uses of *hake* which do not require strictly upward movement: *'alu hake ki mu'a* 'move up to the front' (e.g. of a bus), where *ange* cannot be used with reference to third (*'alu* is a two-place predicate). So there is a continuum of usages between *ange* and *hake*, yet with *hake* being the least grammaticalized, most marked term in the hierarchy.

If we ask why 'UP' is a marked dimension in our everyday perception of space, the answer is quite straightforward: Things which are 'up' cannot readily be dealt with, because we have to fight gravity. Interestingly, 'the upper class' are people the average person does not easily interact with and does not have control over. That 'side' (cf.*ange*) is more marked than 'opposite' (on the *atu* - axis) can also be explained by the fact that one does not normally interact sidewise. Even the time parameter plays a role on the side-dimension: While it is possible to observe different objects at the same time in the ordinary field of vision by looking straight, one cannot see the things on one's extreme left and right at the same time (one would have to move one's head). In other words, space relations are constructed by the same interactive principles as other relations are. So far, we have seen that the basic structuring principles are control, temporal presence vs. absence as well as habituality, entrenchment or topicality[5]. These are general action principles, which are well known in the linguistic encoding of activities. Apart from that, the differentiation of relations as well as the conflation of differentiations along markedness hierarchies takes place along "dimensions" of maximum contrast which are not peculiar to space at all, but correspond to the perceptual capacity of differentiation. Consequently, figure (26a), which does not contain any peculiarly spatial dimensions, but results from interactional principles, is diagrammatically comparable to the purely spatial set-up of figure (1) and (3), respectively.

There is no room here to fully explore the consequences of this state of affairs. However, it seems as if an ultimate account of cognition will be different from a standard representational theory of various concepts by semantic features. Note that our account of *ange*, for instance, has shown that whatever *ange* may refer to, it may be subject to the *time parameter*, though it would be wrong to say that *ange* denotes time. In other words, though time is not part of what is commonly called the "intensional semantics" of the word in question, it is a necessary parameter in explaining the use of *ange* as opposed to *mai* and *atu* [6]. Time as well as the other determinants of the use of directionals may then regarded as the ingredients of an algorithm of conceptualization, i.e. how the mind arrives at or constructs the concepts in question, and it is this algorithm which I think will eventually replace the traditional semantic feature analysis[7].

What makes it so difficult to understand the importance of these (non-spatial) determinants of the use of the words in question, irrespective of their "intensional" meaning, is the fact that they are the backbone of the act of conceptualization itself, and thus they may be called the "blind spot" of conceptualization, which can be arrived at only by indirect evidence, and from this point of view it is only natural that in the process of the grammaticalization of case markers temporal arguments will be expressed "later" than spatial ones, though time is the more fundamental part of the algorithm of conceptualization[8].

---

[5]In this paper there is no time to discuss a further conflation of principles, but from Broschart 1993 and 1994 we can assume that the principles can eventually be reduced to differences in effectivity (modal), in time (temporal) and continued involvement or "interest" (aspectual), along with markedness-distinctions in relation to some (unmarked) point of reference (see below).

[6]I am presently unable to say how the time parameter as one of the determinants of the use of directionals relates to the Wittgensteinian (1953, 43rd. aphorism) idea that meaning is basically identical with the use of the words in question.

[7]A different view is proposed in Bennardo 1995. The general (intensional) meaning of *mai* is assumed to be "to center" (p.14) etc., but this begs the question what enables us to understand the idea of "center" without falling back on spatial metaphor again.

[8]It is Heinz von Foerster who has repeatedly alluded to the "blind spot" of cybernetics and conceptualization, respectively (cf. Baecker 1993:18). In constructivist theory the self-organizing (autopoietic) act of constituting "self" will be the prerequisite of conceptualization, but the concept of (active) "self" is not the first one consciously

But what about the notion of directionality? Surely this must be primarily spatial?

This is not necessarily the case, either. Let us have a look at the following examples:

(28a) 'oku 'ofa 'a Mele 'ia Sione
       PRES love ABS Mele LOC Sione
       "Mele loves Sione"

(28b) 'oku 'ofa 'a Mele kia Sione
       PRES love ABS Mele ALL Sione
       "Mele misses/sends her love to Sione"

(28c) 'Ofa atu!
       love ATU
       "kind regards"

'Ofa 'to love' is constructed with two seemingly local expressions, the simple locative or the allative. With the allative the relation is less "intense" in the sense of sending love or greetings to sb. or in the sense of missing somebody. The allative construction *kiate koe* 'to you' can be substituted by *atu* as in (28c). Even though love does not literally "go" somewhere, it is no doubt a frequent metaphor in the sense of Lakoff/Johnson (1980) that you "send your love" to somebody. But unfortunately, the metaphor does not explain why we understand the metaphor, that is, we still do not know what is the common basis between longing for someone and a movement in space. Note that a metaphor, even a very daring one, must have some "tertium comparationis", otherwise one cannot interpret it[9].

It is Langacker (1987:166-187) who argues that what "moves" is not love but that there is a flow of attention from one participant to the other. This would then be an "abstract" cognitive movement rather than an external movement in space. Unfortunately, that does not work either. Note that "attention" or "interest" are more or less the same kinds of feelings as love (if someone loves somebody, this person is interested in the one he or she loves). So how do we define "abstract" movement? Conversely, supposing we could show that there is really some spatial movement of "interest" in our heads (put your lover in a magnetic resonance tube), then "love" would really move, but then there would be no metaphor.

---

conceptualization, but the concept of (active) "self" is not the first one consciously perceived. In order to eventually do it, a "function must be turned into its own argument" (von Foerster 1993b:129f; trl. JB).

[9]This entails that the concept of a metaphor need not be given up because there is a common denominator between the two areas compared; quite on the contrary, though any metaphor implies that two areas are in principle distinct, there must be some sort of tertium comparationis which guides one's interpretation of the metaphor.

I would suggest a more "animated" idea of direction. Even if we symbolize directions, we usually employ arrows, and real arrows of course are meant to do something to their targets (hit, kill, or make them fall in love). Interestingly, an arrow does not even have to move if people feel that it is directed against them. In other words, directions imply that there is a certain readiness for or probability of achieving some sort of interaction, or that one expects that situation a) (where the arrow is still on the bow string) will change to situation b) (where it might do great harm). As for the feeling of longing for somebody there is also an expectation or hope that somebody's present situation (where the person in question is lonely) will somehow change. Usually, of course, situations only change if one changes one's position, but what really counts as a common basis between "goals" in space and "targets" of desire is that one might expect a certain change of situation resulting in contact or best possible interaction of some kind. This is also true for communication. If the speaker asks the hearer to *talamai* something ('tell me something') the speaker expects that eventually he will possess information which he did not have before. In other words, for every notion of directedness we must assume a certain potential of change in interaction. This potential of change is, in fact, a *modal* idea, and the *modality parameter* is inherently associated with desires (like love) or expectations of all kinds. (This, of course, does not mean that words such as *mai* etc. explicitly "denote" modality in terms of intensional semantics, but I would argue that the concept of directionality cannot be understood without modality, and that it is not possible to explain the various uses of the "directionals" without reference to modality (see also our discussion of the time parameter[10])).

The modal idea mentioned also explains the possibility of using *mai* etc. with one-place non-dynamic predicates as in example (29):

(29)   ko   e   kakai   ena 'oku nau nofo maí
       ESS SP.ART people there PRES 3.PL sit MAI:DEF
       "there are people there who are waiting (lit. sitting) for us"

In this particular situation the people were not even turned towards us (me and my informant). The intended message was simply that there were people sitting with anticipation (for social interaction).

What I wanted to show in this paper is that the use of so-called directionals is governed by parameters of interaction (control, temporality, modality, etc.), and that the "triggers" of these operations are not specifically spatial by nature but relate essentially to the social setting of the situation described and the desire to arrive at an optimum of interaction.

---

[10]The other parameter we were discussing, i.e. the habituality-parameter, is part of aspectual distinctions. Though in this case, too, aspectuality need not be part of the "intensional" meaning of the words in question, Hooper (1994) claims that in Tokelauan *mai* and *atu* are occasionally used to denote aspect-like distinctions. There is no evidence for this in Tongan, however.

Consequently, the information a Tongan native speaker derives from the use of directionals such as *mai, atu, ange* and *hake* usually far exceeds or is even independent of the idea of a movement in space. Rather he or she will be able to tell who considers himself in charge of a situation, whether someone associates himself with a particular household or not, what is his usual or temporary domain of interaction and so on.

By default the question of control relates to one's role in the speech act situation, simply because *mai* is primarily a linguistic unit. Therefore, the prototypical *mai* - referent is the controller of the speech act, which, of course, is also a social act. Yet, under certain conditions *mai* can also point to the controller of the non-linguistic social context (cf. politeness with respect to a client). Potentially, *mai* may simply refer to the controller of perception in general, as in

(30) 'Oku 'asi mai 'a Mele?
      PRES appear MAI ABS Mele
      "Can you (or any observer) see Mele (on the video)?"

And by default, the controller of perception will be the one who looks at a particular spatial set-up of objects in physical space.

Yet in any case, we cannot explain the *mai* - usage without reference to an interactional framework involving an active observer and the phenomena observed. On the other hand, we do not need any "metaphorical" explanations for integrating non-spatial interpretations of the "directionals" in our model. Rather, as we demonstrated above, the control parameter operates in quite different domains, be it the controller of perception in general, the controller of the social act, or the controller of the speech act; these different interpretations can be nested into each other, so that by default a speaker is the controller of a linguistic act, a social act and of perception in general (including localization of persons and things in space):

(31)

control:

a) controller of speech act	+LING	+SOC
b) controller of social act	-LING	+SOC
c) controller of perception	-SOC	+LOC

Similarly, the parameter of entrenchment chooses either the most habitual/familiar participant in the speech act (the speaker) or the most constant/familiar participant in the act narrated (the topic), which by default may coincide again.

(32)

entrenchment:

| a) familiar/habitual participant in speech act |
| b) familiar/constant part. in act (narrated) |

The criterion of temporal presence relates to the desire that a participant (of the speech act or the act narrated) be readily available to interact with him.

(33)

temporal presence:

| a) present in speech act |
| b) present in act (narrated) |

As I emphasized above, there is a clear tendency for social parameters to override the importance of spatial parameters. It is only under conditions where the situation has been "bleached" of social relevance that the interpretation of the directionals is basically spatial and depends on the position of particular participants. This, however, appears to be the least interesting aspect of the use of directionals in language.

Note that I still think that it is worthwhile studying spatial relations, and spatial or rather geometrical models are good ways to formalize relations (for details see Broschart 1993). The many diagrams I have been using in this text are nothing but "geometrical" models for what constitutes "domains" of activity (cf. the conflation of the speaker's and the hearer's "position" in the *mai* - domain under the influence of the solidarity parameter (see (12)). But note that these models cannot be understood in terms of spatial principles as such, and the only reason for being able to devise spatial models of non-spatial phenomena is the fact that there are independent principles structuring domains of any kind in terms of the interaction between observer/perceiver and the observed/perceived. It is not space as such which is fundamentally important for the structuring of human thought. In my opinion, thoughts - which are essentially actions - are structured primarily in terms of action parameters such as temporality and modality etc.. As I tried to demonstrate above, our conception of space is nothing but a side-product of these perceptual activities, which focus mainly on social relations between people.

In any case, if we seriously try to integrate the many non-spatial readings of the directionals into a general theory of their use, we must abandon the traditional habit of regarding the non-spatial uses as "odd" or "derived" from some sort of primary cognition, or of considering space a god-given, unchangeable frame of reference. Neither spatial relations nor relations between people can ever be understood without reference to the interaction between an active observer and the phenomena observed. The rules for applying so-called "directionals" will look totally different if one takes social context and interactional variability into account than if one were to concentrate on standardized physical space, which eventually might turn out to be an artifact, after all.

3. Abbreviations

ABS - absolutive case
ART - article
ALL - allative case
CLASS - classifier
CONTR - control
DEF - definite accent
ESS - essive case
FUT - future tense
H - hearer
HAB - habitual
LING - linguistic
LOC - locative case or localization

PAST - past tense
PERF - perfective aspect
PL - plural
PRES - present (tense)
POSS - possessive
S - speaker
SG - singular
SOC - social
SP - specific ('a certain')
T - time variable
TH - third person
USP - unspecific ('some')

4. References

Anderson, J. 1971. *The Grammar of Case: Towards a Localistic Theory*. London etc. : Cambridge University Press

Baecker, Dirk 1993. 'Kybernetik zweiter Ordnung'. In: S. J. Schmidt (ed.) 1993: 17-23.

Bennardo, G. 1995. 'Linguistic Representation of Spatial Relationships in Tongan'. MS. Cognitive Anthropology Research Group, MPI for Psycholinguistics, Nijmegen.

Bott, E. 1972. 'Psychoanalysis and ceremony'. In: J.S. LaFontaine (ed.). *The Interpretation of Ritual: Essays in Honour of A.J. Richards*. London. 205-37, 277-82.

Bott, E. 1981. 'Power and Rank in the Kingdom of Tonga'. In: *Journal of Polynesian Society* 90/1:7-81.

Bowden, J. 1992. *Behind the Preposition. Grammaticalisation of Locatives in Oceanic Languages*. (Pacific Linguistics Series B-107). Canberra : Australian National University.

Broschart, J. 1993. 'Raum und Grammatik oder: Wie berechenbar ist Sprache? (Mit Beispielen zu Kasusmarkierung, Aspekt, Tempus und Modus)'. In: Th. Müller-Bardey/W. Drossard (eds.). *Aspekte der Lokalisation*. Bochum : Brockmeyer. 1-43.

Broschart, J. 1994. *Präpositionen im Tonganischen. (Zur Varianz und Invarianz des Adpositionsbegriffs)* . Bochum : Brockmeyer.

Brown, R./Gilman, A. 1960. 'The Pronouns of Power and Solidarity'. In: T.A. Sebeok (ed.) *Style in Language*. New York : John Wiley.

Deane, P. 1992. *Grammar in Mind and Brain: Explorations in Cognitive Syntax*. Berlin etc. : de Gruyter.

Fanua, T.P. (undated). *Po Fananga. Folk Tales of Tonga* . Nuku'alofa : Taulua Press.

Heine, B./U.Claudi/F. Hünnemeyer 1991. *Grammaticalization: A Conceptual Framework*. Chicago etc. : Chicago University Press

Helu, F. 1993. 'Identity and Change in Tongan Society since European Contact'. In: *Journal de la Société Océanistes* 97/2: 187-194.

Hooper, R. 1994. 'From directional to aspectual: the polysemy of *mai* and *atu* in Tokelauan'. Talk presented at the Seventh International Conference on Austronesian Linguistics.

Jakobson, R. 1985. 'Mark and Feature'. In: St. Rudy (ed.). *Roman Jakobson. Selected Writings*. Vol. VII. The Hague : Mouton. 122-124.

Kant, I. 1956. *Kritik der reinen Vernunft: Band I-II*. Frankfurt am Main : Suhrkamp (Suhrkamp Taschenbuch Wissenschaft 55).

Köck, W. K. 1986. 'Biosystems theory and empirical aesthetics'. In: *Poetics* 15: 401-437.

Lakoff, G./M. Johnson 1980. *Metaphors we live by*. Chicago etc. : University of Chicago Press.

Langacker, R. 1987. *Foundations of Cognitive Grammar*. Vol I. Palo Alto : Stanford University Press.

Mead, G.H. 1938/1972. 'Perception and the Spatiotemporal'. In: Ch. W. Morris (ed.) *The Philosophy of the Act*. (Vol III of the *Works of George Herbert Mead*). Chicago University Press. 174-204.

Schmidt, S.J. (ed.). 1993. *Heinz von Foerster. Wissen und Gewissen*. Frankfurt am Main : Suhrkamp. (Suhrkamp Taschenbuch Wissenschaft 876).

Smith, B. 1988. *Foundations of Gestalt Theory*. Wien : Philosophia.

Tchekhoff, Cl. 1990. 'Discourse and Tongan *mai, atu, ange*: Scratching the Surface'. In: J. Davidson (ed.). *Pacific Island Languages. Essays in Honour of G.B. Milner*. London: School of Oriental and Asian Studies. 105-110.

Tesnière, L. 1951. *Éléments de Syntaxe Structurale*. Paris : Klincksieck.

von Foerster, H. 1993a. 'Gedanken und Bemerkungen über Kognition'. In: S. J. Schmidt (ed.). 1993: 77-102.

von Foerster, H. 1993b. 'Bemerkungen zu einer Epistemologie des Lebendigen'. In: S.J. Schmidt (ed.) 1993:133.

Wittgenstein, L. 1953/1984. *Philosophische Untersuchungen. Teil I und II*. Frankfurt am Main : Suhrkamp (Suhrkamp Taschenbuch Wissenschaft 501).

Spatial Conceptualizations
of Social Hierarchy in Pohnpei, Micronesia

Elizabeth Keating

Cognitive Anthropology Research Group, Max Plank Institute for Psycholinguistics
PB 310, NL-6500 AH Nijmegen, The Netherlands
phone: 0031-80-521-603, fax: 0031-80-521-300
email: keating@mpi.nl

1 Introduction

This paper looks at the particular way space is organized in Pohnpei, Micronesia and how Pohnpeians map onto physical space a model of their social structure. A status marking feature of the Pohnpeian language affords an empirical way of looking at how spatial resources are used to organize and conventionalize experience. Relationships between space, language use, and hierarchy are examined in this paper in order to analyze how Pohnpeians verbally encode status in space, i.e. set up relative status relations using spatial resources. Pohnpeians encode both vertical and horizontal space with social status such that precise relative rank is constituted by an individual's location in space. The data used are transcripts and video frames of two interactions, one videotaped in Pohnpei in February 1993, and another in 1990. Specific aspects of participants' utterances and non-verbal expressions are analyzed in order to look closely at the link between culture and cognitive mappings in space.

The island of Pohnpei is located in the Caroline Islands of Micronesia, at a latitude of seven degrees north and longitude of 158 degrees east. The circular land mass covers an area of about 190 square kilometers. About 30,000 Pohnpeians live there, not in villages, but scattered about in homesteads along the flatter lands by the shore. The only town or village of Kolonia exists as a seaport and administrative center from colonial days, and was constructed according to Western notions of township spatial organization. A succession of colonial rulers, Spanish, German, Japanese, and American have claimed Pohnpei as part of their sphere of influence. Today Pohnpei is part of the Federated States of Micronesia, which are now independent politically but not economically from the United States. The traditional religio-political structure, a ranked chiefdom, is still quite vital and co-exists harmoniously with an imported democratic form of government. A special vocabulary is used to refer to the activities of chiefly personages as opposed to the rest of society; this special language register will be alluded to more specifically in portions of this paper.

Hierarchy is an underlying principle of organization in Pohnpeian society, constituted in many social practices, such as food sharing. This includes public distributions of food where rank determines whether a person receives any food at all and family meals, where participants eat singly in rank order. Titles are used instead of personal names for adult members of Pohnpeian society; these titles are ranked with respect to each other, and no two have equal significance. Pohnpei is divided into five chiefdoms (*wehi*), and the chiefdoms themselves are hierarchically ranked, with the chiefdom of Madolenihmw ranked highest. Each chiefdom is headed by a paramount chief called the *Nanmwarki* and chieftess *Likend* (chieftesses are called by other titles in other chiefdoms), and by a secondary chief, *Nahnken* and secondary chieftess *Nahnken Iei.* The two chiefs and the two chieftesses head a system of dual lines of hierarchically ranked titles for men and women (for a more complete description of the Pohnpeian polity see Reisenberg 1968; Petersen 1982; Shimizu 1982, 1987). Sub clans within each clan are precisely ranked according to the relative age of their ancestresses, a family of sisters. In Pohnpei power and status flow matrilineally.

Hierarchy is gramaticalized through a particular feature of the Pohnpeian language, its honorific register, a speech register which indexes social status. Speakers can index their own or another's status by choosing specific lexical items. A single utterance can index two separate levels of status aimed at two separate individuals, and one participant's status can be differently constructed by two different speakers in the same interaction. Honorific speech is used in interactions with high ranking participants, in radio announcements, and in religious contexts. This speech register is used by speakers in the two interactions discussed below to inscribe social hierarchy onto physical space.

Space is status marked both vertically and horizontally. The mapping of hierarchy onto space is a visible reflection of the status hierarchy of the island. The

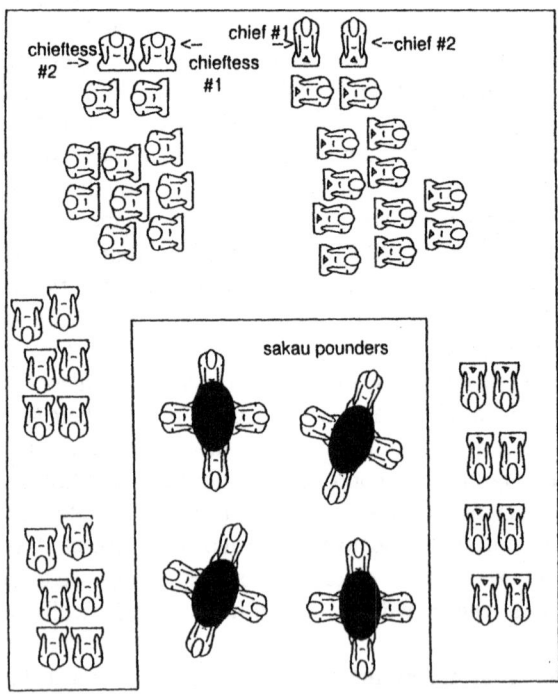

IDEAL CONFIGURATION OF FEAST HOUSE DESCRIBED BY CONSULTANTS

Fig. 1

highest chiefs and chieftesses sit in the highest place, physically and symbolically, whereas the lowest status participants sit in the lower places physically and symbolically. If all five island chiefs and chieftesses come together for an important event, the chief and chieftess of Madolenihmw occupy the space in the feast house which signifies the highest status.

The status marking of all space is based on the layout of the traditional Pohnpeian feasthouse (*nahs*). It is a building that is rectangular in shape, with a floor plan that is U-shaped, that is, one side is completely open to the outside. The U-shaped configuration results from the fact that the floor is raised on three sides above the middle, a bare earth floor. The walls on three sides traditionally extend only half-way to the roof. In most feasthouses, the side platforms are the same height as the rear one. However, the side platforms signify a lower status. The rear platform is reserved for very high status individuals, e.g. the highest chief, the highest chieftess, the secondary chief, and the secondary chieftess. At the very back sit the ancestor deities.

Each part of the feasthouse has a name and a status attached to it; specific locations as well as entryways and posts are named and reserved for certain high status individuals. For example, only a man with the title Keraun en Ledau can lean his back on a certain post and only a man with the title Nahlaimw can use the opposite post. Only specific titled women can sit near the chieftesses. The platform floor as well as the earth floor are portioned into sections, with a name and status. Certain boundaries may not be crossed, for example a person cannot hang her/his legs over the platform (the boundary between where chiefs sit and where pounders sit) or step on the beam which forms the inner edge of a side platform.

Ranked space is reproduced in many contexts outside the feast house, and new, Western influenced aspects of the environment are inscribed with culturally specific meanings. An example is the use of metal folding chairs by the chief and chieftess while all other participants sit on the floor. Space has hierarchical meaning not just in formal settings, but in everyday spatial organization as well. On one occasion, a passerby questioned the appropriateness of my sitting on a higher cement step at the Catholic mission than another woman who was of high status, as we were waiting for a ride.

Historically, on Pohnpei, the highest chief and chieftess were non-participants in social interaction, and strict patterns of avoidance precluded physical association with them. Feast houses were built facing the direction of prevailing winds and fires were strategically built so that smoke would blow inside and obscure the visibility of the high chiefs on the feast house platform. The *manaman*, or sacred power of the chief is considered extremely potent and dangerous. *Mana* is a term with cognates throughout the Pacific (for example, *manaman* on Pohnpei) to describe the power that flows from the deities to the people (cf. Shore 1989). On Pohnpei, *mana* flows matrilineally to descendants within chiefly clans. A chief's successor is not his son but his sister's son. The linkage of the power of *mana* to spatial distance is important for understanding the complexity of the symbolizing of space in Pohnpei, and why bodily movement or location in space is a significant domain for status-marked vocabulary.

2 Status Marking of Spatial Relations

The most frequent site of status marking in language is in expressing verbs of motion and stasis, for example a person's relative position in or path to a position in space. Possessive constructions, which can also be thought of as a form of locative are also status marked, as are certain transitive verbs, and verbs denoting the act of speaking or knowing. Looking at transcripts of interactions, and at recurrent forms, I find the salient way to describe Pohnpeian honorific speech is in terms of a polarity between high and low, with the term for low semantically related to the term for flat, level ground. Because there are no *humiliative* personal pronouns (and only two humiliative nouns, a nominalized form of the verb 'say' and a word for food) verb forms expressing spatial relationships are important sites for the expression of low status (see Keating 1994 for more discussion of this). Typically, all participants except the chief use honorific speech. For chiefs to use the appropriate status elevating verbs and possessive markers toward themselves would be considered immodest (cf. Milner 1961, Duranti 1984). The chiefs' wives do use honorific speech, and are able to lower the status of individuals in the presence of the chief. So only those other than the chief have to make lexical choices based on status.

Space and status are linked in language as well as in seating arrangements in honorific and other contexts. The paramount chief is sometimes referred to as *Wasa Lapalap* (lit. 'place important or place physically large'). Members of the paramount chief's clan are called *sohpeidi*, which literally means 'facing downwards' (they sit facing downwards while others face upwards). Pohnpeians also status mark space through language with the use of a status marked term to refer to 'left' and 'right', *pali meing* and *pali maun*. *Meing* is the term for 'status marked speech.'

Verbs of motion are expressed in honorific register by using one of two stems, *ket-* (for exaltive) or *pato-* (for humiliative) to which directional suffixes are added. Status-marking also includes verbs denoting stasis. The English term stasis does not quite capture the Pohnpeian notion of the *achievement* of position as the result of activity, or the notion of place and status requiring continual attention and activity, rather than an absence of movement. Across languages, honorific vocabularies show a restricted range of lexical resources compared to common speech (see for example Dixon 1971; Haviland 1978, 1979, Agha 1994). Interestingly this is more evident in low status speech than high status in Pohnpei.

The following examples from interactional data illustrate the honorific locative verb stems *pato-* (humiliative) and *ket-* (exaltive) with directional and/or transitive suffixes and without. The construction of stem + directional suffix allows for a range of meanings to be expressed with a minimum of vocabulary. (These represent examples of forms found in interactional data and not the set of all possible forms). Pohnpeian follows the German convention of using 'h' to signify a long vowel.

HUMILIATIVE		EXALTIVE
pato	'locate [stay, be']	ket
pat-pat	'locating [staying']	ket-ket

patoh-do	'locate [come]-here, towards speaker'	ket-do
patoh-la	'locate [go]-there, away from addressee and spkr'	ket-la
patoh-di	'locate [go] downwards' ['sit' 'lie down']	ket-di
patoh-da	'locate [go] up' ['stand up' 'climb up']	ket-da
patoh-long	'locate [go] inwards'	ket-long
patoh-sang	'locate [go] from' ['move from']	ket-sang
patoh-wei	'locate [go] there toward addressee'	ket-wei

Uses of the directional suffixes can be seen in the next excerpt in which the stem *pato-* takes four different endings: *-long* ('inward'), *-sang* ('from'), *-di* ('downwards') and *-wei* ('there towards you'). Each verb phrase also contains an additional deictic reference, such as 'here by me' (*me*) or 'there towards you'(*men*).

01 Chief: ah pwe ma ke mihmi me
              but what if you stay here

02 N: ah pwe ma e **patohlong** me
(man)   but if she goes [she of low status] inside here

03 L: (to another woman)
(man)   ah kowe. **patohsang** men.
              and you. move [you of low status] from there

04 LA: soh i pahn **patohdiwei** men
(woman)   no I will go [I of low status] down there towards you

Note that the chief in line 01 uses *mihmi* ('being','staying') in common speech; all verbs by other speakers are in humiliative speech.

3 Encoding Social Hierarchy onto Space

The first interaction I will discuss does not take place in a traditional feast house, but on the cement porch at the home of the highest ranking chief in the Madolenihmw chiefdom. Neighbors and relatives have come to share in the preparation and drinking of *sakau*, a sacred beverage with mildly soporific effects prepared ceremonially in many Pacific societies (called *kava* in Pacific Islands literature). *Sakau* is made on Pohnpei by pounding the fresh roots of the pepper plant on a large flat stone, adding a small amount of water, and squeezing the root through a strip of soaked hibiscus bark.

In this interaction the participants are structuring the small porch area in order to constitute what they feel is the appropriate hierarchical encoding of space. Using honorific speech, they direct the movements of the chief and chieftess and collaboratively achieve the proper marking of space on the horizontal axis.

Though the interaction does not take place in a traditional feasthouse but on a Western style porch, the hierarchical seating arrangements of the nahs are used as a sort of cognitive model, showing how the spatial relationships according to status can be mapped on to any locale. The chief and chieftess sit on metal folding chairs while other participants sit on the floor or on cement blocks. This constructs the proper *vertical* hierarchy. But vertical is only one axis of hierarchical coding. The other is horizontal. Because this is a porch and not a feasthouse, more explicit instructions are required to achieve the desired goal of hierarchizing space. Verbal status indicators, the honorific verbs, together with non-verbal signals such as gestures, map onto the physical environment the hierarchy of status. Honorific speech marks status lexically as space marks status in the physical domain.

A sequence of positioning and re-positioning of the chief and chieftess is generated by a question posed by the chief's adopted daughter. A certain amount of status accrues to her through her close relationship with her adoptive now chiefly father, however, her title is the counterpart to her husband's title (Nalik). It is a title from another chiefdom since her husband is from a different part of the island. Other participants in the interaction (besides the chief, chieftess, and the daughter) include Nalik (the daughter's husband), Lepen and Sou (middle-aged men) and a teenaged boy, Kadawo.

The chief's daughter has been busy in another part of the house and has been directed by the chieftess to come and join the *sakau*. When she arrives she asks:

Daughter: ah i pahn **pato** iawasa?
               but where will I sit [I of low status]?

The daughter applies the status-lowering verb *pato* to her own actions. Her question draws attention to the fact that the seating map she observes does not clearly indicate appropriate positions. The chief sits farther back (symbolically higher) than the chieftess. The chieftess should be sitting at the same level (horizontal) with the chief. The daughter then issues a series of directives to rearrange the space according to the traditional hierarchical model. In the following excerpt, she directs removal of an ice chest, the movement of the chief, the chieftess, and finally herself. She uses the honorific verb forms appropriate to the status of each: *patohsang* (humiliative 'go from' or 'move from') for the ice chest, and *ketla* (exaltive 'go there' or 'move there') for the chief. In referring to the chieftess she uses the honorific pronoun *komwi* (exaltive 'you'). She uses heavy jabs with her arm and hand towards the high status area as she directs the chief, the chieftess, and herself, dividing up space.

Daughter: Sou pwe ma ice boxo **pato**--
               Sou because if the ice box moves [it of low status]

               ice chest en **patohsang** mwo eri Mwohnsapw
               move the ice chest [it of low status] from there then the chief

               ketla mwo ah komwi
               moves [he of high status] there and you [of high status] (the chieftess)

               ah ngehi
               and me

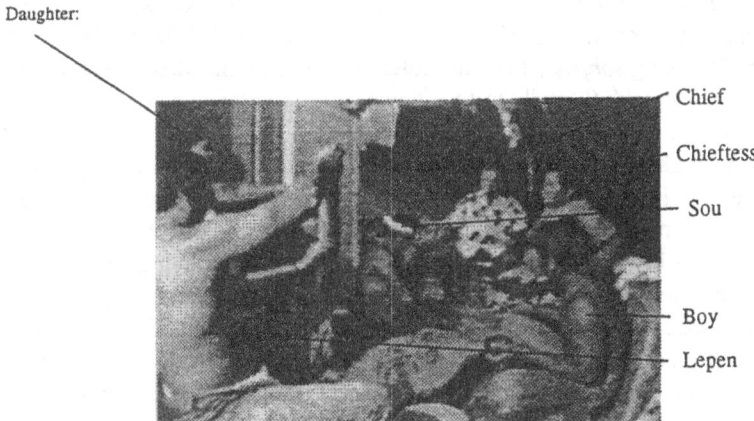

Fig. 2

As soon as Sou moves the ice box, she chooses a more specific directional suffix in again directing the chief's movement (-*wei* 'there towards you'). The directive dividing up space in status order is repeated. In this excerpt it can be seen that the chief resists, claiming that his position in space is already high because he is a member of the high ranking clan, *sohpeidi* ('those who face down').

01 Daughter: eri Mohnsapw **ketwei**
                          then the chief goes [he of high status] there

02   ah **komwi** ah ngehi
                          and you [you of high status] (the chieftess) and me

01 Chief: ah i mwahuer me
                          but I'm fine here

02   sohte me kin kasauada sohpeidi
                          no one habitually moves the high status people

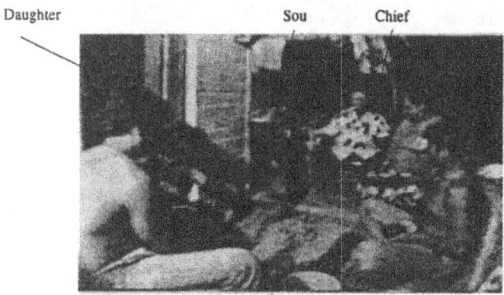

Fig. 3

He objects to being moved, framing his opposition in the formulaic style of many Pohnpeian proverbs, beginning with *sohte me kin* and followed by a disapproved action. By using *sohpeidi* the chief refers to position encoded by spatial reference in language. *Sohpeidi* (literally to face downwards) are those who sit on the feast house platform and look down and out over those of lower status. The chief claims that he is in the highest position no matter where he sits. He's already facing down. But vertical is only one dimension of spatial rank. As I have mentioned earlier, the other important part is horizontal, and horizontal encoding is the only one of significance to persons of lower status. However, though he contests moving, he *does* in fact move to his appropriate position. The chief's comfort is not seen as important as properly defining space to include the spatial hierarchical status marking of those below the chief and chieftess. All of the directives are delivered with status-raising forms of verbs for the chief's action. Below are video frames showing the space before and after the rearrangement.

Initial Seating Positions

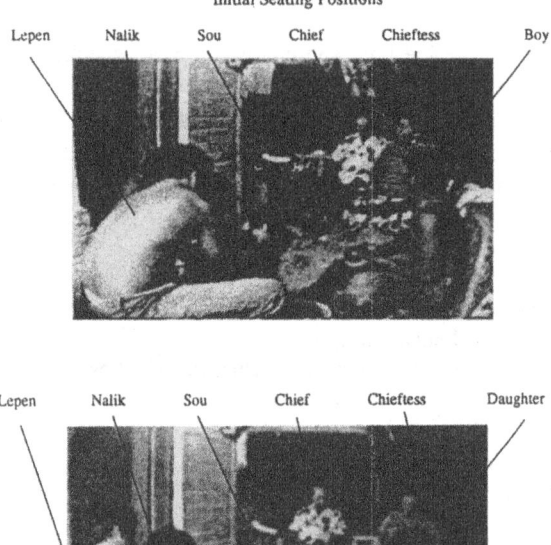

Fig. 4

4 An Interaction with the Second Ranking Chief

In another interaction with the second ranking chief, space is also hierarchically organized using honorific speech. In this interaction, the participants are also preparing themselves and their space in order to prepare *sakau*. This interaction takes place in the traditional feast house.

A Pohnpeian feasthouse is multi-functional, a place for canoe storage, laundry drying, and living quarters. Because of such multiple use patterns, the marking of space for status is often made using honorific speech, as well as gestures and body position, as this interaction will show. This interaction is different from the last one because the participants, surprisingly, are not at first using honorific speech, even though they are in the presence of the chief. They switch register only when one participant is directed to move to a position in space near the chief and high on the platform.

In this excerpt the chief is asking the whereabouts of one of the participants.

Chief: ah ia ih?
            so were is he?

Woman: ie ih
            here he is

Chief: en kohdo e kihsang ah sehten
            tell him to come and take off his shirt

Woman: kohdo kihsang ahmw sehten
            come take off your shirt

Fig. 5

The woman's second utterance uses the common speech form of the verb 'come' (*kohdo*), just as the chief does, and a common possessive classifier. This is not what is expected, given the presence of the chief. Honorific forms co-occur with the reorganization of space on the feasthouse platform to express hierarchy.

A few minutes later, the chief stands up and leaves to go to his private room to get a pack of cigarettes. In the next frame, the chief returns, and stands briefly on the platform before sitting in a new position and pose. Just as he returns from the back room, the *menindei* uses the first honorific words (in boldface type)--verbs of movement:

Menindei: Willy **patohdala** mah dehu
             Willy go up there [you of low status] first and position yourself

             patohwansang ahmw sehten **patohdala** wiada
             take off [you of low status] your shirt go [you of low status] up there

             uhdahn mwohden erir eh
             and truly sit the way a server sits, eh

Fig. 6

The *menindei* uses three humiliative verb forms for the actions he wants Willy to do:
go up there, take off your shirt, and go up there. These verbs contrast with verbs used
earlier before the chief left, e.g. *kihsang ahmw sehten* is now *patohwansang ahmw
sehten*. Honorifics occur when the chief arrives to take up his appropriate sitting
position and when Willy is directed to move *up* to his sitting position on a high status
part of the feast house platform. Prior to this the only moving about by low status
participants was *on the dirt floor*. Though Willy will occupy a high position, it is
symbolically lower than the chief, being closer to the entrance of the feast house.
Willy's lower position is also indicated in the use of humiliative verbs by the *menindei*
to direct his actions.

    In the next frame the chief sits, not in the position he occupied before leaving to
get his cigarettes, but several feet farther back on the platform. The *sakau* pounders
sit lower than the chief, both in their position closer to the entrance of the *nahs* and
in their position lower on the dirt floor. As Willy takes his shirt off the *menindei* tells
him that the chief is going to explain to him the proper way to sit. The *menindei* uses
the exaltive verb *masanih* to refer to the chief's talking.

Menindei: ohlen nek **mahsanihongu**hk dahme
             that man could tell [he of high status] you what is

             pwungen ahmw pahn mwohd
             the correct way for you to sit

Soon, the chief instructs Willy to move up even more (i.e. higher). The chief in assigning position uses common speech, and so does not lower others. Rather, chiefs in this and other contexts seem to engage in acts of *raising* the status of others. Lowering is done by similar status peers.

The final frame shows the position of the four pounders, the *erir*, and the chief as well as the *menindei*. In this frame the *menindei* has given the command to begin pounding the *sakau*, and all talk has ceased.

Nanawa *Willy* *Chief* *Menindei*

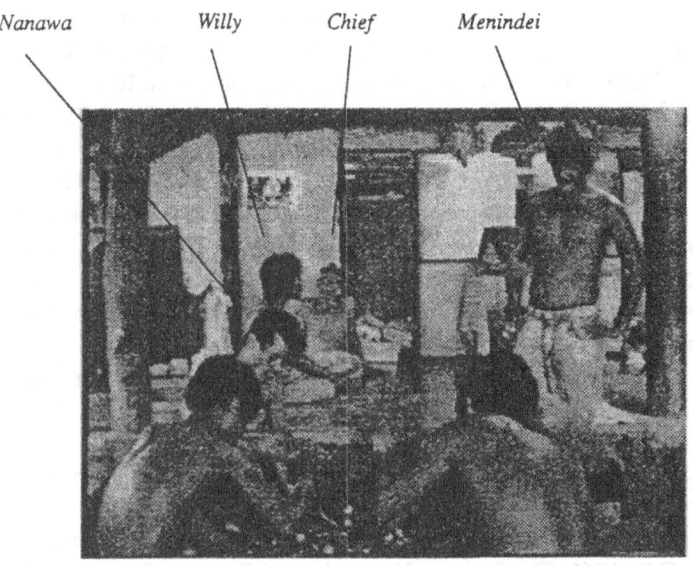

Fig. 7

4 Conclusion

When everyone is finally seated, the relative status of all participants is inscribed on space. This is a more complex encoding of hierarchy than is possible with lexical forms or within the vertical axis. While honorific speech constructs two distinct oppositions between high and low, horizontal spatial organization is able to categorize more finely and sharply between individuals sitting next to each other and sharing and exchanging similar speech patterns. One's proximity to the chief and others in seating arrangements signifies a complex individualized hierarchy of status. Verbs can only reference two different degrees of status (high or low), or in rare cases, three. Humiliative verbs for example reference people as members of a group of low status individuals, not differentiated. The actions of boys and girls as well as titled men and women are status marked as low by the same humiliative verb in the presence of the chief. The status marking of space in the feast house, however, can signal each individual's relative social status. No two positions are equal.

The polarity of high and low contrasts with *multiple* gradations of status constituted by the non-verbal ordering of space. It suggests that verbal symbols may be most powerful when ordered in terms of polar opposites (cf. markedness in languages), whereas visual symbolism is capable of a far greater complexity. This paper has shown how participants make explicit their cognitive map of their hierarchical social structure. It is a highly dialogic creation of hierarchy; non-verbal signs as well as verbal ones inter-relate.

References Cited

Agha, Asif. 1993. Grammatical and Indexical Convention in Honorific Discourse. *Journal of Linguistic Anthropology*, Vol. 3, No. 2 pp. 131-163.

Dixon, R.M.W. 1971. A Method of Semantic Description. In D. Steinberg and L. Jakobovits, ed. *Semantics: An Interdisciplinary Reader in Philosophy, Linguistics, and Psychology*.

Duranti, Alessandro. 1984. Lauga and Talanoaga: Two Speech Genres in a Samoan Political Event. In Brenneis, Donald L. and Fred R. Meyers, eds. 1984. *Dangerous Words: Language and Politics in the Pacific*. New York & London: New York University Press.

Haviland, John. 1978. Guugu-Yimidhir Brother in law Language. *Language in Society* 8:365-93.

_____. 1979. How to Talk to your Brother-in-Law in Guugu Yimidhirr. In *Languages and their Speakers*, Timothy Shopen, ed. Cambridge: Winthrop Publishers.

Keating, Elizabeth. 1994. Unpublished PhD Dissertation. University of California at Los Angeles.

McGarry, William, S.J. Unpublished manuscript. West from Katau.

Milner, G.B. 1961. The Samoan Vocabulary of Respect. *Journal of the Royal Anthropological Institute* 91:296-317.

Petersen, Glenn. 1982. *One Man Cannot Rule a Thousand*. Ann Arbor: University of Michigan Press.

Reisenberg, Saul. 1968. The Native Polity of Ponape. Vol 10 of the Smithsonian Contributions to Anthropology. Washington, D.C.: Smithsonian Insitution Press.

Shimizu, Akitoshi. 1982. Chiefdom and the Spatial Classification of the Life-World: Everyday Life, Subsistence and the Political System on Ponape. In M. Aoyagi, ed., *Islanders and Their Outside World*, pp. 153-215. St. Paul's (Rikkyo) Univ., Tokyo.

_____. 1987. Chieftainships in Micronesia. *Man and Culture in Oceania*, 3 Special Issue: 239-252.

On Drawing Lines on a Map

Barry Smith

Department of Philosophy and Center for Cognitive Science,
State University of New York at Buffalo, NY 14260-1010
phismith@ubvms.cc.buffalo.edu

Abstract

The paper is an exercise in descriptive ontology, with specific applications to problems in the geographical sphere. It presents a general typology of spatial boundaries, based in particular on an opposition between *bona fide* or physical boundaries on the one hand, and *fiat* or human-demarcation-induced boundaries on the other. Cross-cutting this opposition are further oppositions in the realm of boundaries, for example between: crisp and indeterminate, complete and incomplete, enduring and transient, symmetrical and asymmetrical. The resulting typology generates a corresponding categorization of the different sorts of *objects* which (complete) boundaries determine or demarcate. The theory is applied first of all in the areas of geography and of administrative and property law. Indications are then given as to how the typology may be applied also in other fields where physical and fiat boundaries are at work, including the field of cognitive linguistics and the related field of the ontology of truth.

Dividing Reality

Thomas Jefferson famously called into being the states of the so-called Northwest Ordinance by drawing lines on a map.[1] A number of issues are involved in understanding the peculiar creative magic at work in such a performance. These have to do with the nature of Jefferson's politico-geographical authority and with the practical and legal problems of translating ink-lines of a certain thickness on paper into working territorial borders on the ground. To deal in coherent fashion with these issues, however, it will be necessary first of all to consider certain more fundamental ontological questions relating to such creative actions and their products. What sorts of entities are these, which can be brought into being simply by drawing lines on a map? What are the forms and limits of such creativity, and how do the created entities relate to entities of the more humdrum sort? The remarks which follow, offered in answer to these questions, relate to a body of axiomatic work on what has come to be called 'mereotopology', an alliance of topological methods with the ontological theory of part and whole.[2] I shall here confine myself to informal consideration of the

---

[1]

When Jefferson first draw his map in 1874, drawing off 14 neat checkerboard squares between the boundaries of the Atlantic colonies and the Mississippi River, his map was sufficiently inaccurate that it did not even have the Great Lakes in the right place. In the end, 10 states were nonetheless created in this area, having boundaries which follow Jefferson's lines in large degree.

[2]

This work is summarized in Smith 1993, in Eschenbach, *et al.* 1994 and in Casati and Varzi (forthcoming). For a useful overview of related formal work by geographers on these issues, which however does not include a treatment of mereological ideas, see Herring 1991.

basic problems. The topological machinery will allow us to do justice to the fact that Jeffersonian delineations are effective in the geographical sphere only if the boundaries one creates are, in the jargon of topology, Jordan curves (broadly: the boundary of a geopolitical or administrative entity must be free of gaps and must nowhere intersect itself). Constructing topology on a mereological basis, rather than on the basis of set theory as is standardly done, will allow us further to do justice to the fact that there are no (or no obvious) candidate 'atoms' or 'elements' in the geographical world from out of which a universe of sets could be constructed. Rather, geographers deal with fields or regions of different shapes, sizes and functions, with sub-fields of these regions, and with the ways these fields and sub-fields overlap or fail to overlap. They deal, in other words, with a mereologically structured world.

Some of Jefferson's delineations correspond to what we might call *bona fide* boundaries: river-banks, coastlines, and the like. These are boundaries *in the things themselves*. They would exist (and did already exist) even in the absence of all delineating or conceptualizing activity on our or Jefferson's part. *Bona fide* boundaries are boundaries which exist independently of all human cognitive acts – they are a matter of qualitative differentiations or discontinuities in the underlying reality. You, too, possess *bona fide* boundaries of this sort (which correspond, roughly speaking, to the outer surface of your skin and to the boundaries of your internal organs). As is clear, however, if we examine the borders of practically every single political and administrative unit of the North-American continent, there are delineations which correspond to no genuine heterogeneity on the side of the bounded entities themselves. There are, in other words, not only *bona fide* joints in reality, but also pseudo-joints, of a type which are to be found also outside the geographical sphere, for example in the medical divisions, such as that between the upper, middle and lower femur, so extensively documented in atlases of surgical anatomy (see Fig. 1).

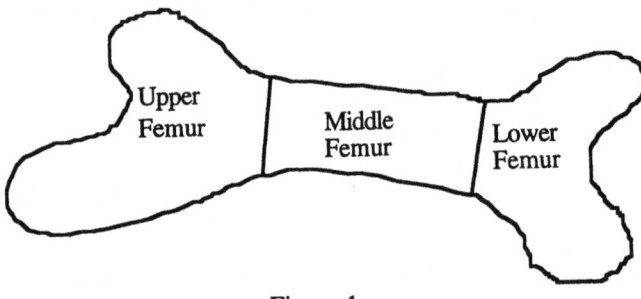

Figure 1.

Fiat Objects

Let us call boundaries of this created sort *fiat* boundaries, a terminology that is designed to draw attention to the sense in which the latter owe their existence to acts of human decision or fiat, to laws or political decrees, or to related human cognitive phenomena. Fiat boundaries as here understood may come into being either via deliberate choice, as in the Jeffersonian case, or as it were automatically, as when, by looking out across the landscape, I create, without further ado, that special type of fiat boundary we call the *horizon*. Clearly, national and state borders, and county- and property-lines provide a wealth of examples of fiat boundaries of the former, deliberate type; we shall see that the realm of human vision is a happy hunting ground for fiat boundaries of the latter, non-deliberate, type.

Fiat boundaries are boundaries which exist only in virtue of the different sorts of demarcations effected cognitively by human beings. Such boundaries may lie entirely skew to all boundaries of the *bona fide* sort (as in the case of the boundaries of Utah and Wyoming). They may also, however (as in the case of Indiana and Pennsylvania), involve a combination of fiat and *bona fide* portions, or indeed they may be constructed entirely out of *bona fide* portions which however, because they are not themselves intrinsically connected, must be glued together out of heterogeneous portions in fiat fashion in order to yield a boundary that is topologically complete. It is my intention that the opposition between fiat and *bona fide* boundaries should be regarded, modulo the existence of these mixed cases, as exhaustive and exclusive. Thus I do not wish to deny that there are types of spatial boundary which are difficult to classify under one or other of the two rubrics:

– exists independently of human cognitive acts
– does not exist independently of human cognitive acts.

And I do not wish to rule out, either, that it may be necessary to introduce at some later stage a categorization more detailed than the simple dichotomy here presented. Since, however, we do have many clear and important cases of boundaries which can be classified unproblematically in terms of this simple dichotomy, I will proceed in what follows as if the dichotomy itself is unproblematic. (And independent evidence for the coherence of this strategy is provided by the fact that almost everything which can be said in terms of the fiat–*bona fide* dichotomy in the spatial realm has an analogue in the realm of temporal objects (events, processes, states: see Smith 1994.)

Once fiat boundaries have been recognized, then we can apply the fiat–*bona fide* dichotomy also to the corresponding (bounded) *objects*. Objects, we can say, come in two sorts, the *bona fide* and the created, the latter being distinguished from the former solely in the fact that their boundaries arise, in whole or in part, through human cognitive operations of certain special sorts, in such a way that both boundaries and objects *exist* only in virtue of these operations.

The fiat–*bona fide* dualism can be contrasted with a range of alternative ontological options of *monistic* flavour which have played a role in the history of ontology and of related disciplines:

1. Some (we shall have occasion only later to press specific charges) would have it that *all* objects are fiat objects (for example that they are the result of human 'conceptual articulations') and that the very idea that there exists an underlying world of *bona fide* objects is merely the expression of an illegitimate 'objectivist' metaphysics, presupposing some notion of a 'God's eye view' that is held to be inappropriate to our post-enlightenment age. (Lakoff 1987)

2. Some, at the opposite extreme – the friends of physics, as we might call them – would have it that *no* objects are fiat objects, that our talk of the latter is *mere talk*, of no further ontological significance. (Friends of *ultimate* physics would insist further that all *bona fide* objects exist on a level way beneath our everyday ken, so that they would reject, too, candidate meso- and macroscopic *bona fide* objects such as people and planets.)

3. Some, finally – we might call them geographical monists – would have it that fiat objects are not created but merely selected from the infinite totality of geometrically possible regions of space.

Against all of these positions (and their many variants) we adopt here a more general, less conceptually constricted, framework which will allow us to express not only what is coherent in the positions mentioned but also what we take to be the correct view, according to which there are not only fiat objects, certain peculiar features of which demonstrate that they are products of genuine creation, but also *bona fide* objects, including *bona fide* objects of human scale such as, for example, you and me. And if it can be accepted that clear examples of fiat objects are provided by the Jeffersonian entities with which we began, then it will follow that not the least important reason for admitting fiat objects into our general ontology will turn on the fact that *most of us live in one* (or in what often turns out to be a nested hierarchy of such objects).

Types of Boundaries

As already pointed out, geographical fiat objects will in general have boundaries which involve a combination of *bona fide* and fiat elements. The shores of the North Sea are *bona fide* boundaries, but we conceive the North Sea as a fiat object nonetheless, because where it abuts the Atlantic it has a boundary of a non-*bona fide* sort. The status of the latter boundary is somewhat peculiar, since there seem to be few practical consequences which turn on the issue as to where, precisely, it lies. The case is similar in regard to many geographical boundaries of what we might call the purely qualitative sort (as contrasted with legal, political and administrative boundaries): consider, for example, the boundary between a hill and an associated valley. As such examples make clear, it is necessary to draw a further opposition between what we might call *crisp* and *indeterminate* boundaries (Cohn and Gotts 1994). For many geographical objects (deserts, valleys, dunes, etc.) are delineated by boundary-like *regions* which are to some degree indeterminate.[3] Moreover, political boundaries were once themselves standardly created in places (mountain ridges, middles of rivers) where there is little human activity and thus little chance or occasion to look into their exact location.[4]

We must bear in mind also that many national and property boundaries do in course of time come to involve boundary-markers: border-posts, watch-towers, barbed-wire-fences, garden-posts, and the like, which will tend in cumulation to convert what is initially a fiat boundary into something more real (tangible, physical). Moreover, there are often reasons of a non-arbitrary sort why these and those fiat objects are

---

[3]Indeterminacy is first of all an epistemological issue, a reflection of the fact we can establish no clear line where the fertile region ends and the desert begins. The difficulties in moving from an epistemological to an ontological concept of indeterminacy are legion, and must here be left aside. A complete treatment of these matters must take account also of the fact that the objects with which we have to deal as cognitive agents are often cognized in terms of fiat boundaries (as inscribed, for example, on maps) which are sharp, even where such sharp boundaries are not genuinely present in the physical world (the world as it exists independently of human cognitive demarcations).

4

See Prescott 1979, e.g. p. 112 on the way in which boundary disputes arise because of incomplete boundary evolution: 'Positional disputes will usually arise at one of two stages. Most of them will arise during the demarcation of the boundary, because the commission will be faced with the problem of matching the boundary definition to the landscape. However, it is also possible that positional disputes will arise at a much later date if the demarcation commission makes an errror.' On the role of maps in boundary-disputes see also *op. cit.,* pp. 127ff. On boundary-impermanence and the histories of boundaries over time, see pp. 171f., 178f.

created rather than others. Thus it seems to have been a complex medley of considerations relating to shipping, trade, harbours, climate, markets, etc., which led our ancestors to create the fiat object "North Sea" in a way which could not, just as well, have motivated them to create, say, a "Middle Sea" stretching between the Bermudas, the Azores, and Gotland. Fiat objects thus in general owe their existence not merely to human fiat but also to associated real properties of the relevant factual material (they are functions of *affordances*, in J. J. Gibson's terms). As demarcated in mesoscopic (geographical) reality they are in every case linked to *bona fide* objects of comparable scale, without which the relevant demarcations could not be effected at all. It is already for this reason a confusion to suppose that *all* objects (or all mesoscopic objects) might be of the fiat type. As the reports of boundary commissions make abundantly clear, the very possibility of fiat demarcation presupposes the existence of *bona fide* landmarks in relation to which fiat boundaries can be initially specified and subsequently re-located.

Note that the admission of fiat objects into our ontology is at least in one respect unproblematic: all fiat objects are supervenient on *bona fide* objects on lower levels in the sense that the fixation of relevant traits at the lower levels suffices to fix the values of traits at higher levels. The interiors of fiat objects are in this sense autonomous portions of autonomous reality. Only the respective external boundaries are created by us; it is these which are the products of our mental and linguistic activity, and of associated conventional laws, norms and habits. The relevant underlying thingly factual material *(unterliegende sachliche Tatbestandsmaterial*, as the German lawyer says) is in every case unaffected thereby.

Some Special Features of Geopolitical Boundaries

Boundaries in general exist as a matter of necessity only in consort with (as dependent parts or moments of) the higher-dimensional entities they bound. (Brentano 1988, Smith 1992) Geopolitical boundaries, or at least the paradigm examples thereof, are distinguished further in being infinitely thin. All political and legal boundaries must, it seems, enjoy in the long run the sort of geometrical perfection that is associated with infinite thinness: they must take up no space. For otherwise disputes would constantly threaten to arise in relation to the no-man's-land which the boundaries themselves would then occupy. If a wall or river separates two distinct portions of land, then either the wall or the river must be split equally down the middle, or it must be assigned as a whole to one or other of the two parties, or it must be declared common property (and then there will exist *two* infinitely thin boundaries separating each of the two distinct parcels of land from the commonly owned region which divides them).

Each adjacent pair of geopolitical boundaries (say: on the Franco-German border) manifests in addition the phenomenon of *coincidence* of boundaries.[5] The boundary of France is not also a boundary of Germany: each points inwards towards its own respective territory. Contrast, in this respect, the Western boundary of the old German Democratic Republic: here, exceptionally, no coincident twin was established, since the Federal Republic did not institute a boundary in that location at all. Moreover, as the case of Texas and the U.S.A. makes clear, distinct geopolitical boundaries may also coincide from within. That is, they may coincide for a part of their length along which they serve as boundaries *on the same side*.

---

[5]See Brentano 1988; compare also Smith 1995 and (forthcoming) for an axiomatic treatment of this notion.

One important reason for conceiving fiat objects and fiat boundaries as created entities (rather then as entities picked out or discovered within the pre-existing totality of all relevant geometrically determined possibilities) turns on the fact that there are fiat boundaries which coincide (occupy an identical spatial location) throughout their total length. The name 'Hamburg' refers on the one hand to a certain German city, on the other hand to one of the constituent states of the German Federal Republic. As it happens the boundaries of Hamburg *Stadt* and of Hamburg *Land* coincide exactly, and both point (serve as boundaries) in the same direction. But they are for all that not identical, as is seen in the fact that the two might in principle diverge (as is currently true, for example, in relation to the analogous case of the city and state of Bremen).

Note that even though political boundaries exist as full-fledged denizens of reality, and even though such boundaries exist always as parts of the things they bound, the coincidence of boundaries yet falls short of identity. France and Germany share no common parts. The border of France is, after all, *French.*

Scattered Objects

The examples of fiat objects mentioned above were in almost all cases examples of proper parts which are delineated or carved out (by fiat) within the interiors of larger (for example continent-sized) *bona fide* wholes. As the case of Japan or New Zealand makes clear, however, the restriction to such cases is by no means necessary. Boundaries (like the things they bound) can be *scattered* (Cartwright 1975); they can be built up mereologically out of separate and disconnected bits. The drawing of fiat boundaries can thus create not merely – Montana-style – fiat parts within larger *bona fide* wholes, but also – Hawaii-style – fiat wholes out of smaller *bona fide* parts. And then, while *bona fide* objects are in general connected, the fiat objects which are circumcluded by fiat boundaries in this way are non-connected.

Interestingly, there are cases where the two distinguished factors – on the one hand the carving out of fiat parts, and on the other hand the gluing together of fiat wholes – operate in tandem, so that geographical objects are created via the fiat unification of disconnected parts within larger *bona fide* wholes: the Holy Roman Empire (of sometimes non-connected principalities, bishoprics, city-states, etc.) will serve as a nice example in this regard, but so will all coastal nations in whose territory islands are included.

Note that there are also scattered fiat objects outside the strictly geographical domain. Examples might be: the Polish nobility, the constellation Orion, the species *cat.* Following Meinong (1899) we might refer to such entities as 'higher-order' fiat objects. Objects of this sort may themselves be unified together modularly into further fiat objects (say: the genus *mammal*, the Union of Pacific Island Nations). Set theory is a general theory of the structures which arise when objects are conceived as being united together *ad libitum* in this fashion on successively higher levels, each object serving as member or element of fiat objects on the next higher level.

Lasting vs. Ephemeral Boundaries

We can distinguish further between enduring and transient boundaries. The boundaries of the Chinese Middle Kingdom and of the Island of Malta are (respectively, fiat and *bona fide*) examples of the former. A great wealth of examples of transient *bona fide* boundaries is provided by non-prognostic weather maps (where we are assuming, realistically, that such boundaries are discovered, and not created, by metereologists.) Examples of transient fiat boundaries, on the other hand – of transient boundaries which are in truth created by human cognitive operations – can be taken from the sphere of visual perception. The psychologist Ewald Hering defines the 'visible field' as the totality or region of real objects imaged at a given moment on the retina of the right or left eye. (1964, p. 226) The visible field is thus a *part* of the ambient environment of the visually perceiving subject. Yet the external boundary of this field is for all that a fiat boundary in the sense set out above – a boundary which exists only as a result of human cognitive activity – and moreover it is a fiat boundary which changes with every movement of the eye and head. Moreover, the interior of this field is itself subject to a complex and subtle fiat organization: it is built out of physical surfaces and other components which are structured in terms of an opposition between (1) entities in the focus of attention and characteristically manifesting determinate boundaries ('figures'), and (2) entities which have indeterminate boundaries and which are experienced as running on (as 'ground') behind them.

Linguistic Fiats

A veritable host of transient fiat boundaries comes to be drawn in reality through our use of language. Such carving out of linguistic fiat objects is in part a matter of sheer grouping together, for example of the sort that is achieved through the use of plural referring expressions such as 'Hannah and her sisters', 'Siouxsie and the Banshees', and so on (see Ojeda 1993). But it is in part also a matter of windowing or foregrounding (Talmy, forthcoming) and in part a matter of the articulation of external reality in terms dictated by our concepts: if I point to a group of irregularly shaped protuberances in the sand and say 'dunes', then the objectual correlate of my expression is a complex plurality (a higher-order fiat object with non-crisp boundaries) divided, via the concept *dune*, into constituent (non-crisp) parts or elements. (Smith 1987, § 15) Cognitive linguists such as Talmy, Langacker and Lakoff have rightly emphasized the degree to which language effects complex and subtle concept-mediated articulations of this sort. Unfortunately, however, they too often draw illegitimate epistemological conclusions from this insight.[6] Moreover they come close to the position mentioned (and rejected) above (a position reminiscent of the fable of King Midas), according to which *all objects to which language refers are fiat objects*. (The error arises through an illegitimate passage from: 'object which we grasp linguistically through concepts' to 'object which exists only in virtue of our linguistically effected demarcations'.)

Certainly an important class of transient fiat boundaries is effected through our use of natural language. As Talmy and others have pointed out, our use of expressions such as 'this' and 'that' in relation to objects in space involves in each case the drawing of an imaginary planar boundary, lying in a plane in front of and parallel to the speaker, which is such that the objects labelled *this* and *that* lie on opposing sides, in roughly the following fashion:

---

[6]As Lakoff writes: "One of the cornerstones of the objectivist paradism is the independence of metaphysics from epistemology. The world is as it is, independent of any concept, belief, or knowledge that people have. Minds, in other words, cannot create reality. I would like to suggest that this is false and that it is contradicted by just about everything known in cultural anthropology." (p. 207) Lakoff goes on to admit that the thesis that 'mind creates reality' does not in fact apply in relation to physical reality; it applies, rather, only in relation to the reality of human institutions. Even in regard to human institutions, however, in contrast to what Lakoff has to say, *our thinking does not make it so*.

```
speaker              this          |          that
```

Figure 2.

It is an interesting feature of this type of transient boundary-creation, that it is effected in exactly the same way independently of order of magnitude, from the tiniest ('this flea') to the grossest ('that galaxy').And as Talmy has also shown (1995), boundaries of the given sort belong to a much larger family which includes also the fictive orientation paths which are created when we assert, for example:

> I aimed the camera into the living room.

(think of an invisible line extending out from the camera into the room). Such orientation paths may further be dynamic in nature:

> I slowly looked towards the door.
> I slowly turned the camera around the room.

And fictive boundaries are at work also in cases of the following sort:

> I offered her the book [creates a virtual sphere around the recipient].
> She accepted the book [she allows the sphere to be broken].
> She rejected the book [she maintains the sphere unbroken].[7]

As should by now be clear, however, it is illegitimate to move from the thesis that such boundary phenomena are pervasive features of our various modes of gaining linguistic access to the world, to the conclusion that the world to which we then have access is a world of fiats only. On the contrary, the very existence of fiat boundaries, here as elsewhere, presupposes a *bona fida* reality consisting of objects of roughly similar scale in and through which such boundaries can be drawn. Moreover, a thesis to the effect that language gives us access only to objects which we ourselves create through our linguistic fiats would imply the impossibility of all scientific investigation of a theory-independent world (including scientific investigation of language itself) and would thus saw off the very hand that feeds it.

One further problem with the work of cognitive linguists such as Lakoff and Talmy is an unclarity as to the question whether the fiat boundaries (including fictive motion paths) created through our uses of natural language are out there in the world (as Talmy's detailed descriptions of his specific examples would suggest) or rather – as the cognitive linguists' favoured methodological pronouncements would have it – somehow such as to exist only in what is referred to as the 'conceptual sphere', so that even space itself can be described as a 'conceptual domain'. If, as I have suggested, the fiat boundaries induced through natural language are of a piece with geographical fiat boundaries, then it is clear how this unclarity is to be resolved: the fiat boundary between things called 'this' and things called 'that' is out there, in the world, in a roughly planar region determined differently from context to context. Cognitive linguists are dealing primarily not, as they themselves often like to suggest, with *conceptual* structures, but rather, like geographers, with structures in the world, albeit with structures of a special, fiat type.

---

[7]A related type of fictive imposition of in this case temporal boundaries is illustrated in the difference between:

> She saw him crossing the road [open interval with indeterminate boundaries].

> She saw him cross the road [closed interval with determinate boundaries].

A Coda on Truth and Against Model-Theoretic Semantics

There is, if one will, a windowing of reality that is effected by our uses of language, especially of those descriptive uses of language which are involved in the making of true empirical judgments. The ephemeral fiat boundaries effected through declarative sentences are indeed, or so I will now argue, analogous to the ephemeral boundaries of the visible fields associated with acts of visual perception. This analogy in its turn suggests a new understanding of that relation between judgment and world we call 'truth'. This relation has classically been understood in terms of a 'correspondence' or isomorphism between a judgment or assertion on the one hand and a certain portion of reality on the other. The central difficulty standing in the way of this classical theory turned on the fact that reality evidently does not come ready-parcelled into judgment- or sentence-shaped portions that would be predisposed to stand in relations of correspondence of the suggested sort. It is for this reason that many practitioners of logical or truth-functional semantics have tended, disastrously, to treat not of truth as such (understood as truth *to* an independent worldly reality), but rather of what they call *truth in a model*, where the model is a specially constructed set-theoretic reality-surrogate whose relation to reality itself is left unspecified.

The theory of ephemeral fiat boundaries and of the windowing of reality in language can help us to avoid the need for this resort to surrogates by allowing us to treat judgment itself as a *sui generis* variety of drawing fiat boundaries around entities in reality of a precisely appropriate (truth-making) sort: veridical judgments then stand to fiat judgment-correlates as acts of veridical perception stand to the visible field. Each true empirical judgment can be seen, in this light, as effecting a division of reality in fiat fashion into two disjoint regions:

– a first, truth-making region, consisting of those entities that are relevant to the truth of the judgment in question,

– a complementary region, consisting of those entities not so involved

Truth itself can then be defined as the relation of correspondence between a judgment and its corresponding truth-making region, in such a way that a true judgment would be something like a map of the corresponding portion of reality. A Jeffersonian view of truth along these lines – for all its superficial strangeness – can be seen on inspection to enjoy a degree of phenomenological, linguistic and ontological adequacy that is higher than standardly available accounts. Its phenomenological adequacy derives from the fact that the account of windowing of reality via language is of a piece with an account of perceptual windowing, so that a theory of evidence, of verification and falsification in perceptual acts, is available from the start. Its linguistic adequacy derives from the fact that the view imposes no unitary logical form (the form of functional application) upon our judgments, but is sensitive, rather, to the wide range of different natural-language sentence forms which are utilized in making true judgments, forms whose corresponding demarcatory effects have been described in detail in the work of cognitive linguists (see especially Langacker 1987/1991). Its ontological adequacy, finally, derives from the fact that the view in question – which after all that has been said we might refer to as the Jeffersonian theory of truth – is able to do justice to the untidy, flesh-and-blood character of the reality to which our judgments are directed.

References

Brentano, Franz 1988 *Philosophical Investigations on Space, Time and the Continuum*, English translation by Barry Smith, London/Sydney: Croom Helm.
Cartwright, Richard 1975 "Scattered Objects", in K. Lehrer (ed.), *Analysis and*

Metaphysics, Dordrecht: Reidel, 153-171, reprinted in Cartwright, *Philosophical Essays*, Cambridge, Mass.: MIT Press, 1987, 171-186.

Casati, Roberto andVarzi, Achille (forthcoming) "The structure of Spatial Localization", *Philosophical Studies*.

Cohn, A. G. and Gotts, N. M. 1994 "A Theory of Spatial Relations with Indeterminate Boundaries" in Eschenbach, Habel and Smith (eds.), 131-150.

Eschenbach, Carola, Habel, Christopher and Smith, Barry (eds.) 1994 *Topological Foundations of Cognitive Science*, Hamburg: Graduiertenkolleg Kognitionswissenschaft.

Hering, Ewald 1964 *Outlines of a Theory of the Light Sense*, trans. L. M. Hurvich and D. Jameson, Cambridge, Mass.: Harvard University Press.

Herring, John R. 1991 "The Mathematical Modeling of Spatial and Non-Spatial Information in Geographic Information Systems", in Mark and Frank, eds., 313-350.

Lakoff, George 1987 *Women, Fire, and Dangerous Things*, Chicago: University of Chicago Press.

Langacker, Ron 1987/1991 *Foundations of Cognitive Grammar*, 2 volumes, Stanford: Stanford University Press.

Mark, David M. and Andrew U. Frank (eds.) 1991 *Cognitive and Linguistic Aspects of Geographic Space*, Dordrecht/Boston/London: Kluwer.

Meinong, Alexius 1899 "Über Gegenstände höherer Ordnung und deren Verhältnis zur inneren Wahrnehmung", *Zeitschrift für Psychologie und Physiologie der Sinnesorgane*, 21, 182-272.

Ojeda, A. 1993 *Linguistic Individuals*, Stanford: CSLI.

Prescott, J. R. V. 1978 *Boundaries and Frontiers*, London: Croom Helm, Totowa, N.J.: Rowman and Littlefield.

Smith, Barry 1987 "On the Cognition of States of Affairs", in K. Mulligan, ed., *Speech Act and Sachverhalt: Reinach and the Foundations of Realist Phenomenology*, Dordrecht/Boston/Lancaster: Nijhoff (1987), 189-225.

Smith, Barry 1992 "Characteristica Universalis", in K. Mulligan, ed., *Language, Truth and Ontology* (Philosophical Studies Series), Dordrecht/Boston/London: Kluwer (1992), 50-81.

Smith, Barry 1993 "Ontology and the Logistic Analysis of Reality", in N. Guarino and R. Poli (eds.), *Proceedings of the International Workshop on Formal Ontology in Conceptual Analysis and Knowledge Representation*, Padova: Institute for Systems Theory and Biomedical Engineering of the Italian National Research Council, 51-68. Revised version in G. Haefliger and P. M. Simons (eds.), *Analytic Phenomenology*, Dordrecht/Boston/London: Kluwer (forthcoming), 223-245.

Smith, Barry 1994 "Fiat Objects", in N. Guarino, L. Vieu and S. Pribbenow (eds.), *Parts and Wholes: Conceptual Part-Whole Relations and Formal Mereology, 11th European Conference on Artificial Intelligence, Amsterdam, 8 August 1994*, Amsterdam: European Coordinating Committee for Artificial Intelligence, 15–23.

Smith, Barry 1995 "Zur Kognition räumlicher Grenzen: Eine mereotopologische Untersuchung", *Kognitionswissenschaft*, 4, 177-184.

Smith, Barry (forthcoming) "Boundaries", in L. H. Hahn (ed.), *The Philosophy of Roderick Chisholm*, LaSalle: Open Court.

Talmy, Leonard 1995 "Fictive Motion in Language and 'Ception'", to appear in P. Bloom, *et al.*, eds., *Language and Space*, Cambridge, Mass.: MIT Press.

Talmy, Leonard (forthcoming) "The Windowing of Attention in Language", in M. Shibatani and S. Thompson (eds.), *Essays in Semantics*, Oxford: Oxford University Press.

How Significant Are Cultural Differences in Spatial Cognition?

Daniel R. Montello

Department of Geography, University of California
Santa Barbara, CA 93106 USA

Abstract. In this essay, I critically discuss ideas about cultural differences in spatial cognition. A critique of the traditional empiricist framework for understanding the development of cognitive structures and processes is described. An evolutionary framework is provided as an alternative. The ambiguity between culture-related and culturally caused differences is also elaborated. Reasons for cultural universality in spatial cognition are then considered. With these ideas as background, some aspects of spatial cognition are enumerated that I believe are largely culturally universal. Following that, aspects that may show more significant cultural variability are discussed. In particular, the relationship of spatial language to spatial thought is addressed in some depth. I conclude that cultural differences in spatial cognition are not nearly as substantial as is often claimed. Finally, the extent to which cultural differences in spatial cognition are important for geographic information system (GIS) research and design is considered.

1 Introduction

It is often proposed that people living in different cultural groups are brought up to perceive and think about the world in very different ways. According to this view, the structures and processes of the mind develop primarily in response to unique experiential and socialization forces that are quite different in different cultures, particularly in widely separated cultures that have had little contact with each other. These include *spatial* structures and processes, of course. To the extent that this view is true, it has important theoretical and practical design consequences for the internationalizing efforts of the spatial and geographic information community. An extreme view of cultural variability in spatial cognition (probably rarely held) implies that such design efforts must be taken up virtually anew, from scratch, for each cultural group that will use these systems. That is, if the "Western" concepts of spatial information and analysis inherent in such systems can even be comprehended by all cultures.

My purpose in this essay is to explore the idea that there are significant cultural differences in spatial cognition. I intentionally approach the idea with an attitude of skepticism. Discussions of cultural differences in spatial cognition typically lack skepticism, in my experience, leading to exaggerated and unbalanced claims about the magnitude and importance of cultural differences. To bolster the skepticism, I will briefly describe a framework for thinking about cognition and its development that differs from the framework commonly associated historically with most ideas about cultural differences. I will critically review some hypotheses and empirical evidence for and against the idea of significant cultural differences. As a way of structuring the essay, I will list aspects of spatial cognition that I think are actually universal or variable across cultures. My conclusion will be that ideas about cultural differences in spatial cognition are often exaggerated, and perhaps not as deserving of careful scrutiny in the GIS research and design communities as is sometimes suggested (e.g., Egenhofer & Golledge, 1994).

Although empirical evidence is discussed below, I will not simply evaluate the significance of cultural differences in spatial cognition according to the evidence provided by specific research studies. Assumptions underlying various positions will also be evaluated, as will issues of conceptual coherency and consistency with existing theories. Empirical evidence is certainly important but cannot be the sole basis for an answer to the question. This is true for several reasons. The issue is broader than a specific theory or hypothesis. To a large extent, it is really a question about a framework for thinking about thinking. Virtually all modern philosophers of science, who otherwise disagree about many things, recognize that scientific debates about such broad theoretical frameworks do not hinge exclusively on an examination of empirical evidence (Kuhn, Feyerabend, Laudan, etc.). In addition, the structure of induction and the *modus tollens* logic used in hypothesis testing result in a great asymmetry between the empirical evaluation of the hypotheses of cultural universality and difference (*ceteris paribus*, supporting the latter is much easier). And of course, much relevant empirical investigation of cultural differences in cognition has yet to be conducted. It is not my intention here to argue for a cessation to such investigation. I am simply proposing that examination of these issues will benefit by looking at them through the lens of an alternative framework, a different set of assumptions. In this way, we will achieve greater balance in our conceptions of cultural universality/difference.

This is certainly not a "yes-or-no" issue I am addressing. There is no question that both cultural universals *and* differences in spatial cognition exist. At the very least, cultural differences in spatial cognition are evident because cultural differences in spatial language plainly exist. But how substantial or superficial are cognitive differences, including those implied by linguistic differences? What are their consequences for GIS researchers and designers? It also bears emphasizing that I have tried to keep my focus on *spatial cognition*. There are many other cultural issues surrounding GIS, especially of a practical nature, that remain important concerns no matter what one's view of the significance of spatial cognitive differences (e.g., Campari and Frank, 1993; A. U. Frank, personal communication, March 20, 1995). These include issues of data quality and standards, administrative procedures, organizational structures, etc.

2 Evolution and the Standard Social Science Model

The thesis of this essay rests partially on a claim that belief in substantial cultural differences in cognition derives, to a large extent, from belief in a framework for understanding learning and cognitive development that is largely incorrect. This framework is essentially a traditional nurture or empiricist framework for thinking about cognition. Historically, the framework has been developed and expressed in the contributions of such notable thinkers as Locke, Durkheim, Marx, Boas, George

Mead, Margaret Mead, Watson, Skinner, and many others. It has been this century's dominant theoretical framework in most of the social and behavioral sciences, especially within the "Anglo-American" tradition.

Tooby and Cosmides (1992) extensively discuss this framework and its continued influence today. They label it the "Standard Social Science Model" (SSSM). Here is a partial list of the characteristics of the SSSM, according to Tooby and Cosmides:

1. Human groups have a culture, which is a virtually group-universal set of beliefs, behavioral practices, symbol systems, etc. These common elements are maintained and transmitted by the group, which has cross-generational continuity.

2. This transmitted culture is the sole explanation for observed within-group similarities and between-group differences.

3. Cultural transmission is maintained through learning, which can be seen as the group-organized process of socialization, imposed by the group on the child. The individual is thus a more or less passive recipient and product of culture. What is organized and contentful in the individual's mind comes from culture and is socially constructed.

4. Features of a particular culture emerge at the sociocultural level of the group and are not given specific shape or content by human biology, nature, or inherited psychological design. The evolved mechanisms of the human mind are content-independent and content-free.

In sum, the SSSM embodies a set of beliefs about virtually unlimited behavioral plasticity as a function of experience, a nearly unbounded range of possible solutions by different cultures to the problems of survival, the critical and overriding importance of childhood experience, and the irrelevance of human evolutionary history. Several of the papers I cite below serve as evidence that some version of the SSSM, usually in a less extreme form, is echoed in the writings of researchers in spatial cognition.

Some writers have recently recognized that the SSSM constitutes a set of assumptions that has served as a barrier to seeing universal aspects of human cognition. Sheets-Johnstone (1990) refers to this consequence of the SSSM as "steadfast cultural relativism". Notably, the SSSM fails to incorporate the insights of Darwin's theory of evolution. This oversight has left the social and behavioral sciences unconnected from biology and from each other, apparently without need for mutual compatibility (Cosmides et al., 1992). Darwin himself applied the theory of evolution by natural selection to mental and behavioral phenomena, not just to physiology and anatomy. And within the last decade or so, a flurry of work has been produced on evolutionary aspects of human personality, mating behavior, kinship relations, and so on (e.g., Buss, 1989; Kenrick et al., 1985).

Of special interest to us, the role of evolution in human cognition has also been recently explicated. Tooby and Cosmides (1992), for instance, provide a model of the human brain as a collection of *domain-specific modules* that evolved in specialized ways to solve particular behavioral problems of enduring importance to humans over hundreds of thousands of years (see also, Fodor, 1983). There are numerous such domains: obtaining food and shelter, construction and repair of material artifacts, metaphysics and ethics, social alliances, communication, mating, status and dominance hierarchies, etc. This contrasts with the SSSM view of the brain as a general all-purpose problem solver that acquires its content and functioning primarily as a result of highly variable exposure to processes of cultural transmission.

3 Cultural Differences, Cultural Universals

Before proceeding, brief definitions of culture and of spatial cognition are in order. *Culture* refers to a body of knowledge and beliefs that is more or less shared between individuals within a group and transmitted across generations. Essentially, culture specifies preferred or accepted patterns of ideation and behavior dealing with religious and value systems, social systems, and material/technological systems. *Cognition* deals with knowledge: its acquisition, storage, retrieval, manipulation, and use by organisms or machines to achieve behavioral goals. Defined in this broad and theoretically neutral way, cognition includes structures and processes involved in perception, learning, thinking, memory, reasoning and problem-solving, and language. *Spatial* cognition refers to these structures and processes when they deal with spatial knowledge: knowledge of location (including distance and direction), size, shape, and pattern, as well as changes in these properties across time.

Cross-cultural research (or just cultural research) is simply research that incorporates cases from two or more cultural groups. Such research implies the possibility of identifying cultural differences *or* universals; it does not necessarily entail a focus on differences. It is critical to remember that two individuals from different cultural groups may differ for a host of reasons aside from their cultural group membership (just as two individuals from the same group may differ a great deal). Any variables upon which the two individuals differ could potentially explain a difference in their spatial cognition. A comprehensive list of possible explanatory variables would essentially be infinite; in addition to culture, there is genetic constitution, physiology and anatomy, sex, gender, age, education, expertise, socioeconomic status, family membership, residential environment, and so on. Frequently, a statistically identified cultural difference is actually due to some other variable that covaries with culture. We might speak of "culturally caused differences", a term referring to differences caused by culturally transmitted ideology, artifacts, rules and mores. Alternatively, we might be dealing with "culture-related differences", differences that are caused by some variable other than cultural group but covary with it (Gentile [1993] discusses a similar distinction with respect to sex and sex-related differences).

When understood in a causal sense, I would argue that cultural differences are typically among the weakest and least important explanations for individuals differing in spatial cognition. Several variables listed above have more substantial effects. Unfortunately, the identification of caused as opposed to culture-related differences is very problematic and difficult to disentangle (D. M. Mark, personal communication, September 21, 1994). This problem is one of the most profound of several thorny interpretive and methodological difficulties in cross-cultural research, such as communication ambiguities between researcher and subject (Hazen, 1983; Pick & Pick, 1978). Covarying factors are often thought to be responsible for observed cultural differences. For instance, Pick and Pick (1978) discussed the likelihood that observed cultural differences in picture perception are really educational differences within cultures. As another example, a difference in the perception of depth due to different residential environments (if it exists; see below) is not a culturally caused difference if it is contained in the visual system and not in the body of culture.

An excellent example of this point for our purposes involves the "amazing feats" of the South Pacific navigators described in the much cited works of Gladwin (1970) and Lewis (1978). As these authors clearly document, the ability to use the subtle and varied cues of the ocean environment, the extensive star patterns, and so on, depends on extensive training. This is true just as competence at reading topographic maps and using compass and sextant depends on training in our own culture (and either system can be readily learned by members of the other culture). In other words, these differences are expertise differences, not culturally caused differences.

3.1 What Aspects of Spatial Cognition Are Culturally Universal?

There are several reasons to expect considerable cultural universality in human spatial cognition. First is the largely common gene pool distribution of the human species and the resulting near universality in nervous system organization and activity. As described above, there is good reason to believe that the nervous system is specialized to solve domain-specific problems that confronted the human species over hundreds of thousands of years of evolution. The problems of learning spaces, navigating in spaces, etc., are excellent candidates for the kinds of pervasive and fundamental problems that we would expect to find reflected in the modularized structure of the nervous system (Lewis-Williams & Dowson, 1988; Neisser, 1987).

A second reason for universality is one that is often overlooked, perhaps in part because of its immediacy and near constant relevance to cognition. That is the fact that humans everywhere have largely the same body structures and processes, including the same motor and sensory systems. Excepting disabilities, humans practice upright bipedality, are bilaterally symmetrical, have heads at the tops of their bodies, have frontally-aligned eyes with similar light sensitivities, have transversely-aligned ears that respond to similar frequency ranges, sense gravity and body movement via proprioceptive senses, can run at about the same speed, are fairly close in body size, have dexterous hands useful for gestures and precise manipulation, can form spoken words with their vocal tracts, and so on. The importance of the concept of "embodied" thinking is eloquently expressed by Lakoff (1987) in his theory of *experiential realism*. "The core of our conceptual systems is directly grounded in perception, body movement, and experience of a physical and social character" (p. xiv). Cultural universals in bodily experience and the use of concrete imagery lead to the universal importance of "kinesthetic image schemas" in language and thought. Apparently independently, Sheets-Johnstone (1990) has organized observations like these in an interesting discussion of the role of the human body in the phylogenesis and ethnogenesis of cognition. She refers to "corporeal thought", siting the origin of human cognition in our shared animate forms and "tactile-kinesthetic correlates".

A third reason for universality is that the learning and socialization to which children are exposed actually exhibits many common features across cultures. The psychological mechanisms that allow various forms of learning to occur are culturally universal (e.g., operant conditioning). Furthermore, although the specific content of transmitted knowledge varies somewhat across cultures, much of it speaks to the same issues or provides solutions to the set of universal human problems reviewed above.

Finally, we might expect universality in human spatial cognition because of the degree of similarity in residential environments. There are in fact many similarities between human environments everywhere and at all times throughout human evolution. In an illuminating essay, Shepard (1984) discusses these "ecological constraints on internal representation". All humans are born into a spatial world that is 3-dimensional and locally Euclidean, is isotropic except for a gravitationally-defined local upright, consists of a prevailing solid terrestrial surface of support and locomotion that is approximately flat (at a suitable scale), and is generally horizontal and normal to the local upright direction. In addition, there is a cycle of directions on the 2-dimensional surface defined for all cultural groups by the earth's structure and pattern of rotation, namely the four cardinal directions (there have never been permanent residents of the poles). Because of these enduring regularities of the physical world, in conjunction with those of the human body and sensory systems, three spatial reference frames are defined for all humans: a body-centered egocentric frame, an object-centered frame, and an environment-centered frame (some additional distinctions can be made). Gibson (1979) discusses other regularities in the terrestrial environments of humans, such as the availability of objects for sitting upon and visual features for remembering as landmarks.

So there are at least four major reasons for cultural universality in spatial cognition: (1) organizational similarities of nervous systems, (2) common body structures and processes, (3) learning and socialization similarities, and (4) similar residential environments. Detailed considerations of the status of these four could obviously constitute several books. The last idea about residential environments calls for further comment here, however, because it has so often been proposed to explain cultural differences in spatial cognition.

The influence of culturally different residential environments on spatial cognition has most specifically been expressed by two ideas: the *carpentered-world hypothesis* and the *ecological hypothesis* (Deregowski, 1980; Liebowitz & Pick, 1972; Pick & Pick, 1978; Segall et al., 1966). Both have been proposed as potential explanations for cultural differences in susceptibility to visual geometric illusions (such as the Müller-Lyer and Ponzo illusions) and for other possible visual perceptual differences (e.g., distortion of angles toward 90°). The carpentered-world hypothesis states that people living in technologically-developed cultures are exposed to many straight lines and right angles prevalent in their "carpentered" built environments. That exposure will ostensibly predispose these people to see the world in "rectilinear" terms. Conversely, people from technologically-undeveloped or "traditional" cultures will not be prone to see the world in rectilinear terms because of the supposed lack of straight lines and right angles in their built environments and the assumption that the natural environment does not provide such stimuli. The related ecological hypothesis focuses on the availability of open vistas that allow perception of long views into the distance. People living in dense jungles or narrow canyons (urban streets?) will presumably not be exposed to opportunities to train their visual systems to interpret depth appropriately (and size constancy, etc.). The classic anecdote is Turnbull's 1961 observation about a BaMbuti Pygmy who temporarily confused distant buffalo for insects. People living in open savanna, tundra, etc., should have had ample opportunity to learn to see depth and not show the effects of restricted vistas.

The empirical evidence pertaining to these ideas is extensively reviewed by the authors cited above. All agree that the evidence is equivocal. In particular, the extensive results of Segall et al., who studied numerous cultural groups from traditional Africa, clearly show that the various illusions are culturally universal, if variable in strength. It is also clear that the various cultural groups studied differ greatly in their exposures to pictorial stimuli. All of the work on cultural differences in visual perception has involved such pictorial stimulus materials. It is well known that even people from technologically-developed cultures have to develop the ability to see a 3-dimensional world in flat pictures (Hagen, 1980). Thus, differences in experience with pictures is the most likely explanation for the modest cultural differences in visual perception that have been found. Hypotheses about residential environments do not enjoy much support (Liebowitz & Pick, 1972).

The carpentered-world hypothesis includes an assumption that natural environments do not provide a stimulus basis for learning rectilinear perception. The implication is that natural, as opposed to built, environments don't contain straight lines and right angles (see also, Couclelis, 1992). How valid is this claim? Does the natural environment contain rectangular stimuli? In response to a comment by her literature professor that "straight lines on the landscape are put there by man", Gail Jensen Sanford composed the following list of straight lines that do exist in nature (reprinted in the February 1995 issue of Harper's Magazine):

> lines along the top of a breaking wave; distant edge of a prairie; paths of hard rain and hail; snow-covered fields; patterns in crystals; lines of white quartz in a granite surface; icicles, stalactites, stalagmites; surface of a calm lake; markings on zebras and tigers; bill of a duck; legs of a sandpiper; angle of migrating birds; dive of a raptor; new frond of a fern; spikes of a cactus; trunks of young, fast-growing trees; pine needles; silk strands woven by spiders; cracks in the surface of ice; strata of metamorphic rock; sides of a volcano; wisp of windblown altocumulus clouds; inside edge of a half-moon

Although one can argue with some of these, it is difficult to avoid the conclusion that straight lines are actually easy to find in natural or noncarpentered environments. Furthermore, the projective geometry of perception provides especially important information of this type. Directions to seen or heard objects are perceived as straight. More explicitly, straight lines are revealed in visual perception when two features are brought into alignment in the visual field. This information occurs in all environments and is available to all residents without sensory impairment. Similar things could be said about right angles, given the universal availability of orthogonal cues from gravity, human body structure (e.g., eyes orthogonal to ears), and postural uprightness with respect the ground.

Given these reasons for spatial cognitive universality as a background, there are several specific ways in which I believe spatial cognition is quite similar across different cultures:

1. Existence and functionality of cognitive maps, as evidenced by the ability to remember features, routes, distances, directions, and make spatial inferences (detours, shortcuts).

2. Salience of the gravitational upright and the perception of a 3-dimensional world with horizon (ground-atmosphere boundary).

3. Differential treatment of spatial information in memory, reasoning, language, as a function of scale.

4. Categorical and hierarchical organization of regions.

5. The prominent role of recognized and labeled visual features in memory organization and problem-solving.

6. The existence and use of multiple frames of reference, varying with respect to the difficulties they engender for reasoning and communication.

7. Orthogonal-oblique differences in the accuracy and precision of angular knowledge.

8. Relative difficulty of apprehending 3-dimensional spatial relations in environmental spaces, as compared to 2-dimensional (e.g., learning 3-dimensional cave structures).

9. The use of spatial metaphors for nonspatial concepts.

Empirical evidence or conceptual arguments for some of these as universals is presented in: Appelle, 1972; Lakoff, 1987; Lynch, 1960; Montello, 1993; Shepard, 1984; Shepard & Hurwitz, 1984; Wallace, 1989. Although there is empirical evidence for the universality of parts of the list, little or none yet exists for other parts. Clearly, more cross-cultural research in spatial cognition would be valuable. The list could provide a guide for such research.

3.2 What Aspects of Spatial Cognition Are Culturally Different?

In this section, I discuss aspects of spatial cognition where I think cultural differences may be more substantial. However, even in these cases, there are important limits on the significance of the differences that should be considered. Because I elaborate on these limits, this list is discussed in more detail than was the previous list of universals. This is especially true in the case of spatial language; no other aspect of spatial cognition has been researched cross-culturally to the degree that language has been.

Spatial Language. A comprehensive discussion of cultural differences in cognition must consider the relationship of language and thought. Presumably, language is one of the most important vehicles of the socialization processes described in the SSSM. Different cultures obviously speak different languages. Furthermore, the history of the behavioral and social sciences contains many claims that thought is "internal language", or at least that thought fundamentally depends on language. J. B. Watson, for instance, proposed that thought was "covert vocalization". More specific to our present concern with spatial cognition, Littlejohn (1963) claimed about the Temne of Africa that "since they have no word for space in general, it is unlikely that they have a concept of it".

The most influential statement about the influence of language on thought is the so-called *Whorfian Hypothesis* (French, cited by Segall et al. [1966], called it the

"Humboldt-Boas-Cassirer-Sapir-Whorf-Lee Hypothesis"). This is the claim that a cultural group's language determines the way its members perceive and think about the world. Weaker forms of *linguistic relativity* suggest only that linguistic differences cause some differences in thought. Linguistic relativity has a rocky intellectual history. Segall et al. (1966) stated that in "100 years of study there has been disagreement as to whether genuine differences in perception [as a function of language] were ever demonstrated" (p. 38). Many writers consider particularly damaging the research by Heider (1972 -- now Rosch) showing that cultures with very few color terms nonetheless perceive the structure of unnamed color categories much as cultures like our's do that have more color terms. Recently, Pinker (1994) unequivocally criticized the Whorfian hypothesis. In his words, "there is no scientific evidence that languages dramatically shape their speaker's ways of thinking" (p. 58). Pinker offers an alternative interpretation of the well-known Kay and Kempton (1984) study with color chips, a study that has more recently been offered as support for a limited Whorfian view.

It is now generally recognized that a strong Whorfian view is untenable. However, Pinker's vigorous claims notwithstanding, it must also be recognized that legitimate questions dealing with the influence of language on thought persist. The validity of some form of linguistic relativism is an ongoing research issue (Mark, 1993; Pederson, in press). In particular, Lakoff (1987) offered an exceptionally detailed, erudite, and subtle discussion in support of some form of Whorfianism.

Turning to the specific topic of this essay, there is no question that spatial information is expressed differently in some languages than in others. Questions about the implications of linguistic differences for aspects of spatial cognition have long been part of the debate over linguistic relativism (see Lakoff, 1987). Recent work continues to attempt to answer these questions (Haviland & Levinson, 1994; Weissenborn & Klein, 1982). How substantial and fundamental are implications of spatial linguistic differences for general cognitive differences?

Although a resolution of the issue is beyond the scope of this essay, there is one phenomenon that I want to consider further. That is research on linguistic differences in the spatial frames-of-reference used in locative expressions. It is well documented (Brown & Levinson, 1993; Haviland & Levinson, 1994; Pederson, 1993) that some cultural groups almost exclusively use absolute frames such as cardinal directions to express location, even at small scales of "figural" or "manipulable" space. This is in stark contrast to cultures like our own that almost exclusively use an egocentric frame (left, right) to express location in small-scale space.

The implications of this provocative difference for the linguistic relativism debate is a little uncertain because both groups can and do shift between both systems (Pederson, 1993). However, I believe it suggests an important point about cultural differences in spatial cognition: Cultural groups are likely to vary in their spatial cognitive processes or structures when no clear "best" solution exists for a problem involving spatial information. As Pederson (1993) discusses, "Each system has its advantages and disadvantages in various communicative contexts" (p. 296). We are all familiar with the ambiguity of using an egocentric frame in communication -- whose left do you mean? And, as Pederson notes, knowing the orientation of the cardinal directions at all times is no trivial or automatic task. In other words, the problem of locative expression in face-to-face communication has no clearly superior best solution. Either you deal with the problem of establishing the deictic frame in communication or with the problem of establishing the orientation of the cardinal directions at a given spot. As a general principle, therefore, we might expect more cultural (and individual) variability with more ambiguous problems. So another way to state the theme of this essay might be: Cultural variation in spatial cognition is relatively minor insofar as most spatial problems do not have a variety of nearly equally effective and efficient solutions.

Pictorial Perception. Clearly people from some traditional cultures have minimal or restricted prior experience with the conventions of pictorial representation (drawings, photos, TV images, etc.), though it is unlikely that any human group does not practice some iconic pictorial representation. As I reviewed above, probably the strongest finding in the cross-cultural literature on perception is that people from some traditional cultures have difficulty interpreting the depiction of 3-dimensional scenery in 2-dimensional media. For people in all cultures, facility at this depends on training and/or practice. The ease with which this skill can be acquired is still contested, as in the case of learning to interpret cartographic "pictures" (Blaut, 1991; Liben & Downs, 1989).

Home Ranges and Activity Spaces. Different cultures practice different forms of economic activity -- they sustain themselves in different ways. This entails, among other things, different temporal and spatial patterns of activity. And within these average cultural group differences, one can also identify variations in patterns of sex-related, age, SES, and other differences; the natures of these variations constitute ongoing research questions (Munroe et al., 1985; Rapoport, 1976). Although, strictly speaking, home ranges and activity spaces constitute examples of behavior rather than cognition, I discuss them here because they are often considered as causes of spatial cognition (e.g., wider travel patterns produce better navigational skills), effects of spatial cognition (e.g., activity space is a subset of "awareness space"), or both.

As I listed above, I believe the regionalization of space is culturally universal, and I expect that all peoples have some concept of territorial control over various regions. Of course, the exact nature of the territories and their control varies. Not all cultures explicitly codify territorial laws or incorporate concepts of sharp boundaries or put up physical barriers, etc. (e.g., Hallowell, 1942). But nothing I have read suggests that there are cultural groups who do not, at a minimum, practice forms of personal or in-group control over access to regions: personal space around the body, hunting territories, home territories, etc.

Formal Measurement of Space. Many traditional groups studied by anthropologists (and presumably the prehistoric groups studied by archeologists) did not have formal, abstracted systems of spatial measurement and manipulation, as in geometry and surveying. In a fascinating article, Hallowell (1942) reviewed his own work with the indigenous Saulteaux of Canada, as well as the work of other anthropologists with other groups. At that time, the Saulteaux used crude spatial quantities arising out of perceptual observation and experience, quantities such as "near" and "far away" that can hardly be called measurement. And like many peoples throughout history and prehistory, to the extent that the Saulteaux did use measurement (i.e., standardized matching quantities), their standards were imprecise, individually and situationally variable, and largely incommensurate. Examples are the expression of larger-scale distances in dynamic activity terms ("3 sleeps away") and that of smaller distances in body-part terms ("3 fingers high") (see also, Littlejohn, 1963). The lack of formal and precise measurement concepts undoubtedly reflected and influenced the way these traditional cultures could think about space. However, it is important to remember that members of technologically-developed cultures like ourselves also reason in our everyday lives in qualitative and fuzzy metric units, temporal activity units, and so on (e.g., our common use of qualitative spatial linguistic terms) (Montello & Frank, in press). As Hallowell stated, "this level of discrimination is undoubtedly a generic human trait" (p. 62).

Environmental Cues. A final important area to look for cultural differences in spatial cognition is in the environmental features or "cues" that are noticed, remembered, and verbally labeled. Hazen (1983) and Rapoport (1976) both offer this as their primary evidence/example of cultural differences in spatial cognition. These authors make much of the variety of features used for orientation by traditional Micronesians, Eskimos, Australian Aborigines, and Saharan Tuareg. But is it a profound difference that Saharan nomads utilize sand dunes, Eskimos snow drifts, and Micronesians ocean swells? It seems more reasonable to see these as superficial and unavoidable variations on the universal ability of humans to take advantage of whatever information is available in their environment to exploit, however subtle. In fact, such cross-cultural research provides evidence for the operation of universal principles of feature selection: Select reliable, perceptible, and informative cues. The celebrated *etak* method of the Micronesians, for instance, is a modest variation on the "celestial navigation system familiar to Western navigators" (Hazen, 1983, p. 4). It is ironic, therefore, that while different cultures do attend to and remember different environmental features for use in navigation and communication, cross-cultural research on the topic may be most valuable because it provides evidence for universal principles.

4 Conclusion: Implications of Cultural Differences in Spatial Cognition for GIS

The theoretical and empirical issues reviewed in this essay suggest to me that the magnitude and pervasiveness of cultural differences in spatial cognition are often exaggerated. Many important aspects of spatial cognitive structures and processes are universally shared by humans everywhere. Furthermore, many of the apparent cultural differences that exist are actually not due to culture but to other factors that vary within cultures, such as professional training and expertise, social class, and so on. Different cultures usually do speak different languages, but it is a matter of ongoing controversy as to whether such differences have substantial effects on spatial perception and thinking. And virtually all of the evidence for substantial cultural differences in spatial cognition suggests that they occur primarily between traditional and technologically-developed cultures, not between different technologically-developed cultures.

This suggests a critical point about the implications of cultural differences in spatial cognition for GIS. The GIS users and designers for whom most such issues would be a concern are not untrained nonspecialists. Nor are they likely to be untrained nonspecialists in the future, except for possibly those using the representational function of GIS (i.e., spatial visualizations). Members of the GIS research and design community constitute a specialized "subculture" requiring training even in the United States and other technologically-developed cultures. The untrained sophomore doesn't understand the nuances of cartography or spatial analysis either; they too need to be taught the distinctions between lagoons, ponds, and marshes. People from traditional groups may not use formal systems of spatial measurement and manipulation, like surveying and geometry, but neither are they using GIS. At such time that members of their culture do use GIS, they will undoubtedly have acquired such formal systems. This is true, I think, because people from all cultural groups can readily learn the spatial concepts and procedures necessary to use GIS even if their cultures did not traditionally use them.

If substantial cultural differences in spatial cognition actually existed, the GIS community should take them into account in GIS research and design. Like other technologies, GIS should not and will not be restricted to European and European-American cultures. And I recognize that there are many significant and important cultural issues surrounding GIS, as I pointed out at the beginning of this essay. However, I do not believe that the topic of cultural differences in *spatial cognition* deserves a great deal of attention in the GIS community. Probably the spatial cognitive issue most deserving of cross-cultural attention is that of feature definitions incorporated into spatial data transfer, as Mark (1993) discusses. The solution may involve the compilation of some cross-cultural dictionaries for geographic features. Alternatively, non feature-based approaches (e.g, raster) may provide an end run around this problem; resident "feature experts" could define locally appropriate features at culturally distinct data processing sites.

None of this is meant to suggest that the GIS community should ignore the cognitive abilities and characteristics of humans. On the contrary, I believe that such considerations continue to be underappreciated and underresearched. The point is that whatever the important cognitive issues are, their culturally universal aspects are much more significant than are their culturally variable aspects.

References

Appelle, S. (1972). Perception and discrimination as a function of stimulus orientation: The "oblique effect" in man and animals. *Psychological Bulletin, 78*, 266-278.

Blaut, J. M. (1991). Natural mapping. *Transactions of the Institute of British Geographers, 16*, 55-74.

Brown, P., & Levinson, S. C. (1993). *Linguistic and Nonlinguistic Coding of Spatial Arrays: Explorations in Mayan Cognition.* Cognitive Anthropology Research Group Working Paper 24. Nijmegen, The Netherlands: Max Planck Institute.

Buss, D. M. (1989). Sex differences in human mate preferences: Evolutionary hypotheses tested in 37 cultures. *The Behavioral and Brain Sciences, 12*, 1-49.

Campari, I., & Frank, A. U. (1993). Cultural differences in GIS: A basic approach. In *Proceedings of Fourth European Conference and Exhibition on Geographical Information Systems* (pp. 10-16).

Cosmides, L., Tooby, J., & Barkow, J. H. (1992). Introduction: Evolutionary psychology and conceptual integration. In J. H. Barkow, L. Cosmides, & J. Tooby (Eds.), *The adapted mind: Evolutionary psychology and the generation of culture* (pp. 3-15). New York: Oxford University Press.

Couclelis, H. (1992). People manipulate objects (but cultivate fields): Beyond the raster-vector debate in GIS. In A. U. Frank, I. Campari, & U. Formentini (Eds.), *Theories and methods of spatio-temporal reasoning in geographic space* (pp. 65-77). Berlin: Springer-Verlag.

Deregowski, J. B. (1980). *Illusions, patterns, and pictures.* London: Academic Press.

Egenhofer, M. J., & Golledge, R. G. (November, 1994). *Time in geographic space: Report on the Specialist Meeting of Research Initiative 10* (Technical Report No. 94-9). National Center for Geographic Information and Analysis.

Fodor, J. A. (1983). *The modularity of mind.* Cambridge, MA: MIT Press.

Gentile, D. A. (1993). Just what are sex and gender, anyway? A call for a new terminological standard. *Psychological Science, 4*, 120-122.

Gibson, J. J. (1979). *The ecological approach to visual perception.* Boston: Houghton Mifflin.

Gladwin, T. (1970). *East is a big bird.* Cambridge, MA: Harvard University Press.

Hagen, M. A. (Ed.). (1980). *The perception of pictures (2 volumes).* New York: Academic Press.

Hallowell, A. I. (1942). Some psychological aspects of measurement among the Saulteaux. *American Anthropologist, 44*, 62-77.

Haviland, J. B., & Levinson, S. C. (1994). Special Issue: Spatial conceptualization in Mayan languages. *Linguistics, 32*.

Hazen, N. L. (1983). Spatial orientation: A comparative approach. In H. L. Pick & L. P. Acredolo (Eds.), *Spatial orientation: Theory, research, and application* (pp. 3-37). New York: Plenum Press.

Heider, E. R. (1972). Universals in color naming and memory. *Journal of Experimental Psychology, 93*, 10-20.

Kay, P., & Kempton, W. (1984). What is the Sapir-Whorf Hypothesis? *American Anthropologist, 86*, 65-79.

Kenrick, D. T., Montello, D. R., & MacFarlane, S. (1985). Personality: Social learning, social cognition, or sociobiology? In R. Hogan & W. H. Jones (Eds.), *Perspectives in personality* (pp. 201-234). Greenwich, CT: JAI Press.

Lakoff, G. (1987). *Women, fire, and dangerous things*. Chicago: University of Chicago Press.

Leibowitz, H. W., & Pick, H. A. (1972). Cross-cultural and educational aspects of the Ponzo perspective illusion. *Perception & Psychophysics, 12*, 430-432.

Lewis, D. (1978). *The voyaging stars: Secrets of the Pacific island navigators*. Sydney: William Collins Publishers.

Lewis-Williams, J. D., & Dowson, T. A. (1988). The signs of all times: Entoptic phenomena in Upper Paleolithic art. *Current Anthropology, 29*, 201-245.

Liben, L. S., & Downs, R. M. (1989). Understanding maps as symbols: The development of map concepts in children. In H. W. Reese (Ed.), *Advances in child development and behavior* (pp. 145-201). San Diego, CA: Academic Press.

Littlejohn, J. (1963). Temne space. *Anthropological Quarterly, 36*, 1-17.

Lynch, K. (1960). *The image of the city*. Cambridge, MA: MIT Press.

Mark, D. M. (1993). Toward a theoretical framework for geographic entity types. In A. U. Frank & I. Campari (Eds.), *Spatial information theory: A theoretical basis for GIS* (pp. 270-283). Berlin: Springer-Verlag.

Montello, D. R. (1993). Scale and multiple psychologies of space. In A. U. Frank & I. Campari (Eds.), *Spatial information theory: A theoretical basis for GIS* (pp. 312-321). Berlin: Springer-Verlag.

Montello, D. R., & Frank, A. U. (in press). Modeling directional knowledge and reasoning in environmental space: Testing qualitative metrics. In J. Portugali (Ed.), *The construction of cognitive maps*. Kluwer Academic.

Munroe, R. H., Munroe, R. L., & Brasher, A. (1985). Precursors of spatial ability: A longitudinal study among the Longoli of Kenya. *Journal of Social Psychology*, *125*, 23-33.

Neisser, U. (1987). A sense of where you are: Functions of the spatial module. In P. Ellen & C. Thinus-Blanc (Eds.), *Cognitive processes and spatial orientation in animals and man* (pp. 293-310). Boston: Martinus Nijhoff.

Pederson, E. (1993). Geographic and manipulable space in two Tamil linguistic systems. In A. U. Frank & I. Campari (Eds.), *Spatial information theory: A theoretical basis for GIS* (pp. 294-311). Berlin: Springer-Verlag.

Pederson, E. (in press). Language as context, language as means: Spatial cognition and habitual language use. *Cognitive Linguistics*.

Pick, A. D., & Pick, H. L. (1978). Culture and perception. In E. C. Carterette & M. P. Friedman (Eds.), *Handbook of perception* (pp. 19-39). New York: Academic Press.

Pinker, S. (1994). *The language instinct*. New York: William Morrow and Company.

Rapoport, A. (1976). Environmental cognition in cross-cultural perspective. In G. T. Moore & R. G. Golledge (Eds.), *Environmental knowing* (pp. 220-234). Stroudsburg, PA: Dowden, Hutchinson & Ross.

Segall, M. H., Campbell, D. T., & Herskovits, M. J. (1966). *The influence of culture on visual perception*. Indianapolis: The Bobbs-Merrill Company, Inc.

Sheets-Johnstone, M. (1990). *The roots of thinking*. Philadelphia: Temple University Press.

Shepard, R. N. (1984). Ecological constraints on internal representation: Resonant kinematics of perceiving, imagining, thinking, and dreaming. *Psychological Review, 91*, 417-447.

Shepard, R. N., & Hurwitz, S. (1984). Upward direction, mental rotation, and discrimination of left and right turns in maps. *Cognition, 18*, 161-193.

Tooby, J., & Cosmides, L. (1992). The psychological foundations of culture. In J. H. Barkow, L. Cosmides, & J. Tooby (Eds.), *The adapted mind: Evolutionary psychology and the generation of culture* (pp. 19-136). New York: Oxford University Press.

Turnbull, C. M. (1961). Some observations regarding the experiences and behavior of the BaMbuti Pygmies. *American Journal of Psychology, 74*, 304-308.

Wallace, R. (1989). Cognitive mapping and the origin of language and mind. *Current Anthropology, 30*, 518-526.

Weissenborn, J., & Klein, W. (Eds.) (1982). *Here and there: Cross-linguistic studies on deixis and demonstration.* Amsterdam: John Benjamins Publishing.

A Formalism for Model-Based Spatial Planning

Janice Glasgow

Department of Computing and Information Science
Queen's University, Kingston, Ontario K7L 3N6
janice@qucis.queensu.ca

Abstract. This paper presents a formalism for spatial planning based
on a symbolic array representation of the world. Entities in the world are
denoted as symbols in the array and relations in the world are modeled
using array inspection functions. A plan is constructed using array transformation functions that correspond to physical transformations in the
world. This model-based approach to planning supports the implementation of heuristic search strategies similar to those applied by humans
when reasoning with cognitive maps.

1 Introduction

Planning is the process of generating a set of actions that can be performed
to achieve a prescribed goal. This paper is concerned with planning problems
where the initial state and goal of a plan are expressed in terms of a spatial configuration. Examples of such problem domains include: a geographic navigation
system, where primitive actions correspond to an agent moving through specified locations in a spatial map; a chess game where a plan consists of a sequence
of moves for a player; and a blocks world where the goal is expressed as spatial
relationships among the blocks. We assume that planning may be reactive, in the
sense that the plan is modified dynamically in order to deal with unanticipated
changes in the state (e.g. reconfiguration of objects or agents in space) as the
plan is being implemented.

In the proposed formalism, a state of the world is represented as a symbolic
array. The array representation corresponds to a model of the world in which spatial relations such as *left-of, on-top-of, north-of, adjacent-to*, etc. are expressed
as corresponding relations in the array. A plan in the formalism is modeled as
a sequence of primitive array function applications, where each array function
corresponds to a physical transformation in the world. The result of applying a
successful plan to an array model of the initial state is an array representation
that satisfies the specified goals. A thesis of this paper is that the symbolic array
representation allows for the design of efficient search strategies for constructing plans in domains involving spatial reasoning. In particular, it supports the
implementation of search strategies that incorporate heuristics similar to those
applied by humans when carrying out spatial planning with cognitive maps. The
paper describes a formal theory of planning based on a semantics for spatial reasoning with symbolic arrays. It also discusses the applicability of the formalism
to hierarchical and reactive planning.

2 Planning with spatial models

This section presents a scheme for model-based spatial planning based on a theory for computational imagery [15, 32]. In this theory, the spatial properties of a world are modeled using an array representation in which entities in the world are denoted as symbols and relations of interest are mapped to corresponding relations in the array. Primitive imagery functions have been designed and implemented for constructing, transforming and inspecting these spatial models. The transformation functions provide a basis for defining the primitive actions in a plan; the inspection functions for computational imagery allow us to determine whether a prescribed goal, or subgoal, has been achieved. The functions described in the paper have been specified using a mathematics of array theory [19, 29] and implemented in the array language Nial [20].

Researchers in cognitive science have studied the role of spatial models in human reasoning and planning. Johnson-Laird [21] describes the representation and reasoning processes involved with several types of mental models. The first, and most fundamental, is a *relational model*, which is a static frame consisting of tokens that represent entities in the world and a set of relations that define the physical relationships among entities. He defines a *spatial model* as a relational model in which the relations of interest are spatial in nature; tokens are located within a symbolic, multi-dimensional space. Spatial models are not necessarily complete or totally accurate: details irrelevant to the current problem solving task are often discarded. Cognitive studies have also suggested that planning may involve mental simulations based on spatial models, where such simulations result in a more efficient set of actions for achieving a goal. In an experiment carried out by Hayes-Roth and Hayes-Roth [18], mental simulations involving a spatial map (model) of a town were used to derive a plan for carrying out a day's errands. During plan generation, a subject would use the spatial map to cluster activities that would occur in neighboring areas of the map. Access to the current and adjacent locations in the map also allowed for reassessment of the plan to improve efficiency. For example, at one point a subject revises his plan after noticing a nearby "short-cut".

A knowledge representation scheme for spatial reasoning with symbolic arrays has previously been developed and its computational advantages argued [10, 11]. This work has also been related to theories of mental models [13], and a formal semantics for model-based deductions using spatial models has been developed [14]. In the remainder of this section, we illustrate how the previously proposed representation scheme can be used to model states and actions in a spatial planning problem.

2.1 State representation

The problem space for a planning system consists of an initial world state, a set of primitive actions for transforming the world and a set of goals. For the domain of spatial planning, we assume that a state (or world) w is denoted as a pair $< S, R >$, such that:

- S is a finite set of symbols corresponding to the entities of interest in the world, and
- R is a set of spatial relations defined over the set of symbols S; each n-ary relation $r \in R$ is denoted as a set of n-tuples containing entities in S.

In the proposed scheme for spatial planning, a world $w = < S, R >$ is modeled using a symbolic array representation $\mathcal{A}$ and a set of array functions $\mathcal{F}$ such that for every n-ary relation $r \in R$, and all symbols $s_1, ..., s_n \in S$, there exists a boolean inspection function $f_r \in \mathcal{F}$ such that if $(s_1, ..., s_n) \in r$ then $f_r(s_1, ..., s_n, \mathcal{A}) = true$. That is, if a spatial relation r holds for the entities denoted by $s_1, ..., s_n$ in the world then a corresponding relation holds for the symbols in array $\mathcal{A}$. Figure 1 illustrates a blocks world configuration and a corresponding array representation.[1] We can construct a model of the world using the array in Figure 1 and the primitive inspection functions f_{left} and f_{on}. The function f_{on} is defined to return the value $true$ if and only if its first parameter is situated directly above the second parameter in the specified array. Thus the function application $f_{on}(a, table)$ would evaluate to $true$, whereas an application $f_{on}(b, c)$ would return $false$. Similarly, f_{left} is defined to correspond to the relation $left$ in the world. Note that the relation $f_{left}(a, c)$ can be derived from inspection of the array representation, although it is not specified in the relation set.

The world described in Figure 1 is $determinate$ in the sense that all relationships among entities are specified (or derivable) from the relation set R. More specifically, a determinate world is one which is representable by a unique array model. Our formalism will assume that the initial state of the world is determinate and that the application of a primitive action to a determinate world will result in a determinate world. We may wish, however, to consider a goal as an indeterminate world. For example, a goal may be specified as a blocks world in which the relation $left$ is defined by the set $\{(a, b)(a, c)\}$. As illustrated in Figure 2, there are two possible configurations that satisfy this goal, thus the world is indeterminate. An indeterminate world is modeled by multiple spatial models, where each model satisfies the relations of the world. A possible worlds semantics for reasoning with spatial models of indeterminate worlds is presented in [14].

Visual information such as shape, distance and size is often discarded in a model. However, if desired, distance and shape attributes can be preserved in the array representation. Figure 3(a) illustrates an island map similar to the one used by Kosslyn and colleagues [22] to study how humans store and inspect mental maps. Much of the information derivable through the visual inspection of the map image can also be inferred from the symbolic array representation

---

[1] For presentation purposes, symbols that occupy two or more adjacent cells of an array are denoted once and separating lines are deleted (as in the case of the symbol "table" in Figure 1). As well, the examples presented consider only relations in two dimensions. However, the theory is not restricted: the representations and functions for spatial imagery have been designed and implemented to accommodate worlds of arbitrary dimensionality.

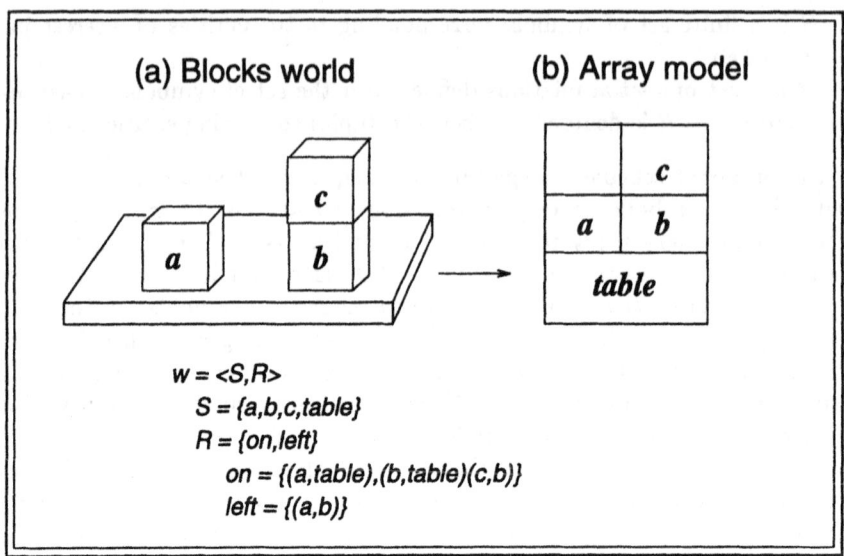

Fig. 1. Array model for spatial world

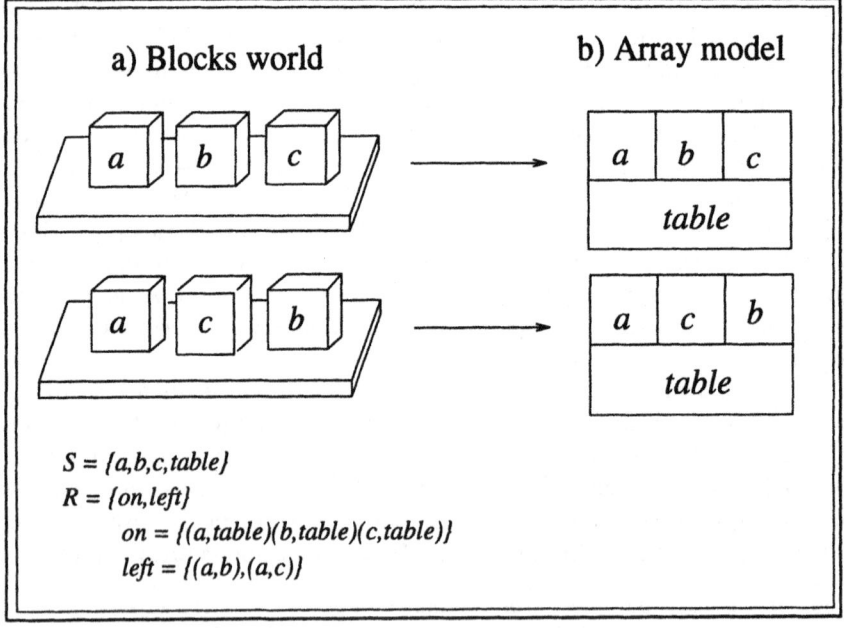

Fig. 2. Model for indeterminate world

in Figure 3(b). Geographic directions are determined in this representation by comparing the relative locations of entities, e.g. the hut is *south-of* the lake and *west-of* the beach. As well, it can be determined that the tree is *near* the lake and that the beach is *closer* to the hut than it is to the lake. Relative size and shape information can be preserved in a representation by increasing the granularity of the array. For example, the shape of the island map is computable from the array representation depicted in Figure 3(c).

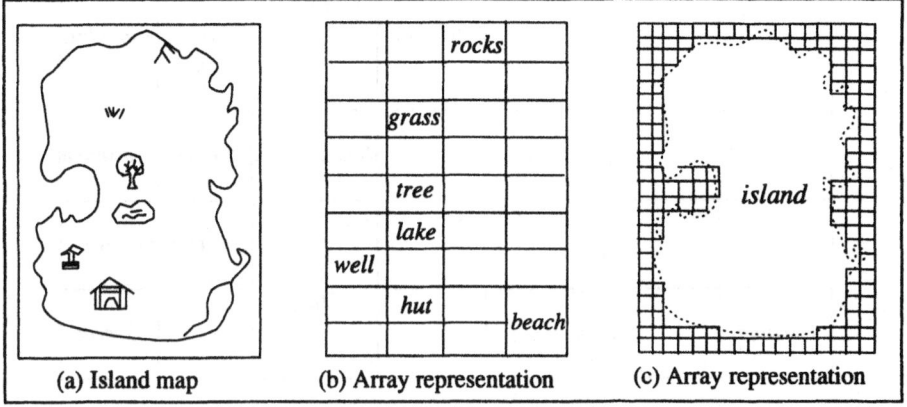

| (a) Island map | (b) Array representation | (c) Array representation |

Fig. 3. Representations of Kosslyn's island map

An array representation is *complete* in the sense that all spatial relationships among entities are made explicit by the the array inspection functions in $\mathcal{F}$. To illustrate the concept of an array model, consider the world described by the symbol set:

$$S = \{Britain, Portugal, Spain, ...\}$$

of major countries in Europe and a set of corresponding geographical relations:

$\mathcal{R} = \{north\text{-}of, west\text{-}of, east\text{-}of, south\text{-}of, borders\text{-}on\}$, such that
$north\text{-}of = \{(Ireland, Portugal), (Britain, Spain), ..\}$,
$west\text{-}of = \{(Ireland, Britain), (Portugal, Spain), ...\}$, etc.

We define an array model $< \mathcal{A}, \mathcal{F} >$ for w such that:

- $\mathcal{A}$ is the array depicted in Figure 4.
- The array inspection functions in $\mathcal{F}$ are defined to model the spatial relations in the world. For example, the relation *west-of* is represented in $\mathcal{A}$ using the function $f_{west\text{-}of} \in \mathcal{F}$, which is defined so that an application of the form $f_{west\text{-}of}(s_1, s_2, \mathcal{A})$ returns the value true if and only if symbol s_1 occurs in a location that is to the left of the left-most occurrence of symbol s_2 in the array data structure $\mathcal{A}$. Similarly the relation *borders-on* is represented in

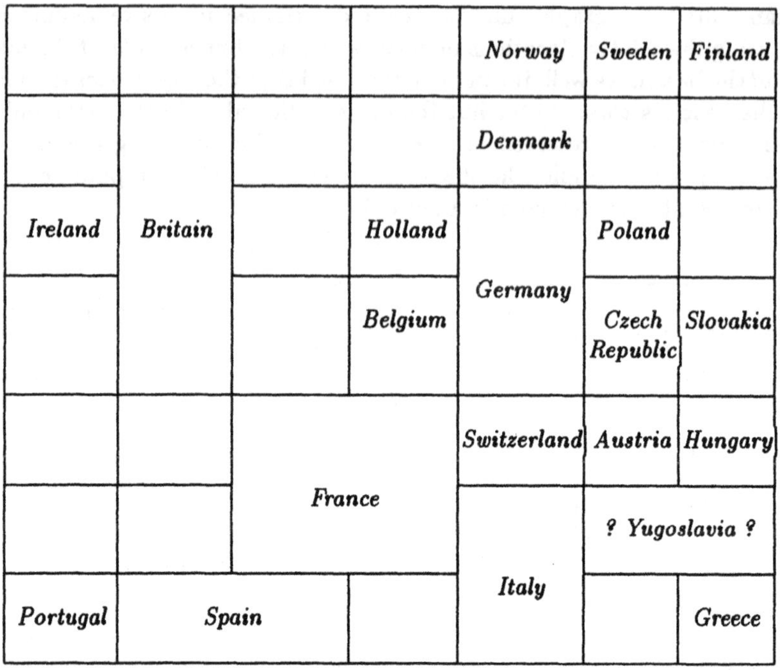

				Norway	Sweden	Finland
				Denmark		
Ireland	Britain		Holland		Poland	
			Belgium	Germany	Czech Republic	Slovakia
				Switzerland	Austria	Hungary
		France			? Yugoslavia ?	
				Italy		
Portugal	Spain					Greece

Fig. 4. Array representation of major countries of Europe

$\mathcal{A}$ using the function $f_{borders\text{-}on}$, such that $f_{borders\text{-}on}(s_1, s_2, \mathcal{A})$ evaluates to true just in the case where symbols s_1 and s_2 are situated in adjacent cells of array $\mathcal{A}$.

The array in Figure 4 does not precisely depict all of the spatial relations among entities in Europe. It is in fact a simplified model that captures the relationships specified in a given world, which we assume itself was a simplification. If more precision is desired, a finer grained array could be constructed that does preserve the exact relations. If the array is fine-grained enough, shape and distance properties could also be captured in the representation.

It is worth noting here that an array representation explicitly depicts the absence, as well as the presence, of entities at relative locations. For example, the symbolic array in Figure 4 can be inspected to infer that there is no country that is directly east of Britain and west of Holland. This property is particularly valuable for problem domains such as planning and navigation.

2.2 Representing actions and plans

The performance of an action generally results in a change to the world state. A primitive action is one that is considered nondecomposable. Primitive actions are often specified as a set of literals denoting the preconditions and postconditions for the action. In our proposed formalism, a primitive action is specified as an array transformation. We assume that a primitive action in the world can be denoted as a transformation t applied to a set of entities. For example the action — *move the table to the left* — is specified using a transformation *move-left* applied to the entity *table*. Such a primitive action is modeled as a primitive array function $f_{move-left}$ applied to the symbol *table*. The result of the function application $f_{move-left}(table)$ to an array model $\mathcal{A}$ is a new array model in which the location of the table is modified to be one cell to the left of where it was previously. If such a physical transformation is impossible (e.g. the location is already occupied) then the function would return an appropriate error message and leave the model unchanged.

Thus, we assume for every primitive transformation t in the world there exists a transformation function $f_t \in \mathcal{F}$ such that if an application of t transforms a world w into a world w' then an application of the function f_t maps an array model of w to an array model for w'. This is illustrated for a blocks world transformation in Figure 5.

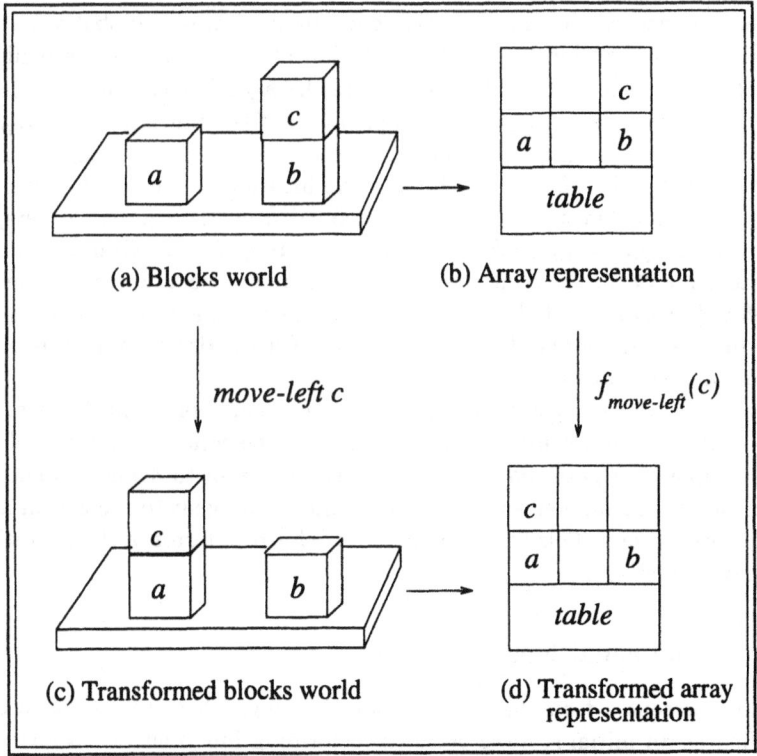

Fig. 5. Transformation of blocks world representation

The primitive actions for a world, and thus the definition of the primitive

array functions, depend on the problem domain. For the blocks world, the primitive actions might be modeled by a set of array functions containing $f_{move-left}$ and $f_{move-right}$. Applying these functions to an entity in an array which is clear (not covered by another block) would result in an array with the entity at a new location directly left or right of its initial position.

A plan $\mathcal{P}$ is defined as an ordered sequence of primitive actions: $\mathcal{P} = < a_1, ..., a_n >$, where each action a_i can be decomposed into a transformation t_i applied to a set of entities. Thus, an action a_i can be modeled as a function application $f_{t_i}(s_1, ..., s_n)$, such that $f_{t_i} \in \mathcal{F}$ is an n-ary function that models transformation t_i and $s_1, ..., s_n$ denote the entities that are specified for the action. A plan can then be modeled as the composition of such function applications.

Cognitive studies suggest that spatial models may be constructed, transformed and inspected from an external or internal perspective [39]. Array functions can be defined for either form of reasoning. For example, in the blocks world of Figure 5 the primitive array function $f_{move-left}$ would be defined with respect to an external perspective of the world. Alternatively, consider a robotic domain where the primitive transformations are $f_{turn-left}, f_{turn-right}$ and $f_{move-forward}$. These functions can be implemented to correspond to an internal, or egocentric, frame of reference by assuming a global variable that records the current location and orientation of the agent. Figure 6 illustrates three time steps for such a domain, where the agent's location and orientation are denoted as a directed arc. The application of the function $f_{turn-right}$ modifies the world by changing the orientation of the agent. Applying the function f_{push} to the symbol *desk* then changes the location of the agent as well as the desk. Inspection functions can also be defined from the egocentric perspective. For example, $f_{in-front}(desk)$ would evaluate to *false* at Time=1, but return *true* at Time=2 for the plan described in Figure 6. The advantage of being able to define functions related to alternative perspectives is that it generally allows for a more straightforward implementation of heuristics that assume either an internal or external frame of reference.

Based on the primitive functions for a particular problem domain, we can also construct more complex functions that correspond to subgoals of a plan. For example, in a path planning system we may wish to define a function *goto* which takes a parameter an entity s and moves an agent to a location adjacent to the location of s. This function could be defined in terms of the primitive turn and move functions.

2.3   Formal definitions

We now present a formalism for spatial planning, where the problem space is specified as an initial state, a set of transforming functions and a goal. A plan is expressed as a sequence of actions; it is successful if the application of its functional model returns an array representation that satisfies the given goal.

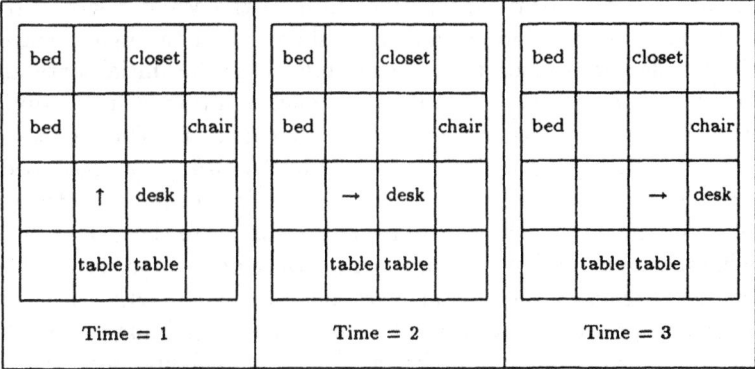

Fig. 6. Results of plan <*turn-right,push table*>

A **spatial planning problem** is specified as a triple $< \mathcal{A}, \mathcal{F}, w >$ such that:

- $\mathcal{A}$ is an array model for the initial state (deterministic world) for the problem;
- $\mathcal{F}$ is a set of array functions for inspecting and transforming spatial array models of the world; and
- $w = < S, R >$ is a world (possibly indeterminate) that describes the goal for the problem.

A **plan** $\mathcal{P}$ is defined as a sequence of primitive actions $< a_1, ..., a_n >$ such that each action a_i is modeled as a function $f_i \in \mathcal{F}$ applied to a set of entity symbols $s_1, ..., s_j \in S$.

A **successful plan** is a plan $\mathcal{P} = < a_1, ..., a_n >$ such that if $f' = f_i(s_1, ..., s_n)$ is the function application that models action a_i and $(f'_n(f'_{n-1}...(f'_1 \ \mathcal{A}))...) = \mathcal{A}'$ then $\mathcal{A}'$ is a model for world w.

Thus, the application of a model (sequence of array function applications) for a successful plan results in an array representation $\mathcal{A}'$ such that for all n-ary relations $r \in R$ and entity symbols $s_1, ..., s_n \in S$ if $(s_1, ..., s_n) \in r$ then $f_r(s_1, ..., s_n, \mathcal{A}') = true$ for inspection function $f_r \in \mathcal{F}$.

2.4 Plan construction

In the previous section we presented a formalism for specifying a spatial planning problem and a solution. In this section, we discuss ways in which the formalism facilitates the design and implementation of a plan generator.

The goal of planning is to construct a set of actions that will transform an initial state of the world into one that satisfies the given goal. As such, it is often viewed as a search problem. Much of the research effort in planning is directed at finding ways of reducing the search space of possible plans. It has been shown that even for simple planners the search space can be exponential [4]. Thus, it is necessary to define heuristic functions to limit the search space of potential plans. For the proposed formalism, we consider search through a space where the initial node corresponds to the array model for the initial state and each subsequent node corresponds to an array model resulting from the application of a primitive transformation function to its parent node. The goal is achieved when a node is reached in which the array model satisfies the prescribed goal.

Array models provide for efficient search strategies in a number of ways. First, they allow for the implementation of heuristics similar to those used by humans employing spatial models of the world. One such strategy is to use neighborhood information. Often a spatial plan involves the inspection of adjacent locations to determined the presence or absence of obstructions or enabling entities. The linear indexing of arrays allows for efficient retrieval of neighborhood information.

Cognitive studies suggest that spatial navigation often depends on specified landmarks, rather than on exact distances and angles [6]. The symbols in the spatial model may correspond to such landmarks; thus, plans can be constructed that specify subplans such as: *drive north until there is a school on your left then turn right.*

Importantly, model-based reasoning addresses the problem of reasoning about change. Modifying a spatial domain by adding knowledge or applying transformations is often problematic in logic-based systems. In traditional planning systems, such as STRIPS [7], an operator corresponding to a primitive action specifies preconditions, post conditions and a list of assertions that need to be deleted. Since modified spatial relations are determined directly through inspection of the array models in our proposed formalism, it is not necessary to examine previous deductions to determine whether relations need to be deleted from the current knowledge base. Being able to manage change in the environment is particularly useful for a reactive planning system. For example, in planning problems related to autonomous mobile robots, the robot agent must continuously react to changes in the environment [9]. While traditional planners may initially construct a complete set of actions, reactive planners construct partial plans that are subject to modification during execution. The model-based approach facilitates reactive planning since unexpected changes to the environment are easily detected and their effects can be evaluated through simulation of the existing plan in the modified model of the world. However, implementing reactive planning in the formalism would require continuous updating of the model through perceptual interaction with the environment or through communication with some external updating procedure.

Another advantage of array models is that they allow for efficient comparison among states. An array model can be easily compared with a model of a goal state to determine the differences that need to be addressed in constructing subgoals for the problem.

2.5 Hierarchical spatial planning

Hierarchical planning involves constructing an initial abstract plan, in which details are left unspecified, then refining each component of the initial plan into a more detailed subplan. Results of cognitive studies suggest that spatial models may be hierarchically organized and that planning and reasoning take place at varying levels of structural decomposition based on a *part-of* relation [28]. For example, when planning a route for a European vacation one might first consider a spatial model of the countries to be visited then later focus in on the details (regions or cities) of the individual countries. Our formalism allows for planning at multiple levels of parts decomposition through the use of nested array data structures: array symbols may denote subarrays that correspond to the subworlds for the structured entities in the world. Figure 7 illustrates a modified representation of the previously presented array for Europe where the symbol *Britain* has been replaced by a subarray that models the world corresponding to this symbol.

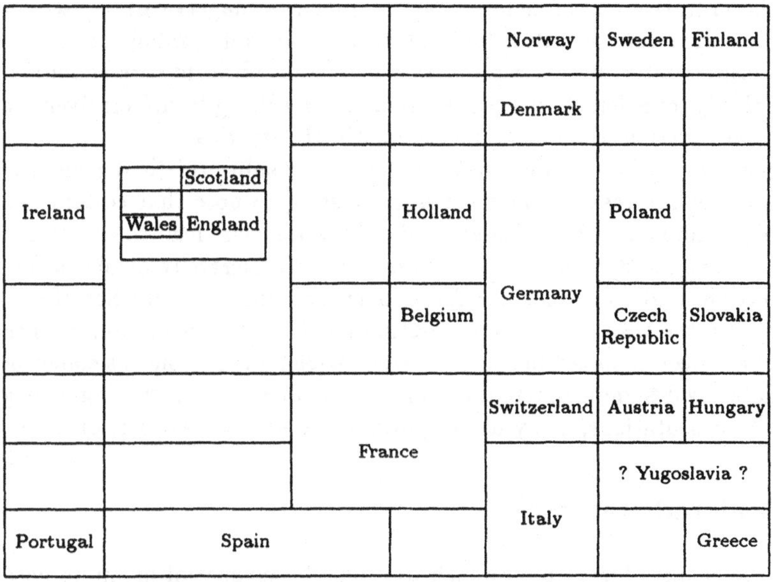

				Norway	Sweden	Finland
				Denmark		
Ireland	Scotland / Wales England		Holland		Poland	
			Belgium	Germany	Czech Republic	Slovakia
				Switzerland	Austria	Hungary
		France			? Yugoslavia ?	
				Italy		
Portugal	Spain					Greece

Fig. 7. Embedded array representation of Europe

A hierarchical representation explicitly depicts relations among entities at multiple levels of the decomposition hierarchy, without having to specify inheritance laws. For example, using the hierarchical representation illustrated in Figure 7 we can compute that Wales is *west-of* Holland, as well as being *west-of* England. This is achieved by extending the implementation of the function $f_{west-of}$ to range over nested data structures. Specifically, if entity s_1 is *west-of* an entity s_2 then all parts of s_1 must be *west-of* s_2, and in turn all parts of s_2. However, relations such as *borders-on* are not necessarily inherited by parts. Other relations may be indeterminate. For example, we can determine the *north-*

of relationships among parts of Britain, but we cannot compare parts of Britain with Holland, or parts of Holland, with regards to this relation.

More formally, a hierarchical world $w =< S, R >$ is a world in which the set S may denote structured entities. For all $s_i \in S$, if s_i denotes a structured entity, then there exists a subworld $w_{s_i} =< S_{s_i}, R_{s_i} >$ of w such that:

- $S_{s_i} \subseteq S$ denotes the set of subentities (parts) of the entity denoted by symbol s_i, and
- R_{s_i} is the set of spatial relations defined for the subworld denoted by s_i.

We assume that each subworld w_{s_i} has an array model. In a hierarchical world w, the set of spatial relations R would contain structural relations corresponding to the part-of hierarchy for the world; that is, *part-of*(s_1, s_2) is specified for all constant symbols s_1 and s_2 such that s_1 denotes a subentity of a structured entity denoted by s_2.

The primitive *focus* function of computational imagery can be used to transform an array representation by replacing a constant symbol corresponding to a structured entity by the representation for the subworld corresponding to the symbol. Figure 8 depicts the successive arrays resulting from iteratively applying the *focus* function to an array model for North America.

The choice of a decomposition hierarchy for subworlds is dependent on the domain and the task demands. It is interesting to note that certain decompositions, combined with course-grained representations, lead to errors analogous to those displayed by humans. For example, the hierarchy displayed in Figure 8 could be used to deduce the false conclusion that the city Seattle, which is situated in the USA, is *south-of* the Canadian city Montreal. Similar misconceptions have been reported in cognitive studies [38]. Such erroneous conclusions are avoidable by designing different decomposition hierarchies, by using finer-grained array representations, or by using spatial collages (see next section).

3 Related Research

This paper introduces an approach to model-based spatial planning based on a symbolic array representation for a world. The concept of a model-based representation is not unique to the symbolic arrays described in this paper. Hayes [17] discusses *direct* representations in which there exist similarities between what is being represented and the medium of the representation. Sloman [36] has also argued the pros and cons of analogical representations, and has concluded that a variety of representation formalisms (including those specialized for spatial reasoning) are important to AI problem solving [37]. Other hybrid approaches have been suggested for visual-spatial and model-based reasoning. Barwise and Etchemendy [1] have proposed a system called *Hyperproof*, which integrates diagrammatic reasoning with sentence-based logics. Hyperproof uses both diagrams and logic notation to teach students how to reason logically. In subsequent work, Barwise and Etchemendy [2] present a formal semantics for reasoning with Hyperproof diagrams. Habel and colleagues [16] have developed

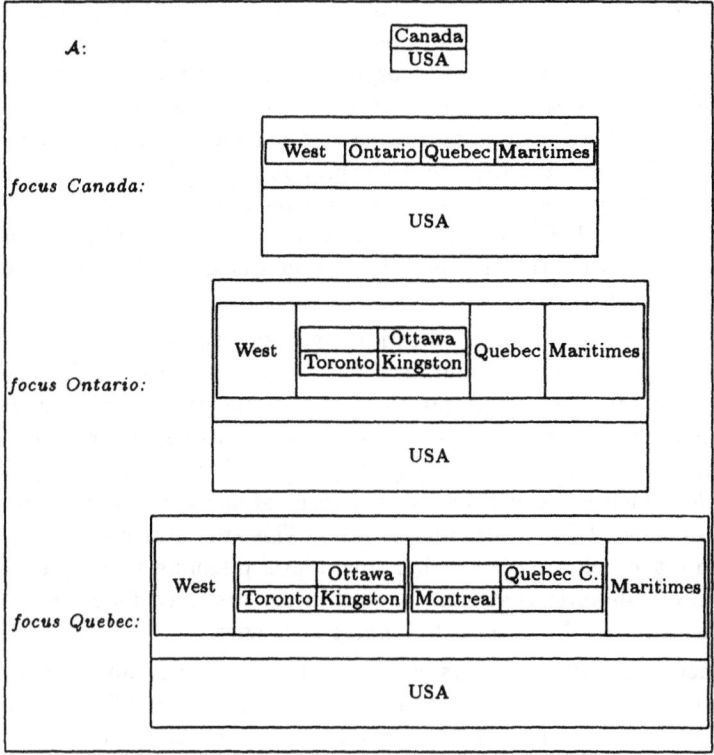

Fig. 8. Successive applications of *focus* function

a hybrid system consisting of a propositional and depictorial partonomy (organization of parts) for reasoning, where the depictorial partonomy reflects the hierarchy proposed in representations for visual processes. They suggest that the advantage of the depictorial representation in their system is that it facilitates an efficient attention-driven method for reasoning. Myers and Konolige [30] treat model-based manipulations as a form of inference within a classical logic system. More specifically, they store partially interpreted sensor data using an analogical representation that interacts with a general-purpose sentential language. A similar approach has been taken by Chandrasekaran and Narayanan [3], who have proposed an architecture where analogical representations derived from visual perception are used in combination with symbolic (propositional) representations. A technique for qualitative spatial reasoning, based on the directional orientation information made available through perceptual processes, has been presented by Freksa and Zimmermann [8]. In this work, orientations in two-dimensional space are defined by the relation between a vector and a point.

Visual-spatial reasoning techniques have also been considered in the context of planning problems. Kuipers [23] has developed a program that determines a path between points by considering a hierarchical network of region representations. Similar to Kuiper's TOUR model [24], our approach can take advantage of problem-solving strategies that incorporate cognitive maps. In his model, however, the relative position of entities is specified as a vector which defines a two-dimensional orientation with respect to a given heading. Kuipers defines a notion of region that allows places to be grouped and referred to collectively. Our ability to embed symbolic arrays within one another also allows for the grouping of entities. The TOUR model has been applied to map learning and navigation in large-scale space – that is, space that is on a larger scale than is available at a single observation [25].

McDermott and Davis [27] describe a general representation for route planning that stores the shapes and locations of entities in the world. Facts in this system are represented as propositions and spatial reasoning is carried out by special-purpose modules that incorporate both theorem proving and numerical computations. Pollack and Riguette [34] have introduced a system called *Tileworld,* which is basically a two-dimensional array containing symbols that denote an agent, tiles, holes and obstacles. This environment was developed to investigate the behaviour of various meta-level planning strategies. The representation of Tileworld is straightforward in our proposed formalism, as are the implementation of the proposed planning strategies.

In their description of a system for qualitative navigation, Levitt and Lawton [26] state three questions that must be addressed in guiding a mobile robot: *Where am I? Where are other places relative to me?* and *How do I get to other places from here?* They present a multi-level theory of spatial representation of the environment based on the observation of landmarks. A topological representation of relative location is integrated with metric knowledge of distance and angles in their approach. Their qualitative approach to path planning also recognizes the need for clustering adjacent regions, which is accomplished in a network formalism.

Tversky [40] supports the notion of mental models for spatial reasoning. She suggests that the representation of cognitive maps allows for inference and perspective-taking, but may discard metric information. Our proposed array representation for planning satisfies many of her findings: it allows for hierarchical representations of space; it tends to align or group closely aligned locations; and it facilitates reasoning with landmarks. Tversky also describes a notion of cognitive collages, where pieces of information may be collected and organized into a single structure. This can be achieved in our formalism through the composition of spatial models. As illustrated in Figure 9, we can compose a symbolic array model of the Pacific Northwest with a symbolic array depicting the cities in Western Canada to generate an image representation that can be inspected to retrieve relations not available in either of the simpler arrays (e.g. Edmonton is north-east of Seattle).

Michel Denis has also considered the use of imagery and spatial models in thought and problem solving [5]. Consistent with our formalism for planning, he proposes that a problem statement should describe objects and physical opera-

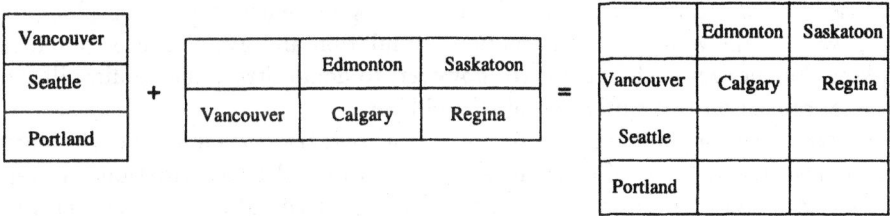

Fig. 9. Composition of spatial models (collages)

tions to be performed on them. Problem solving is then achieved through image or model inspection and transformations simulating the physical processes. Spatial dimensions can be used to denote image classification based on alternative dimensions, as in comparative judgement or transitive inference problems. Denis' research results point to the strong tendency for humans to construct cognitive maps for reasoning in spatial domains.

An application of cognitive maps for navigation was presented in a classic study of Parisian taxi-drivers [31]. The findings in this study suggest a high correlation between the driver's performance (in planning good routes) and the accuracy of their spatial model. They also suggest that the internal representation used by the drivers had two components: the first corresponding to an internalized two-dimensional survey map and a second component corresponding to what they could actually perceive.

Research in geographical information systems and spatial databases has long been concerned with the issue of representing and reasoning with spatial knowledge. Samet [35] has proposed a method for storing geographic knowledge based on the recursive decomposition of space. In this work, the term *quadtree* is used to describe binary array data structures that iteratively subdivide regions into segments until blocks are obtained that consist entirely of 1s or entirely of 0s. These structures (and their three-dimensional counterpart, termed *octrees*) are efficiently stored and implemented as trees, where each node of the tree corresponds to a region in the decomposition hierarchy. An alternative approach to representing geographic systems has been described by Papadias and Sellis [33]. In their work, a symbolic two-dimensional array structure is used to preserve a set of spatial relations among geographic entities. This approach is similar to a model for geographic information systems based on the spatial array models described in this paper [12].

4 Discussion

This paper presents a formalism for model-based spatial planning based on a symbolic array representation for the world. In this representation scheme, primitive array functions are used to model transformations in the world and to retrieve spatial relations of interest. Representing the world using a spatial model has the advantage of allowing for heuristic approaches to planning similar to those used by humans when incorporating a cognitive map for spatial planning. As well, it allows for the construction, simulation and evaluation of potential plans. Further research is required, however, to demonstrate the applicability of the approach on real world planning problems.

Most existing planning systems are logic based: the current state of the world is represented as a set of relations among entities and transformations are represented as logic rules with specified pre and post conditions. As well, current approaches to planning are generally either entirely domain independent or specific to a particular domain, such as robot navigation. The model-based approach to planning described in this paper, however, was designed for a subclass of problems: those that rely on the spatial relations among entities for constructing a plan. A spatial model may be only one component of a more complex representation scheme for reasoning about image or diagrammatic knowledge [21]. Similarly, planning often requires multiple representations and reasoning techniques; it may be necessary to incorporate quantitative data, such as distance or angle information, or logical constraints into the planning system. Consider, for example, the problem of planning the European vacation. Once spatial planning has been used to determine countries and regions to visit, other criteria such as distance, regional geography, etc. may be taken into consideration to construct more detailed or informed plans. Previous research in computational imagery [15] has focussed on how the symbolic array representation can be integrated with visual and declarative representations to form a general scheme for imagery-based reasoning. Future research in this area will focus on extending the planning formalism to incorporate these other forms of image representation.

References

1. J. Barwise and J. Etchemendy. Hyperproof: Logical reasoning with diagrams. In *Working Notes of the AAAI Spring Symposium on Reasoning with Diagrammatic Representations*, 1992.

2. J. Barwise and J. Ethemendy. Heterogeneous logic. In *Working notes of the 1993 IJCAI Workshop on Principles of Hybrid Representation and Reasoning*, 1993.

3. B. Chandrasekaran and N.H. Narayanan. Towards a theory of commonsense visual reasoning. In K.F. Nori and C.E. Veni Madhavan, editors, *Foundations of Software Technology and Theoretical Science*, pages 388–409. Springer-Verlag: Berlin, 1990.

4. D. Chapman. Planning for conjunctive goals. *Artificial Intelligence*, 32:333–377, 1987.

5. M. Denis. Imagery and thinking. In C. Cornoldi and M.A. McDaniel, editors, *Imagery and Cognition*, pages 103–131. Springer-Verlag: New York, 1991.

6. G.W. Evans. Environmental cognition. In *Psychological Bulletin*, volume 88, pages 259–287, 1980.

7. R.E. Fikes and N.J. Nilsson. STRIPS: a new approach to the application of theorem proving to problem solving. *Artificial Intelligence*, 2:189–208, 1971.

8. C. Freksa and K. Zimmermann. On the unilization of spatial structures for cognitively plausible and efficient reasoning. In *Proceedings of the IJCAI '93 Workshop on Spatial and Temporal Reasoning*, pages 61–66, 1993.

9. M. Georgeff and A. Lansky. Reactive reasoning and planning. In *Proceedings of the Sixth National Conference on Artificial Intelligence AAAI '87*, 1987.

10. J.I. Glasgow. Imagery and AI – where do we go from here? *Computational Intelligence*, 9(4):424–435, 1993. Response to taking issue forum.

11. J.I. Glasgow. The imagery debate revisited: A computational perspective. *Computational Intelligence*, 9(4):309–333, 1993. Taking issue paper.

12. J.I. Glasgow. Representation of spatial models for geographic information systems. In N. Pissinou, editor, *Proceedings of the ACM Workshop on Advances in Geographic Information Systems*, pages 112–117, Arlington, VA, 1993.

13. J.I. Glasgow. Array representations for model-based spatial reasoning. In *Proceedings of the Sixteenth Annual Conference of the Cognitive Science Society*, Atlanta, 1994.

14. J.I. Glasgow. Spatial models for indeterminate worlds. In *Proceedings of AAAI '94*, Seattle, 1994.

15. J.I. Glasgow and D. Papadias. Computational imagery. *Cognitive Science*, 16(3):355–394, 1992.

16. C. Habel, S. Pribbenow, and G. Simmons. Propositional and depictorial partonomies: A hybrid approach. In *Proceedings of IJCAI '93 Workshop on Principles of Hybrid Representation and Reasoning*, pages 62–71, 1993.

17. P. Hayes. Some problems and non-problems in representation theory. In *Proceedings of AISB Summer Conference*, pages 63–79, University of Sussex, 1974.

18. B. Hayes-Roth and F. Hayes-Roth. A cognitive model of planning. *Cognitive Science*, 3:275–310, 1979.

19. M.A. Jenkins and J.I. Glasgow. A logical basis for nested array data structures. *Programming Languages Journal*, 14(1):35 – 49, 1989.

20. M.A. Jenkins and W.H. Jenkins. *The Q'Nial Reference Manual*. Nial Systems Ltd., Kingston, Ontario, 1985.

21. P.N. Johnson-Laird. *Mental Models*. Harvard University Press: Cambridge, Mass., 1983.

22. S.M. Kosslyn. *Image and Mind*. Harvard University Press, 1980.

23. B. Kuipers. Modeling spatial knowledge. *Cognitive Science*, 2:129–154, 1978.

24. B.J. Kuipers. Modeling spatial knowledge. In S. Chen, editor, *Advances in spatial reasoning*, pages 171–198. Ablex Publishing Corp., 1990.

25. B.J. Kuipers and T.S. Levitt. Navigation and mapping in large-scale space. In *Advances in spatial reasoning*, volume 2, pages 207–252. Ablex Publishing Corp., 1990.

26. T.S. Levitt and D.T. Lawton. Qualitative navigation for mobile robots. *Artificial Intelligence*, 44:305–360, 1990.

27. D.V. McDermott and E. Davis. Planning routes through uncertain territory. *Artificial Intelligence*, 22, 1984.

28. T.P. McNamara. Mental representation of spatial relations. *Cognitive Psychology*, 18:87 – 121, 1986.

29. T. More. The nested rectangular array as a model of data. In *APL Quote Quad*, 9, 1979. From the proceedings of APL79.

30. K. Myers and K Konolige. Reasoning with analogical representations. In *Proceedings of the Conference on Principles of Knowledge Representations and Reasoning*, Los Altos, CA, 1992. Morgan Kaufmann.

31. J. Pailhous. *La Representation de l'espace urbain*. Paris: Presses Universitaires de France, 1970.

32. D. Papadias and J.I. Glasgow. A knowledge representation scheme for computational imagery. In *Proceedings of the Thirteenth Annual Meeting of the Cognitive Science Society*, pages 48–54, Chicago, 1991. Lawrence Erlbaum Associates: Hillsdale, NJ.

33. D. Papadias and T. Sellis. The semantics of relations in 2d space using representative points: Spatial indices. In A.U. Frank and I. Campari, editors, *Spatial Information Theory: A theretical basis for GIS. Proceedings of the European Conference on Spatial Information Theory (COSIT '93)*, volume 716, pages 234–247. Springer-Verlag Lecture Notes on Computer Science, 1993.

34. M.E. Pollack and M. Ringuette. Introducing the tileworld: experimentally evaluating agent architectures. In *Proceedings of the Eighth National Conference on Artificial Intelligence*, pages 183–189, 1990.

35. H. Samet. *The Design and Analysis of Spatial Data Structures*. Addison-Wesley, 1989.

36. A. Sloman. Afterthoughts on analogical representation. In *Proceedings of Theoretical Issues in Natural Language Processing*, pages 164–168, Cambridge, MA, 1975.

37. A. Sloman. Varieties of fomalisms for knowledge representation. *Computational Intelligence*, 9(4):413–423, 1993.

38. A. Stevens and P. Coupe. Distortions in judged spatial relations. *Cognitive Psychology*, 10:422–437, 1978.

39. H.A. Taylor and B. Tversky. Spatial mental models derived from survey and route descriptions. *Journal of Memory and Language*, 31:261–292, 1992.

40. B. Tversky. Cognitve maps, cognitive collages, and spatial mental models. In A.U. Frank and I. Campari, editors, *Spatial Information Theory, A Theoretical Basis for GIS, Proceedings of the European Conference, (COSIT '93)*, volume 716, pages 14–24. Springer Verlag Lecture Notes in Computer Science, 1993.

Object Localization:
Selection of Optimal Reference Objects

Klaus-Peter Gapp

Cognitive Science Program
Department of Computer Science
University of Saarbrücken
D-66041 Saarbrücken, Germany

gapp@cs.uni-sb.de

Abstract. The quality of an object localization depends essentially on the adequate selection of a suitable reference. In most computational approaches developed so far only the distance between the located object and a potential reference object has been used as a decision criterion. However many other criteria have to be considered for a cognitive plausible selection of adequate reference points. In this paper we investigate how object and context dependent properties, like referentiality, visual salience, functional dependencies, or prior knowledge, influence the quality of a reference object. Each factor is quantitatively determined and scaled by relevance to a certain context. The scaling permits the necessary comparability of the different quality criteria. Finally, on the basis of these factors a computational model is presented which permits a context dependent determination of the optimal reference object in a particular spatial configuration.

1 Introduction

"Where is object A positioned?". The answer to spatial queries like this requires locating a known object A with reference to another (reference) object. The localization of a particular object is an often required procedure in many applications dealing with the domain *space*. The applications range from general spatial information systems, like geographic information systems or driver information systems, and systems using multi-modal instructions, to applications in Virtual Reality. Understanding the use of spatial references permits the development of better interfaces between systems and human users in the spatial domain.

Most algorithms developed so far in this context use very simple mechanisms for the determination of adequate reference objects (e.g., Carsten & Janson, 1985; Wazinski, 1992). The distance from the located object (LO) to the reference object (RO) is often the only criterion used. However many other crucial factors which depend either directly on the objects or on context factors need also to be considered.

The most easily recalled attributes of a region are typically referred to as landmarks. They are used to denote distinguishing features of a route or a region

(Lynch, 1960; Appleyard, 1969; Downs & Stea, 1973; Siegel & White, 1975). In the first case they are used for navigational decisions, whereas in the second case landmarks allow for the maintenance of general geographical orientation. Recognizability (Lynch, 1960), use (Appleyard, 1969), and cultural meaning (Moore, 1979) have been emphasized as being the key factors for the relative landmark status of a place. The metaphor *cognitive map* is usually used to describe mental representations of environments, including landmarks, knowledge about the spatial relations between them, routes and metric survey information (Gärling & Golledge, 1989). However the cognitive map seems to be systematically distorted and potentially contradictory (see e.g., Stevens & Coupe, 1978; Sadalla, Burroughs, & Staplin, 1980; Tversky, 1981; Holyoak & Mah, 1982; Hirtle & Jonides, 1985), and thus not easily reconcilable in a map-like structure (Tversky, 1993). This led Tversky to propose the term *cognitive collage* as often being a more appropriate metaphor for environmental knowledge than cognitive map.

Some landmarks function as spatial reference points, points that serve as the basis for the spatial location of other non-reference points. The concept of spatial reference points implies that the position of a large set of non-reference locations in a particular region is defined in terms of the position of a smaller set of reference locations (Sadalla, Burroughs, & Staplin, 1980). Places, known as reference points, are relatively better known and serve to define the location of adjacent points. According to the findings of Rosch (1975) about the existence of asymmetries in similarity judgements between semantic reference points and non-reference points, Sadalla, Burroughs, and Staplin were able to measure asymmetries of cognitive distances between spatial reference points and non-reference points. For example, subjects were able to indicate the proximity of reference points faster than they could indicate the proximity of equidistant non-reference points, and that the direction of particular target locations is more quickly verified relative to reference locations, than relative to non-reference locations. An experimental cluster analysis of the six clusters size, familiarity, cultural value, form quality, marker salience, and visual salience showed that the attributes of locations designated as reference points are related to the scales of familiarity, visibility from a distance, domination of nearby places, and cultural importance. This gives us an initial idea of how to categorize the attributes which affect the quality of an object, or of a place to serve as a reference object.[1] The first category is related to features specific to the object or place, and the second is determined by additional context dependencies, like functional dependencies or the prior knowledge of the listener.

In this paper we first analyze the factors that govern the choice of reference landmarks in cognitive usage, and how they can be quantitatively measured. Finally, a domain independent model is developed for selecting suitable reference objects to describe a scene to a human user or to answer queries about spatial configurations.

---

[1] Note: *Reference object* in this context means all kinds of spatial objects or places which can be used to locate an object's position.

2 Quality Criteria for Reference Objects

Object localization requires the establishment of a spatial relationship between an object to be located and one or two reference objects (Talmy, 1983; Herskovits, 1986). The localization task can be divided up into three parts: first, the object of interest, the LO, is identified, then an object suitable for use as a referent, the RO, is selected, and finally a linguistic description of the relationship between the objects from some specific point of view is specified. In this paper we will put emphasis on what makes an object a good reference object, and how this can be integrated into a computational theory.

Talmy (1983) categorized the located or primary object, in contrast to the reference or secondary object, as having spatial variables that tend to be: more mobile, smaller, conceived as being geometrically simpler (often point-like), more salient, more recently on the scene, and more in awareness.

The region where an object can be located depends mainly on its size (Habel & Pribbenow, 1988). This means that for large objects, like buildings, the search space for adequate reference objects is significantly larger than for smaller objects, such as a chair or table. Size, as well as many other factors, function as quality criteria in the selection process of adequate reference objects. We first consider features directly related to the potential reference object, and second, factors related to the particular context.

2.1 Relevant Object Features

An important property of a reference object in a general localization task is its visual salience. Visual salience of an object depends on the interaction of basic features like size, shape and color, correlated to the corresponding attributes of the surrounding objects (Treisman, 1988). Objects which are large in size with a salient shape and/or color are therefore preferred reference points.

Visual Salience With an object's size one usually refers to the length, width, and height of the object (cf. Figure 1a). However, for some objects, size refers only to one specific dimension, and this dimension varies between objects. For example the size of a road means its width (cf. Figure 1b), and the size of people, their height (cf. Figure 1c). It seems as if the side which varies most obviously, is usually used for size discrimination.

A phenomenon which is related to size perception, and which makes the exact determination of size more difficult, is that of illusory distortions. Classical illusion configurations such as the Müller-Lyer illusion (Figure 2) give rise to strong misperceptions of size. In the Müller-Lyer illusion the arrowheads on the end of the lines make them appear longer or shorter, depending on the direction in which they point. In most current explanations of size distortions it is assumed that this illusion arises because the arrowheads delineate a set of spans either side of the line, which are either longer or shorter than the line itself.

In Stuart, Bossomaier, and Johnson (1993) the argument was developed that size illusions are the consequence of interference effects in the size domain.

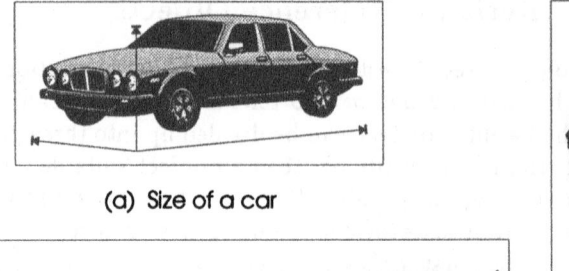

(a) Size of a car

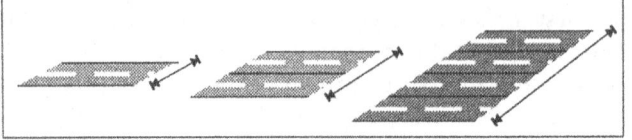

(b) Size of a road

(c) Size of a person

Fig. 1. The size of an object.

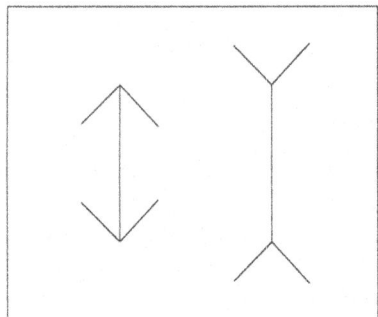

Fig. 2. The Müller-Lyer illusion.

Their empirical work showed that the size domain is coded in parallel, which is an important prerequisite for the assumption of sparse sampling of the size domain.

In the context of reference quality, the size of different kinds of objects must be compared. We therefore propose, as a first approximation, to express each object's size through its visible horizontal and vertical extensions.[2] The relative size of an object compared to the surrounding objects is then determined by comparing the measured differences in each dimension. The result is that object

---

[2] Note: In 3-dimensional space, the perceived objects are projected onto a 2-dimensional plane perpendicular to the line of sight.

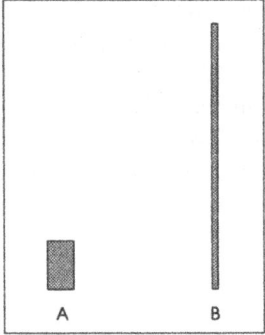

Fig. 3. Object B stands out through its large vertical extension.

B (e.g., a flagpole) presented in Figure 3 is more visually salient than object A because of the bigger difference in vertical size when compared to the difference in horizontal size. It remains to be seen which dimension (probably the vertical) has the bigger influence.

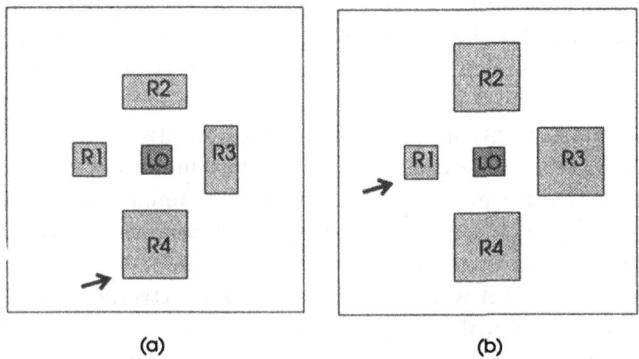

Fig. 4. Size as a feature of relative visual salience.

"What is smaller is preferably located with respect to what is larger" (Levelt, 1989, p 155), but there are exceptions: Size, as well as most other object features, is defined relative to other proximal objects. If in a certain context all potential candidates for a reference object have the same features of visual salience, except that one object is considerably smaller, then the smaller object might be more salient than the others. Consider for example the two spatial situations shown in Figure 4. In Figure 4a the answer to the question *"Where is the LO positioned?"* would be *"above R_4"*, whilst in Figure 4b the object R_1 would be the preferred reference object. In the first case R_4 is visually salient because it is the largest object. However, in the second case R_1 is more salient than the other objects, because its lower size distinguishes it from the surrounding objects.

If objects are of nearly the same size, then the shape of the objects is another feature which helps to make them more distinguishable. The visual salience of a certain shape is again dependent on the shapes of the surrounding objects (cf. Figure 5a,b). Rotation can increase the salience of an object within a group of objects of the same shape (cf. Figure 5c). A car lying on its side is very salient in a parking area, if all the other cars are in normal position.

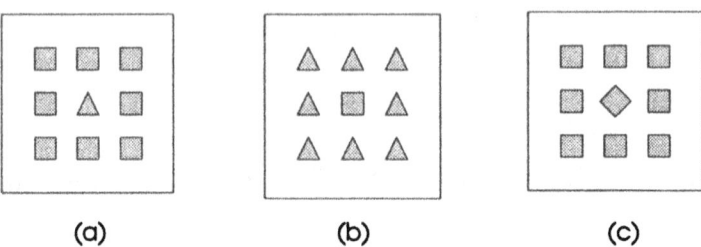

(a)   (b)   (c)

Fig. 5. Shape as a feature of visual salience.

A feature of objects which is easiest to perceive is *color*. Experimental evidence indicates color or, more exactly, color differences as being directly responsible for the conspicuousness of objects (Carter & Carter, 1981). Again there is no absolute measure for the salience of a certain color, because, e.g., a blue dot on a blue background is not salient, whereas a blue dot on a white background is.

Computer systems mainly use the additive, hardware-oriented RGB color theory. In the RGB model each color is divided up into a *red*, a *green*, and a *blue* value. However the use of the RGB color scale is unnatural for man machine communication, since its metrics do not represent color differences on a uniform scale, and colors are not organized in an intuitive manner.

In Johansson (1949) it was stated that the basic attributes by which humans characterize and distinguish the appearances of colors are 1. hue, 2a. chromaticness 2b. saturation, 3a. lightness and 3b. clearness. According to the international lighting vocabulary these values are defined as follows:

Hue is an attribute of a visual sensation according to which an area appears to be similar to one of the perceived colors, red, yellow, green, and blue, or to a combination of two of them.

Chromaticness is an attribute of a visual sensation according to which the perceived color of an area appears to be more or less chromatic.

Saturation is the chromaticness and colorfulness of an area, judged in proportion to its brightness.

Brightness is the attribute of visual sensation according to which a given visual stimulus appears to emit more or less light.

Lightness is the attribute of visual sensation according to which a given visual stimulus appears to emit more or less light in proportions to that emitted by a similar area perceived as having a "white" stimulus.

Johansson proposed that three variables, one from each of the three categories, are necessary to define the appearance of colors unambiguously. In a 3-dimensional geometrical representation of colors, any of these three may be used as major axes. Colors representing constant amounts of these attributes, or constant differences in any of these attributes, in such a geometrical representation are simply illustrated by straight lines.

Systems which describe and classify color based on human color perception are called color appearance systems (Derefeldt, 1991). They are defined by perceptual color coordinates or scales and by a uniform or equal visual spacing of colors according to these scales. Only a few color systems fulfill these criteria. These are the Munsell, the DIN, the NCS, and the OSA/UCS color systems. In many computational applications also the CIELUV and the CIELAB color spaces are also used as approximations to color appearance systems[3]. Despite the fact that none of these systems can claim the advantage of representing all the basic appearance attributes of surface colors and their interrelations, the use of color appearance systems in computer applications always requires the measurement of the chromaticity coordinates of the monitor's RGB phosphors. However in most cases approximations should be precise enough to account for the color when coping with the visual salience of objects.

More salient environmental features may have precedence over less salient ones (Taylor & Tversky, 1992a). Experimental studies in (Mangold, 1986) showed that color dominates size and shape in object identification tasks. Furthermore, size is more easily recognizable than shape. This ordering can be used for a weighted combination of the three features color, size, and shape, related to a single attribute of visual salience.

Mobility A property of an object which is independent from other objects is the distinction between permanently located and movable objects. In most cases, the preferred reference objects are objects with a stationary setting within a certain reference frame (cf. Talmy, 1983). Even if a movable object is located at a particular place for a long time (e.g., a parked car), a permanent stationary reference object is more suitable.

Frame of Reference A last point we want to address in the context of specific object features belongs to the coordination between perception and language required for producing and understanding spatial expressions. Perceptual cues about spatial relations between objects in the perceived environment and expressions that describe these relations must be mapped onto some mental representation of space in order for communication to occur (Miller & Johnson-Laird, 1976). The interaction between perception and language requires the adoption of a frame of reference with respect to which spatial positions can be defined (Carlson-Radvansky & Irwin, 1993). One distinguishes three different frames of reference: the viewer-centered or deictic, the subject-centered or intrinsic, and

---

[3] See (Derefeldt, 1991; Billmeyer, 1985) for an overview to color appearance systems.

the environment-centered or extrinsic frame of reference (Fillmore, 1975; Miller & Johnson-Laird, 1976; Marr & Nishihara, 1978; Talmy, 1983; Levelt, 1984; Pinker, 1985; Retz-Schmidt, 1988).

In Miller and Johnson-Laird (1976) it was argued that the choice of the perspective is dependent on features of the configuration to be described. They proposed the preference of an intrinsic perspective where possible, i.e., if the reference object has an intrinsic front. If a deictic interpretation is intended when an intrinsic interpretation is possible, the speaker will usually add explicitly *"from my point of view"* or *"as I am looking at it"* (Miller & Johnson-Laird, 1976, p 398). In contrast, Levelt showed that the same spatial configuration can be described from both an intrinsic and a deictic perspective. However the speaker has to keep a chosen perspective constant for the whole description (Levelt, 1982). In Ehrich (1985) subjects were asked to describe the arrangement of furniture in a doll's house and to analyze the factors that determined the selection of reference objects and relations. In most descriptions a deictic perspective was used. Ehrich interpreted the dominance of the deictic perspective as the necessity of keeping a selected frame of reference constant. Therefore, it is much easier to describe a complete spatial configuration, such as a room, from a deictic perspective.

Following these results the intrinsic perspective is proposed to be used, if the reference object has an intrinsic front. However, if the task is to completely describe a particular environment with more than three objects, then the deictic perspective is preferred.

2.2 Context Dependencies

The situational context in which an object localization takes place has a great influence on the selected reference point. Although one often tends to see the basic object features like color, shape, and size as the crucial factors when deciding which reference object to choose, context dependencies are sometimes more decisive.

Referentiality Referentiality, for example, is crucial in the process of object localization. It is not possible to refer to an object if one cannot conclude from its visible parts its semantic interpretation, i.e., if the object cannot be unambiguously identified from a certain perspective. The necessity of referentiality was also verified in (Sadalla, Burroughs, & Staplin, 1980), where *visibility from a distance* was rated as one of five important factors for reference points.

Object identification is complicated if an object is not in its intrinsic position or if parts are occluded by other objects. Three different kinds of such *intervening objects (IO)* can be distinguished. An intervening object can appear between the located object and the potential reference object (cf. Figure 6a,b), or it can be located between the viewer and the reference object (cf. Figure 6c). Also the located object can itself function as an intervening object, if it is positioned between the viewer and the reference object (cf. Figure 6d). Depending on its dimensions, an object located between the located and the reference object functions like a barrier and consequently decreases the quality of a potential reference object (Gapp & Maaß, 1994). The same occurs if in 3-dimensional space from a particular point of view, a reference object is fully or partially occluded from another object. If the occlusion is too severe, e.g., referentiality of the potential reference object is no longer given, then either another object has to be selected for reference or, if possible, the point of view must be changed.

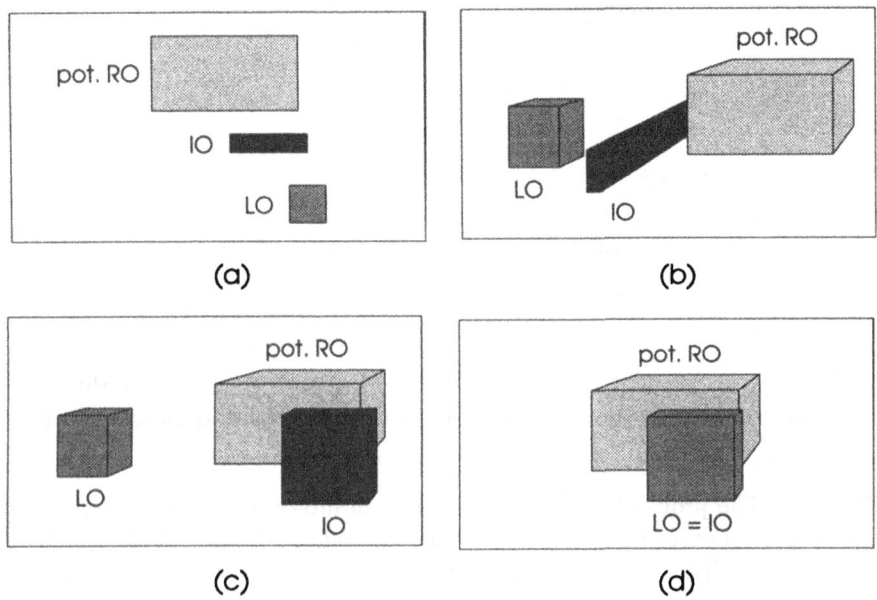

Fig. 6. Intervening objects.

In each case the intervening object occludes parts of the reference object and therefore affects its visual salience. The following method is proposed to account for intervening objects: depending on the location of the intervening object, either from the located object's center of gravity or the point of view, the intervening object is projected onto the reference object (cf. Figure 7a,b). Depending on the degree of occlusion, the hidden parts of the reference object influence the computation of an object's visual salience (e.g., size), as well as its degree of referentiality.

However, it is too simple to make the referentiality of an object only dependent on its degree of visibility. It is, rather, the visibility of an object's prominent parts, which allow for its identification, that defines referentiality. For example, if only the spire of a church is visible, the partially occluded object is still easily perceivable as a being church. However, if only the spire is occluded, the identification of the object as being a church might be much more difficult, although in the latter case the hidden part of the object is smaller than it was in the former.

The minimal set of an object's prominent parts which need to be visible for an unambiguous identification of the object can not yet be computed automatically and need to be marked for each object by hand. It is therefore proposed to use the degree of visibility of an object as a first approximation for its referentiality.

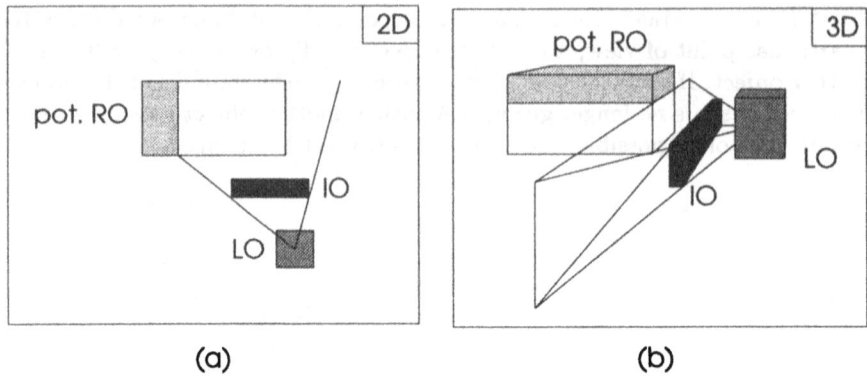

Fig. 7. Projection of the intervening object onto the reference object.

In the examples shown in Figure 7a,b this corresponds to a referentiality factor of about 25% in the 2-dimensional and about 33% in the 3-dimensional case.

Distance The probability of the occurrence of intervening objects depends on the distance between the reference and the located object respective to the point of view. Since distance itself is a decisive criteria for reference objects, the closer a potential reference object is situated to the located object, the better or more precise the localization. This "nearness condition" of objects need not be explicitly mentioned in a localizational phrase, like *"the tree is in front of the church"*. On the contrary, "nearness" is the unmarked case, the default, unless one has evidence to the opposite (Herskovits, 1985). If in a particular localization the reference point is further away than usual, distance information, like *500 feet away*, is usually added.

Common metrics of distances are defined as the minimal distance between the centers or the borders of the objects (cf. Figure 8a,b). To account for the asymmetrical distance judgements between ordinary objects and reference objects (Sadalla, Burroughs, & Staplin, 1980) we propose a new metric which measures the distance from the located object's center of gravity to the closest point of the potential reference object (cf. Figure 8c,d). This results in a larger distance if the "reference object" is an ordinary (and therefore smaller) object.

Functional Dependencies Items that are related are more likely to be remembered together (Tulving, 1962). Spatial distance as a relation between objects is one factor that leads to a grouping in memory (Stevens & Coupe, 1978; Hirtle & Jonides, 1985; McNamara, 1986; Taylor & Tversky, 1992b). Non-spatial organizations may also be used, for example, in remembering items together that are related by function rather than by spatial proximity. In Hirtle and Jonides

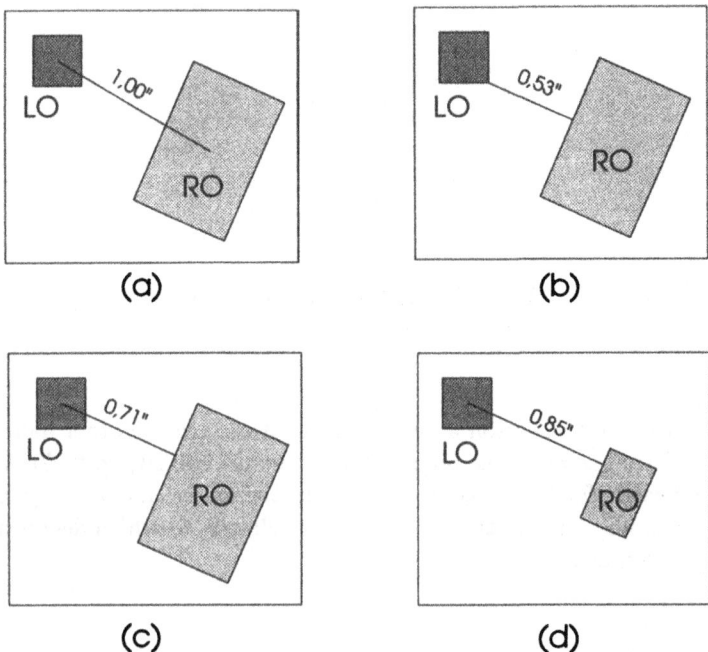

Fig. 8. Simulation of the asymmetrical distance effect.

(1985) it was found that people tended to group commercial buildings with other commercial buildings, and university buildings with other university buildings, despite the fact that the buildings were spatially interspersed. According to this, in Figure 9 it is more likely that people will choose the mail box as a reference object for the letter than the bell. The semantic relationship between the letter and the mail box and their spatial nearness will lead to the formation of a spatial cluster and to a priming effect (Hirtle & Heidorn, 1993). However it is important to note that the priming from non-spatial associations occurs independently from a priming by spatial relations (McNamara & LeSeur, 1989).

Prior Knowledge Grouping and clustering of objects by non-spatial associations is partly dependent on the prior knowledge of the user. The prior knowledge itself can be decisive in the selection process of a reference object. Imagine, for example, a scene with a swimming pool, a soccer field, and a school building as presented in Figure 10. If a pupil is asked where the swimming pool is, he will probably choose the school as the reference. However, a soccer player might tend to refer to the swimming pool as being close to the soccer field.

Prior knowledge influences the localization task in two ways: on the one hand, the chance of an object being selected as a reference object only increases if the

Fig. 9. Functional dependencies.

object is known to the person asked for its location. On the other hand, objects are also preferred as reference objects if they are more likely to be known to the person who asked for the object's location. In the latter case stereotypes can be defined, which help in a particular situation to classify to which degree an object might be known or not.

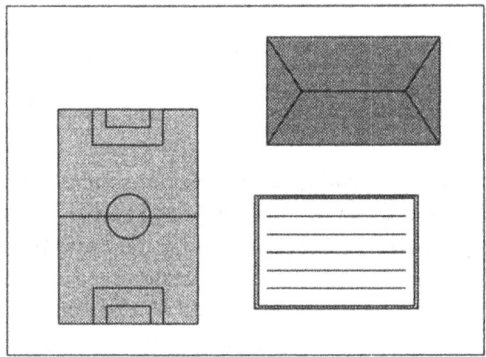

Fig. 10. Prior knowledge: The swimming pool is close to the school/soccer field.

Previously Mentioned Objects A last factor we want to address appears if more than one localization has to be carried out, e.g., in a route description task. In such a situation, objects which have been mentioned as reference objects in previous localizations might be better reference objects than unknown new objects. The importance of the factor "previously mentioned" increases if the localizations are done by inspection of a mental map, which means that the user has no visual image of the scene.

3 Determining the Optimal Reference Object

In this section we focus on how a reference object in a particular situation can be determined on the basis of the criteria discussed above. These criteria are of differing importance and partly dependent on one another. For example, distance, which has a high priority, can be affected by functional dependencies, and referentiality is dependent on visual salience. In Figure 11, a priority sorted list of the selection criteria is presented. However the order is not strict and should only give an idea of which factors are of higher importance than others. The definition of a procedure which determines the best reference object in a particular situation is complex, because the criteria are of differing natures, are object and context dependent, and the values from different factors are not directly comparable per se.

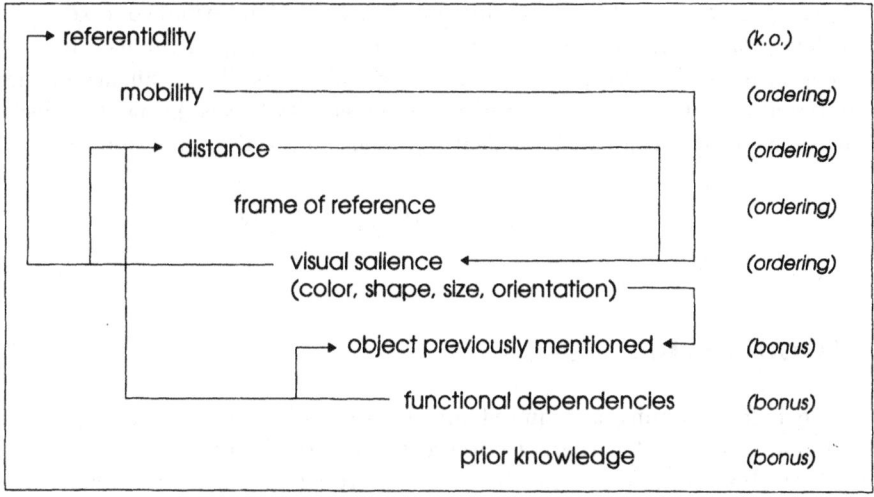

Fig. 11. Priorities of the selection criteria for reference objects.

The input of the algorithm is a set of potential reference object candidates which are proximal to the object to be localized. The size of this proximal region depends on the extensions of the located object. They can be efficiently retrieved by using a spatial indexing method, like quad-trees (Finkel & Bentley, 1974) or range-trees (Lueker, 1978).

It seems reasonable that referentiality, which permits a definite identification of the reference object by the hearer/viewer, is a decisive requirement for successful object localization. If referentiality is not guaranteed, either another object has to be chosen or the point of view has to be changed. We therefore call this property a *k.o.* criterion. Relevant object features function as *ordering* criteria, whereas context dependent factors mainly work as *bonuses* (cf. Figure 11).

The following procedure is used for deciding which object should be selected as the optimal reference object: Each object receives a graded evaluation in the interval $\mathcal{I} = [0,1] \subseteq \mathbb{R}$ for each ordering feature, e.g., 0.0 for a moving object, 0.5 for a movable (but not moving) object, and 1.0 for a permanently located object. Distance can be scaled by intervals and the complete range from 0 to 1 could be used for visual salience. The same is performed for each context dependent factor. If a reference object candidate was previously mentioned, the graduated factor "previously mentioned" for this object depends on the number of objects, located objects as well as reference objects, mentioned since its last occurrence. Functional dependencies are rated in relation to their intensity and prior knowledge is evaluated by the probability that a particular object is known to a listener.

Assume that n factors are used for the evaluation of an object's quality as a reference object. This results in a n-digit feature vector $f = (f_1, \ldots, f_n)$ in the interval space $\mathcal{I}^n = [0,1]^n$. Depending on a particular situation, context factors can change the importance of one or more dimensions, e.g., if the user has only a mental image of the scene and no direct visual access. The influence of these situational aspects is accounted for by the possibility of weighting the affected dimensions using a context dependent n-dimensional scaling function SC_{cxt}:

$$SC_{cxt} : \mathcal{I}^n \to \mathbb{R}^n; (f_1, \ldots, f_n) \mapsto (SC_1(f_1), \ldots, SC_n(f_n))$$

with

$$SC_i(f_i) : \mathcal{I} \to \mathbb{R}, \quad 1 \leq i \leq n$$

The highest scaling is required for the k.o. criterion referentiality. Lower weights are assigned to ordering criteria. If no distinction between the selected reference object candidates can be made according to a feature classified as a bonus criterion, then this particular dimension is dropped from the interval space $\mathcal{I}^n$, and therefore no longer considered in the localization task.

The optimal reference object can now be defined as the $n - x$ dimensional subspace $\mathcal{S}^{n-x} \subseteq \mathcal{I}^n$, $1 \leq x \leq n$, in which x is the sum of factors defined by an interval. This means, that if an optimal reference object is defined by only one single value for each factor f_i, then hence $x = n$ and $\mathcal{S}^{n-x} = \mathcal{S}^0$ denotes a point in $\mathcal{I}^n$.

The context, specific weighting of the single dimensions, makes the factors comparable with each other. Each reference object RO_i is represented as a n-dimensional point $P_{RO_i} \in \mathcal{I}^n$, according to its evaluation of the n quality factors. A scale for the quality of a reference object can now be defined as the shortest distance between P_{RO_i} and $\mathcal{S}^{n-x}$. If $\mathcal{S}^{n-x}$ represents a point $P_{opt} = (f_1^{opt}, \ldots, f_n^{opt}) \in \mathcal{I}^n$ then the quality of a potential reference object $Q(RO_i)$ can be expressed as:

$$Q(RO_i) := |\overline{P_{RO_i} \ P_{opt}}| = \sqrt{\sum_{i=1}^{n}(SC_i(f_i) - f_i^{opt})^2}$$

Hence the best reference object RO_i is defined as the object with the minimal "distance" $Q(RO_i)$. If more than one object have the same distance, direct comparisons on the basis of single feature values can be used for a final decision.

The algorithm presented is not dependent on a certain number of features. The more features used, the better the performance of the algorithm. The performance also depends directly on the quality of the calculation of the individual factors and on their weighting. The individual features of an object can be computed in parallel, however dependencies between some of them can force the reconsideration of already independently computed factors. A parallel processing of all potential candidates is also encouraged. On the one hand this ensures efficient processing, and on the other hand it permits the use of *anytime* architecture (cf. Boddy, 1991) for the system, which allows the generation of answer at "any" time, with increasing quality the longer the computation lasts (resource dependent processing).

4 Discussion

A cognitively plausible description of an object's location in natural language requires the selection of an adequate reference object. As shown in section 2, the factors which determine the suitability of an object to serve as a reference object in a particular situation vary and depend both on the object itself and on the context. Problems are related to the calculation of single factors, which are partly dependent on each other, and to their importance to the quality of a potential reference object, which also differs with the context. An exact definition through the use of empirical studies seems to be unrealistic, given the high complexity of the problem. However, the computational model proposed above permits the computation of an approximated solution, which can be refined by using more factors, modified weights, and a more precise calculation of the single factors. The use of neural nets or genetic algorithms, instead of the distance measure for analyzing an object's reference qualities, also seem to be promising alternatives.

5 Conclusion and Future Work

A cognitively plausible localization of an object requires a context dependent selection of an adequate reference object, which permits the optimal determination of the objects position. We investigated which object and context dependent factors are essential to fulfill this complex task. It turned out that factors which are directly dependent from the potential reference object, like visual salience, have to be defined relative to surrounding objects. On the basis of these investigations a theory for the computation of the optimal reference object was developed. The approach accounts for the different priorities of the single features dependent on the current context and the set of factors considered can be restricted or enhanced without changing the algorithm. An implementation can benefit in performance through the full parallel design of the algorithm. Even anytime architecture can easily be applied.

In future work we will complete the implementation of the algorithm and connect it with the CSR-3D system, which allows for a cognitive plausible computation of spatial relations in 3-dimensional space (Gapp, 1994, 1995a,b). A further part of on-going research is the situation specific selection of spatial relations in object localization tasks.

6 Acknowledgements

The author would like to thank Bernd Andes and Amitabha Mukerjee for valuable discussion. The clarity of this paper was improved thanks to comments from Marc Gurmann and the two anonymous reviewers.

References

Appleyard, D. (1969). Why buildings are known. *Environment and Behaviour, 1*, 131–156.

Billmeyer, F. W. J. (1985). *AIC annotated bibliography on color order systems.* Mimeoform Service, rear, 4805 Prince George's Avenue, Bletsville, MD 20705.

Boddy, M. (1991). Anytime problem solving using dynamic programming. *Proc. of AAAI-91*, Anaheim, CA, 738–743.

Carlson-Radvansky, L. A., & Irwin, D. E. (1993). Frames of reference in vision and language: Where is above? *Cognition, 46*, 223–244.

Carsten, I., & Janson, T. (1985). *Verfahren zur evaluierung räumlicher präpositionen anhand geometrischer szenenbeschreibungen.* Master's thesis, FB Informatik, Univ. Hamburg.

Carter, R. C., & Carter, E. C. (1981). Color and conspicousness. *Optical Society of America, 71*, 723–729.

Derefeldt, G. (1991). Colour appearance systems. In P. Gouras (Ed.), *The perception of colour* (chapter 13, pp. 218–256). Beccles and London, UK: The Macmillan Press.

Downs, R. M., & Stea, D. (1973). Cognitive maps and spatial behaviour: Process and products. In R. M. Downs & D. Stea (Eds.), *Image and environment. cognitive mapping and spatial behaviour* (pp. 8–26). Chicago: Aldine.

Ehrich, V. (1985). Zur linguistik und psycholinguistik der sekundären raumdeixis. In H. Schweizer (Ed.), *Sprache und raum* (pp. 130–161). Stuttgart: Metzler.

Fillmore, C. J. (1975). *Santa cruz lectures on deixis.* Bloomington, IN: Indiana University Linguisic Club. only reproduced, first published in 1971

Finkel, R. A., & Bentley, J. L. (1974). Quad trees: A data structure for retrieval on composite keys. *Acta Informatica, 4*, 1–9.

Gapp, K.-P. (1994). Basic meanings of spatial relations: Computation and evaluation in 3d space. *Proc. of AAAI-94*, Seattle, WA, 1393–1398.

Gapp, K.-P. (1995a). Angle, distance, shape, and their relationship to projective relations. *Proc. of the 17th Annual Conference of the Cognitive Science Society*, Pittsburgh, PA. To appear.

Gapp, K.-P. (1995b). An empirically validated model for computing spatial relations. *Proc. of KI-95*, Berlin, Heidelberg. Springer. To appear.

Gapp, K.-P., & Maaß, W. (1994). Spatial layout identification and incremental descriptions. *Proc. of the AAAI-94 Workshop on Integration of NL and Vision Processing*, Seattle, WA, 145–152.

Gärling, T., & Golledge, R. G. (1989). Environmental perception and cognition. In E. H. Zube & G. T. Moore (Eds.), *Advances in environment, behaviour, and design* (pp. 203–236, Vol. 2). New York: Plenum.

Habel, C., & Pribbenow, S. (1988). *Gebietskonstituierende prozesse*. LILOG-Report 18, IBM Deutschland GmbH, Stuttgart.

Herskovits, A. (1985). Semantics and pragmatics of locative expressions. *Cognitive Science, 9*(3), 341–378.

Herskovits, A. (1986). *Language and spatial cognition. an interdisciplinary study of the prepositions in english*. Cambridge, London: Cambridge Univ. Press.

Hirtle, S. C., & Heidorn, P. B. (1993). The structure of cognitive maps: Representation and processes. In T. Gärling & R. G. Golledge (Eds.), *Behaviour and environment: Psychological and geographical approaches* (chapter 7, pp. 177–189). : Elsevier Science Publishers.

Hirtle, S. C., & Jonides, J. (1985). Evidence of hierachies in cognitive maps. *Memory and Cognition, 13*(208-217).

Holyoak, K. J., & Mah, W. A. (1982). Cognitive reference points in judgement of symbolic magnitude. *Cognitive Psychology, 14*, 328–352.

Johansson, T. (1949). Characteristic properties of colour and colour combinations. *Revue d'Optique Theorique et Instrumentale, 28*, 241–246.

Levelt, W. J. M. (1982). Cognitive styles in the use of spatial direction terms. In R. J. Jarvella & W. Klein (Eds.), *Speech, place, and action* (pp. 251–268). Chichester: Wiley.

Levelt, W. J. M. (1984). Some perceptual limitations on talking about space. In A. J. van Doorn, W. A. van de Grind, & J. J. Koenderink (Eds.), *Limits in perception* (pp. 328–358). Utrecht: VNU Science Press.

Levelt, W. J. M. (1989). *Speaking — from intention to articulation*. Cambridge, MA: MIT Press.

Lueker, G. S. (1978). A data structure for orthogonal range queries. *Proc. of the 19th IEEE Symposium on Foundations of Computer Science*, 28–34.

Lynch, K. (1960). *The image of the city*. Cambridge, MA: MIT Press.

Mangold, R. (1986). *Sensorische faktoren beim verstehen überspezifizierter objektbenennungen*, volume 185. Frankfurt am Main: Lang.

Marr, D., & Nishihara, H. K. (1978). Representation and recognition of the spatial organization of three-dimensional shapes. *Proc. of the Royal Society of London*, 269–294.

McNamara, T. P. (1986). Mental representations of spatial relations. *Cognitive Psychology, 18*, 87–121.

McNamara, T. P., & LeSeur, L. L. (1989). Mental representations of spatial and nonspatial relations. *The Quarterly Journal of Experimental Psychology, 15*, 211–227.

Miller, G. A., & Johnson-Laird, P. N. (1976). *Language and perception.* Cambridge: Cambridge University Press.

Moore, G. T. (1979). Knowing about environmental knowing. *Environment and Behaviour, 11*, 33–70.

Pinker, S. (1985). Visual cognition: An introduction. In S. Pinker (Ed.), *Visual cognition* (pp. 1–63). Cambridge, MA: MIT Press.

Retz-Schmidt, G. (1988). Various views on spatial prepositions. *AI Magazine, 9*(2), 95–105.

Rosch, E. (1975). Cognitive reference points. *Cognitive Psychology, 7*, 532–547.

Sadalla, E., Burroughs, W. J., & Staplin, L. (1980). Reference points in spatial cognition. *Journal of Experimental Psychology: Human Learning and Memory, 6*(5), 516–528.

Siegel, A. W., & White, S. H. (1975). The development of spatial representation of large-scale environments. In W. Reese (Ed.), *Advances in child developement and behaviour* (). New York: Academic Press.

Stevens, A., & Coupe, P. (1978). Distortions in judged spatial relations. *Cognitive Psychology, 10*, 422–437.

Stuart, G. W., Bossomaier, T. R., & Johnson, S. (1993). Preattentive processing of object size: implications for theories of size perception. *Perception, 22*, 1176–1193.

Talmy, L. (1983). How language structures space. In H. Pick & L. Acredolo (Eds.), *Spatial orientation: Theory, research and application* (pp. 225–282). New York, London: Plenum.

Taylor, H., & Tversky, B. (1992a). Descriptions and depictions of environments. *Memory & Cognition, 20*(5), 483–496.

Taylor, H. J., & Tversky, B. (1992b). Spatial mental models derived from survey and route descriptions. *Journal of Memory and Language, 31*, 261–292.

Treisman, A. (1988). Feature and objects: The fourteenth bertlett memorial lecture. *The Quarterly Journal of Experimental Psychology, 40a*(2), 201–237.

Tulving, E. (1962). Subjective organization in free recall of "unrelated" words. *Psychological Review, 69*, 344–354.

Tversky, B. (1981). Distortions in cognitive maps. *Cognitive Psychology, 13*, 407–433.

Tversky, B. (1993). Cognitive maps, cognitive collages, and spatial mental models. *Proc. of the 1ˢᵗ int. Conference on Spatial Information Theory*, Elba, Italy, 14–24. Springer Verlag.

Wazinski, P. (1992). Generating spatial descriptions for cross-modal references. *3rd Conference on Applied Natural Language Processing*, Trento, Italy, 56–63.

Range Queries Involving Spatial Relations:
A Performance Analysis

Yannis Theodoridis[+]

Dimitris Papadias[*]

Dept. of Electrical and Computer Engineering
National Technical University of Athens
GREECE 15773
e-mail: theodor@theseas.ntua.gr

Dept. of Computer Science and Engineering
University of California, San Diego
CA 92093-0114, USA
e-mail: dimitris@cs.ucsd.edu

Abstract: Spatial relations are becoming an important aspect of spatial access methods because of the increasing interest on qualitative spatial information processing. In this paper we show how queries involving spatial relations can be transformed to range queries and implemented in existing DBMSs. We provide a performance analysis of B- and R- tree indexing methods to support such queries and we evaluate the analytical formulas using experimental results. The proposed analytical models for the expected retrieval cost of spatial relations are proved to be good guidelines for a spatial query optimiser.

Keywords: Spatial relations, Performance Analysis, Query Optimisation.

1. Introduction

Spatial information is often processed qualitatively, using spatial relations, rather than absolute coordinates. [Topa95] describes an example of a computerised system for coordinating first-aid vehicles that uses qualitative spatial constraints to navigate vehicles. Additional cases where spatial relations can be used to solve practical problems involving spatial information can be found in [PS94]. As a result, the formalization, representation and processing of spatial relations has become important for user interfaces and query optimization strategies in Geographic Information Systems [PS95, CSE94]. The significance of spatial relations has also been pointed out by a number of researchers in Spatial and Image Databases, [PFK94, SYH94]. The most common types of spatial relations that have been used in geographic applications include *topological*, *direction* and *distance* relations.

[+] Yannis Theodoridis was partially supported by the Department of Research and Technology of Greece (PENED'91).
[*] Dimitris Papadias was partially supported by NSF - IRI 9221276. He is currently with the National Center for Geographic Information and Analysis, University of Maine, Orono ME 04469-5711.

Topological relations deal with concepts of connectedness and inclusion. According to the 4-intersection model [EF91], the most prevalent model in the GIS literature, eight pairwise disjoint relations can be defined using the four intersections between objects' boundaries and interiors: *disjoint, meet, equal, overlap, contains, inside, covers*, and *covered-by*. [Egen91] extended the model by also including intersections among objects' exteriors (9-intersection model). Tests with human subjects have shown that the intersection models have potential for defining cognitively meaningful spatial predicates, a fact that makes the above relations a good candidate for commercial systems [ME94].

Direction relations (*north, northeast*) deal with order in space. Unlike the case of topological relations where the intersection models provide generally accepted definitions, there are no such definitions of direction relations (e.g., is France *north* or *northwest* of Italy?). Although experimental findings from Cognitive and Environmental Psychology can be used as guidelines for the direction relations that people evoke in everyday reasoning, so far the psychological results are too vague to be helpful in defining direction relations in actual systems.

Distance relations (e.g., *near, far*) involve distance concepts. For example, two objects are assumed to be *near* if their distance (however distance is defined) is less than a predefined threshold. A form of queries closely related to distance, is the *nearest neighbour queries* (e.g., find the 3 objects closest to a reference object). The previous types of spatial relations have been studied both independently and in conjunction with each other. [Fran92], for instance, proposes a method for qualitative reasoning that combines direction with distance relations.

Recently the interest about spatial relations has shifted towards Spatial Access Methods. The retrieval of spatial relations in existing DBMSs can be accomplished by maintaining traditional indexes (e.g., B^+-trees), or, alternatively, by incorporating Abstract Data Types (ADTs) with specialised indexes defined by external code (e.g., R-trees). Furthermore, when using extended-relational systems, like Postgres [SR86], both indexing methods are available (or easily included) and application developers can decide which is the most appropriate for their application needs. In particular, B^+-trees have been used for the retrieval of direction relations [TPS95], and R-trees and their variations for direction relations [PTS94], topological relations [PTSE95] and nearest neighbour queries [RKV95].

In this paper we treat all queries involving spatial relations between region objects as range queries, and we provide an analysis for their performance. The advantage of treating spatial relations as range queries is that we can use well known results to estimate the expected performance. Our work constitutes the first attempt to model the performance of spatial relations since previous work has focused on *window queries* (retrieval of objects that share common points with a given object or area). The proposed formulas can be used as guidelines by the query optimisers of database systems that support spatial relations, in order to estimate the cost of spatial queries.

The paper is organised as follows: In section 2 we describe a set of "representative" relations and we demonstrate how they can be retrieved using B- and R-trees. In section 3 we provide analytical models that estimate the performance of each method. Section 4 evaluates the models of section 3 using experimental tests, and section 5 concludes the paper.

2. Retrieval of Spatial Relations

In this paper we will focus on the direction relations *east* and *northeast*, the topological relations *meet* and *inside*, and some distance relations. For these relations we provide a brief description, we demonstrate their retrieval using MBRs and we outline implementations in spatial data structures. The extension of the results to other relations is straightforward.

2.1 Spatial Relations

The relation *east*(p,q) means that the x- coordinates of all points of object p (called *primary object*) are larger than or equal to the x- coordinates of all points of object q (called *reference object*). That is, the primary object (p) must be in the grey area of Figure 1a. Similarly, *northeast*(p,q) means that the x- and y- coordinates of all points of object p are larger than or equal to the x- and y- coordinates of all points of object q (Figure 1b). The relations *meet* and *inside* have their usual meaning according to the 4-intersection model (Figure 1c, 1d).

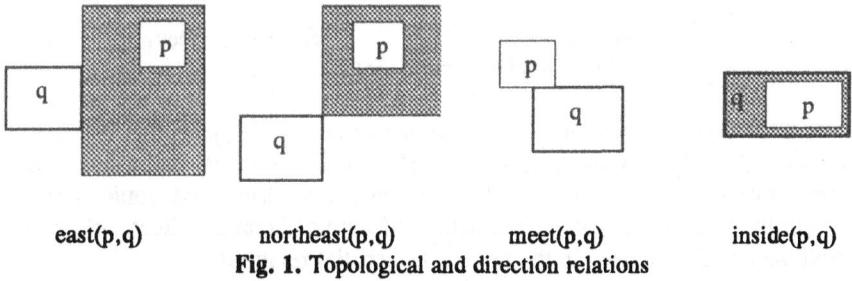

| east(p,q) | northeast(p,q) | meet(p,q) | inside(p,q) |

Fig. 1. Topological and direction relations

In order to define distance relations between objects we will start from distances between points. The distance between two points (*pp_dist*) p_i ($p_i \in p$) and q_j ($q_j \in q$) is defined according to the Euclidean metric: $pp\_dist(p_i,q_j) = \sqrt{(p_{i\_x} - q_{j\_x})^2 + (p_{i\_y} - q_{j\_y})^2}$ (where $p_{i\text{-}x}$ is the x- coordinate of point p_i, $p_{i\text{-}y}$ is the y- coordinate of point p_i, and so on). Using the distance between points, we define the distance between point p_i and object q (*po_dist*) as the minimum distance of p_i from any point of q: $po\_dist(p_i,q)=\min(pp\_dist(p_i,q_j)$, $\forall q_j \in q)$. Finally we define the distance from object p to object q as the maximum of all po_distances from the points of p to q: $oo\_dist(p,q)= \max(po\_dist(p_i,q)$, $\forall p_i \in p)$. Using the above definitions of distances we define the qualitative relation *near* as: $near(p,q,k) \equiv oo\_dist(p,q) \leq k$, that is, *all* points of p must be within k distance from some point of object q (Figure 2a). Similarly, the relation *about*(p,q,k1,k2), where 0<k1<k2, can be defined as: $about(p,q,k1,k2) \equiv near(p,q,k2) \wedge \neg\exists\ p_i$ $(po\_dist(p_i,q) \leq k1)$. That is, according to Figure 2b, all points of p are within distance k1 and k2.

540

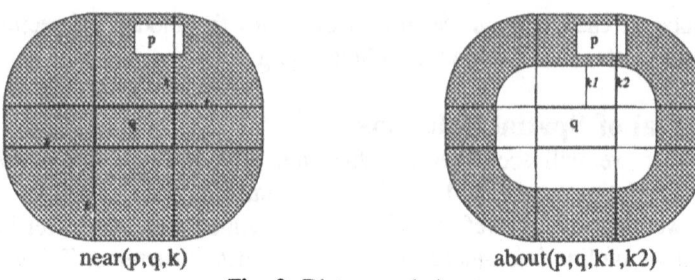

near(p,q,k)   about(p,q,k1,k2)

Fig. 2. Distance relations

We will also consider conjunctions of spatial relations. Figure 3a illustrates a configuration that corresponds to the relation *northeast*(p,q) ∧ *near*(p,q,k), while Figure 3b illustrates a configuration for *east*(p,q) ∧ *meet*(p,q).

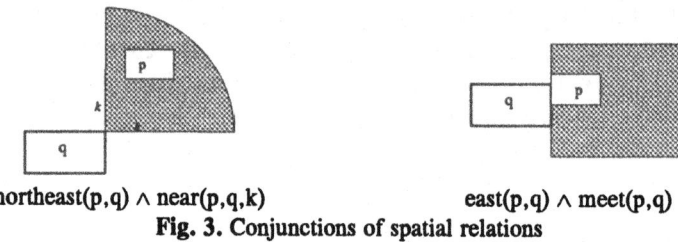

northeast(p,q) ∧ near(p,q,k)   east(p,q) ∧ meet(p,q)

Fig. 3. Conjunctions of spatial relations

Spatial access methods usually store approximations of objects that need only a few points for their representation instead of the objects themselves. Such approximations are used to efficiently retrieve candidates that could satisfy a query. In the next subsection we show how Minimum Bounding Rectangle (MBR) approximations can be used in the retrieval of spatial relations.

2.2 Minimum Bounding Rectangles

MBRs have been used extensively to approximate objects in Spatial Data Structures and Spatial Reasoning because they need only four numbers for their representation; in particular, each object p is represented by the four numbers: p'_{l-x}, p'_{l-y}, p'_{u-x}, p'_{u-y}, where p'_{l-x} stands for the x- coordinate of the lower/left point of MBR p', p'_{l-y} for the y- coordinate of the lower/left point, p'_{u-x} for the x-coordinate of the upper/right point and p'_{u-y} stands for the y- coordinate of the upper/right point.

Since the MBRs are only approximations of the actual objects, the spatial relation between MBRs does not necessarily coincide with the spatial relation between the objects. In most cases the MBRs of objects that satisfy a given relation, should satisfy a number of possible relations with respect to the MBR of the reference object. For example, the MBRs of objects that *meet* a reference object can be related by any topological relation but *disjoint* [PTSE95]. Figure 4 illustrates three configurations where the MBRs satisfy the relations *overlap*, *meet* and *contain* and the actual objects *meet*.

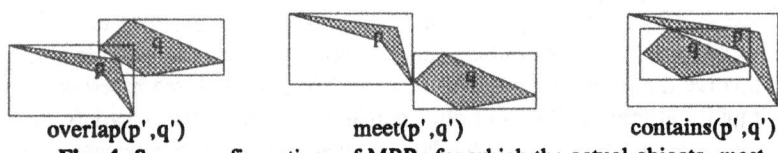

overlap(p',q')   meet(p',q')   contains(p',q')

Fig. 4. Some configurations of MBRs for which the actual objects *meet*

Furthermore, it can be concluded from Figures 1, 2 and 3 that all the relations constrain some or all points of the primary MBR to a *range* (subdivision) of space. In case of *east*, for example, all points of p' must be in the right semi-plane as defined by the vertical line that passes from q's eastern boundary. In some cases all the points of the MBRs to be retrieved must be within the range (e.g., *east*) and this is a sufficient condition for retrieval. In some other cases the range is not a sufficient condition and more specific constraints hold (e.g., *east* ∧ *meet* has the same range as *east* but in this case the MBRs should also *meet*). In general, in order to answer the query "find all objects p that satisfy the relation R with respect to an object q" we have to retrieve all MBRs p' that satisfy certain range constraints with respect to the MBR q' of object q. Table 1 illustrates the mapping from spatial relations R between actual objects to constraints between MBR coordinates. Although the constraints for *east* and *northeast* relations are trivial, the rest require some careful study in order to be clearly understood. For example, the constraints for *meet* state that the two MBRs may not be *disjoint* but permits any other topological relation between them, because any not *disjoint* MBRs may contain objects that *meet* (e.g., Figure 4).

Relation	Constraints on the p'_{l-x}, p'_{l-y}, p'_{u-x}, p'_{u-y} parameters with respect to the reference MBR q'
east(p,q)	$p'_{l-x} \geq q'_{u-x}$
northeast(p,q)	$(p'_{l-y} \geq q'_{u-y}) \wedge (p'_{l-x} \geq q'_{u-x})$
meet(p,q)	$(p'_{l-y} \leq q'_{u-y}) \wedge (p'_{u-y} \geq q'_{l-y}) \wedge (p'_{l-x} \leq q'_{u-x}) \wedge (p'_{u-x} \geq q'_{l-x})$
inside(p,q)	$(q'_{l-y} < p'_{l-y} < q'_{u-y}) \wedge (q'_{l-y} < p'_{u-y} < q'_{u-y}) \wedge$ $(q'_{l-x} < p'_{l-x} < q'_{u-x}) \wedge (q'_{l-x} < p'_{u-x} < q'_{u-x})$
near(p,q,k)	$(q'_{l-y}-k \leq p'_{l-y} \leq q'_{u-y}+k) \wedge (q'_{l-y}-k \leq p'_{u-y} \leq q'_{u-y}+k) \wedge$ $(q'_{l-x}-k \leq p'_{l-x} \leq q'_{u-x}+k) \wedge (q'_{l-x}-k \leq p'_{u-x} \leq q'_{u-x}+k)$
about(p,q,k1,k2)	$(q'_{l-y}-k2 \leq p'_{l-y} \leq q'_{u-y}+k2) \wedge (q'_{l-y}-k2 \leq p'_{u-y} \leq q'_{u-y}+k2) \wedge$ $(q'_{l-x}-k2 \leq p'_{l-x} \leq q'_{u-x}+k2) \wedge (q'_{l-x}-k2 \leq p'_{u-x} \leq q'_{u-x}+k2) \wedge$ $\neg ((p'_{l-y} \leq q'_{u-y}+k1) \wedge (p'_{u-y} \geq q'_{l-y}-k1) \wedge$ $(p'_{l-x} \leq q'_{u-x}+k1) \wedge (p'_{u-x} \geq q'_{l-x}-k1))$
near(p,q,k)∧ northeast(p,q)	$(q'_{u-y} \leq p'_{l-y} \leq q'_{u-y}+k) \wedge (q'_{u-y} \leq p'_{u-y} \leq q'_{u-y}+k) \wedge$ $(q'_{u-x} \leq p'_{l-x} \leq q'_{u-x}+k) \wedge (q'_{u-x} \leq p'_{u-x} \leq q'_{u-x}+k)$
meet(p,q) ∧ east(p,q)	$(p'_{l-x} = q'_{u-x}) \wedge (p'_{l-y} < q'_{u-y}) \wedge (p'_{u-y} > q'_{l-y})$

Table 1. Constraints for the retrieval of spatial relations using MBRs

Because MBRs differ from the actual objects they enclose, they are not always adequate to express the relation between the actual objects. For this reason, spatial queries involve the following two step strategy: First a *filter step* based on MBRs is used to rapidly eliminate MBRs of objects that could not possibly satisfy the query and select a set of potential candidates. Then during a *refinement step* each candidate is examined (by using computational geometry techniques) and false hits are detected and eliminated. The relations *meet* and *inside* require a refinement step [PTSE95], while *east* and *northeast* do not [PTS94], that is, all MBRs enclose objects that satisfy the query.

The distance relations may also need a refinement because near(p',q',k) does not necessarily imply near(p,q,k) and vice versa (the same is true for *about*). If it is not the case that near(p',q',k), but three out of four vertices of the primary MBR p' are within k distance (*po_dist*) from q' (and one vertex in po_dist>k), then p' may enclose an object that satisfies the relation near(p,q,k) (see Figure 5a). Obviously, the same MBR may enclose an object that is not *near* and a refinement step is needed to make the distinction. If however, two vertices of p' are further than k distance, then p cannot be *near* (see Figure 5b). This conclusion is based on the observation that each edge of the MBR coincides with at least one point of the enclosed object. If two vertices are further than distance k, it means that at least a whole edge (and therefore some point of the enclosed object) is further than k. In some cases it is also possible that the entire MBR is within distance k, but the objects are not *near* because some (or all) points of p are further than k from any point of q (Figure 5c). Thus, the refinement step is needed for all MBRs p' for which there exist points that are further than k from any point of q'. Similarly, conjunctions of relations in which one relation needs a refinement step (*northeast* ∧ *near*, *meet* ∧ *east*), also require refinement.

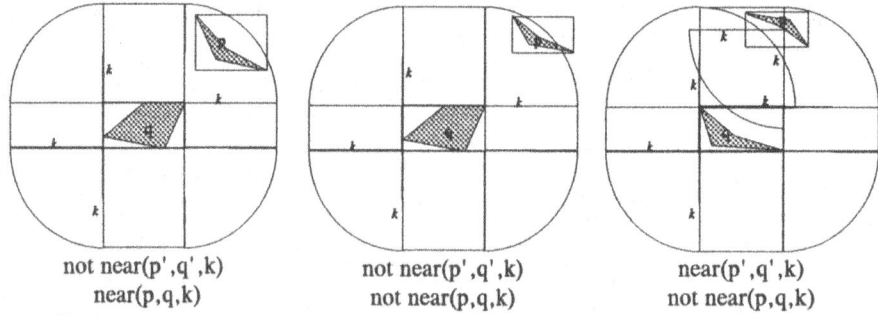

not near(p',q',k)   not near(p',q',k)   near(p',q',k)
near(p,q,k)   not near(p,q,k)   not near(p,q,k)

Fig. 5. Configurations for which a refinement step is needed in the case of *near*

In the rest of the section we will demonstrate how we can use existing data structures to retrieve spatial relations.

2.3 Implementation of Spatial Relations
The first solution for the retrieval of direction relations includes the maintenance of four B-tree indexes (in the rest of the paper we will refer to B⁺-trees using the

general term "B-trees"). Each index corresponds to one of the four numbers: p'_{l-x}, p'_{l-y}, p'_{u-x}, p'_{u-y}. Obviously, some relations imply search on one B-tree while others imply search on more B-trees. For instance, the query "find all objects p that are *east* of object q" is transformed to the constraint $p'_{l-x} \geq q'_{u-x}$ (see Table 1) which is a simple range query in the corresponding B-tree. On the other hand, most queries need to search two or more B-trees and, in a second phase, to compute the intersection of the intermediate answer sets (for details see [TPS95]).

In general, the processing of a query of the form "find all objects p that satisfy a given spatial relation with respect to object q" using B-trees involves the following steps:

Step 1. Depending on the relation to be retrieved, select the B-trees to be searched from the set of four indexes. This procedure involves Table 1.

Step 2. Search each index involved to find the corresponding answer sets.

Step 3. If multiple indexes are involved, find the intersection set. A "realistic" assumption is that this procedure is executed in main memory.

Step 4. If necessary, follow a refinement step for the selected object IDs.

The performance of the retrieval mechanism using B-trees depends significantly on the particular relation because the number of B-trees to be searched relies on the number of constraints that are involved in the definition of the relation. *East*, for example, involves only one constraint $(p'_{l-y} \geq q'_{u-y},)$, while *near* contains four constraints. As it will be shown later this fact significantly affects the efficiency of retrieval.

Another data structure that is efficient for spatial relations is the R-tree. The R-tree [Gutt84] is a height-balanced tree, which consists of intermediate and leaf nodes. The MBRs of the actual data objects are assumed to be stored in the leaf nodes of the tree. Intermediate nodes are built by grouping rectangles at the lower level. An intermediate node is associated with some rectangle which encloses all rectangles that correspond to lower level nodes. Improved variations of R-trees include the R^+-trees [SRF87] and the R^*-trees [BKSS90]. In this paper we use R^*-trees because we found them to have consistently better performance in the retrieval of spatial relations than both R- and R^+- trees.

In order to retrieve objects that satisfy a spatial relation with respect to a reference object we have to specify the MBRs that could enclose such objects using Table 1 and then to search the intermediate nodes that contain these MBRs. For instance, the intermediate nodes P that could contain MBRs p' that satisfy the relation *east* with respect to q' $(p'_{l-x} \geq q'_{u-x})$ should satisfy the constraint $P_{u-x} \geq q'_{u-x}$. Table 2 presents the constraints for the intermediate nodes for each direction relation of Table 1. Notice that the same relation between intermediate nodes and the reference MBR holds for all the levels of the tree structure.

Relation	Constraints for the intermediate Nodes P to be Searched with respect to the reference MBR q'
east(p,q)	$P_{u-x} \geq q'_{u-x}$
northeast(p,q)	$(P_{u-x} \geq q'_{u-x}) \wedge (P_{u-y} \geq q'_{u-y})$
meet(p,q)	$(P_{l-y} \leq q'_{u-y}) \wedge (P_{u-y} \geq q'_{l-y}) \wedge (P_{l-x} \leq q'_{u-x}) \wedge (P_{u-x} \geq q'_{l-x})$
inside(p,q)	$(P_{l-y} < q'_{u-y}) \wedge (P_{u-y} > q'_{l-y}) \wedge (P_{l-x} < q'_{u-x}) \wedge (P_{u-x} > q'_{l-x})$
near(p,q,k)	$(P_{l-y} \leq q'_{u-y}+k) \wedge (P_{u-y} \geq q'_{l-y}-k) \wedge$ $(P_{l-x} \leq q'_{u-x}+k) \wedge (P_{u-x} \geq q'_{l-x}-k)$
about(p,q,k1,k2)	$(P_{l-y} \leq q'_{u-y}+k2) \wedge (P_{u-y} \geq q'_{l-y}-k2) \wedge (P_{l-x} \leq q'_{u-x}+k2) \wedge$ $(P_{u-x} \geq q'_{l-x}-k2) \wedge \neg ((q'_{l-y}-k1 \leq P_{l-y} \leq q'_{u-y}+k1) \wedge$ $(q'_{l-y}-k1 \leq P_{u-y} \leq q'_{u-y}+k1) \wedge (q'_{l-x}-k1 \leq P_{l-x} \leq q'_{u-x}+k1) \wedge$ $(q'_{l-x}-k1 \leq P_{u-x} \leq q'_{u-x}+k1))$
near(p,q,k)$\wedge$ northeast(p,q)	$(q'_{l-y}-k \leq P_{l-y} \leq q'_{u-y}+k) \wedge (P_{u-y} \geq q'_{u-y}) \wedge$ $(q'_{l-x}-k \leq P_{l-x} \leq q'_{u-x}+k) \wedge (P_{u-x} \geq q'_{u-x})$
meet(p,q)$\wedge$east(p,q)	$(P_{l-y} < q'_{u-y}) \wedge (P_{u-y} > q'_{l-y}) \wedge (P_{l-x} \leq q'_{u-x}) \wedge (P_{u-x} > q'_{u-x})$

Table 2 Constraints for intermediate nodes of R-trees

In general, the processing of a query of the form "find all objects p that satisfy a given spatial relation with respect to object q" using R-trees involves the following steps:

Step 1. Starting from the top node, exclude the intermediate nodes P which could not enclose MBRs that satisfy the spatial relation and recursively search the remaining nodes. This procedure involves Table 2.

Step 2. Among the leaf nodes retrieved, select the ones that satisfy the spatial relation. This procedure involves Table 1.

Step 3. If necessary, follow a refinement step for the selected MBRs.

Intuitively, R-trees perform better than B-trees in cases where many constraints are involved in the definition of the direction relation of interest. The next section provides a mathematical analysis that supports this argument.

3. Cost Analysis

In this section we provide analytical formulas that estimate the performance of B-trees and R-trees on the retrieval of spatial relations. Existing formulas for the expected performance of the above structures focus on traditional retrieval (matching queries on B-trees [Yao78, Come79, Bato81] and overlap queries on R-trees [FSR87, PSTW93, FK94]). Our work extends previous work and estimates the expected cost (i.e., number of disk accesses) for the retrieval of several types of spatial relations and combinations. In this discussion we assume that both data and query rectangles are uniformly distributed over the unit square address space.

3.1. Analysis of B-trees

As explained in section 2, searching between one and four B-trees is necessary depending on the relation we want to retrieve. Constraints can be grouped in two categories:

(a) *exact matching* constraints (e.g., $p'_{l-y} = q'_{u-y}$) and

(b) *partial matching* constraints (e.g., $p'_{l-y} > q'_{u-y}$, $q'_{l-y} < p'_{l-y} < q'_{u-y}$) which are characterised by a range r ($0 \leq r \leq 1$).

If we suppose that the data keys are stored in a B-tree index of height h with L leaf nodes then the average cost $C(r)$ for the retrieval of a constraint with range r is [Come79]

$$C(r) = h + r \cdot L \tag{1}$$

It is obvious that the exact matching constraint is a special case of partial matching constraint ($r = 0$). It is also clear that the cost for the retrieval of a relation is the sum of the costs for each constraint involved.

In order to compute the expected cost $C(r)$ for the retrieval of a constraint characterised by a range r we need to provide equations for the parameters of Eq. 1, namely h, L, r. Suppose now that m is the maximum number of entries in a B-tree node, c is the average capacity of a node, and N is the total number of keys stored in the leaf nodes. We have the following equations [Bato81] for the average h and L values in order to use them in Eq. 1:

$$h = 1 + \left\lceil \log_{c \cdot m} \frac{N}{m} \right\rceil \tag{2}$$

$$L = \frac{N}{c \cdot m} \tag{3}$$

What remains in order to have a complete expression for Eq. 1 is the value of parameter r which depends on the particular constraint. If we assume that the size of a data object MBR is $p_x \cdot p_y$ and the size of a query object MBR is $q_x \cdot q_y$ then we can provide in Table 3 the values of parameters r according to possible constraints (the constraints refer only to x- coordinate since constraints for y- coordinate can be expressed in a similar way).

Constraint	Average range r
$p'_{l-x} < q'_{l-x}$ or $p'_{u-x} > q'_{u-x}$	$r = (1 - q_x) / 2$
$q'_{l-x} \leq p'_{l-x} < q'_{u-x}$ or $q'_{l-x} < p'_{u-x} \leq q'_{u-x}$	$r = q_x$
$p'_{l-x} > q'_{u-x}$ or $p'_{u-x} < q'_{l-x}$	$r = (1 - (2 \cdot p_x + q_x)) / 2$
$p'_{l-x} < q'_{u-x}$ or $p'_{u-x} > q'_{l-x}$	$r = (1 + q_x) / 2$

Table 3 Average values for range r of a constraint

Using information from Table 3 and Eq. 2 and 3 we can estimate the expected cost for each constraint (see Eq. 1). Summing up, the expected cost $C(R,k)$ for the retrieval of a spatial relation R with k constraints is:

$$C(R,k) = \sum_{i=1}^{k} C(r_i) = \sum_{i=1}^{k} (h + r_i \cdot L) \tag{4}$$

3.2. Analysis of R-trees

Most of the work in the literature has dealt with the expected performance of R-trees for processing overlap queries i.e., the retrieval of data objects p that share common area with a query window q. More particularly, let N be the total number of data objects indexed in a R-tree, h be the height of the tree, c the average node capacity at every level of the tree and m the maximum number of entries in a node. If we assume that the average node size at level j is $n_{j,x} \cdot n_{j,y}$ (the root is assumed at level $j=h$ and the leaf-nodes at level $j=1$) and the average size of a query object MBR is $q_x \cdot q_y$ then the expected retrieval cost (number of disk accesses) of an overlap query using R-trees is [PSTW93, FK94]

$$C(q_x,q_y)=\sum_{j=1}^{h}\left\{\frac{N}{(c\cdot m)^j}\cdot\left(n_{j,x}+q_x\right)\cdot\left(n_{j,y}+q_y\right)\right\} \qquad (5)$$

The expression for computing the height h of the R-tree is similar to that of B-trees (Eq. 2). If we name N_j the number of nodes at level j, and d_j the density (i.e., the sum of the nodes' areas divided by the global area) of these nodes then the average node sizes $n_{j,x}$ and $n_{j,y}$ are given by the following equations [TS95]:

$$n_{j,x} = n_{j,y} = \left(\frac{d_j}{N_j}\right)^{\frac{1}{2}} \qquad (6)$$

where

$$d_j = \left\{1+\frac{(d_{j-1})^{\frac{1}{2}}-1}{(c\cdot m)^{\frac{1}{2}}}\right\}^2 \qquad (7)$$

and

$$N_j = \frac{N_{j-1}}{c\cdot m} \qquad (8)$$

Therefore, d_j and N_j can be computed recursively using d_0 and N_0 which denote the density d and the amount N respectively of the data object MBRs. Qualitatively, this means that we can estimate the retrieval cost of a window query just with the knowledge of the data set and the query window.

Since Eq. 5 expresses the expected performance of R-trees on overlap queries using a query window q, in order to estimate the retrieval cost of a spatial relation $R(p,q)$ we need the following transformation: $R(p,q) \Rightarrow overlap(p',Q)$. In other words, the retrieval of a spatial relation using R-trees is equivalent (in terms of cost) to the retrieval of an overlap query using an appropriate query window Q. The necessary transformation Q for each spatial relation R should take into consideration the corresponding constraint of the intermediate nodes because only these nodes are important when estimating the retrieval cost [PTSE95].

For the spatial relations that we consider in this paper, the appropriate query windows Q are illustrated in Figure 6. Each query window is an appropriate transformation of the corresponding constraints presented in Table 2. Notice that *meet* and *inside* correspond to the same query window Q. This is a property that can be extracted by examining the constraints of Table 2. The same property holds in our tests for *near* and *about* (if k=k2) because, according to Table 2, the constraints for the intermediate nodes to be searched for about(p,q,k1,k2) are identical to the ones for near(p,q,k2) plus some extra constraints which can be evaluated during the refinement step.

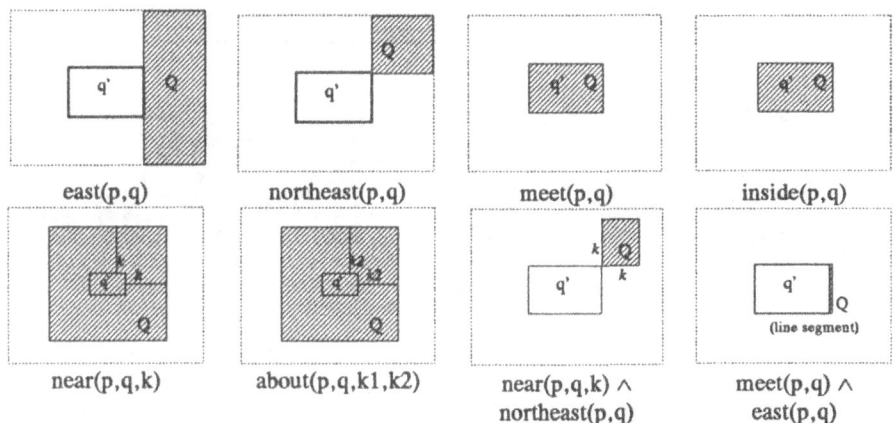

Fig. 6. Query windows for the estimation of the retrieval cost

4. Experimental Results

In order to evaluate the quality of the proposed analytical approximations, we compare the expected and the experimental cost for the retrieval of several representative spatial relations using both the B-tree and the R-tree indexing mechanisms. For the experimental tests we used several *data files* that contained 10,000 data object MBRs with small up to large data sizes (p_x and p_y ranged from 0.5% up to 5% of global side size respectively). The sizes of the reference object MBRs used for the retrieval of spatial relations were equal to the corresponding data object MBRs (i.e., $p_x=p_y=q_x=q_y$). For the distance relations we set $k=3 \cdot q_x$ (*near*), $k1=q_x$ and $k2=3 \cdot q_x$ (*about*).

The expected cost using B-trees was computed by using Eq. 4, information from Tables 1 and 3, and the following typical values:
- average capacity at leaf nodes $c = 0.67$,
- maximum number of entries in a node $m = 126$ (1 page of 1024 bytes includes 126 keys · 4 bytes + 127 pointers · 4 bytes + 12 bytes node-overhead),
- total number of keys $N = 10,000$.

On the other hand, the expected cost using R-trees was computed by using Eq. 5, information from Figure 6 and the following typical values:
- average node capacity $c = 0.67$,
- maximum number of entries in a node $m = 50$ (1 page of 1024 bytes includes 50 entries · 4 values per entry · 4 bytes + 50 pointers · 4 bytes + 24 bytes node-overhead),
- total number of data $N = 10,000$.

The results for each spatial relation are illustrated in Figure 7. The experimental results are illustrated with columns and the analytical results with lines.

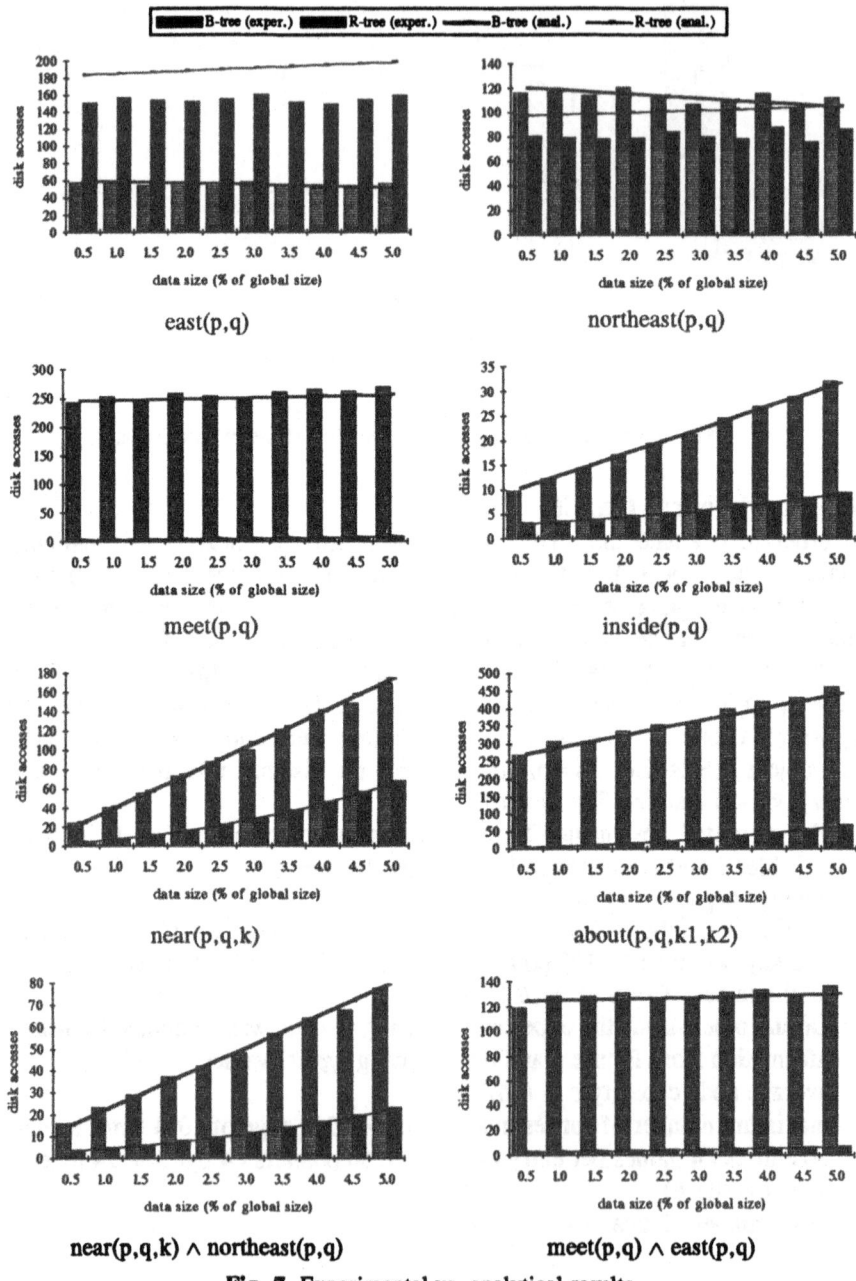

Fig. 7. Experimental vs. analytical results

The common observation in all the graphs of Figure 7 is that the analytical estimate is very close to the experimental results. With the exception of the R-tree estimation for *east* and *northeast*, the relative error is usually below 5%. The weakness of the R-tree model for *east* and *northeast* is, more or less, expected because of the very large window queries Q (see Figure 6) that make R-trees an unstable index for these relations. On the other hand, the range of a constraint in a B-tree query does not affect the B-tree retrieval mechanism and, therefore, the B-tree estimate is always very close to the experimental results.

The comparison of the B- and R- tree indexing mechanism depends on two factors: the number of constraints involved (for B-trees) and the size of the query window Q (for R-trees). For example, B-trees perform better than R-trees when one constraint and large Q are involved (*east*). When the opposite happens, R-trees perform better than B-trees. Following these guidelines, a spatial query optimiser can predict the efficiency of the one or the other indexing mechanism on the support of several spatial queries.

According to our experimental tests, the analytical models of section 3 are proved to be efficient for the estimation of the particular spatial queries that we have implemented using transformation to range queries. Following the same procedure, the performance of other spatial queries can also be estimated with similar accuracy.

5. Conclusion

Relations in space are becoming an important aspect of access methods as a result of the increasing interest on qualitative spatial information processing. In this paper we focus on the retrieval of spatial relations using classic alphanumeric (B-trees) and spatial (R-trees) indexing methods. First we transform queries involving spatial relations into range queries, then we provide analytical formulas for their expected performance (extending previous work on analysis of indexing methods), and finally we evaluate the analytical model. In most cases we found the analytical estimate almost identical to the actual results, a fact that leads to the conclusion that the derived formulas can be used successfully in query optimisers of Geographic Databases in order to estimate the cost of spatial queries.

Although we have worked with a small set of representative topological, direction and distance relations, the results are directly applicable to other spatial relations and combinations. Future work can be done:
(a) to apply the analytical models on specialised spatial relations such as nearest-neighbour or furthest-neighbour and combinations,
(b) to provide analytical models for other spatial indexing methods, such as Grid files [NHS84] or K-D-B-trees [Robi81], in order to evaluate their efficiency with respect to the particular spatial relations and
(c) to derive analytical formulas assuming a general (non-uniform) distribution of objects over the work space in order to efficiently support any kind of spatial information.

Acknowledgements

We thank Emmanuel Stefanakis for implementing and testing B-trees. We also thank Timos Sellis for providing useful comments.

References

[Bato81]   Batory, D.S., "B+ Trees and Indexed Sequential Files: A Performance Comparison", In the Proceedings of ACM SIGMOD Conference, 1981.

[BKSS90]   Beckmann, N., Kriegel, H.P. Schneider, R., Seeger, B., "The R*-tree: an Efficient and Robust Access Method for Points and Rectangles", In the Proceedings of ACM SIGMOD Conference, 1990.

[CSE94]   Clementini, E., Sharma, J., Egenhofer, M., "Modeling Topological Spatial Relations: Strategies for Query Processing", International Journal of Computer and Graphics, 18(6), 815-822.

[Come79]   Comer, D., "The Ubiquitous B-Tree", ACM Computing Surveys, Vol. 11(2), pp. 121-137, 1979.

[Egen91]   Egenhofer, M., "Reasoning about Binary Topological Relations", In the Proceedings of the Second Symposium on the Design and Implementation of Large Spatial Databases (SSD), Springer Verlag LNCS, 1991.

[EF91]   Egenhofer, M., Franzosa R., "Point-Set Topological Spatial Relations", International Journal of Geographic Information Systems, Vol 5(2), pp. 160-174, 1991.

[FSR87]   Faloutsos, C., Sellis, T., Roussopoulos, N., "Analysis of Object Oriented Spatial Access Methods", In the Proceedings of ACM SIGMOD Conference, 1987.

[FK94]   Faloutsos, C., Kamel, I., "Beyond Uniformity and Independence: Analysis of R-trees Using the Concept of Fractal Dimension", In the Proceedings of the 13th ACM PODS Symposium, 1994.

[Fran92]   Frank, A.U., "Qualitative Spatial Reasoning about Distances and Directions in Geographic Space", Journal of Visual Languages and Computing, Vol. 3, pp. 343-371, 1992.

[Gutt84]   Guttman, A., "R-trees: A Dynamic Index Structure for Spatial Searching", In the Proceedings of ACM SIGMOD Conference, 1984.

[Knut73]   Knuth, D., "The Art of Computer Programming, vol.3: Sorting and Searching", Addison-Wesley, 1973.

[ME94]   Mark, D., Egenhofer, M., "Calibrating the Meaning of Spatial Predicates from Natural Language: Line Region Relations", In the Proceedings of the 6th International Symposium on Spatial Data Handling (SDH), Taylor Francis, 1994.

[NHS84]   Nievergelt, J., Hinterberger, H., Sevcik, K.C., "The Grid File: An Adaptable, Symmetric Multikey file Structure", ACM Transactions on Database Systems, Vol 9(1), pp. 38-71, 1984.

[PSTW93]   Pagel, B., Six, H., Toben, H., Widmayer, P., "Towards an Analysis of Range Query Performance", In the Proceedings of the 12th ACM PODS Symposium, 1993.

[PFK94]   Papadias, D., Frank, A.U., Koubarakis, M., "Constraint-Based Reasoning in Geographic Databases: The Case of Symbolic Arrays", In the Proceedings of the 2nd ICLP Workshop on Deductive Databases, 1994.

[PS94]   Papadias, D., Sellis, T., " Qualitative Representation of Spatial Knowledge in two-dimensional Space", Very Large Data Bases Journal, Special Issue on Spatial Databases, Vol 3(4), pp. 479-516, 1994.

[PTS94]   Papadias, D., Theodoridis, Y., Sellis, T., "The Retrieval of Direction Relations Using R-trees", In the Proceedings of the 5th Conference on Database and Expert Systems Applications (DEXA), Springer Verlag LNCS, 1994.

[PS95]   Papadias, D., Sellis, T., "A Pictorial Query-by-Example Language", Journal of Visual Languages and Computing, Special Issue on Visual Query Systems, 6(1), pp 53-72, 1995.

[PTSE95]   Papadias, D., Theodoridis, Y., Sellis, T., Egenhofer, M., "Topological Relations in the World of Minimum Bounding Rectangles: a Study with R-trees", In the Proceedings of ACM SIGMOD Conference, 1995.

[Robi81]   Robinson, J.T., "The K-D-B-Tree: A Search Structure for Large Multidimensional Dynamic Indexes", In the Proceedings of ACM SIGMOD Conference, 1981.

[RKV95]   Roussopoulos, N., Kelley, F., Vincent, F., "Nearest Neighbor Queries", In the Proceedings of ACM SIGMOD Conference, 1995.

[SRF87]   Sellis, T., Roussopoulos, N., Faloutsos, C., "The R^+-tree: A Dynamic Index for Multi-Dimensional Objects", In the Proceedings of the 13th Very Large Data Bases Conference, 1987.

[SR86]   Stonebraker, M., Rowe, L., "The Design of Postgres", In the Proceedings of ACM SIGMOD Conference, 1986.

[SYH94]   Sistla, P., Yu, C., Haddad, R., "Reasoning about Spatial Relationships in Picture Retrieval Systems", In the Proceedings of the 20th Very Large Data Bases Conference, 1994.

[Topa95]   Topaloglou, T., "Spatial Databases with Partial Information: Representation and Reasoning", Forthcoming Ph.D Thesis, University of Toronto, Canada, 1995.

[TPS95]   Theodoridis, Y., Papadias, D., Stefanakis, E., "Supporting Direction Relations in Spatial Database Systems", Technical Report, KDBSLAB-TR-95-02, National Technical University of Athens, Athens, Greece, 1995.

[TS95]   Theodoridis, Y., Sellis, T., "Indexing Point and Non-point Spatial Data: A Performance Analysis", Technical Report, KDBSLAB-TR-95-03, National Technical University of Athens, Athens, Greece, 1995.

[Yao78]   Yao, A.C., "On Random 2-3 Trees", Acta Informatica, Vol. 9(2), pp. 159-168, 1978.

[PSZ1] Papadias, D., Sellis, T., " Qualitative Representation of Spatial Knowledge in two-dimensional Space," Very Large Data Base Journal, Special Issue on Spatial Databases, Vol 3(4), pp. 479-516, 1994.

[PTSM] Papadias, D., Theodoridis, Y., Sellis, T., "The Retrieval of Direction Relations Using R-trees", In the Proceedings of the 5th Conference on Database and Expert Systems Applications (DEXA), Springer Verlag, 1994.

Evaluating and Refining
Computational Models of Spatial Relations
Through Cross-Linguistic Human-Subjects Testing

David M. Mark[1], David Comas[2], Max J. Egenhofer[3],
Scott M. Freundschuh[4], Michael D. Gould[5], Joan Nunes[6]

Abstract. Human-subjects testing based on a sound formal model is outlined as an effective way to evaluate and refine computational models of spatial relations. The cognitive response of a subject (person) to a real-world situation depends on the characteristic of the situation and the characteristics of the subject. Spatial entities can be characterized by topological relations, metric, scale (scope), kinds of phenomena (semantics), motion (time), and (for maps and displays) graphic presentation. Subjects' responses may vary according to their natural languages, their cultures, and individual characteristics such as gender, age, and handedness. Given this conceptual background, seven human-subjects testing protocols are introduced. Each test is described, and results of applications of these tests are summarized where possible. The tests are compared with their requirements, and the different aspects of human spatial cognition that they might test. Lastly, a program for applying these tests and refining the formal models based on test results is presented.

KEYWORDS: Spatial relations, spatial language, spatial cognition, formal models, human-subjects testing, geographic information systems, GIS.

---

[1]   National Center for Geographic Information and Analysis, Department of Geography, State University of New York at Buffalo, Buffalo, NY 14261-0023, U.S.A. Email: dmark@geog.buffalo.edu.

[2]   GIS Centre, Universitat de Girona, Pl. Ferrater Mora, 1, 17071 Girona, Spain. Email: comas@skywalker.udg.es.

[3]   National Center for Geographic Information and Analysis, Department of Spatial Information Science and Engineering, and Department of Computer Science, University of Maine, Orono, ME 04469-5711, U.S.A.   Email: max@mecan1.maine.edu.

[4]   Department of Geography, University of Minnesota-Duluth, Duluth, Minnesota 55812, U.S.A. Email: sfreunds@d.umn.edu.

[5]   Departamento de Geografía y Ordenación del Territorio, Universidad de Extremadura, Avda. Quijotes, s/n, E-10071 Cáceres, Spain.   Email: ghhum02@sis.ucm.es.

[6]   Departament de Geografia, Edifici B, Universitat Autònoma de Barcelona, 08193 Bellaterra, Spain. Email: ILGE5@cc.uab.es.

1. Introduction

Development and enhancement of a general theory of spatial relations are critical to several fields. Computational models of spatial relations are essential for the design of geographic information systems (GISs). Spatial relations are used internally by GISs in various overlay procedures. They also may be provided to end-users of the GIS as spatial predicates to constrain database retrievals and logical selections. But a general theory of spatial relations is also needed in cognitive science, especially in cognitive and computational linguistics. Spatial language appears in a wide range of linguistic areas, and is also a source domain for metaphors that people use to structure and talk about more abstract domains. Also, reasoning about spatial relations is very important in robotics and in computational vision. And lastly, a general theory of spatial relations would be of central interest to geography itself, for what it reveals about the nature of geographic space and spatial relations.

There has been much progress toward theories of spatial relations, notably the works of Talmy (1983) and Herskovits (1986) in linguistics and cognitive science, of Cohn and his colleagues in artificial intelligence (Randell *et al.*, 1992; Cohn *et al.*, 1994; Cohn and Gotts, 1995), and of Egenhofer and his colleagues in geographic information science (Egenhofer and Franzosa, 1991; Egenhofer and Herring, 1994; Egenhofer and Mark, 1995). However, all the above work, and other related work in linguistics, artificial intelligence, mathematics, and GIS has been based on the intuition of the researchers. A model is accepted and published if it is internally consistent, intuitive, and makes plausible predictions or distinctions about spatial relations.

Recently, some of us have felt the need to conduct human-subjects experiments to confirm these formal models of spatial relations, or in some cases, to refine them. The resulting research approach seems powerful. The formal model is used to define things that can be distinguished in objective descriptions of situations, and experimental stimuli are designed to span the full range of situations that the formal model can distinguish. Test results may show that all of those distinctions are actually made by subjects, or that only some are made, or that some are made more frequently than others. What is learned can be fed back into the further development of the formal models. Basically, we believe that a person's reaction to a situation can be predicted in part from a formal description of a situation, and in part from characteristics of the subject (Figure 1). Our goal is to build such a predictive model, or suite of models, for spatial relations, with emphasis on relations in space of a geographic scale, and for how the relations are described or expressed in natural language.

If this research project is successful, the results would provide the basis for computer programs that could solve such problems as these:

- take an objective description of the geometry and attributes of two or more spatial entities, and the general characteristics of some person who is expected to understand the sentence, and determine an appropriate natural-language sentence or phrase that describes the spatial relation between the entities.

- take any sentence involving locatives, any objective description of a situation involving spatial relations, and any general class of person, and determine how well the sentence describes the situation.

The computational ability to do the above would in turn allow, for example, retrieval from a spatial database of all those examples that fit a given sentence up to some specified degree. And, in the grand scheme of this project, this will be possible for all the world's natural languages.

In this paper, we present in some detail an outline of the factors that may be needed to describe spatial entities and their spatial relationships, using formal models that have sufficient expressive power to encompass everything that we need for building the models described above. Then, we will outline some characteristics of human subjects and eventual computer users, that also will have to be modeled. And third, we present a portfolio of human-subjects tests that we are performing, to provide the evidence to confirm, refine, and perhaps extend the formal models of spatial relations already developed.

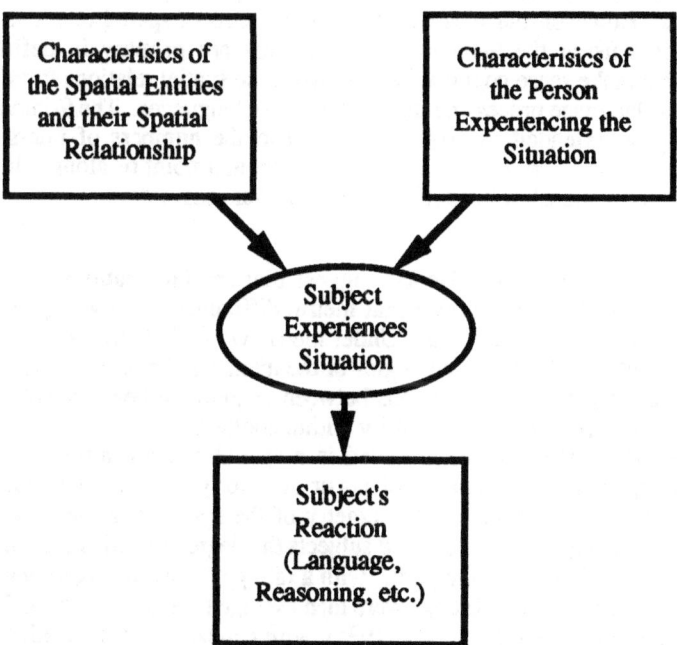

Figure 1: Diagram representing the framework for this research.

2. Characterizing Spatial Entities and Their Relationships

2.1. Topology

The term topology refers to properties that are invariant under continuous transformations such as rotation, re-scaling, and stretching. Containment and coincidence are two of the more basic topological relations. In the research described in this paper, we define topological spatial relations as described by Egenhofer and Franzosa (1991) and Egenhofer and Herring (1994). Specifically, we use the 9-intersection model (Egenhofer and Herring, 1994), in which each entity is considered to have an interior, a boundary, and an exterior. In its simple form, the 9-intersection considers which of the three parts, for one entity intersect which parts of the other. Each of these 9 intersections can be either empty or non-empty. This would appear to lead to 512 distinct relations, but if each entity is simply connected, most combinations of part intersections are impossible. Egenhofer and Herring presented proofs on which combinations are impossible: the 9-intersection can distinguish in R^2 only eight spatial relations between regions in the plane, but 19 relations between an unbranched line and a region, and 33 between two unbranched lines.

The testing protocols described in this paper are not bound to the 9-intersection, but could be applied to stimuli chosen according to other models of topological differences. However, our thesis is that those aspects of spatial relations that can be differentiated using the 9-intersection topology are primary in defining spatial relations that make sense cognitively, and that these spatial relations form the basis of both natural-language processing and spatial query languages. The 9-intersection also can readily be extended by paying attention to the numbers of intersections, the dimensions of the intersections (e.g., intersection at a point or along a line), and the directions (orientations) of lines (Egenhofer and Franzosa, 1995).

2.2. Metric

Whereas both theory and evidence suggest that spatial relations will be defined primarily by topology, we expect that metric differences also will play a role, and may dominate in some situations. Under metric we include the shapes and relative sizes of the entities, the sizes and angles of the intersections or of parts formed by the intersections, distances and directions between disjoint entities, relative and absolute orientations, and positions of one entity within another.

The shape of both entities involved in a spatial relationship can influence the relationship as judged by human subjects or even for purely computational purposes. More important may be the exact geometry of the spatial relation between the two features. For example, we have asked subjects the degree to which they agreed that in each case, "the road crosses the park," with a rating of 5 indicating strong agreement, and 1 strong disagreement. Ratings were then averaged across subjects and re-scaled to a 0 to 1 scale (Mark and Egenhofer, 1995), with a mean of 1.0 indicating unanimous strong agreement, and 0.0 unanimous strong disagreement. Six of the 60 stimuli that we have presented to about 120 English-speaking subjects were topologically-identical according to the 9-intersection model, but showed metric differences (Table 1). The three cases in which the road traversed the park between two opposing sides had mean agreement ratings of 0.88 and 0.91 that the road did "cross" the park, whereas for other stimuli, topologically identical to these, mean agreement ratings were as low as 0.28.

Table 1: Mean Agreement that "The Road Crosses the Park"
for Six Topologically-Identical Stimuli Over a Range of
Geometries.

Stimulus	Agreement ("crosses")
	0.914
	0.886
	0.882
	0.781
	0.367
	0.285

The orientation of the Figure with respect to the Ground also may be important in defining spatial relations. The best example of "cross" may have a line running perpendicular to the long axis of a region that the line crosses, whereas a line that parallels the long axis may not fit the concept of "crosses" quite as well. Also, in testing based on diagrams, the orientation of the Figure and Ground with respect to the testing medium, be it a piece of paper or a computer screen may effect results, since a line that runs across the medium may have a stronger rating for "crosses" than one running up and down the page, other things being equal (Wagner, 1993). Geographic reference frames (north-south, east-west, etc.) may have similar effects, although this conjecture is completely untested.

2.3. Scale/Scope

There is considerable evidence that people reason about small spaces such as table-tops, and the manipulable objects in them, differently than they reason about geographic spaces. Mark and Freundschuh (1995) recently reviewed the evidence that there are different kinds of cognitive spaces, based on the way people interact with them, but defined largely by scale (size, scope). Manipulable spaces are populated with three-dimensional entities that move without changing shape or other properties. Entities mostly have physical existence, and thus two things cannot occupy the same space. Geographic spaces, on the other hand, are mostly two-dimensional, and entities in them are viewed in a more abstract way, so that the same place can be occupied by, say, a province and a mountain at the same time.

Talmy (1983) has noted that almost all spatial relation terms in all languages apply across all scales. This is almost certainly true for the lexical elements, but the meanings of these words or case structures or whatever may change, both formally and informally, with scale or scope.

Testing using maps is somewhat problematic in this regard, since the map itself and its elements are clearly manipulable, but the map normally represents a geographic space and geographic entities. If there is a difference in the details of the meanings of spatial relations with scale, then results of map-based tests might relate to either space, or a mixture. In the long term, the research program must include experiments in which stimuli with identical geometries and topological relations, but where some subjects get them labeled as maps, and others get them labeled as diagrams or table-top-objects.

2.4. Kinds of Features and Phenomena

We believe that the nature of the phenomena involved can have an influence on how the spatial relations are perceived and characterized. We do not have experimental evidence to support this conjecture, as all experiments performed so far have involved a "road" and a "park." Furthermore, our conjecture conflicts with a generalization presented by Talmy (1983), who claimed that spatial relations in natural language tend to be both magnitude (size) neutral and material neutral. An example, based on introspection, was presented by Mark and Egenhofer (1994a):

> [T]he following pair of sentences would seem to mean about the same:
> The road goes through the park.
> The road goes across the park.
> The near-identity of "through" and "across" also seems to hold if we are talking about a man walking though or across a field. There may be a slight shade of difference in the meanings of the next pair of sentences, but again, the meanings seem very similar:
> The highway goes through Boston.
> The highway goes across Boston.
> But now consider this pair of sentences:
> When we went to Seattle, we drove through Canada.
> When we went to Seattle, we drove across Canada.
> The first sentence would be true if any part of the route is in Canada. It would, for example, be reasonable to say that if we had driven from Boston to Buffalo, then cut through southern Ontario,

and then driven on U.S. roads from Detroit to Seattle. However, the second sentence implies that most of the route was in Canada, that is, from one "side" of Canada to the other.

We also expect that the nature of the boundaries of regions may play a role in the grouping or similarity of spatial relations. Recently, interest has focused on the fact that many geographic entities have indistinct boundaries (Frank, 1995; Burrough and Frank, 1995). Experiments performed so far, with a "road" and a "park," involved entities with crisp or distinct boundaries, and data suggest that subjects often made distinctions based on exact details of whether, say, the end of the road (line) fell precisely on the boundary of the park (region). Although we have not yet tested other phenomena, we suspect that if boundaries are thought by the subjects to be uncertain or imprecise, then the exact relations between two boundaries may not be of high salience.

2.5. Motion (Dynamic/Static)

Some languages make a major distinction between static and dynamic spatial relations. For example, Korean typically expresses spatial relation through trajectory verbs, rather than through static locative particles (Bowerman, 1993). Pilot data in this research have shown small differences in subjects' responses to "the road crosses the park" and "the road goes across the park," with mean agreement by subjects being weaker for the second sentence in the case of 48 of 60 stimuli, although differences were statistically significant (at the 5 percent level) for only three of the 60 stimuli (exactly five percent of them). Such a grammatical distinction may make a more substantial difference in some languages.

2.6. Presentation

It is fairly common to perform tests using maps or map-like drawings as stimuli, and then to draw conclusions about geographic terminology from the results. This may work, reasonably well, most of the time, but it must be remembered that experiments conducted in this way really only reveal things about how people talk about maps and the spatial relations shown on them. In fact, the situation may be worse than that, since most tests of spatial relations, spatial knowledge acquisition, or navigation have used as stimuli schematic drawings that do not look much like cartographic maps. It seems possible that subjects may lose track of the map metaphor during the task, and end up making judgments about points, lines, and polygons. For example, Mark and Egenhofer (1994b) showed cards with a closed polygon (shaded) and a solid line on each, saying that these represented a road and a state park. The subjects grouped the cards and then described the spatial relation in each group. In the 97 descriptions composed by 11 English-speaking subjects, 93 did use the terms "road" and "park" for the line and region, respectively. However, in the remaining six descriptions, coming from two of the subjects, the "road-park" (map) context broke down, and the subjects referred to the entities as the "line" or the "area." Thus, issues of the medium of presentation, the style or graphic design of the stimuli, and the presence or absence of context or base-map information, all are important variables in research designs in this area.

3. Characteristics of Subjects that may Influence Their Reactions to Spatial Relationships

As noted in the introduction to this paper, objective formal description of a spatial relation is only one side of the equation. A subject's reaction to a situation, and by inference, reactions of people to spatial relations in everyday life, also is influenced by the characteristics of the person. At present, we divide the individual characteristics into language spoken, culture, and individual characteristics.

3.1. Language Spoken

Language spoken has an obvious surface effect, but whether language has an effect on spatial reasoning and judgments is less certain and more controversial (see Montello, 1995). Benjamin Lee Whorf and Edward Sapir proposed that people think in language, and that speakers of different languages think and reason differently (Whorf, 1940; for a recent discussion in a geographic context, see Mark, 1993). This conjecture, known now as the Sapir-Whorf hypothesis is very controversial, because if such differences do exist, then it could be argued that speakers of one language are better equipped to think about certain topics than are speakers of some other language. Whereas there do seem to be differences in how languages categorize geographic features (Mark 1993), it is not at all clear whether such effects would extend to the definitions and groupings of spatial relations. Much data, using a variety of experimental protocols, and for many languages from as many language families as possible, will be needed before such effects can be confirmed or laid to rest.

3.2. Culture

"It is impossible to comprehend a culture without taking into account its language, probably its single most important element. It is also impossible to completely understand a language independent of its cultural context. ... But despite the many ways in which culture and language influence each other, their integration is not absolute. Each has many properties uniquely its own that are not directly, or even indirectly, influenced by the other." (Howard, 1989, p. 77)

Oft-cited examples of cultural influences on language are the "Eskimo words for snow" idea, in which the Inuit are claimed to have many words for kinds of snow, and no single word for everything that an English-speaker would consider to be snow. Whereas the facts regarding snow and the language of the Inuit seem unclear, the idea that some cultures have more detailed vocabularies than others seems to be true at least to some degree. The skiing "sub-culture" or "speech community" within North American English also has many words for snow, some of which might not be familiar to some non-skiing native speakers of North American English from snow-free localities. Whether there is any analogous variation in words for spatial relations is not known at this time, but is something to be looked for in interpretations of test results. We must be careful not to generalize about English speakers from undergraduate student subjects in Buffalo, New York, or about Spanish from office workers in San José, Costa Rica.

3.3.   Individual Characteristics

Sex (gender) appears to be correlated with both verbal and spatial abilities (see Halpern, 1986, for an excellent summary). Since the tasks outlined in this paper involve judgments of spatial relations, or sentences, or both, we expect that both verbal and spatial abilities may influence results, and thus a gender effect is possible. We have asked all subjects in all the experiments so far to report their age, sex, and native (first) language spoken, although we do not expect significant sex-related differences in the results of these particular experiments. Direct testing of subjects' spatial and verbal abilities would be desirable, to see if that correlates with any aspects of their responses. Since Halpern (1986) noted significant interactions between cognitive abilities, sex, and handedness, it might also be worthwhile to record the handedness of subjects, although we have not done that so far.

4.   Human-subjects Experiments for Assessing Spatial Relations

We have designed seven human-subjects protocols for exploring, evaluating, or refining computational models of spatial relations in natural language. All of these have been approved by the Human Subjects Review Board at the State University of New York at Buffalo. We have decided to use a road and a park as the entities in the spatial relationship for all of the testing in the early phase of our research, although eventually will extend the research to other kinds of geographic features. As illustrated in Figure 2, four tasks require road-park drawings, and two tasks produce such drawings. Also, two tasks require road-park spatial relation sentences, and three others produce them. We have developed a standard set of 60 road-park drawings for use with the verbal description (4), grouping (5) and agreement (7) tasks; these have been printed in an appendix to a technical report (Egenhofer et al., 1994), and can be made available in electronic form to interested researchers.

4.1.   Verbal Examples Task

In the Verbal Examples Task, subjects are asked to list as many different road-park sentences as they can, in their native language. Sentence frequency will be counted by subject, and sentence sequences can be studied to see whether subjects list approximately synonymous sentences consecutively, or use some other recall strategy.

4.2.   Graphic Examples Task

This task is very similar to the previous one, except that subjects are asked to draw examples of as many different spatial relations as they can. Results will be classified topologically according to the distinctions of the 9-intersection model, and metric properties of the examples might also be characterized.

4.3.   Graphic Comparison Task

This protocol presents road-park drawings three at a time, and asks subjects to judge whether the middle example of the three is more similar to the one to the left, or to the one to the right. This is a "forced choice" task, and we expect results to indicate the relative importance of topology and metric. Since both this and the previous task do not involve language explicitly, it will be very interesting to see whether there are any systematic differences in responses for speakers of different languages. The stimuli that we will use first involve relations between rectangles and straight lines.

Some subjects will be judging geometric figures with no physical interpretation, whereas others will be told that the line represents a road and the rectangle represents a park. Any difference between those two conditions would tentatively be attributed to the semantics of the phenomena involved.

4.4. Verbal Description Task

The Verbal Description Task asks subjects to write a sentence that describes the spatial relation shown in each of the 60 diagrams mentioned above. In future, we may design other sets of diagrams.

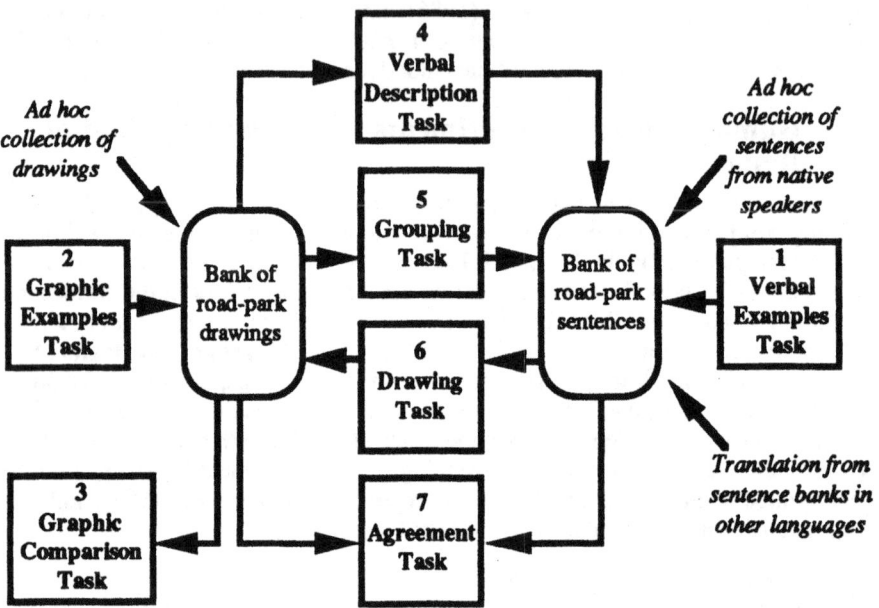

Figure 2: Information requirements and products from the seven experimental protocols defined thus far.

4.5. Grouping Task

In the Grouping Task, the 60 drawings are printed on individual cards, and shown to subjects, who are asked to arrange the drawings into groups such that a single natural-language phrase in their native language can describe each drawing in the group. To the maximum extent possible, the instructions are given and responses recorded in the native language of the subjects. In English, the instructions are:

> Here are 60 different sketches of a road and a State Park. Please arrange the sketches into several groups, such that you would use the same verbal description for the spatial relationship between the road and the Park in each case.

When the subject completes the task, the experimenter records the groups, and elicits a descriptive phrase for the spatial relation in each group. Lastly, the subject is asked to pick a "best example" from the stimuli in each group.

4.6. Drawing Task

The Drawing Task presents subjects with outlines of a park, with a sentence under each drawing. In English, the instructions are:

> On each of the following diagrams, the shaded polygon represents a state park. Please draw a line on each diagram to represent a road that has the spatial relationship to the park that is described by the sentence under the diagram. Try to draw a road that makes the diagram a "best example" of the relationship described by the sentence. If you think two sentences indicate the same spatial relation, you can draw the road in the same place to exemplify each.

Responses are categorized to determine which of the 19 possible spatial relations has been drawn. We also will study the geometry of the examples produced, noting whether arrows were added to "direct" the roads, etc.

4.7. Agreement Task

In the Agreement Task, subjects are presented with sentences, each of which describes a spatial relation between a "road" and a "park." The first page includes the following instructions:

> Examine each of the 60 maps, and determine how well you think the sentence printed at the top of side 1 of your answer form fits the spatial (geographic) relationship between the thicker dark road and the park. Your judgment should be on the scale of (a) "strongly disagree" to (e) "strongly agree."

This instruction page is followed by 8 pages, and each but the last of these pages had 8 road-park diagrams.

4.8 Comparison of the Tasks

The tasks described in sections 4.1 to 4.7 clearly have the potential to reveal different aspects of how subjects think about spatial relations. A good example of the difference between tasks is the contrast between the responses to the Drawing Task and the Agreement Task for the sentence "the road crosses the park" (Figure 3). In a sense, the results agree, since the topological class with the highest agreement rating for the sentence from the Agreement Task also was the class drawn by 29 of 32 subjects (over 90%) as an example of the sentence. However, although other topological classes had ratings from the Agreement Task that were almost as high, only 3 subjects drew any of these other patterns. Agreement values span the range from 0 to 1, showing that the category represented by "the road crosses the park" has graded or perhaps fuzzy membership. But the Drawing Task misses this, since the subject produces a prototypical or best example.

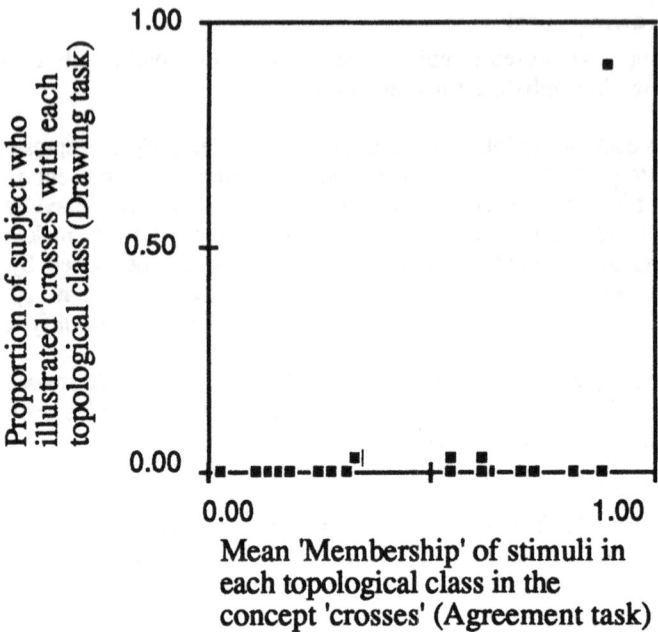

Figure 3: Plot of proportion of subjects who drew examples which each of the nineteen 9-intersection topologies as examples of "the road crosses the park" (vertical), plotted against mean agreement that "the road crosses the park" for examples of each of those topological classes.

4.9 Progress and Preliminary Results

Of the seven experimental protocols described in this paper, three have already been applied to groups of subjects.

The Grouping (5) Task was the first one that we used in this research program, asking 12 native speakers of English, 12 native speakers of Chinese, and four others, to group our first forty road-park stimuli into groups, such that every example within each group could be described by a simple sentence in their native language (Mark and Egenhofer, 1994b). The results showed a great deal of individual differences. Data derived from the results also were used to evaluate probable information loss that would result if the 9-intersection model were simplified to ignore certain topological distinctions (Egenhofer and Mark, 1995). Clearly, data from more subjects, and especially from more languages, are needed.

The Drawing (6) Task has been used to compare prototypical meanings of 64 sentences in English and 43 sentences in Spanish, with 32 and 19 subjects, respectively (Mark and Egenhofer, 1995). It seems that this task would be a useful early experiment in the exploration of a language new to these experimental procedures. The geometry of the drawings provides a rich set of data for exploring geometric factors in prototypical meanings.

So far, the Agreement (7) Task has been applied to several sentences in English, Norwegian, and Spanish. Mark and Egenhofer (1994a, 1994b) reported results for four sentences in the English language, and four others have been tested subsequently. These are listed in Table 2. Abrahamson (1994) reported results for the Agreement Task for five sentences in Norwegian, and also we have tested one sentence in Spanish.

Table 2: English-language Sentences tested Using the
Agreement Task, as of June 1995

---

the road crosses the park
the road goes across the park
the road goes through the park
the road enters the park
the road goes into the park
the road goes along the park
the road goes by the park
the road goes to the park

---

5. Summary and Further Work

In this paper, we have described a battery of human-subjects testing protocols that we feel will produce valuable insights into how people think about spatial relations and how they express these relations in natural language. Subject responses are expected to be associated with objective descriptions of the spatial entities and relationships, and on subject characteristics. Experimental design is based on the assumption that topology is primary in defining spatial relations, and that a particular formal model of topology, the 9-intersection, highlights the same distinctions that people make when judging and talking about spatial relations. Other aspects that may influence subjects' responses include metric, scale (scope), kinds of phenomena (semantics), motion (time), and graphic presentation. In the experiments conducted thus far, most of these have been held constant, but future work will involve systematic variation of test stimuli along these and perhaps other dimensions. Subjects' responses are expected to be influenced by native language, culture, and individual characteristics such as gender, age, and handedness.

Experiments that we have performed so far have been based on a line representing a road and a region representing a state park. Our plans for further testing employ a strategy that works outward from a core provided by the road-park situation in English. One dimension is to extend the tests to other sentences in English. Another is to test similar sentences in as many other languages as possible. And a third dimension is to change the phenomena involved from a road and a park to other geographic entity types. In each dimension, results can be compared back to the road-park-English core sentences. We will try to avoid the temptation to jump ahead into completely novel combinations of language, spatial relation, and phenomena, but instead attempt to build in an orderly fashion from the foundation provided by the formal model of topology and the tests performed thus far.

6. Acknowledgments

This paper is part of a collaborative research project funded in part by the Scientific and Environmental Affairs division of the North Atlantic Treaty Organization (NATO). The paper also is a part of Research Initiative 10, "Spatio-Temporal Reasoning in GIS," of the U.S. National Center for Geographic Information and Analysis (NCGIA), supported by the National Science Foundation (SBE-88-10917); support by NSF is gratefully acknowledged. Max Egenhofer's research is also supported by NSF under grant IRI-9309230, and by grants from Intergraph Corporation, Space Imaging Inc., and Environmental Systems Research Institute. Ann Deakin assisted with experiments at Buffalo and provided comments on the manuscript, and Jan Terje Bjorke coordinated the experiments with Norwegian.

7. References

Abrahamson, F., 1994. Uklarhet og Noyaktighet i Geografiske Informasjonssystemer (Fuzziness and Accuracy in Geographical Information Systems), Unpublished Master's thesis, Division of Surveying and Mapping, University of Trondheim (Norway) and Norwegian Institute of Technology.

Bowerman, M., 1991. The Origins of Children's Spatial Semantic Categories: Cognitive vs. Linguistic Determinants. In J. J. Gumperz and S. C. Levinson, Editors, *Rethinking Linguistic Relativity*. Cambridge: Cambridge University Press.

Burrough, P., and Frank, A. U., (Editors) 1995. *Geographic Objects with Indeterminate Boundaries*. London, Taylor and Francis, in press.

Cohn, A. G., and Gotts, N. M., 1995. The "Egg-Yolk" Representation of Regions with Indeterminate Boundaries. In P. Burrough and A. U. Frank, Editors, *Geographic Objects with Indeterminate Boundaries*. London, Taylor and Francis, in press.

Cohn, A. G., Randell, D. A., and Cui, Z., 1994. Taxonomies of Logically Defined Qualitative Spatial Relations. In N. Guarino and R. Poli, Editors, *Formal Ontology in Conceptual Analysis and Knowledge Representation*. Kluwer, in press.

Egenhofer, M., and Franzosa, R., 1991. Point-Set Topological Spatial Relations. *International Journal of Geographical Information Systems* 5(2): 161-174.

Egenhofer, M., and Franzosa, R., 1995. On the Equivalence of Topological Relations. *International Journal of Geographical Information Systems* 9(2): 133-152.

Egenhofer, M., and Herring, J., 1994. Categorizing Topological Spatial Relations Between Point, Line, and Area Objects. In M. J. Egenhofer, D. M. Mark, and J.

R. Herring, 1994. *The 9-Intersection: Formalism and its Use For Natural-Language Spatial Predicates.* Santa Barbara, CA: National Center for Geographic Information and Analysis, Report 94-1.

Egenhofer, M. J., and Mark, D. M., 1995. Modeling Conceptual Neighborhoods of Topological Relations. *International Journal of Geographical Information Systems* , in press.

Egenhofer, M. J., Mark, D. M., and Herring, J. R. 1994. *The 9-Intersection: Formalism and its Use For Natural-Language Spatial Predicates.* Santa Barbara, CA: National Center for Geographic Information and Analysis, Report 94-1.

Frank, A. U., 1995. The Prevalence of Objects with Sharp Boundaries in GIS. In P. Burrough and A. U. Frank, Editors, *Geographic Objects with Indeterminate Boundaries.* London, Taylor and Francis, in press.

Halpern, D. F., 1986. *Sex Differences in Cognitive Abilities.* Hillsdale, N. J.: Erlbaum Associates.

Herskovits, A., 1986. *Language and Spatial Cognition: A Interdisciplinary Study of the Prepositions in English.* Cambridge, England: Cambridge University Press.

Howard, M. C., 1989. *Contemporary Cultural Anthropology* (Third Edition). Harper Collins Publishers.

Mark, D. M., 1993. Toward a Theoretical Framework for Geographic Entity Types. In A. U. Frank and I. Campari, Editors, *Spatial Information Theory: A Theoretical Basis for GIS*, Berlin: Springer-Verlag, *Lecture Notes in Computer Sciences* Vol. 716, pp. 270-283.

Mark, D. M., and Egenhofer, M. J., 1994a. Calibrating the Meanings of Spatial Predicates From Natural Language: Line-Region Relations. *Sixth International Symposium on Spatial Data Handling* 1994, Vol. 1, pp. 538-553.

Mark, D. M., and Egenhofer, M. J., 1994b. Modeling Spatial Relations Between Lines and Regions: Combining Formal Mathematical Models and Human Subjects Testing. *Cartography and Geographic Information Systems* 21(3), in press.

Mark, D. M., and Egenhofer, M. J., 1995. Topology of Prototypical Spatial Relations Between Lines and Regions in English and Spanish. *AutoCarto 12*, Charlotte, NC, March 1995, pp. 245-254.

Mark, D. M., and Freundschuh, S. M., 1995. Spatial Concepts and Cognitive Models for Geographic Information Use. In T. L. Nyerges, D. M. Mark, R. Laurini, and M. Egenhofer, Editors, *Cognitive Aspects of Human-Computer Interaction for Geographic Information Systems.* Dordrecht: Kluwer Academic Publishers.

Montello, D., 1995. Cultural Differences in Spatial Cognition? *COSIT'95, Lecture Notes in Computer Science,Lecture Notes in Computer Science*, Berlin: Springer-Verlag, in press.

Randell, D. A., Cohn, A. G., and Cui, Z., 1992. A Spatial Logic Based on Regions and Connection. *3rd International Conference on Knowledge Representation and Reasoning.* San Francisco: Morgan Kaufmann, pp. 156-176.

Wagner, L., 1993. *The Mental Representation of the Meaning of Across.* Unpublished Undergraduate Thesis, Wellesley College.

Talmy, L., 1983. How Language Structures Space. In H. Pick and L. Acredolo, Editors, *Spatial Orientation: Theory, Research and Application.* Plenum Press.

Whorf, B. L., 1940. Science and Linguistics. *Technology Review (M.I.T.)* 42(6). reprinted in J. B. Carroll, Editor, 1956. *Language, Thought, and Reality: Selected Writings of Benjamin Lee Whorf.* Cambridge, MA: The M.I.T. Press, pp. 207-219.

Posters

Specification of Geographic Applications: Are General-Purpose Object-Oriented Methods Sufficient?
Gracia Abrantes (INESC, Portugal)

Universal Analytical GIS Operations: A Framework for a Task-Oriented User Interface
Jochen Albrecht (ISPA Vechta, Germany)

Hierarchical Spatial Reasoning
Adrijana Car (TU Vienna, Austria)

Verbal and Nominal Expressions in Route Descriptions
Agnes Gryl (LIMSI-CRNS, France)

Measurement-Based GIS
Peter Haunold (TU Vienna, Austria)

An Object-Oriented Framework for Representing Map Data and Examples of Problem Solving in this Framework
Marie-Therese Kelly (University of Wales, United Kingdom)

Time-Space Behaviour of Tourists in a Cross Border Region
Wim van der Knaap (Centre for Recreation and Tourism Studies, The Netherlands)

A Common-Sense Model of Flow Processes in a Spatially Distributed Parameter
Monika Lundell (Swiss Federal Institute of Technology, Switzerland)

Intelligently Zooming?
Sabine Timpf (TU Vienna, Austria)

Authors' Index

Springer-Verlag
and the Environment

We at Springer-Verlag firmly believe that an international science publisher has a special obligation to the environment, and our corporate policies consistently reflect this conviction.

We also expect our business partners – paper mills, printers, packaging manufacturers, etc. – to commit themselves to using environmentally friendly materials and production processes.

The paper in this book is made from low- or no-chlorine pulp and is acid free, in conformance with international standards for paper permanency.

Lecture Notes in Computer Science

For information about Vols. 1–912

please contact your bookseller or Springer-Verlag

Vol. 949: D.G. Feitelson, L. Rudolph (Eds.), Job Scheduling Strategies for Parallel Processing. Proceedings, 1995. VIII, 361 pages. 1995.

Vol. 950: A. De Santis (Ed.), Advances in Cryptology - EUROCRYPT '94. Proceedings, 1994. XIII, 473 pages. 1995.

Vol. 951: M.J. Egenhofer, J.R. Herring (Eds.), Advances in Spatial Databases. Proceedings, 1995. XI, 405 pages. 1995.

Vol. 952: W. Olthoff (Ed.), ECOOP '95 - Object-Oriented Programming. Proceedings, 1995. XI, 471 pages. 1995.

Vol. 953: D. Pitt, D.E. Rydeheard, P. Johnstone (Eds.), Category Theory and Computer Science. Proceedings, 1995. VII, 252 pages. 1995.

Vol. 954: G. Ellis, R. Levinson, W. Rich. J.F. Sowa (Eds.), Conceptual Structures: Applications, Implementation and Theory. Proceedings, 1995. IX, 353 pages. 1995. (Subseries LNAI).

VOL. 955: S.G. Akl, F. Dehne, J.-R. Sack, N. Santoro (Eds.), Algorithms and Data Structures. Proceedings, 1995. IX, 519 pages. 1995.

Vol. 956: X. Yao (Ed.), Progress in Evolutionary Computation. Proceedings, 1993, 1994. VIII, 314 pages. 1995. (Subseries LNAI).

Vol. 957: C. Castelfranchi, J.-P. Müller (Eds.), From Reaction to Cognition. Proceedings, 1993. VI, 252 pages. 1995. (Subseries LNAI).

Vol. 958: J. Calmet, J.A. Campbell (Eds.), Integrating Symbolic Mathematical Computation and Artificial Intelligence. Proceedings, 1994. X, 275 pages. 1995.

Vol. 959: D.-Z. Du, M. Li (Eds.), Computing and Combinatorics. Proceedings, 1995. XIII, 654 pages. 1995.

Vol. 960: D. Leivant (Ed.), Logic and Computational Complexity. Proceedings, 1994. VIII, 514 pages. 1995.

Vol. 961: K.P. Jantke, S. Lange (Eds.), Algorithmic Learning for Knowledge-Based Systems. X, 511 pages. 1995. (Subseries LNAI).

Vol. 962: I. Lee, S.A. Smolka (Eds.), CONCUR '95: Concurrency Theory. Proceedings, 1995. X, 547 pages. 1995.

Vol. 963: D. Coppersmith (Ed.), Advances in Cryptology - CRYPTO '95. Proceedings, 1995. XII, 467 pages. 1995.

Vol. 964: V. Malyshkin (Ed.), Parallel Computing Technologies. Proceedings, 1995. XII, 497 pages. 1995.

Vol. 965: H. Reichel (Ed.), Fundamentals of Computation Theory. Proceedings, 1995. IX, 433 pages. 1995.

Vol. 966: S. Haridi, K. Ali, P. Magnusson (Eds.), EURO-PAR '95 Parallel Processing. Proceedings, 1995. XV, 734 pages. 1995.

Vol. 967: J.P. Bowen, M.G. Hinchey (Eds.), ZUM '95: The Z Formal Specification Notation. Proceedings, 1995. XI, 571 pages. 1995.

Vol. 968: N. Dershowitz, N. Lindenstrauss (Eds.), Conditional and Typed Rewriting Systems. Proceedings, 1994. VIII, 375 pages. 1995.

Vol. 969: J. Wiedermann, P. Hájek (Eds.), Mathematical Foundations of Computer Science 1995. Proceedings, 1995. XIII, 588 pages. 1995.

Vol. 970: V. Hlaváč, R. Šára (Eds.), Computer Analysis of Images and Patterns. Proceedings, 1995. XVIII, 960 pages. 1995.

Vol. 971: E.T. Schubert, P.J. Windley, J. Alves-Foss (Eds.), Higher Order Logic Theorem Proving and Its Applications. Proceedings, 1995. VIII, 400 pages. 1995.

Vol. 972: J.-M. Hélary, M. Raynal (Eds.), Distributed Algorithms. Proceedings, 1995. XI, 333 pages. 1995.

Vol. 973: H.H. Adelsberger, J. Lažanský, V. Mařík (Eds.), Information Management in Computer Integrated Manufacturing. IX, 665 pages. 1995.

Vol. 974: C. Braccini, L. DeFloriani, G. Vernazza (Eds.), Image Analysis and Processing. Proceedings, 1995. XIX, 757 pages. 1995.

Vol. 975: W. Moore, W. Luk (Eds.), Field-Programmable Logic and Applications. Proceedings, 1995. XI, 448 pages. 1995.

Vol. 976: U. Montanari, F. Rossi (Eds.), Principles and Practice of Constraint Programming — CP '95. Proceedings, 1995. XIII, 651 pages. 1995.

Vol. 977: H. Beilner, F. Bause (Eds.), Quantitative Evaluation of Computing and Communication Systems. Proceedings, 1995. X, 415 pages. 1995.

Vol. 978: N. Revell, A M. Tjoa (Eds.), Database and Expert Systems Applications. Proceedings, 1995. XV, 654 pages. 1995.

Vol. 979: P. Spirakis (Ed.), Algorithms — ESA '95. Proceedings, 1995. XII, 598 pages. 1995.

Vol. 980: A. Ferreira, J. Rolim (Eds.), Parallel Algorithms for Irregularly Structured Problems. Proceedings, 1995. IX, 409 pages. 1995.

Vol. 981: I. Wachsmuth, C.-R. Rollinger, W. Brauer (Eds.), KI-95: Advances in Artificial Intelligence. Proceedings, 1995. XII, 269 pages. (Subseries LNAI).

Vol. 982: S. Doaitse Swierstra, M. Hermenegildo (Eds.), Programming Languages: Implementations, Logics and Programs. Proceedings, 1995. XI, 467 pages. 1995.

Vol. 983: A. Mycroft (Ed.), Static Analysis. Proceedings, 1995. VIII, 423 pages. 1995.

Vol. 985: T. Sellis (Ed.), Rules in Database Systems. Proceedings, 1995. VIII, 373 pages. 1995.

Vol. 986: Henry G. Baker (Ed.), Memory Management. Proceedings, 1995. XII, 417 pages. 1995.

Vol. 987: P.E. Camurati, H. Eveking (Eds.), Correct Hardware Design and Verification Methods. Proceedings, 1995. VIII, 342 pages. 1995.

Vol. 988: A.U. Frank, W. Kuhn (Eds.), Spatial Information Theory. Proceedings, 1995. XIII, 571 pages. 1995.

Vol. 989: W. Schäfer, P. Botella (Eds.), Software Engineering — ESEC '95. Proceedings, 1995. XII, 519 pages. 1995.

Vol. 990: C. Pinto-Ferreira, N.J. Mamede (Eds.), Progress in Artificial Intelligence. Proceedings, 1995. XIV, 487 pages. 1995. (Subseries LNAI).